Lecture Notes
in Control and Information Sciences   369

Editors: M. Thoma, M. Morari

Michael J. Hirsch, Panos M. Pardalos,
Robert Murphey, Don Grundel (Eds.)

Advances in Cooperative Control and Optimization

Proceedings of the 7th International Conference on Cooperative Control and Optimization

 Springer

Editors

Dr. Michael J. Hirsch

Raytheon, Inc.
P.O. Box 12248
St. Petersburg, FL 33733 USA
Email: Michael_J_Hirsch@Raytheon.com

Dr. Robert Murphey

GNC Branch, Munitions Directorate
AFRL
101 W. Eglin Blvd., Ste. 331
Eglin AFB, FL 32542 USA
Email: Murphey@eglin.af.mil

Dr. Panos M. Pardalos

Department of Industrial and Systems Engineering
University of Florida
P.O. Box 116596
Gainesville, FL 32611 USA
Email: Pardalos@ufl.edu

Dr. Don Grundel

AAC/ENA
101 W. Eglin Blvd., Ste. 383
Eglin AFB, FL 32542 USA
Email: Don.Grundel@eglin.af.mil

Library of Congress Control Number: 2007932965

ISSN print edition: 0170-8643
ISSN electronic edition: 1610-7411
ISBN-10   3-540-74354-5 Springer Berlin Heidelberg New York
ISBN-13   978-3-540-74354-5 Springer Berlin Heidelberg New York

Springer is a part of Springer Science+Business Media
springer.com
© Springer-Verlag Berlin Heidelberg 2007

Typesetting: by the authors and SPS using a Springer LATEX macro package

Printed on acid-free paper   SPIN: 11864295   89/SPS   5 4 3 2 1 0

Preface

Optimal control of cooperative systems continues to be at the forefront of research initiatives in the military sciences. Recently, cooperative system research has expanded from the military domain to other engineering disciplines, including drug design and disaster recovery. While there exist many powerful techniques for optimal cooperative control problems, this area is still considered one of the most difficult in the applied sciences. Thus, there must be continual improvements and new insight directed to the modeling and analysis of optimal cooperative control problems. This present volume, as well as volumes from previous years, clearly illustrate novel solutions from some of the best and brightest optimal cooperative control researchers.

This volume represents the most recent in a series of publications discussing recent research and challenges in the field of optimal cooperative control. Most of the chapters in this book were presented at the Seventh International Conference on Cooperative Control and Optimization, which took place in Gainesville, Florida, January 31 – February 2, 2007. It is our belief that this book will be an invaluable resource to faculty, researchers, and students in the fields of optimization, control theory, computer science, and applied mathematics.

We gratefully acknowledge the financial support of the Air Force Research Laboratory, The Center for Applied Optimization at The University of Florida, and Raytheon, Inc. We thank the contributing authors, the anonymous referees, and Springer Publishing for making the conference so successful and the publication of this book possible.

Michael J. Hirsch
Panos M. Pardalos
Robert Murphey
Don Grundel

Table of Contents

Locating RF Emitters with Large UAV Teams

Paul Scerri, Robin Glinton, Sean Owens, and Katia Sycara

School of Computer Science
Carnegie Mellon University
Pittsburgh, PA 15213, USA
{pscerri, rglinton, owens, katia}@cs.cmu.edu
Gerald Fudge and Joshua Anderson
L-3 Communications Integrated Systems
Greenville, TX 75402, USA
{Gerald.L.Fudge, Joshua.D.Anderson}@L-3com.com

Abstract. This chapter describes a principled, yet computationally efficient way for a team of UAVs with Received Signal Strength Indicator (RSSI) sensors to locate radio frequency emitting ground vehicles in a large environment. Such a capability has a range of both civilian and military applications. RSSI sensor readings are noisy and multiple emitters will cause ambiguous, overlapping signals to be received by the sensor. Generating a probability distribution over emitter locations requires integrating multiple signals from different UAVs into a Bayesian filter, hence requiring cooperation between the UAVs. To build a coherent distributed picture given communication limitations, the UAVs share only those sensor readings that induce the largest changes in their local filter. Each UAV translates its probability distribution into a map of information *entropy* and then plans a path that will maximize the reduction in entropy (or conversely provides the highest information gain.) Planned paths are shared with a subset of other UAVs to minimize overlapping search. Experiments in a medium fidelity simulation environment show the approach to be lightweight and effective. Live flight results with lightweight Class I UAVs validate our approach.

1 Introduction

The rapidly improving availability of small, unmanned aerial vehicles (UAVs) and their ever decreasing cost is leading to considerable interest in multi-UAV applications. However, while UAVs have become smaller and cheaper, there is a lack of sensors that are light, small and power efficient enough to be used on a small UAV yet are capable of taking useful measurements of objects often several hundred meters below them. Static or video cameras are one option, however image processing normally requires human input or at least computationally intensive offboard processing, restricting their applicability to very small UAV teams. In this chapter, we look at how teams of UAVs can use very small Received Signal Strength Indicator (RSSI) sensors whose only capability is to detect the approximate strength of a Radio Frequency (RF) signal, to search for and

M.J. Hirsch et al. (Eds.): Adv. in Cooper. Ctrl. & Optimization, LNCIS 369, pp. 1–20, 2007.
springerlink.com   © Springer-Verlag Berlin Heidelberg 2007

accurately locate such sources. RSSI sensors give at most an approximate range to an RF emitter and will be misleading when signals overlap. Applications of such UAV teams range from finding lost hikers or skiers carrying small RF beacons to military reconnaissance operations. Moreover, the core techniques have a wider applicability to a range of robotic teams that rely on highly uncertain sensors, e.g., search and rescue in disaster environments.

Many of the key technologies required to build a UAV team for multi-UAV applications have been developed and are reasonably mature and effective [1,2]. However, for large UAV teams with very noisy sensors, key problems remain, specifically, much previous work is formally grounded but impractical [3]. Often the coordination and planning algorithms and the representations of the environment are not appropriate for more than two or three UAVs and targets. For example, some solutions require a UAV to know the planned paths of all other UAVs in order to plan its own path [8], but this is infeasible (both in terms of communication and computation) when the number of UAVs is large. Other approaches only solve part of the problem, e.g., estimating locations from sensor readings [12] or planning cooperative paths [11], but do not combine these elements in an integrated solution, although there are some exceptions [4]. Signal processing techniques for creating probability distributions from noisy signals have been extensively studied, but rarely have distributed filters versions been created and those that have been do not scale to larger teams [9].

Our approach to this problem has three key elements that enable locating RF emitters with large teams of lightweight UAVs. The first key element is a distributed filter to localize RF emitters in the environment. Each UAV has a Binary Bayesian Grid Filter [7] where a value of a cell in the grid represents the probability that there is an emitter in the corresponding location on the ground. Due to limitations on available communication bandwidth, it is infeasible for UAVs to share their entire distribution, instead they share a small subset of their sensor readings with others in the team. Hence, departing from previous approaches that elicited a model of what teammates know in order to choose what to send [9], we started from the assumption that if some information leads to large local information gain, it will probably do so for much of the team. We investigated two information gain based heuristics for choosing which readings to share with teammates. The first heuristic is to send sensor readings that have the greatest impact on the UAV's local probability distribution. The second heuristic is to create a parallel probability distribution based purely on readings received from teammates and send sensor readings that have the biggest impact on that distribution. Intuitively, the first heuristic sends readings that were most important for the local UAV, while the second sends sensor readings that are most important to the team, given a local model of what the team knows. Experiments show that the first heuristic results in better team behavior than sending random messages, but the second heuristic performs worse than random.

The second element of the approach is to tightly couple estimates of the current locations of the emitters to the UAV path planning process. Specifically, a probability distribution over emitter locations is translated into a map of the information *entropy* in the environment. UAVs plan paths through areas of maximum entropy, hence maximizing expected information gain. The UAVs plan only a relatively short distance ahead in each planning cycle. This approach allows the UAVs to be reactive to new information, which is critical when sensors are highly uncertain and the domain is dynamic. For example, if a UAV traverses an area, but the sensor readings do not provide an accurate picture of that area, the entropy will remain high and the UAV will consider re-traversing the area. Notice that the entropy map coupled with the path planner looking to maximize information gain provides an integrated way for trading off between going to the locations where there will be most information gain and locations that can be quickly reached.

The third key element of the approach is a very lightweight, computationally inexpensive method for cooperative path planning. The important application feature underlying the approach is that due to the high uncertainty and dynamicism in the environment, some overlap of paths is acceptable (or even desirable), provided that the UAVs mainly spread out and search areas of maximum entropy. Our approach is for each UAV to share its planned path with some other members of the team. When planning, each UAV estimates the change in entropy that would be induced by those paths being flown by others and plans on the resulting entropy map. If the most current path of a particular UAV is not known the most recent location is used to roughly estimate where that UAV might be searching.

2   Problem

This chapter presents a method for localizing an unknown number of RF emitters using a team of UAVs. UAVs are outfitted wth RSSI sensors which detect the power of an RF signal at a position in space. The UAVs must maintain a belief over the state of all emitters in the environment in a decentralized manner.

The emitters are represented by the set: $E = \{e_1 \dots e_n\}$ where n is not known to the team of UAVs. Emitters are all assumed to be emitting at a single known frequency.[1] Emitters are mobile and emit intermittently. The homogeneous UAVs are represented by the set: $U = \{u_1 \dots u_m\}$. Each u_i flies a path given by $\boldsymbol{u}^i(t)$. During flight a UAV takes sensor readings, $z_t(\boldsymbol{loc})$ which are the received signal power at a location $\boldsymbol{loc} = \{x, y, z\}$ where $\{x, y, z\}$ gives the Euclidean coordinates of a point in space relative to a fixed origin. The power of the signal received is a result of three components. The first component, $\Gamma(\boldsymbol{loc}, e_i) = \frac{e_{const}}{dist(\boldsymbol{loc}, e_i)^2}$, where $dist(loc, e_i)$ is the Euclidean distance between loc and e_i and e_{const} is a constant that gives the power at $dist(loc_{e_1}, e_i) = 0$, is due to the sources themselves. The second component, $EN(\boldsymbol{loc}, E)$, is due to multi-path and attenuation of the signal due to environmental factors. Multi-path occurs when a reflected component of the signal arrives at a receiver and in combination with an attenuated direct signal results in a perturbed source

---

[1] This will be relaxed in future work.

location estimate. Finally ϵ gives typical zero-mean normally distributed sensor noise. The total power received at a location (loc) in space is then given by:

$$z_t(\boldsymbol{loc}) = \sum_{e_i \in E} \Gamma(\boldsymbol{loc}, e_i) + EN(\boldsymbol{loc}, E) + \epsilon \sim \mathcal{N}(0, \sigma)$$

Figure 1 shows some signals that will be received at different distances from a single emitter (i.e., no overlap). This is the basic signal model used in the simulation results below and closely represents real data collected from RSSI sensors on a physical UAV. The x-axis shows the distance and the y-axis shows the signal strength in dB (which is a log scale.) There are two important things to notice about this signal. First, it is very noisy, with high variation at all distances from the emitter, with some background noise high enough to represent being close to the emitter. Second, it has a very long "tail", i.e., at a reasonable distance from the emitter there is still useful information in the signal. Figure 2 shows the sensor readings when the UAV flies near one emitter and then another. Notice the overlap in the signals between the emitters, which are about 350m apart.

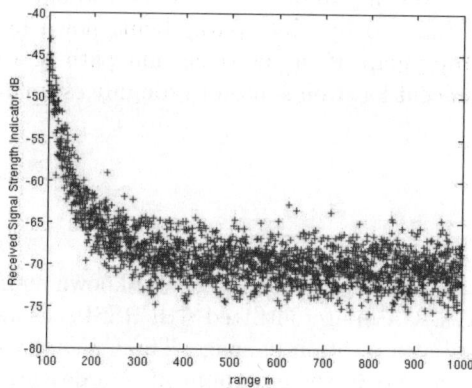

Fig. 1. Sensor readings taken from different distances from an RF emitter

The sensor readings taken by the ith UAV, up until time t are $z^i_{t_0} \dots z^i_t$. Each UAV maintains a posterior distribution P over emitter locations given by $P^i_t(e_1 \dots e_n | z^i_{t_0} \dots z^i_t)$. The UAVs proactively share sensor readings to improve each other's posterior distribution. At time t each u_i can send some subset of locally sensed readings: $\boldsymbol{z}^i_t \subset z^i_{t_0} \dots z^i_t$.

The true configuration of the emitters in the environment at time t is represented as a distribution Q such that

$$Q_t(e_1 \dots e_n) = 1$$

when $e_1 \dots e_n$ gives the true configuration of the emitters at t. The objective is to minimize the divergence between the team belief and the true state of the emitters, while minimizing the cost of UAV flight path, and minimizing the total

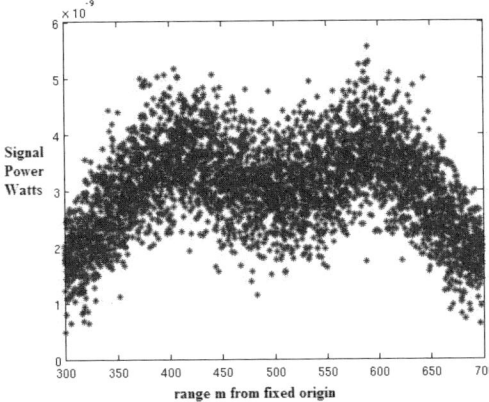

Fig. 2. Sensor readings taken when flying between two emitters, first near one, then near the other

number of messages shared between UAVs. The following function expresses this mathematically:

$$\min_{\boldsymbol{u}^i} \sum_t \sum_{u_i \in U} \beta_1 Cost(\boldsymbol{u}^i(t)) + \beta_2 D_{\mathrm{KL}}(P_t^i \| Q) + \beta_3 |\boldsymbol{z}_t^i|$$

where D_{KL} denotes the Kullback Leibler divergence and $\beta_{1\ldots3}$ are weights which control the importance of the individual factors in the optimization process.

3   Algorithm

The most important feature of the overall algorithm is the tight integration of all the key elements to maximize performance at a reasonable computational and communication cost. A Binary, Bayesian Grid Filter (BBGF) maintains an estimate of the current locations of any RF emitters in the environment. This distribution is translated into a map of the entropy in the environment. The entropy is captured in a *cost map*. UAVs plan paths with a modified Rapidly-expanding Randomized Tree (RRT) planner that maximize the expected change in entropy that will occur due to flying a particular path. The most important incoming sensor readings, as computed by the KL information gain they cause, are forwarded to other members of the team for integration into the BBGFs of other UAVs. Planned paths are also shared so that other UAVs can take into account the expected entropy gain of other UAVs when planning their own paths. The paths of other UAVs are also captured in a *cost map*. Additional cost maps, perhaps capturing results of terrain analysis or no-fly zones, can be easily added to the planner.

3.1   Implementation

The overall, integrated process aims to balance the desire to have a principled, formally grounded approach, yet be lightweight and robust enough to be practical for a team of UAVs. The hardware independent components (planners, filters, etc.) are isolated from the hardware specific components (sensor drivers, autopilot) to allow the approach to be quickly integrated with different UAVs or moved from simulation to physical UAVs. The hardware independent components are encapsulated in a *proxy* which will either be on the physical UAV or on a UAV ground station, depending on the vehicle. In the experiments below, the simulations use *exactly* the same proxy code as the live flight experiments with physical UAVs. Figure 3 shows the main components and information flows from the perspective of one UAV-proxy.

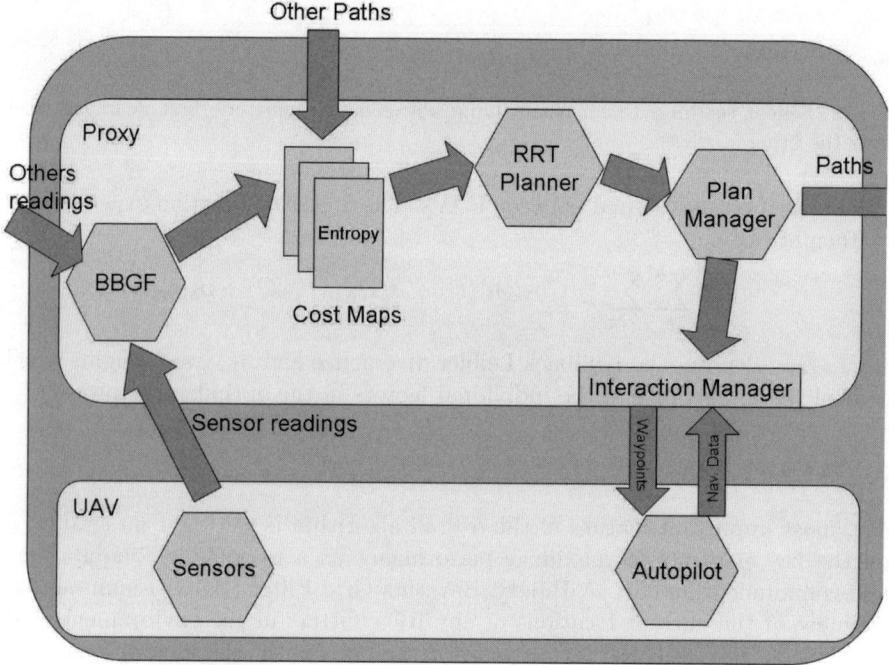

Fig. 3. Block diagram of architecture

4   Distributed State Estimation

In this section, we describe the filter used to estimate the locations of the emitters and the decisions individual UAVs make about sending information to one another.

4.1   Binary, Bayesian Grid Filter

The filter uses a grid representation, where each cell in the grid represents the probability that there is an emitter in the area on the ground corresponding to that location.[2] For a grid cell c the probability that it contains an emitter is written $P(c)$. The grid as a whole acts as the posterior $P_t^i(e_1 \ldots e_n | z_{t_0}^i \ldots z_t^i)$.

To make calculations efficient, we represent probabilities in *log odds* form, i.e., $l_t = logP(i)$. Updates on grid cells are done in a straightforward Bayesian manner.

$$l_t = l_{t-1} + log\frac{P(e_i|z_t)}{1 - P(e_i|z_t)} - log\frac{P(e_i)}{1 - P(e_i)}$$

where $P(e_i|z_t)$ is a inversion of the signal model, with the standard deviation extended for higher powered signals, i.e.,

$$P(e_i|z_t) = \begin{cases} \frac{1}{\sqrt{2\pi(\sigma_1^2)}}e^{-\frac{1}{2}(z_t - \Gamma)^2} & \text{if } z_t \geq \Gamma \\ \frac{1}{\sqrt{2\pi(\sigma_2^2)}}e^{-\frac{1}{2}(z_t - \Gamma)^2} & \text{otherwise} \end{cases}$$

where $\sigma_1 > \sigma_2$ scales the standard deviation on the noise to take into account structural environmental noise and overlapping signals. Intuitively, overlapping and other effects might make the signal stronger than expected, but they are less likely to make the signal weaker than expected. Figure 4 shows a plot of the (log) probability (y-axis) of a signal of a particular strength (x-axis) when the emitter is 500 m from the sensor.

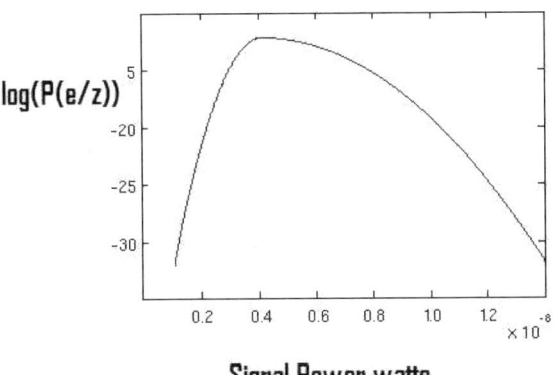

Fig. 4. Mapping between probability and signal strength

Notice that there is no normalization process across the grid because the number of emitters is not known. If the number of emitters were known, a normalization process might be able to change the probability of emitters even in areas where no sensor readings had been taken. Initial values of grid cells are set to values reflecting any prior knowledge or some small uniform value if no knowledge is available.

---

[2] A quad-tree or other representation might reduce memory and computational requirements in very large environments, but the algorithmic complexity is not justified for reasonable sized domains.

Entropy. The UAVs will fly to areas of maximum entropy, hence the probability distribution has to be translated into an entropy distribution. We assume independence between grid cells, so entropy can be calculated on a grid cell by grid cell basis. Specifically, the entropy, H, of a grid cell i is:

$$H(i) = P(i)log(P(i)) + (1 - P(i))log(1 - P(i))$$

Figure 5 shows how probability and entropy are related.

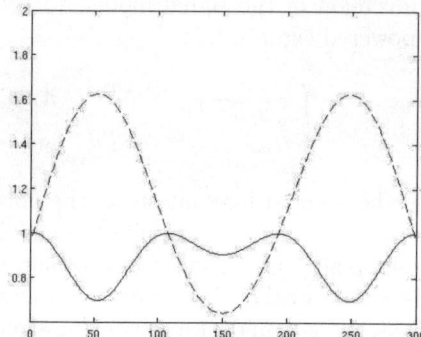

Fig. 5. Mapping between probability of an emitter and entropy. The broken line shows the probability and the unbroken line the entropy.

4.2   Information Sharing Approaches

For UAVs to plan the best possible paths, i.e., ones that lead to the greatest information gain for the team, it is important that each member of the team have an accurate picture of the distribution. Hence, UAVs must share local sensor readings with other members of the team. However, it is not scalable to simply send all sensor readings, nor is it likely to be particularly useful since some readings will not change the distribution very much. In this section, we describe a number of heuristics that are used to decide which sensor readings to pass around the team.

There are two reasons why we choose to share sensor readings rather than sharing probability distributions. First, for arbitrary probability distributions it is difficult to find concise representations that can be easily sent. Second, each UAV will have different confidence in different parts of its distribution and this confidence would need to be calculated and communicated with the distribution. While these problems are not insurmountable, they justify first trying the simpler approach of sending raw sensor readings.

Rosencrantz and Thrun [9] developed an approach to distributed particle filters that relied on teammates providing some information about what they know, so that the most appropriate information can be sent to each teammate. While such an approach clearly has some benefits in terms of getting the right information to the right team members, it does not scale to larger teams, from either a computational or communication perspective.

Instead, by using only local information to determine what readings to send and allowing other team members to decide whether those readings are forwarded, we reduce computational and communication complexity. Specifically, each agent looks at each sensor reading they get, either directly from their sensors or from other agents. If they think that this reading is sufficiently useful, they will create a *token* containing both the sensor reading and a time to live, TTL. The TTL is initially set to some small number (see 4.2). The token is randomly forwarded to a teammate[3]. The team member receiving the reading will integrate it into its own probability distribution. Each token has a unique identifier which is used to ensure sensor readings are only incorporated into the filter once. If the receiving agent finds the reading useful, it will increase the TTL on the token, otherwise it will decrease the TTL. While $TTL > 0$ and not all team members have been visited by the token, it will continue to be passed around the team, but as soon as $TTL = 0$ propagation stops. In this way, readings that are widely useful to the team are widely shared because many UAVs will increment the TTL, but those that are not widely useful will either be not shared at all or shared with only a small portion of the team (see 4.2).

The UAVs increment the TTL on tokens with sensor readings that lead to a new distribution with a KL-difference from the original distribution above some threshold α. Formally, the UAV increases the TTL on a token containing, $z_t(\boldsymbol{l})$, iff:

$$D_{\mathrm{KL}}(P_t^i(e_1 \ldots e_n | z_{t_0}^i \ldots z_t^i)) || P(e_1 \ldots e_n | z_{t_0} \ldots z_{t-1}^i)) =$$

$$\sum_i P_t^i(e_1 \ldots e_n | z_{t_0}^i \ldots z_t^i) \log \frac{P_t^i(e_1 \ldots e_n | z_{t_0}^i \ldots z_t^i)}{P_t^i(e_1 \ldots e_n | z_{t_0} \ldots z_{t-1}^i)} > \alpha$$

Intuitively, the UAVs are sending the most important readings from their perspective. In the results below, we refer to this heuristic as H_LOCAL_KL.

However, in some situations, information that does not seem important locally, may be important to the rest of the team. For example, a sequence of readings might slowly change the local perspective, with none of the readings having large enough KL information to send, but overall having high value. A second KL-difference heuristic utilizes a second probability distribution over emitters locations, but created *only* from sensor readings received on incoming tokens or sent on out-going tokens. Intuitively, this second distribution models the team's perspective of the BBGF. The heuristic H_TEAM_KL increases the TTL on tokens where the sensor reading leads to a KL-difference greater than the threshold on this model of the team's perspective on the BBGF. In experiments below we baseline the approach by sending random readings, denoted H_RAND.

Analysis. In this section we describe an analytical approach to modeling the propagation of sensor readings using the H_LOCAL_KL or H_TEAM_KL

---

[3] More intelligent approaches than completely random can be envisioned, but random sending minimizes computational requirements at the UAV and works effectively.

heuristics. Let p denote the probability that an agent will compute a new distribution with KL-divergence greater than the threshold α, given a new sensor reading. We assume that for a given sensor reading, p is identical for all m agents, and all agents will make a decision independently of the others. We also assume that an agent will never forward a reading to another agent that has already seen it; this can be implemented by attaching a history of recipients to the token.

Let c be the TTL increment. Because no agent ever receives the same reading twice, the total number of agents that ultimately receive a token always has the form $T = ic + 1$, where $i \in \{0, 1, 2, \dots\}$. The distribution of T for values less than m is given exactly by

$$\Pr(T = ic + 1) = p^i (1 - p)^{ic - i + 1} \sum_{x_1=1}^{c} \sum_{x_2=x_1+1}^{2c}$$

$$\cdots \sum_{x_{i-2}=x_{i-3}+1}^{(i-2)c} (i-1)c - x_{i-2} \tag{1}$$

and the expected value of T can be calculated directly by

$$\langle T \rangle = \left(\sum_{i=0}^{\lceil (m-1)/c \rceil} (ic + 1) \Pr(T = ic + 1) \right)$$

$$+ m \left(1 - \sum_{i=0}^{\lceil (m-1)/c \rceil} \Pr(T = ic + 1) \right) \tag{2}$$

Calculating $\langle T \rangle$ from Eq. 2 can be cumbersome for large teams, but fortunately significant insight into the behavior of the system can be obtained without resorting to brute calculation. An agent receiving a sensor reading will forward it to pc other agents on average. When $pc < 1$ and $m \gg c$, $\Pr(T = m) \approx 0$ and the expected value of T can be approximated by the geometric series $\langle T \rangle \approx \sum_{j=0}^{\infty} (pc)^j = 1 + pc/(1 - pc)$. When $pc > 1$, on average each forwarding of a token will result in even more agents forwarding the sensor reading, and hence $\Pr(T = m) > 0$ even for very large m. As p increases from $1/m$ to 1, $\langle \frac{T}{m} \rangle$ increases to 1, primarily because $\Pr(T = m)$ increases toward 1. Intuitively, when $pc > 1$, if enough of the team receives a reading, it becomes very likely that eventually all of the agents will receive the reading. Mathematically this is shown by the the fact that the probability of a token stopping before reaching all of the team decreases exponentially with the accumulated TTL. The use of an initial TTL greater than 1 takes advantage of this fact and greatly increases $\langle T \rangle$ for $p > 1/c$, although it has a much lesser effect when $p < 1/c$.

The dramatic change in behavior at $p = \frac{1}{c}$ offers a promising way to choose c. Suppose that sensor readings are of two types, either useful to the team or useless to the team, and that agents correctly classify useful readings with probability p (and thus forward them) and incorrectly classify useless readings with probability

$1 - q$ (and thus forward them with probability q). This causes both useful and useless readings to be forwarded through the team, and we wish to choose c such that the fraction of useful messages passed is maximized. Since the number of messages passed for a reading is equal to the number of agents that receive the reading, we wish to maximize the ratio $\langle T_p \rangle / (\langle T_p \rangle + \langle T_q \rangle)$, where T_p, T_q are the number of agents that ultimately receive useful and useless readings, respectively. As long as $p > q$, this can be accomplished by choosing c such that $q < \frac{1}{c} < p$. This is quite powerful because as long as agents are correct more often than they are wrong (a quite reasonable assumption), then $q < 1/2 < p$, and so $c = 2$ suffices to dramatically reduce the fraction of useless messages. Figure 6 shows the effect of different values of c on $\langle T_p \rangle / (\langle T_p \rangle + \langle T_q \rangle)$ for $p = 0.8$ and $q = 0.3$ and $m = 500$; for these settings the optimal choice is $c = 2$.

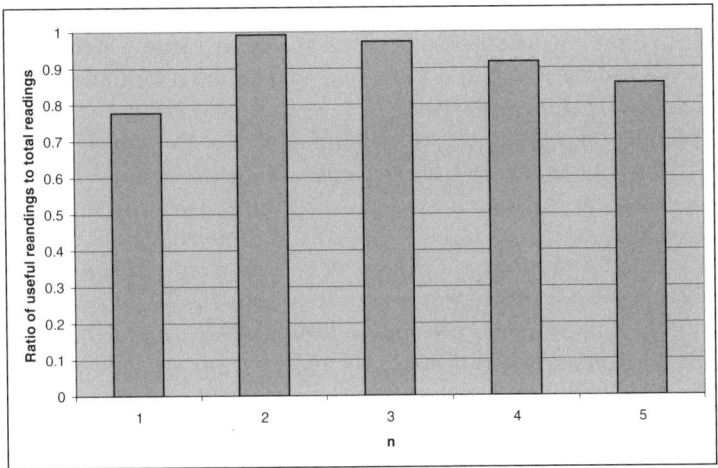

Fig. 6. The ratio of useful messages to total messages for a team of 500 agents that forward useful readings with probability $p = 0.8$ and forward useless readings with probability $q = 0.3$. The fraction of useful messages is maximized at $c = 2$.

5   Cooperative Search

In this section, we describe the cooperative path planning for maximizing the team's expected information gain and, hence, its estimate of emitter locations.

Shortly before traversing a path, the UAV plans its next path, using an RRT planner as described below. The path is encapsulated in a token and forwarded to some of the other team members. It is not critical for the token to reach all other team members, although team performance will be better if it does. UAVs store all the paths they receive via tokens. When planning new paths a change in entropy due to other UAVs flying their planned paths is assumed by the planner. Effectively, the entropy is reduced in areas where other UAVs plan to fly, reducing the incentive for flying in those areas. If the UAV does not know the current planned path of a particular UAV, it takes the last known location of that UAV, i.e., typically the last point on the last plan from that UAV, and assumes that the UAV moved randomly from there.[4] Using this technique, the UAVs mostly search different parts of the environment, but will sometimes have overlapping paths. Importantly, the approach is computationally and communication efficient, scalable and very robust to message loss.

5.1   Modified RRT Planner

Once the UAV has the entropy map and knowledge of the paths of other UAVs, it needs to actually plan a path that maximizes the team's information gain. We chose to apply an RRT planner [6,5] because it is fast, capable of handling large, continuous search spaces and able to handle non-trivial vehicle dynamics.

However, efficient RRT planners typically rely on using a goal destination to guide which points in the space to expand to. In this case, there is no specific goal, the UAV should just find a path that maximizes information gain. Initial tests with an RRT planner showed them to be inefficient in such cases. Moreover, the RRT planner did not handle the subtle features of the entropy map well. To make the planner more efficient for this particular problem, it was necessary to change a key step in the algorithm. Specifically, instead of picking a new point in space to expand the nearest node towards, a promising node is selected and expanded randomly outwards in a number of directions. This modified search works something like a depth first search, but with the RRT qualities of being able to quickly handle large, continuous search spaces and vehicle dynamics. Notice that this change also eliminates the most computationally expensive part of a normal RRT planner, the nearest neighbor computation, making it much faster.

Algorithm 5.1 shows the modified RRT planning process. Input to the algorithm includes a cost map encoding the goals of the vehicle and another cost map with the known paths of other vehicles. Lines 1-5 initialize the algorithm, creating a priority queue (*plist*) and initial node (*n*). The ordering of the priority queue is very important for the functioning of the algorithm, since the highest priority node will be expanded. The function COMPUTEPRIORITY uses both the cost of the node and the number of times it has been expanded to determine a priority. Intuitively, the algorithm works best if good nodes that have not been expanded too many times previously are expanded. The main search loop is lines 6-17 and is repeated 20,000 times (about 10ms on a standard desktop.) The highest priority node is taken off the queue (then added again with new priority). This node, representing the most promising path, is expanded 10 times in the inner loop, lines 10-17. The expansion creates a new node, representing the next point on a path, extending the previous best path by a small amount. The Expand function is designed so that all new nodes lead to kinematically feasible paths. The function COMPUTECOST then determines the cost for the new

---

[4] In future work, we may take into account that the other UAV will also be attempting to maximize entropy and thereby create better models of what it intends to do.

search node, taking into account the cost of the node it succeeds and the *cost maps*. The cost map representing other paths will return positive infinity if the new node leads to a path segment that would lead to a collision. The expanded nodes are added to the priority list for possible future expansion and the process continues. Finally, the node with the lowest cost is returned. The best path is found by iterating back over the *prev* pointers from the best node.

Algorithm 1. RRT Planning Process

RRTPLANNER($x, y, CostMaps, time, state$)

(1) $plist \leftarrow []$
(2) $n \leftarrow \langle x, y, t, cost = 0, prev = \emptyset, priority = 0 \rangle$
(3) $n \leftarrow$ COMPUTEPRIORITY(n)
(4) $plist.insert(n)$
(5) $best = n$
(6)   **foreach** 20000
(7) $n \leftarrow plist.removeFirst()$
(8) $n.priority \leftarrow$ COMPUTEPRIORITY(n)
(9) $plist.insert(n)$
(10)   **foreach** 10
(11) $n' \leftarrow$ EXPAND(n)
(12) $n'.prev = n$
(13) $n'.cost =$ COST($n, CostMaps$)
(14) $n'.priority \leftarrow$ COMPUTEPRIORITY(n')
(15) $plist.insert(n')$
(16)   **if** $n'.cost < best.cost$
(17) $best \leftarrow n$
(18) Return $best$

The planning process plans several kilometers and takes less then 0.5s on a standard desktop machine, even with other proxy processes continuing in parallel.

Using the Planner. If the UAV only plans a short distance ahead, it can fail to find plans that lead it to high value areas that are a long distance away. However, if the UAV plans long paths, it loses reactivity to new information (both sensor readings and plans of others). Our approach is to allow the UAV to plan long paths, but only use the first small piece of the path. In this way, the UAV will reach high value, distant areas by repeatedly creating plans to that area and executing part of the plan, but it can also react quickly to new information.

6   Experiments

In this section, we present empirical simulation results of the approach described above. The approach is implemented with the Machinetta proxy [10] framework integrated with either the Sanjaya UAV simulation environment or with an OP-NET simulation environment. The signal model is derived from real data from an RSSI sensor flown on a real UAV. The code used is exactly the same code as being used in an ongoing flight test, with the exception of the code between the proxy and the autopilot. Unless otherwise noted, the simulated environment is 50km by 50km and there were four RF emitters in the environment. The results below represent several hundred hours of simulated flying time, with each data point an average of five runs. The simulator and proxies are spread out over up to 15 desktop computers and communication is via multi-cast UDP resulting in around 3% message loss. These experiments are conducted in simulation due to the practical difficulty of conducting experiments with large numbers of physical UAVs. This approach was validated at L-3 Communications Integrated Systems in a series of live flight tests in late 2006 involving up to four Class I UAVs under autonomous control by the Machinetta proxies.

Information Sharing Experiments. In the first experiment, we looked at the three different information sharing heuristics. Figure 7 shows the average KL-divergence from the ground truth over time. Ground truth is modeled as tight $\frac{1}{r^2}$ probability around the real emitter location. The figure shows that all the information sharing algorithms were effective at determining the location of the emitters, but that H_LOCAL_KL was substantially better than the other heuristics. Interestingly, sending random sensor readings, H_RAND, around the team was clearly better than H_TEAM_KL, sending readings according to a model of the team. Figure 8 gives one possible reason for this, i.e., that H_TEAM_KL sent very few readings around. H_LOCAL_KL gives a low number of messages along with its good KL-divergence, showing it to be clearly the best heuristic.

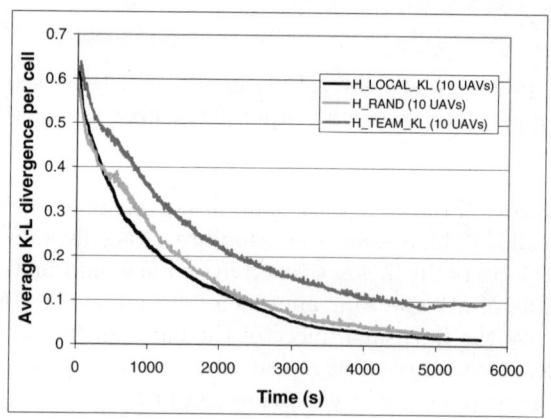

Fig. 7. The KL-divergence over time for three different information sharing algorithms

Number of UAVs and Number of Emitters. The second experiment varied both the number of emitters and number of UAVs in the environment. Figure 9 shows that more UAVs led to a faster decrease in the KL-divergence, showing that the additional UAVs were useful. Interestingly, more UAVs actually made

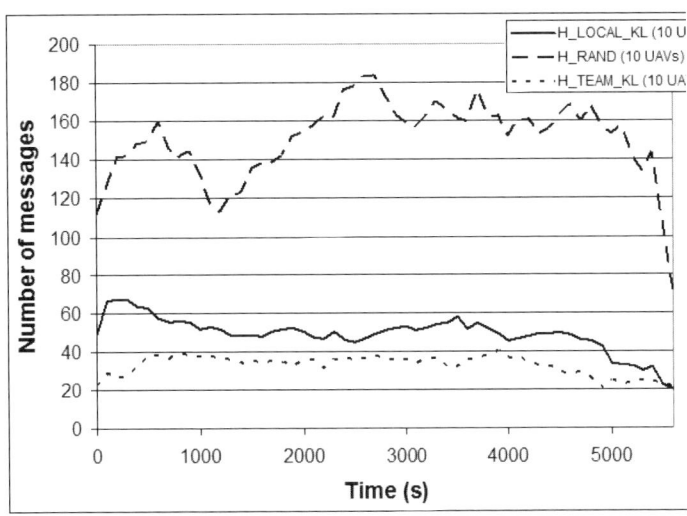

Fig. 8. The number of messages sent between UAVs for three different information sharing algorithms

reducing the KL-divergence faster. We hypothesize that this was because the UAVs were able to use the additional signals in the environment to quickly identify RF emitter locations.

Intermittent Signals. The next experiment varied how often the RF emitters were giving off a signal that could be detected, see Figure 10. The four emitters had periods ranging from 5 seconds to 30 minutes, then the percentage of that period that they were on for was varied from 25% to 100%. Curiously, the KL-divergence appears better when the emitter is *off* more. However, this is only due to a quirky interaction between the KL-divergence measurement and the very noisy sensors. Specifically, the noisy sensors do not allow the UAVs to very precisely locate the emitters, so believing that they were not there at all could actually lower the KL-divergence. Figure 11 shows an oscillation in the number of messages sent between UAVs as emitters turn on and off.

Probability of Collision. One of the issues that must be addressed in any practical autonomous UAV system is the possibility of collision. In this experiment, we examine how the probability of intra-system collision (collision between UAVs within the autonomous UAV system) varies with the number of UAVs. We note that our path planning approach will naturally tend to avoid collisions because of the tendency for the UAVs to spread out in order to maximize entropy gain as they coordinate their path planning. Nonetheless, because of software time delay, communications bandwidth, wind, navigation error, etc., the probability of collision is non-zero, especially if a large number of UAVs is covering the same region. For the simulation illustrated in Figure 12, the search area is

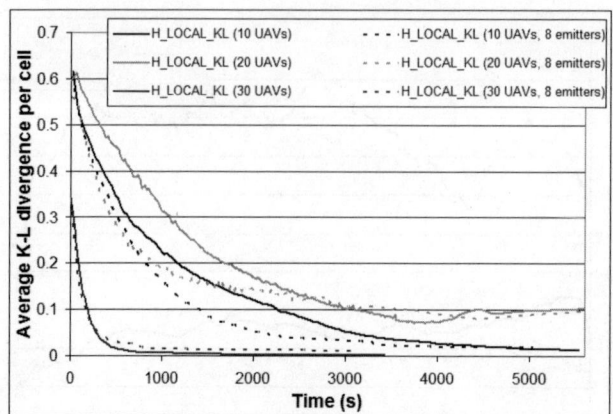

Fig. 9. The impact on KL-divergence of changing the number of UAVs and the number of RF emitters

1 square kilometer, the UAVs are all flying at the same altitude, and simulated wind is 5 m/sec, with wind gust standard deviation of 2 m/sec. For the purposes of this simulation, a collision is defined to occur anytime the distance between a pair of UAVs is less than 5 meters. The time interval for the probability of collision metric is from the start of the mission until the entropy map is 80

Live Flight Experiments. Live flight experiments with teams of up to 4 UAVs demonstrate the validity of the overall approach. The RSSI sensor, shown in Figure 13 and weighing only a few ounces, was selected to be suitable for integration on a lightweight Class I UAV, such as the Procerus UAV shown in Figure 14. In the experiments, the Procerus UAVs are hand launched, then once all are in the air, the control is handed over to the Machinetta proxies. The proxy software is hosted on a ground station rather than the on-board processor in order to minimize on-board processing requirements for the experimental system. As discussed earlier, the simulations and live flight tests use the same proxy software. This joint simulation / live flight development environment not only minimizes simulation artifacts, but also accelerates the overall development by allowing a rapid cycle of algorithm and code development, simulation, live flight tests, post-mission analysis, and back to algorithm and code development. Figure 15 illustrates some of the preliminary live flight results in a side-by-side comparison with simulation results for a 1 square kilometer region with three UAVs. The bottom half of Figure 15 shows the entropy map (left-hand side) and BBGF resulting from a live flight test at roughly the 20 minute point. The top half of Figure 15 shows the corresponding simulation results. From this, we can see that the emitter localization (shown in light green in the BBGFs on the right hand side) is, not surprisingly, somewhat better for the simulation than for the live flight test. Analysis of a series of live flight tests and simulations indicate that one significant factor is sporadic loss of communications during the live flights.

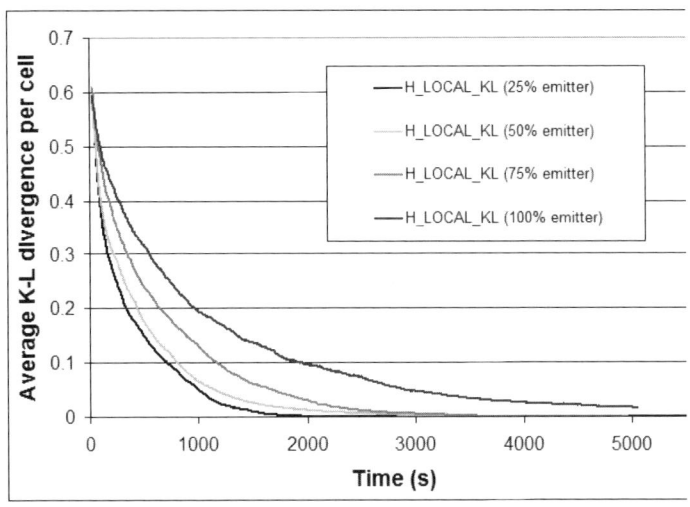

Fig. 10. The KL-divergence over time for four different percentages of time the RF emitters emit

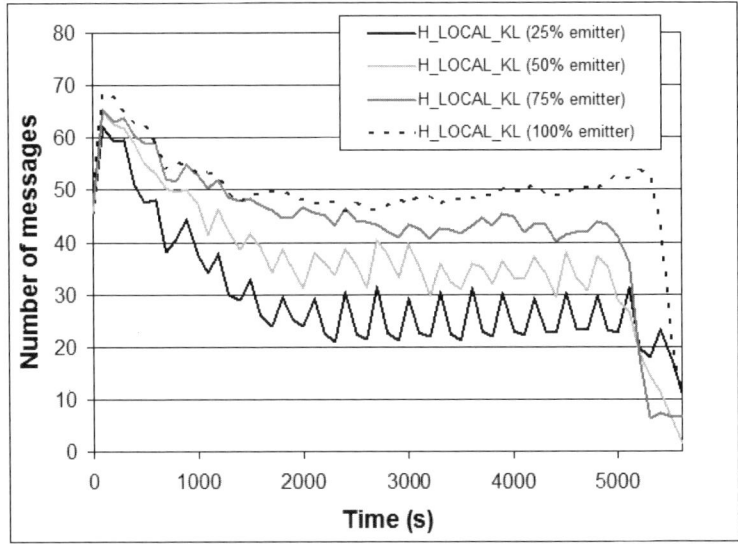

Fig. 11. The number of messages over time, for different levels of intermittency

The UAV autopilot is programmed to circle anytime during loss of communications in order to allow for quick visual identification of data link problems and to provide for taking manual control of the UAV if necessary. Upon restoration of communications, the autopilot automatically goes back to control of its proxy.

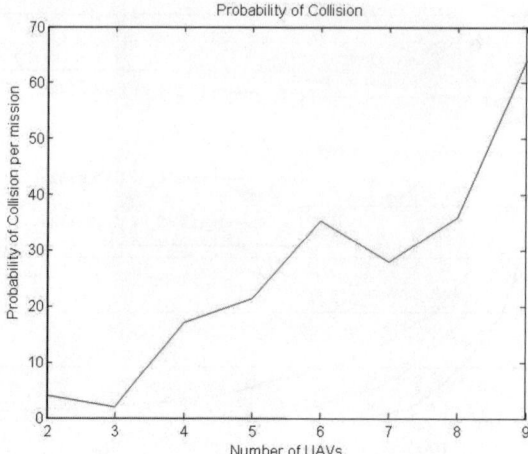

Fig. 12. Probability of collision versus number of UAVs in 1 square kilometer search area

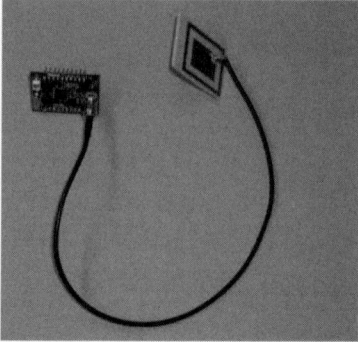

Fig. 13. RSSI sensor used for live flight experiments

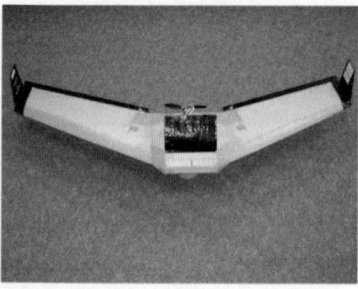

Fig. 14. Procerus UAV with 60" wingspan used for live flight experiments

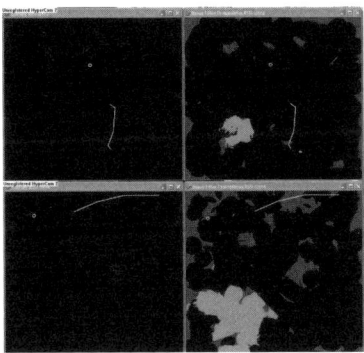

Fig. 15. Example simulation and live flight comparison of entropy and BBGF maps for similar scenario

7   Related Work

The application presented in this chapter builds on a large body of previous work spanning a range of areas.

In [9] a method for distributed probabilistic state estimation is presented. In this work agents share local beliefs with neighbors through a query-answer protocol. There are several difficulties with this approach for our application. Firstly, for a UAV team the concept of a neighbor is problematic, since UAVs move so quickly neighbors change often and simply keeping up with who a UAVs neighbors are can be expensive. Secondly, information exchanges are strictly between pairs. This fact, in combination with the KL-divergence criteria for determining the importance of information to be shared, puts excessive responsibility on local agents to determine the importance of local information to the team. Furthermore, each shared reading is only shared with a single neighbor. This limits the potential of shared readings to contribute to the entire team and therefore limits team search optimization. In contrast, our approach enables readings with high utility to the team to propagate to the entire team. Consider for example if a UAV flies directly over an emitter, clearly such a reading should be shared with the entire team.

In [12,13] the locations of sources are detected using information theoretic techniques. This work depends on a fixed array of receivers and as such does not contend with the added complexity of incorporating moving sensors into the formulation or proactive path planning for sensors to improve source localization.

In [4] a multiple UAV team is used to localize a group of emitters. In that work, a single UAV broadcasts all sensor readings to teammates. To ameliorate the exponential cost of this sharing paradigm, UAVs form sub-teams which each maintain a separate subteam posterior. The main drawback of this approach is that it is not possible to optimize search paths over the entire team. In fact, with this approach all optimization occurs within small sub-teams.

8 Conclusions

This chapter presented an integrated approach to finding RF emitters with a large team of UAVs. Simulation and live flight experiments show the approach to be effective, light-weight and robust. The key to the scalability and robustness was to find algorithms that can exploit information provided by the team but not rely on it. The most critical aspect of this was to design algorithms where local knowledge was exploited to make coordination decisions. While local knowledge was not always accurate, many local decisions make for good overall behavior because on average local decisions are good.

Acknowledgements

This research has been sponsored in part by AFOSR FA9550-07-1-0039, AFOSR FA9620-01-1-0542, L3-Communications (4500257512) and NSF ITR IIS-0205526.

References

1. R. Beard, T. Mclain, D. Nelson, and D. Kingston. Decentralized cooperative aerial surveillance using fixed-wing miniature uavs. IEEE Proceedings: Special Issue on Multi-Robot Systems, to appear.
2. L. Bertuccelli and J. How. Search for dynamic targets with uncertain probability maps. In IEEE American Control Conference, 2006.
3. S. Butenko, R. Murphey, and P. Pardalos, editors. Recent Developments in Cooperative Control and Optimization. Springer, 2003.
4. P. DeLima, G. York, and D. Pack. Localization of ground targets using a flying sensor network. In SUTC (1), pages 194199, 2006.
5. N. Kalra, D. Ferguson, and A. Stentz. Constrained exploration for studies in multirobot coordination. In Proc. IEEE International Conference on Robotics and Automation, 2006.
6. S. LaValle and J. Kuffner. Randomized kinodynamic planning. International Journal of Robotics Research, 20(5):378400, 2001.
7. H. P. Moravec. Sensor fusion in certainty grids for mobile robots. AI Magazine, 9(2):6174, 1998.
8. D. Pack, G. York, and G. Toussaint. Localizing mobile RF targets using multiple unmanned aerial vehicles with heterogeneous sensing capabilities. In IEEE International Conference on Networking, Sensing, and Control, 2005.
9. M. Rosencrantz, G. Gordon, and S. Thrun. Decentralized sensor fusion with distributed particle filters. In Proceedings of the Conference on Uncertainty in AI (UAI), 2003.
10. P. Scerri, D. V. Pynadath, L. Johnson, P. Rosenbloom, N. Schurr, M. Si, and M. Tambe. A prototype infrastructure for distributed robot-agent-person teams. In The Second International Joint Conference on Autonomous Agents and Multiagent Systems, 2003.
11. T. Schouwenaars, J. P. How, and E. Feron. Multi-vehicle path planning for nonline of sight communication. In IEEE American Control Conference, 2006.
12. M. Wax. Detection and estimation of superimposed signals. PhD thesis, Stanford Univ, 1985.
13. M. Wax and T. Kailath. Detection of signals by information theoretic criteria. IEEE Trans. Acoust., Speech, Signal Processing, 33, 1985.

Out-of-Order Sigma-Point Kalman Filtering for Target Localization Using Cooperating Unmanned Aerial Vehicles

Gregory L. Plett[*], Dimitri Zarzhitsky, and Daniel J. Pack

Department of Electrical and Computer Engineering
United States Air Force Academy, USAFA, CO 80840, USA

Abstract. This chapter outlines our research efforts toward developing a cooperative target localization method based on multiple autonomous unmanned aerial vehicles (UAVs) that are outfitted with heterogeneous sensors. The current focus of the research includes (1) optimizing the UAV trajectories to place them at desired locations at desired times to capture target locations, (2) cooperative sensor scheduling, and (3) intelligent fusing of multiple sensor measurements to accurately estimate the position and velocity of a target. The focus of this paper is the sensor-fusion task. One might consider addressing this problem using some form of Kalman filter. However, a complicating factor in the present application is that sensor readings arrive out-of-sequence to the sensor-fusion process. For example, there is non-deterministic latency in the inter- and intra-UAV communication channels. We address this problem by developing an out-of-order sigma-point Kalman filter (O^3SPKF).

1 Introduction

Detecting and localizing targets using multiple cooperative heterogeneous sensors is a challenging problem that can directly impact military and law-enforcement applications such as intelligence, surveillance, and reconnaissance as well as civilian applications such as search-and-rescue and forest-fire early detection. The particular solution of our interest must address a number of challenging requirements: (a) covert/passive sensing must be used; (b) the dynamic characteristics of the target are unknown; (c) the target is episodically mobile; and (d) the target is intermittently occluded from particular sensing mechanism(s). The cooperative method proposed in this paper plays an important role in our larger overall goal to develop a multiple cooperative UAV system that can autonomously search, detect, and localize multiple targets [1,2]. Our solution addresses these requirements using a flight of small autonomous UAVs with heterogeneous sensing capabilities. Multiple autonomous UAVs offer certain advantages over other conventional sensor platforms. They offer robustness in the presence of a loss of members; can quickly search a large area; can operate

---

[*] Corresponding author. Visiting Research Professor (Associate Professor on sabbatical from the University of Colorado at Colorado Springs, CO 80918, USA)

M.J. Hirsch et al. (Eds.): Adv. in Cooper. Ctrl. & Optimization, LNCIS 369, pp. 21–43, 2007.
springerlink.com   © Springer-Verlag Berlin Heidelberg 2007

using a decentralized but cooperative control algorithm, requiring minimal human intervention; and they are small and relatively inexpensive, allowing quick and easy deployment.

Each UAV carries a suite of one or more sensors including perhaps: radio frequency (RF) sensors to detect and determine direction-of-arrival (DOA), time-difference-of-arrival (TDOA) and/or frequency-difference-of-arrival (FDOA); infrared sensors to detect heat signatures; and optical image sensors. Due to cost considerations and payload constraints, it is not desirable for every UAV to have a full complement of sensing capability. In our implementation, when a target is initially detected by one UAV, a small formation of UAVs comprising complementary heterogeneous sensors is autonomously assembled to localize the target. The UAVs then cooperatively locate the target by combining the sensor information collected by heterogeneous sensors onboard the UAVs. The output of the localization process gives an estimate of the target's position and velocity and provides error bounds on the estimate.

We integrate dynamic sensor fusion and target localization using a modified sigma-point Kalman filter (SPKF). SPKFs are a generalization of the ubiquitous Kalman filter [3,4] to problems with nonlinear descriptions.[1] The problem specifically addressed in this chapter is that sensor readings may arrive out-of-sequence to the fusion process due, for example, to non-deterministic communication-channel latency between UAVs. A similar issue was solved in [11] where the latency was known and deterministic. However, in our case, the latency is not known a priori and is variable, so we take quite a different approach, which we have named the "out-of-order sigma-point Kalman filter" (O³SPKF).

This chapter proceeds by first outlining several approaches that may be taken to handle out-of-order measurements. Sigma-point Kalman filters are then reviewed to provide background for the development of the O³SPKF. To illustrate the results, we then give an example model of target dynamics, the overall simulation system used, and some results. Finally, we close with some concluding remarks.

2   Approaches to Handling Out-of-Sequence Measurements

There are a number of straightforward approaches to handling out-of-sequence measurements that might be considered in a target-localization application. These include the methods that we call the "simple approach" and the "buffered approach," which will be described below. The O³SPKF method is perhaps not quite as straightforward, but has performance advantages, as will be shown. All of these methods are based on sigma-point Kalman filtering, which is itself explained in Section 3. We limit ourselves to this one technology to make a valid

---

[1] One variety of SPKF is the unscented Kalman filter (UKF) [5,6,7], which has been used to locate targets using TDOA measurements [8]. We use the central difference Kalman filter (CDKF) [9,10] here since it has slightly higher theoretic accuracy [10] and requires fewer algorithm parameters be tuned.

comparison of results.[2] The differences between the methods we describe in this chapter reside in either (1) how the data is made available to the SPKF, as in the "simple" approach and in the "buffered" approaches, or (2) how the SPKF is modified to be able to accommodate out-of-sequence measurements, as in the O^3SPKF approach.

Before we outline the methods themselves, it is instructive to consider the method for evaluating their performance. A number of important metrics might be considered: What is the computational complexity of the algorithm? What is the required amount of auxiliary memory/storage for the algorithm? What is the estimation accuracy of the algorithm? For the methods described in this chapter, the first two questions are quite straightforward to answer. The third requires more discussion. The issue lies in the question "How does one compute a localization error estimate at an arbitrary time, since the filter is only updated at random, asynchronous points in time?"

To clarify this explanation, we define some notation. Let t_m be the time a particular measurement is taken, t_a be the time that measurement arrives for processing ($t_a = t_m + d_t$, where d_t is the transport/processing delay), t_f be the time the measurement arrives at the filter ($t_f = t_a + d_q$, where d_q is the queueing delay), t_x be the time associated with the filter's most recent state estimate, and t be the present (real) time. Further, let $x(t)$ be the true state at time t and $\hat{x}(t)$ be an estimate of the state corresponding to time t.

Control decisions need to be made based on $\hat{x}(t)$; however, the filter only "knows" $\hat{x}(t_x)$ corresponding to the timestamp of the measurement most recently incorporated into the filter's estimate. Since (with probability one) $t \neq t_x$, we must define a means to propagate the filter's most recent estimate $\hat{x}(t_x)$ forward in time to predict $\hat{x}(t)$. This can be done using the target's motion-model state equation. Note that $\hat{x}(t_x)$ has incorporated all measurements with $t_f < t$, but has not necessarily incorporated all measurements with $t_a < t$ or $t_m < t$ due to the delays involved. We begin to suspect that there will be a definite cost to transmission latency: not all measurements taken prior to the present time will be included in the target position estimate, thus degrading accuracy of the state prediction. Furthermore, there will be a cost to queueing latency since the greater the difference $t - t_x$, the greater length of time we need to predict over, further degrading accuracy. The ad-hoc methods that we describe in this chapter to process out-of-sequence measurements take the approach of trying to manage queueing latency to improve the accuracy of the estimates. We will see that the O^3SPKF approach improves on these ad-hoc methods by adding no latency, and making better use of all available measurements.

The Simple Approach to Handling Out-of-Sequence Measurements. Since we suspect that latency will degrade our real-time estimate of target state, our first ad-hoc approach is to simply discard all measurements that arrive at the filter

---

[2] We have found the SPKF to give a very good tradeoff between computational complexity and accuracy of location estimates, but we recognize that if additional computational resources were available, a technology such as a particle filter might produce even more accurate estimates.

out-of-sequence. That is, if a particular measurement has $t_m < t_x$, that measurement is discarded. We call this the "simple" approach.

The Buffered Approach to Handling Out-of-Sequence Measurements. The simple approach has the beneficial property that it adds no latency ($d_q = 0$). However, many potentially useful measurements are discarded. We will see later, that if the simple approach is used with UAVs having multiple sensors, it is possible that only the measurements taken by one of these sensors are used if the processing time required by the sensors is quite different. Therefore, we seek means whereby d_q may be adjusted to provide improved target state estimates without introducing significant complexity.

The method we call the "buffered" approach is one way to do this. We form a buffer of N measurements. The oldest measurement in the buffer has timestamp denoted $t_{\min}$. When a new measurement arrives, we compare its timestamp t_m against $t_{\min}$. If $t_m < t_{\min}$, the measurement is discarded, as in the simple approach. However, if $t_m \geq t_{\min}$, the oldest buffer measurement is removed from the buffer and is used to update the SPKF, and the presently received measurement is added to the buffer. The filter time t_x will be updated to $t_{\min}$. If the buffer is very large, few measurements will be discarded—which we expect will result in near-optimal estimation of $\hat{x}(t_x)$—but since the difference between t and t_x will generally be large—we expect poorer estimation of $\hat{x}(t)$. If the buffer is very small, many measurements will be discarded—resulting in poor estimation performance of $\hat{x}(t_x)$ and by extension of $\hat{x}(t)$—but the difference between t and t_x will generally be small. In the limit as the buffer size goes to zero, the buffered approach becomes the simple approach, and there will be some size N that optimizes the estimation performance of $\hat{x}(t)$.

When determining the buffer size N, one might consider modeling the inter-UAV transmission latency (the dominant effect) as an exponential random variable, as is common in communication theory. Then, if μ is the expected latency, 68% of measurements are received with $d_t < \mu$, 86% of measurements are received with $d_t < 2\mu$, and 95% of measurements are received with $d_t < 3\mu$. We might then choose $N = \lceil k\mu/T_s \rceil \times$ number of UAVs $\times$ number of sensors per UAV, where $k \in \{1, 2, 3\}$ selects the average percentage of measurements queued and used in the filter, and T_s is the inter-sample interval for each sensor.

The O^3SPKF Approach to Handling Out-of-Sequence Measurements. The simple and buffered approaches both result in measurements arriving at the SPKF in order (either by discarding all out-of-sequence measurements, or using the buffering mechanism to sort the great majority of measurements in-order, while still discarding a few out-of-sequence measurements). The approach we propose in this chapter is fundamentally different from either of these in that it modifies the SPKF itself to be able to accommodate the out-of-sequence measurements. If a new measurement arrives with $t_m \geq t_x$, it is incorporated in the filter state using the standard SPKF steps and t_x is updated to t_m. However, if a new measurement arrives with $t_m < t_x$, an alternate sequence of steps is executed to update the filter state estimate using the novel information regarding the system state at time t_x in this stale measurement, and the filter time remains at t_x.

The O^3SPKF is not a buffering method, so incurs no additional latency ($d_q =$ 0). Furthermore, it never discards measurements; therefore, we expect that it will give better estimates than the other two approaches. Also, the O^3SPKF is of the same computational complexity as SPKF, so we incur no processing penalty for using it. However, because the O^3SPKF does not discard any measurements (as do the simple and buffered methods), it will execute more often. Finally, since the O^3SPKF does not buffer measurements it has no additional auxiliary storage requirements.

We note in passing that there are some other approaches to fusing out-of-order measurements that bear some similarity to O^3SPKF, which we did not consider while preparing this chapter. One method maintains a buffer of sensor data, but unlike the simple buffered methods proposed above, updates the SPKF immediately upon receiving a new measurement. The SPKF state and co-variance estimates are stored along with the corresponding measurement in the buffer. When an out-of-order measurement arrives, the filter state is updated by first "rolling back" to the estimate immediatly prior to that measurement, and the SPKF algorithm is applied repeatedly to all following measurements in the buffer—potentially resulting in many SPKF update steps per new data point. This method gives the best achievable results, but we did not consider it due to its requirements of a potentially large buffer and challenges in a real-time implementation. A second method is similar—upon receiving an out-of-order measurement, a Kalman smoother is run backward in time from t_x to t_m (potentially requiring many iterations of the smoother steps as each data point is considered again) to make a smoothed estimate of the state at time t_m. The present measurement is then incorporated into $\hat{x}(t_m)$ via a measurement update equation, and the state estimate is re-propagated to time t_x [12]. The O^3SPKF has similar steps to this method: It also propagates the present state back in time, but it always does this in one step (not requiring a buffer of measurements). Additionally, it does not update the state estimate at time t_m; rather, covariance calculations are performed using the present and prior data that allow direct updating of the state estimate at time t_x using the old measurement.

We now present the SPKF and the O^3SPKF. Readers familiar with either the UKF or SPKF might skim the next section to discover the notation that we use, and to see how we partition the SPKF into six steps which are then mirrored in the O^3SPKF in Section 4.

3   Sigma-Point Kalman Filters (SPKF)

Kalman filters are an intelligent (and sometimes optimal) way to estimate the unmeasurable "state" $x(t)$ of some dynamic system given measurements of a signal $u(t)$ possibly affecting that state (the dynamic "input", sometimes called a forcing function), and measurements $y(t)$ (the dynamic "output") related to linear or nonlinear combinations of members of that state and $u(t)$. Here, we assume that the state of the target to be estimated comprises its position and velocity: that is, $x(t) = [p_x(t), p_y(t), v_x(t), v_y(t)]^T$, where $p_x(t)$ is the "x" position coordinate of the target, $p_y(t)$ is the "y" position coordinate of the target, $v_x(t)$ is the "x" velocity of the target and $v_y(t)$ is the "y" velocity of the target. (We could easily extend this to three dimensions by simply adding extra state elements $p_z(t)$ and $v_z(t)$.) The state vector $x(t)$ is assumed to have dynamics that can be modeled in a "state-space" form. For example, a relationship that can be used with some kinds of Kalman filter is:

$$\dot{x}(t) = f(x(t), u(t), w(t), t) \tag{1}$$
$$y(t) = h(x(t), u(t), v(t), t), \tag{2}$$

where $w(t)$ is an unmeasurable "process noise" that is often modeled as a zero-mean white Gaussian random process, $v(t)$ is unmeasurable "sensor noise" that is also modeled as a zero-mean white Gaussian random process, $f(\cdot)$ is the "state equation" function that captures the dynamics of the state, and $h(\cdot)$ is the "measurement equation" or "output equation" function that describes how the sensor measurements relate to the state.

Creating an optimum estimate $\hat{x}(t)$ of the true state $x(t)$ is a very challenging problem in general. Very close approximations to the optimum estimate can be made using particle filters, but these are too computationally intensive for our application. Alternative suboptimal solutions can be derived by assuming that the state estimation error always retains a Gaussian probability density function—this assumption is the basis of the original Kalman filter, the extended Kalman filter, and the sigma-point Kalman filters to be discussed. Then, rather than having to propagate the entire density function through time, we need only to evaluate the conditional mean and covariance of the state vector once each sampling interval and make updates to the estimates using the following two relationships:

$$\hat{x}^+(t_x) = \hat{x}^-(t_x) + L(t_x, t_m)\big(y(t_m) - \hat{y}(t_m)\big) \tag{3}$$
$$\Sigma^+_{\tilde{x}(t_x)} = \Sigma^-_{\tilde{x}(t_x)} - L(t_x, t_m)\Sigma^-_{\tilde{y}(t_m)}L(t_x, t_m)^T, \tag{4}$$

where the superscript T is the matrix/vector transpose operator, and

$$\hat{x}^+(t_x) = \mathbb{E}\left[x(t_x) \mid \mathbb{Y}^+\right] \tag{5}$$
$$\hat{x}^-(t_x) = \mathbb{E}\left[x(t_x) \mid \mathbb{Y}^-\right] \tag{6}$$
$$\hat{y}(t_m) = \mathbb{E}\left[y(t_m) \mid \mathbb{Y}^-\right] \tag{7}$$
$$\Sigma^-_{\tilde{x}(t_x)} = \mathbb{E}\left[(x(t_x) - \hat{x}^-(t_x))(x(t_x) - \hat{x}^-(t_x))^T\right] = \mathbb{E}\left[\tilde{x}^-(t_x)\tilde{x}^-(t_x)^T\right] \tag{8}$$
$$\Sigma^+_{\tilde{x}(t_x)} = \mathbb{E}\left[(x(t_x) - \hat{x}^+(t_x))(x(t_x) - \hat{x}^+(t_x))^T\right] = \mathbb{E}\left[\tilde{x}^+(t_x)\tilde{x}^+(t_x)^T\right] \tag{9}$$
$$\Sigma^-_{\tilde{y}(t_m)} = \mathbb{E}\left[(y(t_m) - \hat{y}(t_m))(y(t_m) - \hat{y}(t_m))^T\right] = \mathbb{E}\left[\tilde{y}(t_m)\tilde{y}(t_m)^T\right]$$
$$= \mathbb{E}\left[\tilde{y}(t_m)\tilde{y}(t_m)^T\right] \tag{10}$$

$$L(t_x, t_m) = \mathbb{E}\left[(x(t_x) - \hat{x}^-(t_x))(y(t_m) - \hat{y}(t_m))^T\right]\left(\Sigma^-_{\tilde{y}(t_m)}\right)^{-1}$$
$$= \Sigma^-_{\tilde{x}(t_x)\tilde{y}(t_m)}\left(\Sigma^-_{\tilde{y}(t_m)}\right)^{-1}. \tag{11}$$

While this is a linear recursion, we have not directly assumed that the system model is linear. In the notation we use, time t_x indicates the timestamp of the measurement closest in time to the present and t_m indicates the timestamp of the most recently received measurement. (In the literature, no distinction is usually made between t_x and t_m. However, they will not be equal if measurements arrive at the Kalman filter out-of-sequence, which is the topic of this chapter). The decoration "circumflex" indicates an estimated quantity (e.g., $\hat{x}$ indicates an estimate of the true quantity x). A superscript "−" indicates an a priori estimate (i.e., an estimate of a quantity's value at some point in time based on all sensor data except the most recently received measurement) and a superscript "+" indicates an a posteriori estimate (i.e., an estimate of a quantity's value at some point in time based on all sensor data including the most recently received measurement). The decoration "tilde" indicates the error of an estimated quantity (e.g., $\tilde{x}$ is the difference between x and $\hat{x}$). The symbol $\Sigma_{xy} = \mathbb{E}\left[xy^T\right]$ indicates the auto- or cross-correlation of the variables in its subscript. (Note that often these variables are zero-mean, so the correlations are identical to covariances). Also, for brevity of notation, we often use Σ_x to indicate the same quantity as Σ_{xx}. The symbol $\mathbb{Y}^+$ indicates the set of all sensor readings taken up to and including the most recently received measurement, while $\mathbb{Y}^-$ indicates the set of all sensor readings excluding the most recently received measurement.

Equations (3) through (11) (and approximations thereof) may be used to derive either the Kalman filter, the extended Kalman filter, or the sigma-point Kalman filter. All members of this family of filters comply with a structured sequence of six steps per iteration, as outlined here.

General step 1: State estimate time update. For each measurement received, first assign $t_p = t_x$ (the prior value of the filter time), then set $t_x = \max(t_x, t_m)$ (the updated value of the filter time). For in-order measurements this results in $t_x = t_m > t_p$; for out-of-order measurements, this results in $t_x = t_p > t_m$. An updated state prediction $\hat{x}^-(t_x)$ of the value of $x(t_x)$ is then made, based on a priori information and the system model using (1) and (6).

General step 2: Error covariance time update. The second step is to determine the predicted state-estimate error covariance matrix $\Sigma_{\tilde{x}(t_x)}^-$ based on a priori information and the system model using (8).

General step 3: Estimate system output $y(t_m)$. The third step is to estimate the system's output corresponding to the timestamp of the most recently received measurement using present a priori information and (2) and (7).

General step 4: Estimator gain matrix $L(t_x, t_m)$. The fourth step is to compute the estimator gain matrix by evaluating (11). We again emphasize that the literature generally makes no distinction between t_x and t_m. However, this distinction is key to the O³SPKF developed herein, of which the most important aspect is the ability to compute the correct value for $L(t_x, t_m)$.

General step 5: State estimate measurement update. The fifth step is to compute the a posteriori state estimate by updating the a priori estimate using the estimator gain and the output prediction error using (3).

General step 6: Error covariance measurement update. The final step computes the a posteriori error covariance matrix using (4). The estimator output comprises $\hat{x}^+(t_x)$ and $\Sigma^+_{\tilde{x}(t_x)}$. The estimator then waits until the next measurement is received, and returns to Step 1.

The standard Kalman filter is obtained by replacing (1) and (2) with linear state-space equations using a fixed sampling interval, resulting in closed-form equations for Steps 1–6. When the system equations are nonlinear, however, we must make some approximations to evaluate the expectation operators, and might consider the extended Kalman filter. A better alternative is the sigma-point Kalman filter, which has the same computational complexity as the extended Kalman filter, but more accurately approximates these steps.

SPKF computes estimates of the mean and covariance of the output of a nonlinear function using a small fixed number of function evaluations. A set of points (*sigma points*) is chosen as input to the function so that the (possibly weighted) mean and covariance of the points exactly matches the a priori mean and covariance of the input random variable being modeled. These points are then passed through the nonlinear function, resulting in a transformed set of output points. The a posteriori mean and covariance that are sought are then approximated by the mean and covariance of these points. Note that the sigma points comprise a fixed small number of vectors that are calculated deterministically—unlike particle filter methods.

Specifically, if the input random vector x has mean $\bar{x}$ and covariance $\Sigma_{\tilde{x}}$, then $p + 1 = 2 \times \dim(x) + 1$ sigma points are generated as the set

$$\mathcal{X} = \left\{\bar{x}, \bar{x} + \gamma\sqrt{\Sigma_{\tilde{x}}}, \bar{x} - \gamma\sqrt{\Sigma_{\tilde{x}}}\right\},$$

with members of $\mathcal{X}$ indexed from 0 to p, where γ is a tuning parameter (see below for an example), and where the matrix square root $R = \sqrt{\Sigma}$ computes a result such that $\Sigma = RR^T$. Typically, the efficient *Cholesky decomposition* [13,14] is used, resulting in a lower-triangular R. The reader can verify that the weighted mean and covariance of $\mathcal{X}$ equal the original mean and covariance of random vector x for a specific set of $\{\gamma, \alpha^{(m)}, \alpha^{(c)}\}$ if we define the weighted mean as $\bar{x} = \sum_{i=0}^{p} \alpha_i^{(m)} \mathcal{X}_i$, the weighted covariance as $\Sigma_{\tilde{x}} = \sum_{i=0}^{p} \alpha_i^{(c)} (\mathcal{X}_i - \bar{x})(\mathcal{X}_i - \bar{x})^T$, $\mathcal{X}_i$ as the ith column of $\mathcal{X}$, and both $\alpha_i^{(m)}$ and $\alpha_i^{(c)}$ as real scalars with the necessary (but not sufficient) conditions that $\sum_{i=0}^{p} \alpha_i^{(m)} = 1$ and $\sum_{i=0}^{p} \alpha_i^{(c)} = 1$. The various sigma-point methods differ only in the choices taken for these weighting constants. The two most common methods are the *unscented Kalman filter* (UKF) [5,6,7,15] and the *central difference Kalman filter* (CDKF) [9,10]. The CDKF has only one "tuning parameter" h, which makes implementation simpler. It also has marginally higher theoretic accuracy than the UKF [10], so we focus on this method in the application sections later. Using the CDKF,

$$\gamma = h, \quad (h = \sqrt{3} \text{ for Gaussian distributions})$$
$$\alpha_0^{(m)} = \alpha_0^{(c)} = \frac{h^2 - \dim(x)}{h^2}$$
$$\alpha_i^{(m)} = \alpha_i^{(c)} = \frac{1}{2h^2}, \quad i \neq 0$$

To use SPKF in a general estimation problem, with nonlinear state and output equations, we first define an augmented random vector x^a that combines the randomness of the state, process noise, and sensor noise. This augmented vector is then used as the state in the estimation process. However, we will assume a linear state equation with zero-mean process noise and additive zero-mean sensor noise to the otherwise nonlinear output equation. This allows the SPKF steps to be somewhat simplified. The model we use is:

$$x(t) = A(t_0)x(t - t_0) + w_{t_0}(t), \qquad t_0 > 0$$
$$y(t) = h(x(t), u(t)) + v(t),$$

where $A(t_0)$ is a state-transition matrix that represents the homogeneous dynamics of the state over a generic time interval t_0. For a given t_0, $A(t_0)$ is a constant matrix; however, in this work we receive measurements at random times, so we must treat t_0 and therefore $A(t_0)$ as variable in the development of the algorithm.

In this section, we will assume that measurements arrive in-sequence, such that when the measurement arrives at the sensor-fusion process $t_m \geq t_x$. We will call this case the "In-order SPKF."

In-order SPKF step 1: State estimate time update. First, assign $t_p = t_x$ (the prior value of the filter time), then set $t_x = \max(t_x, t_m) = t_m$. We desire to estimate $\hat{x}^-(t_x)$ using prior information regarding $x(t_p)$ and the state equation. To do so, we compute sigma points $\mathcal{X}^+(t_p)$ corresponding to the prior state and covariance estimates. These $p + 1$ vectors are

$$\mathcal{X}^+(t_p) = \left\{ \hat{x}^+(t_p), \hat{x}^+(t_p) + \gamma \sqrt{\Sigma_{\tilde{x}(t_p)}^+}, \hat{x}^+(t_p) - \gamma \sqrt{\Sigma_{\tilde{x}(t_p)}^+} \right\}.$$

Sigma points corresponding to a prediction of the state at time t_x are generated by evaluating the process equation $f(\cdot)$ using all $\mathcal{X}_i^+(t_p)$ (where the subscript i denotes that the ith vector is being extracted from the original set), yielding the a priori sigma points $\mathcal{X}_i^-(t_x)$. The state prediction is a weighted average of the $\mathcal{X}_i^-(t_x)$. In general,

$$\hat{x}^-(t_x) = \mathbb{E}\left[f(x(t_p), u(t_x), w(t_x)) \mid \mathbb{Y}^- \right]$$
$$\approx \sum_{i=0}^{p} \alpha_i^{(m)} \mathcal{X}_i^-(t_x).$$

However, this simplifies for our linear state equation. If $t_0 = t_m - t_p$,

$$\hat{x}^-(t_x) = \mathbb{E}\left[A_{t_0}x(t_p) + w_{t_0}(t_x) \mid \mathbb{Y}^-\right]$$

$$= \sum_{i=0}^{p} \alpha_i^{(m)} \mathcal{X}_i^-(t_x) = \sum_{i=0}^{p} \alpha_i^{(m)} A(t_0)\mathcal{X}_i^+(t_p)$$

$$= A(t_0)\hat{x}^+(t_p).$$

In-order SPKF step 2: Error covariance time update. Using the a priori sigma points from Step 1, the a priori covariance estimate is computed as

$$\Sigma_{\tilde{x}(t_x)}^- = \sum_{i=0}^{p} \alpha_i^{(c)} \left(\mathcal{X}_i^-(t_x) - \hat{x}^-(t_x)\right)\left(\mathcal{X}_i^-(t_x) - \hat{x}^-(t_x)\right)^T + \Sigma_{w_{t_0}}.$$

For our linear state equation, this again simplifies:

$$\Sigma_{\tilde{x}(t_x)}^- = A(t_0)\Sigma_{\tilde{x}(t_p)}^+ A(t_0)^T + \Sigma_{w_{t_0}}.$$

In-order SPKF step 3: Predict system output $y(t_m) = y(t_x)$. The system output is predicted by evaluating the model output equation using the sigma points describing the state at time t_m. The in-order case has $t_m = t_x$, so we first compute the points $\mathcal{Y}_i^-(t_m) = h(\mathcal{X}_i^-(t_m), u(t_m)) = h(\mathcal{X}_i^-(t_x), u(t_m))$. The output estimate is then

$$\hat{y}(t_m) = \mathbb{E}\left[h(x(t_x), u(t_m)) + v(t_m) \mid \mathbb{Y}^-\right]$$

$$\approx \sum_{i=0}^{p} \alpha_i^{(m)} h(\mathcal{X}_i^-(t_x), u(t_m))$$

$$= \sum_{i=0}^{p} \alpha_i^{(m)} \mathcal{Y}_i^-(t_m).$$

In-order SPKF step 4: Estimator gain matrix $L(t_x, t_m)$. To compute the estimator gain matrix, we must first compute the required covariance matrices.

$$\Sigma_{\tilde{y}(t_m)}^- = \sum_{i=0}^{p} \alpha_i^{(c)} \left(\mathcal{Y}_i^-(t_m) - \hat{y}(t_m)\right)\left(\mathcal{Y}_i^-(t_m) - \hat{y}(t_m)\right) + \Sigma_v$$

$$\Sigma_{\tilde{x}(t_x)\tilde{y}(t_m)}^- = \sum_{i=0}^{p} \alpha_i^{(c)} \left(\mathcal{X}_i^-(t_x) - \hat{x}^-(t_x)\right)\left(\mathcal{Y}_i^-(t_m) - \hat{y}(t_m)\right).$$

Then, we simply compute $L(t_x, t_m) = \Sigma_{\tilde{x}(t_x)\tilde{y}(t_m)}^- \left(\Sigma_{\tilde{y}(t_m)}^-\right)^{-1}$.

In-order SPKF step 5: State estimate measurement update. The a posteriori state estimate is computed using (3).

Table 1. Summary of variable sample period in-order SPKF using linear state equation and additive noises

---

Nonlinear state-space model:

$$x(t) = A(t_0)x(t - t_0) + w_{t_0}(t)$$
$$y(t) = h(x(t), u(t)) + v(t),$$

where $w_{t_0}(t)$ and $v(t)$ are independent, zero-mean Gaussian noise processes of covariance matrices $\Sigma_{w_{t_0}}$ and Σ_v, respectively.

Definition: Let $p = 2 \times \dim(x(t))$.
Initialization: At time zero, set $t_x = 0$ and

$$\hat{x}^+(0) = \mathbb{E}\left[x(0)\right]$$
$$\Sigma_{\tilde{x}(0)}^+ = \mathbb{E}\left[(x(0) - \hat{x}^+(0))(x(0) - \hat{x}^+(0))^T\right]$$

Computation: For each sample occurring in-order, (i.e., $t_m \geq t_x$) compute:

Initialize time pointers:	$t_p = t_x$, $t_x = t_m$, and $t_0 = t_x - t_p$.
State est. time update:	$\hat{x}^-(t_x) = A(t_0)\hat{x}^+(t_p)$.
Error cov. time update:	$\Sigma_{\tilde{x}(t_x)}^- = A(t_0)\Sigma_{\tilde{x}(t_p)}^+ A(t_0)^T + \Sigma_{w_{t_0}}$.
Output estimate:	$\mathcal{X}^-(t_x) = \left\{\hat{x}^-(t_x), \hat{x}^-(t_x) + \gamma\sqrt{\Sigma_{\tilde{x}(t_x)}^-},\right.$
	$\left.\hat{x}^-(t_x) - \gamma\sqrt{\Sigma_{\tilde{x}(t_x)}^-}\right\}$.
	$\mathcal{Y}_i(t_m) = h(\mathcal{X}_i^-(t_x), u(t_m))$.
	$\hat{y}(t_m) = \sum_{i=0}^{p} \alpha_i^{(m)} \mathcal{Y}_i(t_m)$.
Estimator gain matrix:	$\Sigma_{\tilde{y}(t_m)}^- = \sum_{i=0}^{p} \alpha_i^{(c)}\left(\mathcal{Y}_i(t_m) - \hat{y}(t_m)\right)\left(\mathcal{Y}_i(t_m) - \hat{y}(t_m)\right)^T$
	$+ \Sigma_v$.
	$\Sigma_{\tilde{x}(t_x)\tilde{y}(t_m)}^- = \sum_{i=0}^{p} \alpha_i^{(c)}\left(\mathcal{X}_i^-(t_x) - \hat{x}^-(t_x)\right)\left(\mathcal{Y}_i(t_m) - \hat{y}(t_m)\right)^T$.
	$L(t_x, t_m) = \Sigma_{\tilde{x}(t_x)\tilde{y}(t_m)}^-\left(\Sigma_{\tilde{y}(t_m)}^-\right)^{-1}$.
State est. meas. update:	$\hat{x}^+(t_x) = \hat{x}^-(t_x) + L(t_x, t_m)\left(y(t_m) - \hat{y}(t_m)\right)$.
Error cov. meas. update:	$\Sigma_{\tilde{x}(t_x)}^+ = \Sigma_{\tilde{x}(t_x)}^- - L(t_x, t_m)\Sigma_{\tilde{y}(t_m)}^- L^T(t_x, t_m)$.

In-order SPKF step 6: Error covariance measurement update. The state estimate error covariance matrix is updated directly from the optimal formulation:
$$\Sigma_{\tilde{x}(t_x)}^+ = \Sigma_{\tilde{x}(t_x)}^- - L(t_x, t_m)\Sigma_{\tilde{y}(t_m)}^- L(t_x, t_m)^T.$$

For reference, the in-order SPKF optimized for a linear state equation and additive sensor noise is summarized in Table 1.

4   Out-of-Order Sigma-Point Kalman Filters (O^{3}SPKF)

This chapter introduces a novel variant of the SPKF that allows the filter to be updated using out-of-sequence sensor data. That is, the filter state estimate may already be updated to time t_x using data sensed at t_x when a new piece of information arrives that is the result of a sensor reading taken at time $t_m < t_x$. The most common reasons for such out-of-sequence data include: inter- and intra-UAV communication latency, and processing latency. Ideally, this old sensor data should not be discarded, since it still contains information related to the target's present state, but its impact should be discounted appropriately, based on how stale the measurement is.

A similar problem was treated in [11], where a SPKF needed to be updated based on time-lagged sensor data from a global positioning system (GPS) unit. In their work, however, the time lag was constant and known a priori. In our case, the time lag is not constant, neither is it known a priori. However, we do assume that sensor data has a time-stamp on it so that we can calculate the time lag. Nevertheless, reference [11] gives us a clue as to how to modify the SPKF to our purposes.

In this section, we will assume that measurements arrive out-of-sequence; that is, $t_m < t_x$.

Out-of-order SPKF steps 1 and 2: State estimate time update: First, assign $t_p = t_x$ (the prior value of the filter time), then set $t_x = \max(t_x, t_m) = t_x$. We desire to estimate $\hat{x}^-(t_x)$ using prior information regarding $x(t_p)$ and the state equation. However, since t_x has not been changed by this measurement, we simply retain the prior values of $\hat{x}^-(t_x) = \hat{x}^+(t_x)$ and $\Sigma^-_{\tilde{x}(t_x)} = \Sigma^+_{\tilde{x}(t_x)}$.

Out-of-order SPKF step 3: Estimate system output $y(t_m) \neq y(t_x)$: When using out-of-sequence measured data to update the SPKF, the state update equation maintains the same linear form $\hat{x}^+(t_x) = \hat{x}^-(t_x) + L(t_x, t_m)(y(t_m) - \hat{y}(t_m))$. The key insight from [11] is that in such a case, $L(t_x, t_m)$ should be calculated via Eq. (11) instead of using the standard SPKF formulation where $t_x = t_m$. In order to compute this update, we require an estimate $\hat{y}(t_m)$ and the covariances required to compute $L(t_x, t_m)$. These in turn require sigma points representing $\hat{x}^-(t_x)$ and $\hat{y}(t_m)$. The first are easily computed:

$$\mathcal{X}^-(t_x) = \left\{ \hat{x}^-(t_x), \hat{x}^-(t_x) + \gamma \sqrt{\Sigma^-_{\tilde{x}(t_x)}}, \hat{x}^-(t_x) - \gamma \sqrt{\Sigma^-_{\tilde{x}(t_x)}} \right\}.$$

It remains to calculate the sigma points to represent the distribution of $\hat{y}(t_m)$—we can do so using the output equation $h(\cdot)$ to find these output sigma points if we are able to calculate the sigma points representing $\hat{x}^-(t_m)$. To do so, consider the following specific form of a state equation where we define the time interval $t_0 = t_x - t_m$[3]

---

[3] Note that the particular form of a linear state equation given above is not necessary for this general idea to work; however, if the equation is nonlinear, it must be locally Lipschitz. Sigma points representing $x(t_x)$ and $w_{t_0}(t_x)$ must be propagated backward in time to compute the mean and covariance estimates of $x(t_m)$.

$$x(t_x) = A(t_0)x(t_m) + w_{t_0}(t_x)$$
$$x(t_m) = A(t_0)^{-1}x(t_x) - A(t_0)^{-1}w_{t_0}(t_x)$$
$$\hat{x}^-(t_m) = \mathbb{E}\left[x(t_m)|\mathbb{Y}^-\right] = A(t_0)^{-1}\hat{x}^-(t_x),$$

where $\mathbb{Y}^-$ is the history of all measurements, excluding the "new" measurement taken at time t_m. Therefore, we can "predict" a prior state estimate given the present state estimate. The prior covariance can also be computed, and is found to be

$$\Sigma^-_{\tilde{x}(t_m)} = A(t_0)^{-1}\left(\Sigma^-_{\tilde{x}(t_x)} + \Sigma_{w_{t_0}}\right)A(t_0)^{-T}.$$

Using these two quantities, we can compute the desired sigma points representing $\hat{x}^-(t_m)$ as

$$\mathcal{X}^-(t_m) = \left\{\hat{x}^-(t_m), \hat{x}^-(t_m) + \gamma\sqrt{\Sigma^-_{\tilde{x}(t_m)}}, \hat{x}^-(t_m) - \gamma\sqrt{\Sigma^-_{\tilde{x}(t_m)}}\right\}.$$

These sigma points are passed through the output equation $h(\cdot)$ to first form $\mathcal{Y}_i^-(t_m) = h(\mathcal{X}_i^-(t_m), u(t_m))$ and then $\hat{y}^-(t_m)$ as before.

Out-of-order SPKF steps 4–6:   The remaining steps are straightforward now that we have a means for calculating the sigma points corresponding to $y(t_m)$. The entire O³SPKF for a linear state equation and additive sensor noise is summarized in Table 2.

5   An Example Model of Motion

In order to use any Kalman filtering technique to localize a target, we require a model of the target's dynamics. Due to the non-cooperative nature of the target we wish to localize in our present research, this model cannot be known a priori; therefore, we must employ an approximate model. Here, we use a "nearly constant velocity" (NCV) model of dynamics. For a state vector $x(t) = [p_x(t), p_y(t), v_x(t), v_y(t)]^T$, where $p_x(t)$ is the "x" position coordinate of the target, $p_y(t)$ is the "y" position coordinate of the target, $v_x(t)$ is the "x" velocity of the target and $v_y(t)$ is the "y" velocity of the target, we have:

$$\dot{x}(t) = \underbrace{\begin{bmatrix} 0 & 0 & 1 & 0 \\ 0 & 0 & 0 & 1 \\ 0 & 0 & 0 & 0 \\ 0 & 0 & 0 & 0 \end{bmatrix}}_{A} x(t) + w(t)$$

$$y(t) = h(x(t), u(t)) + v(t),$$

where the stochastic signals $w(t)$ and $v(t)$ are assumed to be Gaussian and white, and sensor noise $v(t)$ has covariance matrix Σ_v and process noise $w(t)$ has covariance matrix $\Sigma_w(t) = \text{diag}(0, 0, \sigma^2, \sigma^2)$. The output equation depends on $h(\cdot)$, which itself depends on the sensor being used to produce a measurement

Table 2. Summary of variable sample period out-of-order SPKF using linear state equation and additive noises

---

Nonlinear state-space model:

$$x(t) = A(t_0)x(t - t_0) + w_{t_0}(t)$$
$$y(t) = h(x(t), u(t)) + v(t),$$

where $w_{t_0}(t)$ and $v(t)$ are independent, zero-mean Gaussian noise processes of covariance matrices $\Sigma_{w_{t_0}}$ and Σ_v, respectively.

Definition: Let $p = 2 \times \dim(x(t))$.
Initialization: At time zero, set $t_x = 0$ and

$$\hat{x}^+(0) = \mathbb{E}\left[x(0)\right]$$
$$\Sigma_{\tilde{x}(0)}^+ = \mathbb{E}\left[(x(0) - \hat{x}^+(0))(x(0) - \hat{x}^+(0))^T\right]$$

Computation: For each sample occurring out-of-order, (i.e., $t_m < t_x$) compute:

Initialize time pointers:	$t_0 = t_x - t_m.$
Output estimate:	$\hat{x}^-(t_m) = A(t_0)^{-1}\hat{x}^+(t_x).$

$$\Sigma_{\tilde{x}(t_m)}^- = A(t_0)^{-1}\left(\Sigma_{\tilde{x}(t_x)}^+ + \Sigma_{w_{t_0}}\right)A(t_0)^{-T}.$$

$$\mathcal{X}^-(t_m) = \left\{\hat{x}^-(t_m), \hat{x}^-(t_m) + \gamma\sqrt{\Sigma_{\tilde{x}(t_m)}^-},\right.$$
$$\left.\hat{x}^-(t_m) - \gamma\sqrt{\Sigma_{\tilde{x}(t_m)}^-}\right\}.$$

$$\mathcal{Y}_i(t_m) = h(\mathcal{X}_i^-(t_m), u(t_m)).$$

$$\hat{y}(t_m) = \sum_{i=0}^p \alpha_i^{(m)}\mathcal{Y}_i(t_m).$$

Estimator gain matrix:
$$\mathcal{X}^+(t_x) = \left\{\hat{x}^+(t_x), \hat{x}^+(t_x) + \gamma\sqrt{\Sigma_{\tilde{x}(t_x)}^+},\right.$$
$$\left.\hat{x}^+(t_x) - \gamma\sqrt{\Sigma_{\tilde{x}(t_x)}^+}\right\}.$$

$$\Sigma_{\tilde{y}(t_m)}^- = \sum_{i=0}^p \alpha_i^{(c)}\left(\mathcal{Y}_i(t_m) - \hat{y}(t_m)\right)\left(\mathcal{Y}_i(t_m) - \hat{y}(t_m)\right)^T$$
$$+ \Sigma_v.$$

$$\Sigma_{\tilde{x}(t_x)\tilde{y}(t_m)}^- = \sum_{i=0}^p \alpha_i^{(c)}\left(\mathcal{X}_i^+(t_x) - \hat{x}^+(t_x)\right)\left(\mathcal{Y}_i(t_m) - \hat{y}(t_m)\right)^T.$$

$$L(t_x, t_m) = \Sigma_{\tilde{x}(t_x)\tilde{y}(t_m)}^-\left(\Sigma_{\tilde{y}(t_m)}^-\right)^{-1}.$$

State est. meas. update:	$\hat{x}^+(t_x) = \hat{x}^-(t_x) + L(t_x, t_m)\left(y(t_m) - \hat{y}(t_m)\right).$
Error cov. meas. update:	$\Sigma_{\tilde{x}(t_x)}^+ = \Sigma_{\tilde{x}(t_x)}^- - L(t_x, t_m)\Sigma_{\tilde{y}(t_m)}^- L^T(t_x, t_m).$

---

and perhaps a measurable input signal $u(t)$. This model says, in effect, that the target velocity is generally constant except via perturbations to its acceleration through $w(t)$, and that measurements may be taken that somehow relate to the position and velocity of the target.

We will be updating the Kalman filter at non-deterministically separated discrete points in time. Therefore, we need to be able to integrate the effect of $w(t)$

on $x(t)$ over a variable period t_0 (the integral essentially performing a convolution) and result in a variable-sample-rate discrete-time model of the form:

$$x(t + t_0) = A(t_0)x(t) + w_{t_0}(t)$$
$$y(t) = h(x(t), u(t)) + v(t).$$

(We see that the output equation is unchanged). We compute $A(t_0) = e^{At_0}$, where $e^{(\cdot)}$ is the matrix-exponential function. We further compute [16]

$$\Sigma_{w_{t_0}} = \int_0^{t_0} e^{A(t_0 - \tau)} \Sigma_w e^{A^T(t_0 - \tau)} \, d\tau$$

to evaluate the equivalent discrete-time noise covariance based on the continuous-time noise covariance. For the NCV model, the state equation becomes

$$x(t + t_0) = \underbrace{\begin{bmatrix} 1 & 0 & t_0 & 0 \\ 0 & 1 & 0 & t_0 \\ 0 & 0 & 1 & 0 \\ 0 & 0 & 0 & 1 \end{bmatrix}}_{A(t_0)} x(t) + w_{t_0}(t)$$

where $\Sigma_{w_{t_0}}$ can be evaluated analytically, and is found to be

$$\Sigma_{w_{t_0}} = \begin{bmatrix} \frac{t_0^3 \sigma^2}{3} & 0 & \frac{t_0^2 \sigma^2}{2} & 0 \\ 0 & \frac{t_0^3 \sigma^2}{3} & 0 & \frac{t_0^2 \sigma^2}{2} \\ \frac{t_0^2 \sigma^2}{2} & 0 & t_0 \sigma^2 & 0 \\ 0 & \frac{t_0^2 \sigma^2}{2} & 0 & t_0 \sigma^2 \end{bmatrix}.$$

This model is suitable for use with either the in-order SPKF or the out-of-order SPKF as developed in this chapter.

6   Performance Comparisons

Results indicative of the performance of the simple, buffered, and O³SPKF approaches were generated via simulation of multiple UAVs locating a mobile target. Target and UAV trajectories were generated using the United States Air Force Academy (USAFA) multiple-UAV simulator, to be described next. The overall methodology for generating results will then be discussed.

6.1   The USAFA Multiple UAV Simulation System Control Architecture

In this section, we briefly present the distributed control architecture we developed to search, detect, and locate ground targets using multiple UAVs. The purpose is to provide readers the proper context in which the out-of-order Sigma

Point Kalman Filter technique allows us to achieve our overall goal. As mentioned earlier, the overall goal is to develop a cooperative multiple heterogeneous UAVs system for military applications. In particular, we are interested in developing a distributed control architecture for each UAV that can optimize transmitted sensor information obtained by nearby UAVs. Collectively, the multiple UAVs cooperate to search, detect, and locate ground targets. To that end, we have developed the following control architecture.

The control architecture is made of a behavior-based state machine with four different states shown in Fig. 1: Global Search (GS), Approach Target (AT), Locate Target (LT), and Target Re-acquisition (TR). Each UAV operates in one of the four states at a time. The switch between two operating states is based on the current state of a UAV, sensor values obtained from the UAV and other neighboring UAVs, and state data transmitted from other UAVs in the mission area. For details of the switching conditions, see Table 3.

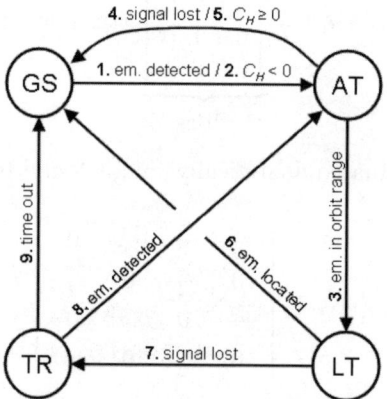

Fig. 1. Decision state machine for UAV state selection. The numbered events that trigger each particular directional connector are listed in Table 3.

Each UAV starts in the GS state when launched. During this state, a UAV uses a set of heuristic rules to guide its movement. The rules include visiting locations with little or no recent history, moving away from other nearby UAVs to maximize the search coverage, and moving straight, if possible, to reduce fuel use. Once a target is detected, the UAV that detected the target switches to the AT state and approaches the target, while other UAVs that receive the target detection information will independently decide whether to approach the detected target or continue to operate in the current operating state based on the estimated distance of the target, the number of UAVs that are approaching the target, and the estimated number of targets in the mission area that have not been detected. Once a UAV is within a region from which it can safely locate the target, it switches the operating state from AT to LT and flies around the target with a pre-determined orbit. Other UAVs that committed to help locate

Table 3. List of events that trigger decisions to change states in the decision state machine of each UAV

#	From	To	Event
1	GS	AT	New target (e.g., RF emitter) detected by UAV's sensor.
2	GS	AT	Decision to cooperate with an ongoing localization effort.
3	AT	LT	UAV arrives at orbit range from target's estimated position.
4	AT	GS	Target becomes occluded (e.g., emitter stops transmitting).
5	AT	GS	Decision to abandon an ongoing localization effort.
6	LT	GS	Target successfully located.
7	LT	TR	Target becomes occluded (e.g., emitter stops transmitting).
8	TR	AT	Target detected by UAV's sensor.
9	TR	GS	Maximum time for TR reached.

the target also switch to the LT state as they enter the orbit. As the UAVs fly around the target, they position themselves to maximize collective sensing capabilities while combining sensor data among the UAVs on the orbit. It is this state of our operation where the current work on O^3SPKF is used to combine multiple sensor information obtained by the UAVs. A target may disappear from the sensors before it is localized within a desired accuracy. For example, for a radio frequency signal emitting target, it may stop emitting before it can be located. For such situations, UAVs who are operating in the LT state switch to the TR state. During this state, a UAV engages in a search pattern similar to a global search except in a smaller scale to continue to look for the lost target. The UAV will continue in this state either until the target reappears at which time it switches back to the LT state or when a pre-determined time interval has elapsed at which time it switches to the GS state.

Figure 2 shows a screenshot of the simulator in action. The emitter location is indicated by a white cross in the center of the large circle—the emitter leaves behind it a fading trail of crosses showing its path. The large circle indicates the desired orbiting radius of the UAVs—this is not possible in practice due to the random motion of the targets, but is approximated by the UAV control algorithm. The small squares denote the UAV positions (two in this case). Lines are drawn between the UAVs and their target-position estimates, with the SPKF state three-sigma uncertainty denoted by the ellipse centered on the target position estimate.

6.2   The Simulation Process

The following methodology was employed

- First, the USAFA Multiple UAV Simulation System was used to generate the trajectories of a mobile target and the UAVs tracking it. UAV locations were initialized by randomly placing them within a 2 km radius of the target. The UAV flight paths were then controlled to converge to an orbit of 0.5 km distance from the target, with inter-UAV angles of 90° for a two-UAV simulation and ±120° for a three-UAV simulation. Locations of the target

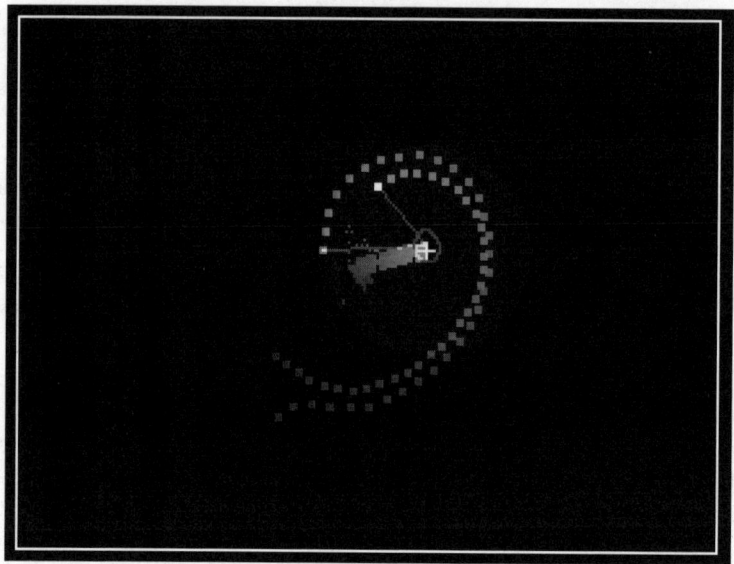

Fig. 2. Screenshot of simulation system in action

and UAVs were recorded once per second (of simulated time). Targets moved according to the NCV model with $\sigma = 2 \times 10^{-4}$, with maximum velocity limited to 20 km/h. Nominal UAV velocity was 100 km/h.
- Secondly, randomness was applied to the simulated data. Measurement requests were made to each sensor at a nominal sample rate of 1 Hz, corresponding to the original data. Random clock timing jitter uniformly distributed between −5 ms and 5 ms was applied to each measurement time (locations of target and UAV were interpolated at these instants from the original data). Two sensors per UAV were simulated:
  - One sensor providing target emission DOA information, with measurement timestamp of 0.05 s after the measurement was requested, and a sensor-data processing time uniformly distributed between [0.06 s, 0.061 s] (the randomness accounts for timing uncertainty in a multi-threaded processing system), and Gaussian sensor noise with zero mean and standard deviation of 6 deg. These characteristics correspond roughly to a radio-frequency (RF) DOA sensor that we are currently building for a prototype UAV.
  - A second sensor also providing DOA information, with measurement timestamp of 0.01 s after the measurement was requested, and a sensor-data processing time uniformly distributed between [0.2 s, 0.22 s], and Gaussian sensor noise with zero mean and standard deviation of 3 deg. These characteristics correspond roughly to a camera-based DOA sensor that we are building for a prototype UAV. The measurement process is much faster than for the RF sensor, but requires a longer processing time.

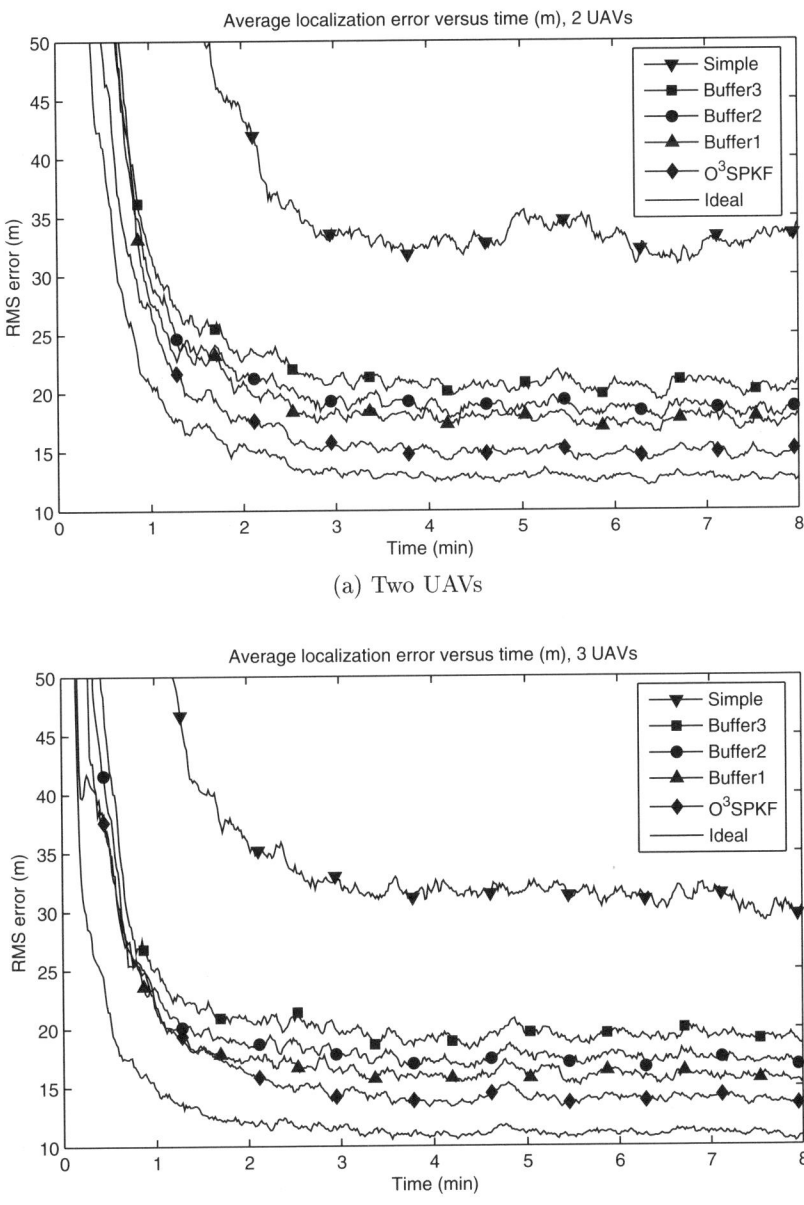

Fig. 3. Example of localization error improvement over time. "Ideal" = no-latency SPKF result.

We note that the "simple approach" will never make use of the data from the second sensor. We therefore expect that results using this approach will be

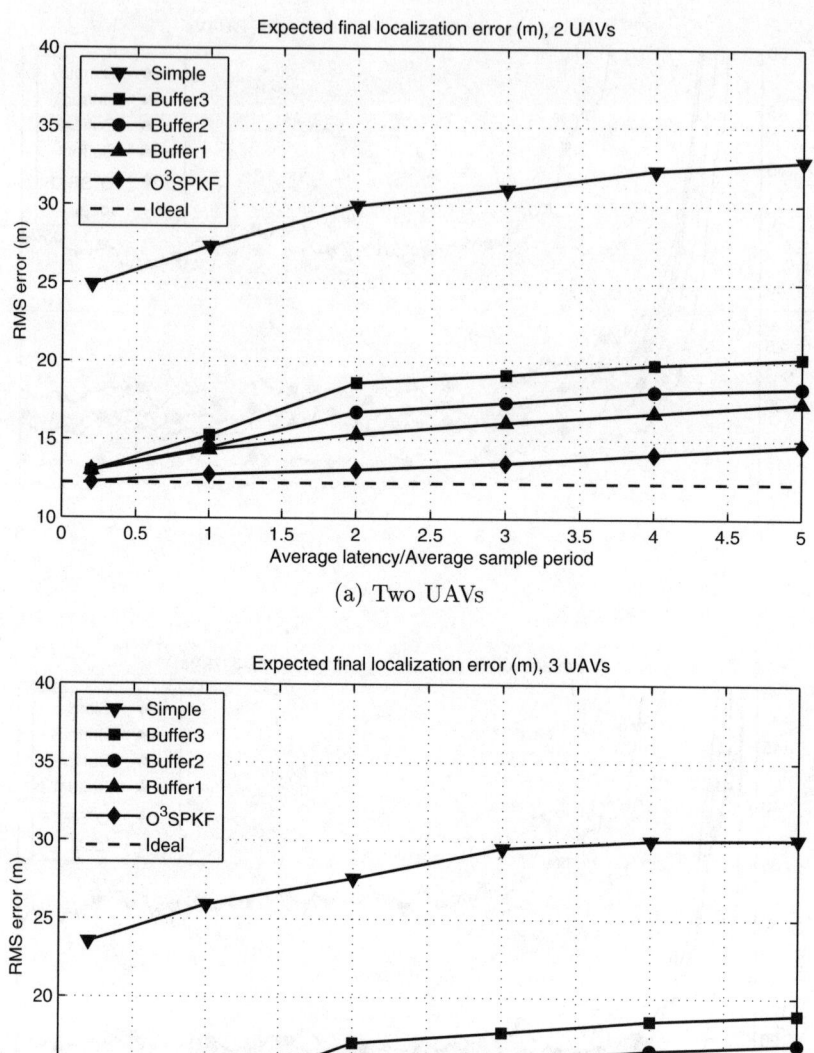

(a) Two UAVs

(b) Three UAVs

Fig. 4. Simulation summary performance plots. "Ideal" = no-latency SPKF result.

quite poor, and that even a buffer of one sample could improve performance significantly.

Random latency was also added to those data traveling from one UAV to another. This was computed using an exponential random variable with differing means, as reported in the results sections below.

– The output sensor data and UAV position data from the second step were then used as input to the ideal method, the simple method, the buffered method with differing buffer lengths, and the O^3SPKF (the "ideal method" is a post-processing method that sorts all data in order by t_m, then applies an SPKF to it. This provides a performance bound that is not achievable in practice since the sorting process is non-causal). Target state estimates and uncertainties were output whenever they were updated.
– True position data from the first step and estimated position data from the third step were compared.

6.3 Results

Simulations using the above methodology were run for scenarios comprising two UAVs localizing a target and three UAVs localizing a target. In each case 500 simulations were run, with the data being processed by each of the methods, and the results averaged in a root-mean-square (RMS) sense. Figure 3 shows a representative plot of the average localization error versus time for an expected latency of 5 s. The ideal (non-achievable) case is best, as expected, followed by O^3SPKF, the buffered method with three different buffer lengths, and then the simple method. Note that the the "Buffer$k$" method denotes a buffer length $N = \lceil k\mu/T_s \rceil \times$ number of UAVs $\times$ number of sensors per UAV. The poor results of the simple method are not difficult to explain since the majority of the sensor readings are discarded. Perhaps surprisingly, the buffered method with the smallest buffer performed best—it appears that discarding data is not as costly as a stale state estimate (due to a larger prediction time) when the target is moving.

Various latencies were simulated, and summary results are presented in Fig. 4. Communication latencies of 0.2 s (the value we expect in our prototype UAV system), and 1 s through 5 s were simulated (again, the RMS average of the final localization error of 500 simulations per data point are plotted). We see that all methods degrade quite gently with increasing latency, and that the O^3SPKF performs best of all.

7 Conclusions

In this chapter we address the problem of sensor fusion to localize a target when some of the measurements arrive at the sensor-fusion process out-of-order. We propose that sigma-point Kalman filtering is a good approach to sensor fusion, but is not able to handle the out-of-order measurements directly. Several simple remedies are presented, which include discarding out-of-order measurements, or alternately buffering a number of measurements before presenting them to the sensor-fusion process so that the majority of them may be retained and sorted in order. The problem with simply discarding out-of-order measurements is that they do contain useful data regarding the target position, and it is not wise to throw away this information. The problem with buffering measurements is that the fusion process has added delay built into it, so that its state estimate is stale—this estimate must then be propagated forward in time to the present in order to make control decisions, and the propagation step adds error that increases with the amount of time required for the propagation.

We present an alternate approach to either of these ad-hoc methods. We re-derive the sigma-point Kalman filter such that the modified filter, which we call the out-of-order sigma-point Kalman filter (O^3SPKF), is able to directly incorporate the out-of-order measurements without buffering and without discarding measurements. If a measurement arrives at the sensor-fusion process in-order, the standard SPKF steps are executed. If a measurement arrives at the sensor-fusion process out-of-order, the modified O^3SPKF steps are executed. Since no measurements are discarded by the O^3SPKF, it must execute its steps more frequently than the methods that do discard sensor data. (For example, the buffered methods for $k \in \{1, 2, 3\}$ retain on average 68%, 86%, and 95% of the measurements received (respectively), so the number of iterations of the SPKF required by the buffered methods are similarly that fraction of the number of iterations required by the O^3SPKF.) However, per iteration, the computational complexity of O^3SPKF is the same as SPKF, it does not require memory overhead for buffering, and it gave the best simulation results of all the methods attempted. In conclusion, the O^3SPKF works very well, and is an excellent candidate for sensor fusion for the application of locating targets using multiple UAVs.

References

1. York, G., Pack, D.J.: Comparative study of time-varying target localization methods using multiple unmanned aerial vehicles: Kalman estimation and triangulation techniques. In: Proc. IEEE International Conference on Networking, Sensing and Control, (Tuscon, AZ) (March 2005) 305–10
2. Pack, D.J., York, G.: Developing a control architecture for multiple unmanned aerial vehicles to search and localize RF time-varying mobile targets: Part I. In: Proc. IEEE International Conference on Robotics and Automation, (Barcelona, Spain) (April 2005) 3965–70
3. Kalman, R.: A new approach to linear filtering and prediction problems. Transactions of the ASME—Journal of Basic Engineering **82, Series D** (1960) 35–45
4. Kalman, R.: The Seminal Kalman Filter Paper (1960). http://www.cs.unc.edu/~welch/kalman/kalmanPaper.html Accessed 20 May 2004.
5. Julier, S., Uhlmann, J., Durrant-Whyte, H.: A new approach for filtering nonlinear systems. In: Proceedings of the American Control Conference. (1995) 1628–32
6. Julier, S., Uhlmann, J.: A new extension of the Kalman filter to nonlinear systems. In: Proceedings of the 1997 SPIE AeroSense Symposium, SPIE (Orlando, FL, April 21–24, 1997)
7. Julier, S., Uhlmann, J.: Unscented filtering and nonlinear estimation. Proceedings of the IEEE **92**(3) (March 2004) 401–22

8. Savage, C., Cramer, R., Schmitt, H.: TDOA geolocation with the unscented Kalman filter. In: Proc. 2006 IEEE International Conference on Networking, Sensing and Control. (April 2006) 602–6
9. Nørgaard, M., Poulsen, N., Ravn, O.: Advances in derivative-free state estimation for nonlinear systems. Technical rep. IMM-REP-1998-15, Dept. of Mathematical Modeling, Tech. Univ. of Denmark, 28 Lyngby, Denmark (April 2000)
10. Nørgaard, M., Poulsen, N., Ravn, O.: New developments in state estimation for nonlinear systems. Automatica **36**(11) (November 2000) 1627–38
11. van der Merwe, R.: Sigma-Point Kalman Filters for Probabilistic Inference in Dynamic State-Space Models. PhD thesis, OGI School of Science & Engineering at Oregon Health & Science University (April 2004)
12. Hirsch, M., Ortiz-Pena, H., Cooperman, R.: Smoothing techniques using an IMM and multi-sensor tracking with time-late data. In: Proceedings of the National Symposium on Sensor and Data Fusion. (Laurel, MD, June 2004)
13. Press, W., Teukolsky, S., Vetterling, W., Flannery, B.: Numerical Recipes in C: The Art of Scientific Computing. second edn. Cambridge University Press (1992)
14. Stewart, G.: Matrix Algorithms. Volume I: Basic Decompositions. SIAM (1998)
15. Julier, S., Uhlmann, J.: A general method for approximating nonlinear transformations of probability distributions. Technical report, RRG, Department of Engineering Science, Oxford University (November 1996)
16. Burl, J.B.: Linear Optimal Control: H_2 and H_∞ Methods. Addison-Wesley, Menlo Park, CA (1999)

Multi-cumulant Control
for Zero-Sum Differential Games:
Performance-Measure Statistics
and State-Feedback Paradigm

Khanh D. Pham

Space Vehicles Directorate
Air Force Research Laboratory
Kirtland AFB, NM 87117 U.S.A
khanh.pham@kirtland.af.mil

Abstract. This chapter presents an extension of cost-cumulant control theory over a finite horizon for a class of stochastic zero-sum differential games wherein the evolution of the states of the game in response to decision strategies selected by two players from sets of admissible controls is described by a stochastic linear differential equation and a standard integral-quadratic cost. A direct dynamic programming approach for the Mayer optimization problem is used to solve for a multi-cumulant based solution when both players measure the states and minimize the first finite number of cumulants of the standard integral-quadratic cost associated with this special class of differential games. This innovative decision-making paradigm is proposed herein to provide not only a mechanism in which the conflicting interests of noncooperative players can be optimized, but also an analytical tool which is used to provide a complete statistical description of the global performance of the stochastic differential game.

1 Introduction

This chapter considers a closed-loop two-person zero-sum linear-quadratic game wherein the dynamics of the game in response to control variables selected by both players from a class of linear-feedback controllers is described by a stochastic linear differential equation. In seeking optimal control strategies whose respective objectives are minimization and maximization of a finite linear combination of the first k cost cumulants of an integral-quadratic random cost associated with the class of linear stochastic systems over a finite horizon, the recently developed statistical control theory [7]-[17] is extended herein. The extension which is manifested through the resulting cumulant-generating equations, now allows the incorporation of classes of linear feedback controllers to affect and predict more accurately the effects of non-Gaussian perturbations on the accuracy of system performance via a complete statistical description. In other words, using these high-order cost cumulants, it is possible to obtain an approximation of the system performance distribution.

M.J. Hirsch et al. (Eds.): Adv. in Cooper. Ctrl. & Optimization, LNCIS 369, pp. 45–64, 2007.
springerlink.com © Springer-Verlag Berlin Heidelberg 2007

Since the formulation of multi-cumulant and zero-sum games is parameterized both by the number of cost cumulants and by the scalar coefficients in the linear combination, it may be viewed both as a generalization of linear-quadratic Gaussian control, when the first cost cumulant is optimized and of two-person zero-sum differential games when a certain denumerable linear combination of cost cumulants is optimized. The set of coupled matrix Riccati differential equations is introduced, whose solvability leads to the existence of the closed-loop feedback saddle points for the corresponding multi-cumulant and zero-sum game under some additional mild conditions. It is worth mentioning that the multi-cumulant and zero-sum game is an initial cost problem, in contrast with the more traditional terminal cost class of investigations. One may address an initial cost problem by introducing changes of variables which convert it to a terminal cost problem. However, this modifies the natural context of cost cumulants, which it is preferable to retain. Instead, one may take a more direct dynamic programming approach to the initial cost problem. Such an approach is illustrative of the more general concept of the principle of optimality, an idea tracing its roots back to the 17th century.

2   Problem Formulation

Let's consider a zero-sum stochastic differential game with two noncooperative players, identified as u_1 and u_2. Suppose $(t_0, x_0) \in [t_0, t_f] \times \mathbb{R}^n$ is fixed and a system input noise $w(t) \triangleq w(t, \omega) : [t_0, t_f] \times \Omega \mapsto \mathbb{R}^p$ is an p-dimensional stationary Wiener process defined with $\{\mathcal{F}_t\}_{t \geq 0}$ being its natural filtration on a complete filtered probability space $(\Omega, \mathcal{F}, \{\mathcal{F}_t\}_{t \geq 0}, \mathcal{P})$ over $[t_0, t_f]$ with the correlation of increments

$$E\left\{[w(\tau) - w(\xi)][w(\tau) - w(\xi)]^T\right\} = W|\tau - \xi|, \quad W > 0 .$$

Also, decision sets $\mathcal{U}_1 \in L^2_{\mathcal{F}_t}(\Omega; \mathcal{C}([t_0, t_f]; \mathbb{R}^{m_1}))$ and $\mathcal{U}_2 \in L^2_{\mathcal{F}_t}(\Omega; \mathcal{C}([t_0, t_f]; \mathbb{R}^{m_2}))$ are assumed to be the subsets of Hilbert space of $\mathbb{R}^{m_1}$-valued and $\mathbb{R}^{m_2}$-valued, square integrable processes on $[t_0, t_f]$ that are adapted to the σ-field $\mathcal{F}_t$ generated by $w(t)$, respectively. Associated with each $(u_1, u_2) \in \mathcal{U}_1 \times \mathcal{U}_2$ is a standard finite-horizon integral-quadratic form (IQF) random cost $J : [t_0, t_f] \times \mathbb{R}^n \times \mathcal{U}_1 \times \mathcal{U}_2 \mapsto \mathbb{R}^+$ (for which the first player, u_1 tries to minimize, while the second player, u_2 attempts to maximize it) such that

$$J(t_0, x_0; u_1, u_2) = x^T(t_f)Q_f x(t_f)$$
$$+ \int_{t_0}^{t_f} \left[x^T(\tau)Q(\tau)x(\tau) + u_1^T(\tau)R_{11}(\tau)u_1(\tau) - u_2^T(\tau)R_{22}(\tau)u_2(\tau)\right] d\tau , \quad (1)$$

where the system states of the game, $x(t) \triangleq x(t, \omega) : [t_0, t_f] \times \Omega \mapsto \mathbb{R}^n$ belong to the Hilbert space $L^2_{\mathcal{F}_t}(\Omega; \mathcal{C}([t_0, t_f]; \mathbb{R}^n))$ with $E\left\{\int_{t_0}^{t_f} x^T(\tau)x(\tau)d\tau\right\} < \infty$ and evolve according to the stochastic differential equation

$$dx(t) = (A(t)x(t) + B_1(t)u_1(t) + B_2(t)u_2(t))\,dt + G(t)dw(t) \tag{2}$$
$$x(t_0) = x_0 \ .$$

The system coefficient matrices $A \in \mathcal{C}([t_0, t_f]; \mathbb{R}^{n \times n})$, $B_1 \in \mathcal{C}([t_0, t_f]; \mathbb{R}^{n \times m_1})$, $B_2 \in \mathcal{C}([t_0, t_f]; \mathbb{R}^{n \times m_2})$ and $G \in \mathcal{C}([t_0, t_f]; \mathbb{R}^{n \times p})$ are deterministic bounded matrix-valued functions. The terminal penalty weighting $Q_f \in \mathbb{R}^{n \times n}$, the state weighting $Q \in \mathcal{C}([t_0, t_f]; \mathbb{R}^{n \times n})$ and control weightings $R_{11} \in \mathcal{C}([t_0, t_f]; \mathbb{R}^{m_1 \times m_1})$, and $R_{22} \in \mathcal{C}([t_0, t_f]; \mathbb{R}^{m_2 \times m_2})$ are deterministic bounded matrix-valued functions with properties of symmetry and positive semi-definiteness. In addition, $R_{11}(t)$ and $R_{22}(t)$ are invertible.

To put this stochastic differential game in a class of closed-loop feedback control, it is observed that the system (2) is linear and the performance measure (1) is quadratic. Therefore, it is reasonable to assume that the players choose control actions that are optimal within the class of memoryless perfect-state strategies, $\gamma_1 : [t_0, t_f] \times L^2_{\mathcal{F}_t}(\Omega; \mathcal{C}([t_0, t_f]; \mathbb{R}^n)) \mapsto L^2_{\mathcal{F}_t}(\Omega; \mathcal{C}([t_0, t_f]; \mathbb{R}^{m_1}))$ and $\gamma_2 : [t_0, t_f] \times L^2_{\mathcal{F}_t}(\Omega; \mathcal{C}([t_0, t_f]; \mathbb{R}^n)) \mapsto L^2_{\mathcal{F}_t}(\Omega; \mathcal{C}([t_0, t_f]; \mathbb{R}^{m_2}))$

$$u_1(t) = \gamma_1(t, x(t)) = K_1(t)x(t) \ , \tag{3}$$
$$u_2(t) = \gamma_2(t, x(t)) = K_2(t)x(t) \ , \tag{4}$$

where the admissible gains $K_1 \in \mathcal{C}([t_0, t_f]; \mathbb{R}^{m_1 \times n})$ and $K_2 \in \mathcal{C}([t_0, t_f]; \mathbb{R}^{m_2 \times n})$ are deterministic bounded matrix-valued functions defined in appropriate senses.

For a given initial condition $(t_0, x_0) \in [t_0, t_f] \times \mathbb{R}^n$ and subject to strategies (3)-(4), the dynamics of the game (2) is given by

$$dx(t) = [A(t) + B_1(t)K_1(t) + B_2(t)K_2(t)]\,x(t)dt + G(t)dw(t) \ , \tag{5}$$
$$x(t_0) = x_0 \ ,$$

and its IQF cost in the form of a Chi-square random variable, follows

$$J(t_0, x_0; K_1, K_2) = x^T(t_f)Q_f x(t_f)$$
$$+ \int_{t_0}^{t_f} x^T(\tau) \left[Q(\tau) + K_1^T(\tau)R_{11}(\tau)K_1(\tau) - K_2^T(\tau)R_{22}(\tau)K_2(\tau) \right] x(\tau)d\tau \ . \tag{6}$$

It is necessary to develop a procedure for generating cost cumulants of the two-player zero-sum differential game by adapting the parametric method in [5] to characterize a moment-generating function. These cost cumulants are then used to form performance index in the cost-cumulant control optimization. This approach begins with a replacement of the initial condition (t_0, x_0) by any arbitrary pair (α, x_α). Thus, for the given admissible feedback gains K_1 and K_2, the cost functional (6) is seen as the "cost-to-go", $J(\alpha, x_\alpha)$

$$J(\alpha, x_\alpha) \triangleq x^T(t_f)Q_f x(t_f)$$
$$+ \int_{\alpha}^{t_f} x^T(\tau) \left[Q(\tau) + K_1^T(\tau)R_{11}(\tau)K_1(\tau) - K_2^T(\tau)R_{22}(\tau)K_2(\tau) \right] x(\tau)d\tau \ .$$

The moment-generating function of the vector-valued random process (5) is given by the definition

$$\varphi\left(\alpha, x_\alpha; \theta\right) \triangleq E\left\{\exp\left(\theta J\left(\alpha, x_\alpha\right)\right)\right\} , \tag{7}$$

where the scalar $\theta \in \mathbb{R}^+$ is a small parameter. What follows next is the cumulant-generating function

$$\psi\left(\alpha, x_\alpha; \theta\right) \triangleq \ln\left\{\varphi\left(\alpha, x_\alpha; \theta\right)\right\} , \tag{8}$$

in which $\ln\{\cdot\}$ denotes the natural logarithmic transformation of an enclosed entity.

Theorem 1. *Cost Cumulant Generating Function.*
For all $\alpha \in [t_0, t_f]$ and the small parameter $\theta \in \mathbb{R}^+$, define

$$\varphi\left(\alpha, x_\alpha; \theta\right) \triangleq \varrho\left(\alpha; \theta\right) \exp\left(x_\alpha^T \Upsilon(\alpha; \theta) x_\alpha\right) , \tag{9}$$

$$v\left(\alpha; \theta\right) \triangleq \ln\{\varrho\left(\alpha; \theta\right)\} . \tag{10}$$

Then, the cost-cumulant generating function is expressed by

$$\psi\left(\alpha, x_\alpha; \theta\right) = x_\alpha^T \Upsilon(\alpha; \theta) x_\alpha + v\left(\alpha; \theta\right) , \tag{11}$$

where the scalar solution $v\left(\alpha; \theta\right)$ solves the time-backward differential equation with the terminal boundary condition $v\left(t_f; \theta\right) = 0$

$$\frac{d}{d\alpha}v\left(\alpha; \theta\right) = -\operatorname{Tr}\left\{\Upsilon(\alpha; \theta)G\left(\alpha\right)WG^T\left(\alpha\right)\right\} , \tag{12}$$

and the matrix-valued solution $\Upsilon(\alpha; \theta)$ satisfies the time-backward differential equation together with its terminal-valued condition $\Upsilon(t_f; \theta) = \theta Q_f$

$$\begin{aligned}
\frac{d}{d\alpha}\Upsilon(\alpha; \theta) = &-[A(\alpha) + B_1(\alpha)K_1(\alpha) + B_2(\alpha)K_2(\alpha)]^T \Upsilon(\alpha; \theta) \\
&- \Upsilon(\alpha; \theta)[A(\alpha) + B_1(\alpha)K_1(\alpha) + B_2(\alpha)K_2(\alpha)] \\
&- 2\Upsilon(\alpha; \theta)G(\alpha)WG^T(\alpha)\Upsilon(\alpha; \theta) \\
&- \theta\left[Q(\alpha) + K_1^T(\alpha)R_{11}(\alpha)K_1(\alpha) - K_2^T(\alpha)R_{22}(\alpha)K_2(\alpha)\right] .
\end{aligned} \tag{13}$$

In addition, the auxiliary solution $\varrho(\alpha; \theta)$ is satisfying the time-backward differential equation with the terminal boundary condition $\varrho\left(t_f; \theta\right) = 1$

$$\frac{d}{d\alpha}\varrho\left(\alpha; \theta\right) = -\varrho\left(\alpha; \theta\right)\operatorname{Tr}\left\{\Upsilon(\alpha; \theta)G\left(\alpha\right)WG^T\left(\alpha\right)\right\} . \tag{14}$$

Proof. For any given θ, let $\varpi\left(\alpha, x_\alpha; \theta\right) \triangleq \exp\left(\theta J\left(\alpha, x_\alpha\right)\right)$. The moment-generating function becomes

$$\varphi\left(\alpha, x_\alpha; \theta\right) = E\left\{\varpi\left(\alpha, x_\alpha; \theta\right)\right\} ,$$

with time derivative of

$$\frac{d}{d\alpha}\varphi\left(\alpha, x_\alpha; \theta\right) = -\varphi\left(\alpha, x_\alpha; \theta\right)\theta x_\alpha^T \left[Q(\alpha) + K_1^T(\alpha)R_{11}(\alpha)K_1(\alpha)\right.$$
$$\left. - K_2^T(\alpha)R_{22}(\alpha)K_2(\alpha)\right]x_\alpha \ .$$

Using the standard Ito's formula in [1], one gets

$$d\varphi\left(\alpha, x_\alpha; \theta\right) = E\left\{d\varpi\left(\alpha, x_\alpha; \theta\right)\right\},$$
$$= E\left\{\varpi_\alpha\left(\alpha, x_\alpha; \theta\right)d\alpha + \varpi_{x_\alpha}\left(\alpha, x_\alpha; \theta\right)dx_\alpha\right.$$
$$\left. + \frac{1}{2}\mathrm{Tr}\left\{\varpi_{x_\alpha x_\alpha}(\alpha, x_\alpha; \theta)G(\alpha)WG^T(\alpha)\right\}d\alpha\right\},$$
$$= \varphi_\alpha\left(\alpha, x_\alpha; \theta\right)d\alpha$$
$$+ \varphi_{x_\alpha}\left(\alpha, x_\alpha; \theta\right)\left[A(\alpha) + B_1(\alpha)K_1(\alpha) + B_2(\alpha)K_2(\alpha)\right]x_\alpha d\alpha$$
$$+ \frac{1}{2}\mathrm{Tr}\left\{\varphi_{x_\alpha x_\alpha}\left(\alpha, x_\alpha; \theta\right)G\left(\alpha\right)WG^T\left(\alpha\right)\right\}d\alpha \ ,$$

when combined with (9) leads to

$$- \varphi\left(\alpha, x_\alpha; \theta\right)\theta x_\alpha^T\left[Q(\alpha) + K_1^T(\alpha)R_{11}(\alpha)K_1(\alpha) - K_2^T(\alpha)R_{22}(\alpha)K_2(\alpha)\right]x_\alpha$$
$$= \frac{\frac{d}{d\alpha}\varrho\left(\alpha; \theta\right)}{\varrho\left(\alpha; \theta\right)}\varphi\left(\alpha, x_a; \theta\right) + \varphi\left(\alpha, x_\alpha; \theta\right)x_\alpha^T\frac{d}{d\alpha}\varUpsilon(\alpha; \theta)x_\alpha + \varphi\left(\alpha, x_\alpha; \theta\right)\left\{x_\alpha^T\left[A(\alpha)\right.\right.$$
$$\left. + B_1(\alpha)K_1(\alpha) + B_2(\alpha)K_2(\alpha)\right]^T\varUpsilon(\alpha; \theta)x_\alpha$$
$$\left. + x_\alpha^T\varUpsilon_a(\alpha; \theta)\left[A(\alpha) + B_1(\alpha)K_1(\alpha) + B_2(\alpha)K_2(\alpha)\right]x_\alpha\right\}$$
$$+ \varphi\left(\alpha, x_\alpha; \theta\right)\left\{2x_\alpha^T\varUpsilon(\alpha; \theta)G(\alpha)WG^T(\alpha)\varUpsilon(\alpha; \theta)x_\alpha + \mathrm{Tr}\left\{\varUpsilon(\alpha; \theta)G(\alpha)WG^T(\alpha)\right\}\right\}.$$

To have constant and quadratic terms independent of x_α, it is required that

$$\frac{d}{d\alpha}\varUpsilon(\alpha; \theta) = -[A(\alpha) + B_1(\alpha)K_1(\alpha) + B_2(\alpha)K_2(\alpha)]^T\varUpsilon(\alpha; \theta)$$
$$- \varUpsilon(\alpha; \theta)[A(\alpha) + B_1(\alpha)K_1(\alpha) + B_2(\alpha)K_2(\alpha)]$$
$$- 2\varUpsilon(\alpha; \theta)G(\alpha)WG^T(\alpha)\varUpsilon(\alpha; \theta)$$
$$- \theta\left[Q(\alpha) + K_1^T(\alpha)R_{11}(\alpha)K_1(\alpha) - K_2^T(\alpha)R_{22}(\alpha)K_2(\alpha)\right] \ ,$$
$$\frac{d}{d\alpha}\varrho\left(\alpha; \theta\right) = -\varrho\left(\alpha; \theta\right)\mathrm{Tr}\left\{\varUpsilon(\alpha; \theta)G\left(\alpha\right)WG^T\left(\alpha\right)\right\} \ ,$$

with the terminal conditions $\varUpsilon(t_f; \theta) = \theta Q_f$ and $\varrho\left(t_f; \theta\right) = 1$. Finally, the remaining time-backward differential equation satisfied by $\upsilon\left(\alpha; \theta\right)$ is given by

$$\frac{d}{d\alpha}\upsilon\left(\alpha; \theta\right) = -\mathrm{Tr}\left\{\varUpsilon(\alpha; \theta)G\left(\alpha\right)WG^T\left(\alpha\right)\right\} \ , \quad \upsilon\left(t_f; \theta\right) = 0 \ .$$

$\square$

Now cost cumulants can be generated for the zero-sum stochastic differential game by looking at a MacLaurin series expansion of the cumulant-generating function

$$\psi\left(\alpha, x_\alpha; \theta\right) = \sum_{i=1}^{\infty} \kappa_i(\alpha, x_\alpha) \frac{\theta^i}{i!} = \sum_{i=1}^{\infty} \left. \frac{\partial^{(i)}}{\partial\theta^{(i)}} \psi(\alpha, x_\alpha; \theta) \right|_{\theta=0} \frac{\theta^i}{i!} \ , \tag{15}$$

in which $\kappa_i(\alpha, x_\alpha)$'s are called the cost cumulants. Note that the series coefficients can be computed using (11)

$$\left. \frac{\partial^{(i)}}{\partial\theta^{(i)}} \psi(\alpha, x_\alpha; \theta) \right|_{\theta=0} = x_\alpha^T \left. \frac{\partial^{(i)}}{\partial\theta^{(i)}} \Upsilon(\alpha; \theta) \right|_{\theta=0} x_\alpha + \left. \frac{\partial^{(i)}}{\partial\theta^{(i)}} \upsilon(\alpha; \theta) \right|_{\theta=0} \ . \tag{16}$$

In view of results (15) and (16), cost cumulants for the stochastic differential game problem can be obtained as

$$\kappa_i(\alpha, x_\alpha) = x_\alpha^T \left. \frac{\partial^{(i)}}{\partial\theta^{(i)}} \Upsilon(\alpha; \theta) \right|_{\theta=0} x_\alpha + \left. \frac{\partial^{(i)}}{\partial\theta^{(i)}} \upsilon(\alpha; \theta) \right|_{\theta=0} \ , \tag{17}$$

for any finite $1 \le i < \infty$. For notational convenience, the following definitions are introduced:

$$H(\alpha, i) \triangleq \left. \frac{\partial^{(i)}}{\partial\theta^{(i)}} \Upsilon(\alpha; \theta) \right|_{\theta=0} \quad \text{and} \quad D(\alpha, i) \triangleq \left. \frac{\partial^{(i)}}{\partial\theta^{(i)}} \upsilon(\alpha; \theta) \right|_{\theta=0} \ . \tag{18}$$

The next theorem yields an attractive method of generating cost cumulants in time domain. This computational method is preferred to that of (16) in the formulation of cost-cumulant control problems.

Theorem 2. *Cost-Cumulants in Zero-Sum Stochastic Differential Games. Suppose that (A, B_1) and (A, B_2) are uniformly stabilizable. The players choose control strategies $(u_1(t), u_2(t)) = (K_1(t)x(t), K_2(t)x(t))$ for the zero-sum differential game characterized by (5) and (6). For $k \in \mathbb{Z}^+$ fixed and $1 \le i \le k$, the kth cost cumulant in the zero-sum stochastic game is given by*

$$\kappa_k(t_0, x_0; K_1, K_2) = x_0^T H(t_0, k)x_0 + D(t_0, k) \ , \tag{19}$$

in which the cumulant variables $\{H(\alpha, i)\}_{i=1}^k$ and $\{D(\alpha, i)\}_{i=1}^k$ evaluated at $\alpha = t_0$ satisfy the following differential equations (with the dependence of $H(\alpha, i)$ and $D(\alpha, i)$ upon the admissible gains K_1 and K_2 suppressed)

$$\begin{aligned} \frac{d}{d\alpha} H(\alpha, 1) = &- \left[A(\alpha) + B_1(\alpha)K_1(\alpha) + B_2(\alpha)K_2(\alpha)\right]^T H(\alpha, 1) \\ &- H(\alpha, 1)\left[A(\alpha) + B_1(\alpha)K_1(\alpha) + B_2(\alpha)K_2(\alpha)\right] \\ &- Q(\alpha) - K_1^T(\alpha)R_{11}(\alpha)K_1(\alpha) + K_2^T(\alpha)R_{22}(\alpha)K_2(\alpha) \ , \end{aligned} \tag{20}$$

and, for $2 \leq i \leq k$

$$
\begin{aligned}
\frac{d}{d\alpha} H(\alpha, i) = & - \left[A(\alpha) + B_1(\alpha) K_1(\alpha) + B_2(\alpha) K_2(\alpha) \right]^T H(\alpha, i) \\
& - H(\alpha, i) \left[A(\alpha) + B_1(\alpha) K_1(\alpha) + B_2(\alpha) K_2(\alpha) \right] \\
& - \sum_{j=1}^{i-1} \frac{2i!}{j!(i-j)!} H(\alpha, j) G(\alpha) W G^T(\alpha) H(\alpha, i-j) \;,
\end{aligned}
\tag{21}
$$

together with $1 \leq i \leq k$

$$
\frac{d}{d\alpha} D(\alpha, i) = -\mathrm{Tr}\left\{ H(\alpha, i) G(\alpha) W G^T(\alpha) \right\} \;,
\tag{22}
$$

where the terminal conditions $H(t_f, 1) = Q_f$, $H(t_f, i) = 0$ *for* $2 \leq i \leq k$ *and* $D(t_f, i) = 0$ *for* $1 \leq i \leq k$.

Proof. The cost-cumulant expression in (19) is readily justified by using the result (17) and the definitions (18). What remains is to show that the solutions $H(\alpha, i)$ and $D(\alpha, i)$ for $1 \leq i \leq k$ indeed satisfy (20)-(22). Note that the equations (20)-(22) are satisfied by the solutions $H(\alpha, i)$ and $D(\alpha, i)$ and can be obtained by repeatedly taking the derivative with respect to θ of (12)-(13) together with the assumption $A(\alpha) + B_1(\alpha) K_1(\alpha) + B_2(\alpha) K_2(\alpha)$ is stable for all $\alpha \in [t_0, t_f]$.   □

In the subsequent development, the subset of symmetric matrices of the vector space of all $n \times n$ matrices with real elements is denoted by $\mathbb{S}^n$. Now, let the k-tuple variables $\mathcal{H}$ and $\mathcal{D}$ be defined as follows

$$
\mathcal{H}(\cdot) \triangleq (\mathcal{H}_1(\cdot), \ldots, \mathcal{H}_k(\cdot)) \text{ and } \mathcal{D}(\cdot) \triangleq (\mathcal{D}_1(\cdot), \ldots, \mathcal{D}_k(\cdot)) \;,
$$

for each element $\mathcal{H}_i \in \mathcal{C}^1([t_0, t_f]; \mathbb{S}^n)$ of $\mathcal{H}$ and $\mathcal{D}_i \in \mathcal{C}^1([t_0, t_f]; \mathbb{R})$ of $\mathcal{D}$ having the representations

$$
\mathcal{H}_i(\cdot) \triangleq H(\cdot, i) \text{ and } \mathcal{D}_i(\cdot) \triangleq D(\cdot, i)
$$

with the right members satisfying the dynamic equations (20)-(22) on the horizon $[t_0, t_f]$. For ease of presentation, the following mappings are introduced:

$$
\begin{aligned}
\mathcal{F}_i &: [t_0, t_f] \times (\mathbb{S}^n)^k \times \mathbb{R}^{m_1 \times n} \times \mathbb{R}^{m_2 \times n} \mapsto \mathbb{S}^n \\
\mathcal{G}_i &: [t_0, t_f] \times (\mathbb{S}^n)^k \mapsto \mathbb{R}
\end{aligned}
$$

where the actions are given by

$$\mathcal{F}_1(\alpha, \mathcal{H}, K_1, K_2) \triangleq - [A(\alpha) + B_1(\alpha)K_1(\alpha) + B_2(\alpha)K_2(\alpha)]^T \mathcal{H}_1(\alpha)$$
$$- \mathcal{H}_1(\alpha) [A(\alpha) + B_1(\alpha)K_1(\alpha) + B_2(\alpha)K_2(\alpha)]$$
$$- Q(\alpha) - K_1^T(\alpha)R_{11}(\alpha)K_1(\alpha) + K_2^T(\alpha)R_{22}(\alpha)K_2(\alpha)$$

$$\mathcal{F}_i(\alpha, \mathcal{H}, K_1, K_2) \triangleq - [A(\alpha) + B_1(\alpha)K_1(\alpha) + B_2(\alpha)K_2(\alpha)]^T \mathcal{H}_i(\alpha)$$
$$- \mathcal{H}_i(\alpha) [A(\alpha) + B_1(\alpha)K_1(\alpha) + B_2(\alpha)K_2(\alpha)]$$
$$- \sum_{j=1}^{i-1} \frac{2i!}{j!(i-j)!} \mathcal{H}_j(\alpha)G(\alpha)WG^T(\alpha)\mathcal{H}_{i-j}(\alpha) , \quad 2 \le i \le k$$

$$\mathcal{G}_i(\alpha, \mathcal{H}) \triangleq -\mathrm{Tr}\left\{\mathcal{H}_i(\alpha)G(\alpha)WG^T(\alpha)\right\} , \quad 1 \le i \le k .$$

For a compact formulation, the product mappings are established as such

$$\mathcal{F}_1 \times \cdots \times \mathcal{F}_k : [t_0, t_f] \times (\mathbb{S}^n)^k \times \mathbb{R}^{m_1 \times n} \times \mathbb{R}^{m_2 \times n} \mapsto (\mathbb{S}^n)^k$$
$$\mathcal{G}_1 \times \cdots \times \mathcal{G}_k : [t_0, t_f] \times (\mathbb{S}^n)^k \mapsto \mathbb{R}^k$$

along with the corresponding notations $\mathcal{F} \triangleq \mathcal{F}_1 \times \cdots \times \mathcal{F}_k$ and $\mathcal{G} \triangleq \mathcal{G}_1 \times \cdots \times \mathcal{G}_k$. Thus, the dynamic equations of motion (20)-(22) can be rewritten as

$$\frac{d}{d\alpha}\mathcal{H}(\alpha) = \mathcal{F}(\alpha, \mathcal{H}(\alpha), K_1(\alpha), K_2(\alpha)) , \quad \mathcal{H}(t_f) = \mathcal{H}_f \qquad (23)$$

$$\frac{d}{d\alpha}\mathcal{D}(\alpha) = \mathcal{G}(\alpha, \mathcal{H}(\alpha)) , \qquad \mathcal{D}(t_f) = \mathcal{D}_f , \qquad (24)$$

where the terminal values $\mathcal{H}_f = (Q_f, 0, \ldots, 0)$ and $\mathcal{D}_f = (0, \ldots, 0)$.

Note that the product system uniquely determines $\mathcal{H}$ and $\mathcal{D}$ once the admissible feedback gains K_1 and K_2 are specified. Hence, $\mathcal{H}$ and $\mathcal{D}$ are considered as $\mathcal{H}(\cdot, K_1, K_2)$ and $\mathcal{D}(\cdot, K_1, K_2)$, respectively. The performance index in cost-cumulant control can now be formulated in the admissible feedback gains K_1 and K_2.

Definition 1. *Performance Index in Cost-Cumulant Control.*
Fix $k \in \mathbb{Z}^+$ and the sequence $\mu = \{\mu_i \ge 0\}_{i=1}^k$ with $\mu_1 > 0$. Then for the given initial condition (t_0, x_0), the performance index $\phi_0 : [t_0, t_f] \times (\mathbb{S}^n)^k \times \mathbb{R}^k \mapsto \mathbb{R}^+$ of the finite-horizon cost-cumulant control is defined by

$$\phi_0 (t_0, \mathcal{H}(t_0, K_1, K_2), \mathcal{D}(t_0, K_1, K_2)) \triangleq \sum_{i=1}^k \mu_i \kappa_i(K_1, K_2)$$

$$= \sum_{i=1}^k \mu_i \left[x_0^T \mathcal{H}_i(t_0, K_1, K_2)x_0 + \mathcal{D}_i(t_0, K_1, K_2) \right] , \qquad (25)$$

where additional parametric design freedom μ_i mutually chosen by players represent different levels of influence as they deem important to the overall cost distribution. Symmetric solutions $\{\mathcal{H}_i(t_0, K_1, K_2) \ge 0\}_{i=1}^k$ and $\{\mathcal{D}_i(t_0, K_1, K_2) \ge 0\}_{i=1}^k$ evaluated at $\alpha = t_0$ satisfy (23)-(24).

For the given terminal data $(t_f, \mathcal{H}_f, \mathcal{D}_f)$, the classes $\mathcal{K}^1_{t_f, \mathcal{H}_f, \mathcal{D}_f; \mu}$ and $\mathcal{K}^2_{t_f, \mathcal{H}_f, \mathcal{D}_f; \mu}$ of admissible feedback gains may be defined as follows.

Definition 2. *Admissible Feedback Gain Strategies.*
Let the compact subsets $\overline{K}_1 \subset \mathbb{R}^{m_1 \times n}$ and $\overline{K}_2 \subset \mathbb{R}^{m_2 \times n}$ be the sets of allowable gain values. For the given $k \in \mathbb{Z}^+$ and the sequence $\mu = \{\mu_i \geq 0\}_{i=1}^k$ with $\mu_1 > 0$, the sets of admissible control strategies $\mathcal{K}^1_{t_f, \mathcal{H}_f, \mathcal{D}_f; \mu}$ and $\mathcal{K}^2_{t_f, \mathcal{H}_f, \mathcal{D}_f; \mu}$ are assumed to be the classes of $\mathcal{C}([t_0, t_f]; \mathbb{R}^{m_1 \times n})$ and $\mathcal{C}([t_0, t_f]; \mathbb{R}^{m_2 \times n})$ with values $K_1(\cdot) \in \overline{K}_1$ and $K_2(\cdot) \in \overline{K}_2$ for which solutions to the dynamic equations of motion (23)-(24) exist on the finite horizon $[t_0, t_f]$.

Then one may state the cost-cumulant control optimization problem for the zero-sum stochastic differential game.

Definition 3. *Optimization Problem.*
Suppose that $k \in \mathbb{Z}^+$ and the sequence $\mu = \{\mu_i \geq 0\}_{i=1}^k$ with $\mu_1 > 0$ are fixed. Then, the cost-cumualnt control optimization problem over $[t_0, t_f]$ is given by

$$\min_{K_1(\cdot) \in \mathcal{K}^1_{t_f, \mathcal{H}_f, \mathcal{D}_f; \mu}} \max_{K_2(\cdot) \in \mathcal{K}^2_{t_f, \mathcal{H}_f, \mathcal{D}_f; \mu}} \phi_0\left(t_0, \mathcal{H}(t_0, K_1, K_2), \mathcal{D}(t_0, K_1, K_2)\right) \quad (26)$$

subject to the dynamic equations (23)-(24) for $\alpha \in [t_0, t_f]$.

Next, the fundamental theorem of calculus and stochastic differential rules is utilized to derive the existence of a saddle point.

Theorem 3. *Existence of a Saddle Point.*
Consider the linear-quadratic zero-sum stochastic differential game

$$dx(t) = [A(t) + B_1(t)K_1(t) + B_2(t)K_2(t)]\, x(t)dt + G(t)dw(t) \ ,$$
$$x(t_0) = x_0 \ ,$$

which in turn, is associated with the finite-horizon IQF cost

$$J(t_0, x_0; K_1, K_2) = x^T(t_f) Q_f x(t_f)$$
$$+ \int_{t_0}^{t_f} x^T(\tau) \left[Q(\tau) + K_1^T(\tau) R_{11}(\tau) K_1(\tau) - K_2^T(\tau) R_{22}(\tau) K_2(\tau) \right] x(\tau) d\tau \ .$$

For any given $k \in \mathbb{Z}^+$ and the sequence $\mu = \{\mu_i \geq 0\}_{i=1}^k$ with $\mu_1 > 0$, there exists a saddle point $(K_1^*, K_2^*) \in \mathcal{K}^1_{t_f, \mathcal{H}_f, \mathcal{D}_f; \mu} \times \mathcal{K}^2_{t_f, \mathcal{H}_f, \mathcal{D}_f; \mu}$ such that there hold

$$\phi_0\left(t_0, \mathcal{H}(t_0, K_1^*, K_2), \mathcal{D}(t_0, K_1^*, K_2)\right) \leq \phi_0\left(t_0, \mathcal{H}(t_0, K_1^*, K_2^*), \mathcal{D}(t_0, K_1^*, K_2^*)\right)$$
$$\phi_0\left(t_0, \mathcal{H}(t_0, K_1^*, K_2^*), \mathcal{D}(t_0, K_1^*, K_2^*)\right) \leq \phi_0\left(t_0, \mathcal{H}(t_0, K_1, K_2^*), \mathcal{D}(t_0, K_1, K_2^*)\right) \ .$$

It is now concluded that the existence of a saddle point yields both necessary and sufficient conditions for the minimax problem to be equivalent to the corresponding maximin problem. In other words, the Issacs condition holds according to [3]. The value function, $\mathcal{V}(\varepsilon, \mathcal{Y}, \mathcal{Z})$ for the game starting at the time-states triple $(\varepsilon, \mathcal{Y}, \mathcal{Z})$ is defined as follows.

Definition 4. *Value Function.*
The value function $\mathcal{V} : [t_0, t_f] \times (\mathbb{S}^n)^k \times \mathbb{R}^k \mapsto \mathbb{R}^+ \cup \{+\infty\}$ *associated with the Mayer problem is defined by*

$$\mathcal{V}(\varepsilon, \mathcal{Y}, \mathcal{Z}) \triangleq \min_{K_1(\cdot) \in \mathcal{K}^1_{\varepsilon, \mathcal{Y}, \mathcal{Z}; \mu}} \max_{K_2(\cdot) \in \mathcal{K}^2_{\varepsilon, \mathcal{Y}, \mathcal{Z}; \mu}} \phi_0 \left(t_0, \mathcal{H}(t_0, K_1, K_2), \mathcal{D}(t_0, K_1, K_2) \right)$$

$$= \max_{K_2(\cdot) \in \mathcal{K}^2_{\varepsilon, \mathcal{Y}, \mathcal{Z}; \mu}} \min_{K_1(\cdot) \in \mathcal{K}^1_{\varepsilon, \mathcal{Y}, \mathcal{Z}; \mu}} \phi_0 \left(t_0, \mathcal{H}(t_0, K_1, K_2), \mathcal{D}(t_0, K_1, K_2) \right) ,$$

for any $(\varepsilon, \mathcal{Y}, \mathcal{Z}) \in [t_0, t_f] \times (\mathbb{S}^n)^k \times \mathbb{R}^k$.

Conventionally, set $\mathcal{V}(\varepsilon, \mathcal{Y}, \mathcal{Z}) = \infty$ when either $\mathcal{K}^1_{\varepsilon, \mathcal{Y}, \mathcal{Z}; \mu}$ or $\mathcal{K}^2_{\varepsilon, \mathcal{Y}, \mathcal{Z}; \mu}$ is empty. The development in the sequel is motivated by the excellent treatment in [4], and is intended to follow it closely. Unless otherwise specified, the dependence of trajectory solutions $\mathcal{H}$ and $\mathcal{D}$ on the admissible gains K_1 and K_2 is omitted for notational clarity.

Theorem 4. *Necessary Conditions.*
The value function evaluated along any trajectory corresponding to a pair of control strategy gains feasible for its terminal states is a non-increasing function of time. The value function evaluated along any optimal trajectory is constant.

It is important to note that these properties are necessary conditions for optimality. The next theorem shows that these conditions are also sufficient for optimality.

Theorem 5. *Sufficient Condition.*
Let $\mathcal{W}(\varepsilon, \mathcal{Y}, \mathcal{Z})$ *be an extended real-valued function defined on*

$$[t_0, t_f] \times (\mathbb{S}^n)^k \times \mathbb{R}^k$$

such that $\mathcal{W}(\varepsilon, \mathcal{Y}, \mathcal{Z}) = \phi_0(\varepsilon, \mathcal{Y}, \mathcal{Z})$.

Let t_f, $\mathcal{H}_f$, $\mathcal{D}_f$ *be given terminal conditions, and suppose that, for each trajectory pair* $(\mathcal{H}, \mathcal{D})$ *corresponding to a control strategy pair* (K_1, K_2) *in* $\mathcal{K}^1_{t_f, \mathcal{H}_f, \mathcal{D}_f; \mu} \times \mathcal{K}^2_{t_f, \mathcal{H}_f, \mathcal{D}_f; \mu}$, $\mathcal{W}(\alpha, \mathcal{H}(\alpha), \mathcal{D}(\alpha))$ *is finite and non-increasing on* $[t_0, t_f]$.

If (K_1^*, K_2^*) *is a control strategy pair in* $\mathcal{K}^1_{t_f, \mathcal{H}_f, \mathcal{D}_f; \mu} \times \mathcal{K}^2_{t_f, \mathcal{H}_f, \mathcal{D}_f; \mu}$ *such that for the corresponding trajectory pair* $(\mathcal{H}^*, \mathcal{D}^*)$, $\mathcal{W}(\alpha, \mathcal{H}^*(\alpha), \mathcal{D}^*(\alpha))$ *is constant then the pair* (K^*, K_2^*) *is a saddle point and* $\mathcal{W}(t_f, \mathcal{H}_f, \mathcal{D}_f) = \mathcal{V}(t_f, \mathcal{H}_f, \mathcal{D}_f)$.

Corollary 1. *Restriction of Strategy Gains.*
Let (K_1^*, K_2^*) *be an optimal control strategy pair in* $\mathcal{K}^1_{t_f, \mathcal{H}_f, \mathcal{D}_f; \mu} \times \mathcal{K}^2_{t_f, \mathcal{H}_f, \mathcal{D}_f; \mu}$ *and* $(\mathcal{H}^*, \mathcal{D}^*)$ *the corresponding trajectory pair of dynamic equations*

$$\frac{d}{d\alpha}\mathcal{H}(\alpha) = \mathcal{F}(\alpha, \mathcal{H}(\alpha), K_1(\alpha), K_2(\alpha)) , \quad \mathcal{H}(t_f) = \mathcal{H}_f$$

$$\frac{d}{d\alpha}\mathcal{D}(\alpha) = \mathcal{G}(\alpha, \mathcal{H}(\alpha)) , \quad \mathcal{D}(t_f) = \mathcal{D}_f .$$

Then, the restriction of the pair (K_1^*, K_2^*) *to* $[t_0, \alpha]$ *is an optimal control strategy pair for the control problem with the terminal-valued condition* $(\alpha, \mathcal{H}^*(\alpha), \mathcal{D}^*(\alpha))$ *when* $t_0 \leq \alpha \leq t_f$.

Both necessary and sufficient conditions implied by these properties for a control gain to be optimal give hints that one may find a function $\mathcal{W}(\varepsilon, \mathcal{Y}, \mathcal{Z}) : [t_0, t_f] \times (\mathbb{S}^n)^k \times \mathbb{R}^k \mapsto \mathbb{R}^+$ such that $\mathcal{W}(\varepsilon, \mathcal{Y}, \mathcal{Z}) = \phi_0(\varepsilon, \mathcal{Y}, \mathcal{Z})$, $\mathcal{W}(\varepsilon, \mathcal{Y}, \mathcal{Z})$ is constant on the corresponding trajectory pair, and $\mathcal{W}(\varepsilon, \mathcal{Y}, \mathcal{Z})$ is non-increasing on other trajectories.

Note that the value function $\mathcal{V}(\varepsilon, \mathcal{Y}, \mathcal{Z})$ is supposed to be continuously differentiable in $(\varepsilon, \mathcal{Y}, \mathcal{Z})$ which then results in the uniqueness of a saddle point (K_1^*, K_2^*). Formally speaking, the result regarding the differentiability of the value function, which is adapted from [4], is stated as follows.

Theorem 6. *Differentiability of Value Function.*
Let admissible feedback gains $K_1^*(\alpha, \mathcal{H}, \mathcal{D})$ and $K_2^*(\alpha, \mathcal{H}, \mathcal{D})$ constitute a saddle point. Further, let $t_0(\varepsilon, \mathcal{Y}, \mathcal{Z})$ and $(\mathcal{H}(t_0(\varepsilon, \mathcal{Y}, \mathcal{Z}); \varepsilon, \mathcal{Y}), \mathcal{D}(t_0(\varepsilon, \mathcal{Y}, \mathcal{Z}); \varepsilon, \mathcal{Z}))$ be the initial time and initial states for the trajectories of

$$\frac{d}{d\alpha}\mathcal{H}(\alpha) = \mathcal{F}(\alpha, \mathcal{H}, K_1^*(\alpha, \mathcal{H}, \mathcal{D}), K_2^*(\alpha, \mathcal{H}, \mathcal{D})) \ ,$$

$$\frac{d}{d\alpha}\mathcal{D}(\alpha) = \mathcal{G}(\alpha, \mathcal{H}) \ ,$$

with the terminal-valued condition $(\varepsilon, \mathcal{Y}, \mathcal{Z})$. Then, the value function $\mathcal{V}(\varepsilon, \mathcal{Y}, \mathcal{Z})$ is differentiable at each point at which $t_0(\varepsilon, \mathcal{Y}, \mathcal{Z})$ and $\mathcal{H}(t_0(\varepsilon, \mathcal{Y}, \mathcal{Z}); \varepsilon, \mathcal{Y})$ and $\mathcal{D}(t_0(\varepsilon, \mathcal{Y}, \mathcal{Z}); \varepsilon, \mathcal{Z})$ are differentiable with respect to $(\varepsilon, \mathcal{Y}, \mathcal{Z})$.

As a tenet of transition from the principle of optimality, a family of games based on different starting points is now considered. Let's begin with an interlude of time, ε in mid-play. At its commencement, the path has reached some definitive points. Consider all possible $(\mathcal{H}, \mathcal{D})$ which may be reached at the end of the interlude for all possible choices of (K_1, K_2). Suppose that for each endpoint, the game beginning there has already been solved. Then the value function $\mathcal{V}(\varepsilon, \mathcal{H}, \mathcal{D})$ resulting from each choice of (K_1, K_2) is known, and they are to be so chosen as to render it minimax. As the duration of the interlude approaches t_f, this leads to a sufficient condition to Hamilton-Jacobi-Isaacs (HJI) equation. By adapting to the initial-cost problem and the terminologies present in the cost-cumulant control, the HJI equation satisfied by the value function $\mathcal{V}(\varepsilon, \mathcal{Y}, \mathcal{Z})$ is then given.

Definition 5. *Playable Set.*
Let the playable set $\mathcal{Q}$ be defined as

$$\mathcal{Q} \triangleq \left\{ (\varepsilon, \mathcal{Y}, \mathcal{Z}) \in [t_0, t_f] \times (\mathbb{S}^n)^k \times \mathbb{R}^k \text{ such that } \mathcal{K}_{\varepsilon, \mathcal{Y}, \mathcal{Z}; \mu}^1 \times \mathcal{K}_{\varepsilon, \mathcal{Y}, \mathcal{Z}; \mu}^2 \neq 0 \right\} \ .$$

Theorem 7. *HJI Equation-Mayer Problem.*
Let $(\varepsilon, \mathcal{Y}, \mathcal{Z})$ be any interior point of the playable set $\mathcal{Q}$ at which the value function $\mathcal{V}(\varepsilon, \mathcal{Y}, \mathcal{Z})$ is differentiable. Then $\mathcal{V}(\varepsilon, \mathcal{Y}, \mathcal{Z})$ satisfies the partial differential inequality

$$0 \geq \frac{\partial}{\partial \varepsilon} \mathcal{V}(\varepsilon, \mathcal{Y}, \mathcal{Z}) + \frac{\partial}{\partial \operatorname{vec}(\mathcal{Y})} \mathcal{V}(\varepsilon, \mathcal{Y}, \mathcal{Z}) \cdot \operatorname{vec}(\mathcal{F}(\varepsilon, \mathcal{Y}, K_1, K_2))$$

$$+ \frac{\partial}{\partial \operatorname{vec}(\mathcal{Z})} \mathcal{V}(\varepsilon, \mathcal{Y}, \mathcal{Z}) \cdot \operatorname{vec}(\mathcal{G}(\varepsilon, \mathcal{Y})) \ ,$$

for all $(K_1, K_2) \in \overline{K}_1 \times \overline{K}_2$.

If there exists a saddle point $(K_1^*, K_2^*) \in \mathcal{K}_{\varepsilon, \mathcal{Y}, \mathcal{Z}; \mu}^1 \times \mathcal{K}_{\varepsilon, \mathcal{Y}, \mathcal{Z}; \mu}^2$, then the partial differential equation of differential games

$$0 = \min_{K_1 \in \overline{K}_1} \max_{K_2 \in \overline{K}_2} \left\{ \frac{\partial}{\partial \varepsilon} \mathcal{V}(\varepsilon, \mathcal{Y}, \mathcal{Z}) + \frac{\partial}{\partial \operatorname{vec}(\mathcal{Y})} \mathcal{V}(\varepsilon, \mathcal{Y}, \mathcal{Z}) \cdot \operatorname{vec}(\mathcal{F}(\varepsilon, \mathcal{Y}, K_1, K_2)) \right.$$

$$\left. + \frac{\partial}{\partial \operatorname{vec}(\mathcal{Z})} \mathcal{V}(\varepsilon, \mathcal{Y}, \mathcal{Z}) \cdot \operatorname{vec}(\mathcal{G}(\varepsilon, \mathcal{Y})) \right\} \quad (27)$$

is satisfied together with $\mathcal{V}(t_0, \mathcal{H}_0, \mathcal{D}_0) = \phi_0(t_0, \mathcal{H}_0, \mathcal{D}_0)$ and $\operatorname{vec}(\cdot)$ the vectorizing operator of enclosed entities. The optimum in (27) is achieved by the left limit $(K_1^*(\varepsilon)^-, K_2^*(\varepsilon)^-)$ of the optimal strategy pair at ε.

The construction of a scalar-valued function which is a candidate for the value function is discussed in the following theorem.

Theorem 8. *Verification Theorem.*
Fix $k \in \mathbb{Z}^+$. Let $\mathcal{W}(\varepsilon, \mathcal{Y}, \mathcal{Z})$ be a continuously differentiable solution of the HJI equation

$$0 = \min_{K_1 \in \overline{K}_1} \max_{K_2 \in \overline{K}_2} \left\{ \frac{\partial}{\partial \varepsilon} \mathcal{V}(\varepsilon, \mathcal{Y}, \mathcal{Z}) + \frac{\partial}{\partial \operatorname{vec}(\mathcal{Y})} \mathcal{V}(\varepsilon, \mathcal{Y}, \mathcal{Z}) \cdot \operatorname{vec}(\mathcal{F}(\varepsilon, \mathcal{Y}, K_1, K_2)) \right.$$

$$\left. + \frac{\partial}{\partial \operatorname{vec}(\mathcal{Z})} \mathcal{V}(\varepsilon, \mathcal{Y}, \mathcal{Z}) \cdot \operatorname{vec}(\mathcal{G}(\varepsilon, \mathcal{Y})) \right\}$$

and satisfy the boundary condition

$$\mathcal{W}(t_0, \mathcal{H}_0, \mathcal{D}_0) = \phi_0(t_0, \mathcal{H}_0, \mathcal{D}_0) \ , \qquad \text{for } (t_0, \mathcal{H}_0, \mathcal{D}_0) \in \mathcal{M} \ , \qquad (28)$$

where $\mathcal{M} = \{t_0\} \times (\mathbb{S}^n)^k \times \mathbb{R}^k$.

Let $(t_f, \mathcal{H}_f, \mathcal{D}_f)$ be a point of $\mathcal{Q}$, (K_1, K_2) a control strategy pair in $\mathcal{K}_{t_f, \mathcal{H}_f, \mathcal{D}_f; \mu}^1 \times \mathcal{K}_{t_f, \mathcal{H}_f, \mathcal{D}_f; \mu}^2$ and $\mathcal{H}$ and $\mathcal{D}$ the corresponding solutions of the equations

$$\frac{d}{d\alpha} \mathcal{H}(\alpha) = \mathcal{F}(\alpha, \mathcal{H}(\alpha), K_1(\alpha), K_2(\alpha)) \ , \quad \mathcal{H}(t_f) = \mathcal{H}_f$$

$$\frac{d}{d\alpha} \mathcal{D}(\alpha) = \mathcal{G}(\alpha, \mathcal{H}(\alpha)) \ , \quad \mathcal{D}(t_f) = \mathcal{D}_f \ .$$

Then, $\mathcal{W}(\alpha, \mathcal{H}(\alpha), \mathcal{D}(\alpha))$ is a non-increasing function of α. If (K_1^*, K_2^*) is a control strategy pair in $\mathcal{K}_{t_f, \mathcal{H}_f, \mathcal{D}_f; \mu}^1 \times \mathcal{K}_{t_f, \mathcal{H}_f, \mathcal{D}_f; \mu}^2$ defined on $[t_0, t_f]$ with corresponding solution, $\mathcal{H}^*$ and $\mathcal{D}^*$ of the above equations such that for $\alpha \in [t_0, t_f]$

$$
\begin{aligned}
0 = {} & \frac{\partial}{\partial \varepsilon} \mathcal{W}(\alpha, \mathcal{H}^*(\alpha), \mathcal{D}^*(\alpha)) \\
& + \frac{\partial}{\partial \operatorname{vec}(\mathcal{Y})} \mathcal{W}(\alpha, \mathcal{H}^*(\alpha), \mathcal{D}^*(\alpha)) \cdot \operatorname{vec}(\mathcal{F}(\alpha, \mathcal{H}^*(\alpha), K_1^*(\alpha), K_2^*(\alpha))) \\
& + \frac{\partial}{\partial \operatorname{vec}(\mathcal{Z})} \mathcal{W}(\alpha, \mathcal{H}^*(\alpha), \mathcal{D}^*(\alpha)) \cdot \operatorname{vec}(\mathcal{G}(\alpha, \mathcal{H}^*(\alpha))) \ , \qquad (29)
\end{aligned}
$$

then (K_1^*, K_2^*) is a saddle-point strategy pair in $\mathcal{K}_{t_f, \mathcal{H}_f, \mathcal{D}_f; \mu}^1 \times \mathcal{K}_{t_f, \mathcal{H}_f, \mathcal{D}_f; \mu}^2$ and

$$
\mathcal{W}(\varepsilon, \mathcal{Y}, \mathcal{Z}) = \mathcal{V}(\varepsilon, \mathcal{Y}, \mathcal{Z}) \ , \qquad (30)
$$

where $\mathcal{V}(\varepsilon, \mathcal{Y}, \mathcal{Z})$ is the value function.

It is observed that to have a saddle-point solution (K_1^*, K_2^*) in $\mathcal{K}_{t_f, \mathcal{H}_f, \mathcal{D}_f; \mu}^1 \times \mathcal{K}_{t_f, \mathcal{H}_f, \mathcal{D}_f; \mu}^2$ defined and continuous for all $\alpha \in [t_0, t_f]$, the solution $\mathcal{H}(\alpha)$ to (23) when evaluated at $\alpha = t_0$ must also exist. Therefore, it is necessary that $\mathcal{H}(\alpha)$ is finite for all $\alpha \in [t_0, t_f)$. Moreover, the solution of (23) exists and is continuously differentiable in a neighborhood of t_f. Applying the results from [2], these solutions can further be extended to the left of t_f as long as $\mathcal{H}(\alpha)$ remains finite. Hence, the existence of unique and continuously differentiable solutions to (23) are certain if $\mathcal{H}(\alpha)$ are bounded for all $\alpha \in [t_0, t_f)$. As the result, the candidate value functions $\mathcal{V}(\alpha, \mathcal{H}, \mathcal{D})$ are continuously differentiable as well.

Theorem 9. *Necessary and Sufficient Conditions for a Saddle-Point Solution.* (K_1^*, K_2^*) *is a saddle-point strategy if and only if* $\mathcal{H}(\alpha)$ *is bounded for all* $\alpha \in [t_0, t_f)$.

3   Multi-cumulant Saddle-Point Solution

Recall that the optimization problem being considered herein is in "Mayer form" and can be solved by applying an adaptation of the Mayer form verification theorem of dynamic programming given in [4]. In the framework of dynamic programming, it is often required to denote the terminal time and states of a family of optimization problems as $(\varepsilon, \mathcal{Y}, \mathcal{Z})$ rather than $(t_f, \mathcal{H}_f, \mathcal{D}_f)$. That is, for $\varepsilon \in [t_0, t_f]$ and $1 \le i \le k$, the states of the system (23)-(24) defined on the interval $[t_0, \varepsilon]$ have terminal values denoted by $\mathcal{H}(\varepsilon) \equiv \mathcal{Y}$ and $\mathcal{D}(\varepsilon) \equiv \mathcal{Z}$. Since the cumulant-based performance index (25) is quadratic affine in terms of arbitrarily fixed x_0, this observation then suggests a solution to (27) may be sought in the form

$$
\mathcal{W}(\varepsilon, \mathcal{Y}, \mathcal{Z}) = x_0^T \sum_{i=1}^{k} \mu_i (\mathcal{Y}_i + \mathcal{E}_i(\varepsilon)) x_0 + \sum_{i=1}^{k} \mu_i (\mathcal{Z}_i + \mathcal{T}_i(\varepsilon)) \ , \qquad (31)
$$

where the parametric functions of time $\mathcal{E}_i \in \mathcal{C}^1([t_0, t_f]; \mathbb{S}^n)$ and $\mathcal{T}_i \in \mathcal{C}^1([t_0, t_f]; \mathbb{R})$ are yet to be determined. The next theorem shows how the partial differential equation in the notation of $\mathcal{W}(\varepsilon, \mathcal{Y}, \mathcal{Z})$ looks like using inverse vectorizing transformation.

Corollary 2. *Time Derivative of a Candidate Function.*
Fix $k \in \mathbb{Z}^+$ and let $(\varepsilon, \mathcal{Y}, \mathcal{Z})$ be any interior point of the reachable set $\mathcal{Q}$ at which the real-valued function (31) is differentiable. Then, the time derivative of $\mathcal{W}(\varepsilon, \mathcal{Y}, \mathcal{Z})$ is found to be

$$\frac{d}{d\varepsilon}\mathcal{W}(\varepsilon, \mathcal{Y}, \mathcal{Z}) = \sum_{i=1}^{k} \mu_i \left(\mathcal{G}_i(\varepsilon, \mathcal{Y}) + \frac{d}{d\varepsilon}\mathcal{T}_i(\varepsilon) \right)$$

$$+ x_0^T \sum_{i=1}^{k} \mu_i \left(\mathcal{F}_i(\varepsilon, \mathcal{Y}, K_1, K_2) + \frac{d}{d\varepsilon}\mathcal{E}_i(\varepsilon) \right) x_0 . \quad (32)$$

The substitution of this hypothesized solution (31) into (27) and making use of the result (32) yield

$$0 = \min_{K_1 \in \overline{K}_1} \max_{K_2 \in \overline{K}_2} \left\{ \frac{\partial}{\partial \varepsilon}\mathcal{W}(\varepsilon, \mathcal{Y}, \mathcal{Z}) + \frac{\partial}{\partial \operatorname{vec}(\mathcal{Y})}\mathcal{W}(\varepsilon, \mathcal{Y}, \mathcal{Z}) \cdot \operatorname{vec}(\mathcal{F}_i(\varepsilon, \mathcal{Y}, K_1, K_2)) \right.$$

$$\left. + \frac{\partial}{\partial \operatorname{vec}(\mathcal{Z})}\mathcal{W}(\varepsilon, \mathcal{Y}, \mathcal{Z}) \cdot \operatorname{vec}(\mathcal{G}_i(\varepsilon, \mathcal{Y})) \right\}$$

$$= \min_{K_1 \in \overline{K}_1} \max_{K_2 \in \overline{K}_2} \left\{ x_0^T \left(\sum_{i=1}^{k} \mu_i \frac{d}{d\varepsilon}\mathcal{E}_i(\varepsilon) \right) x_0 + \sum_{i=1}^{k} \mu_i \frac{d}{d\varepsilon}\mathcal{T}_i(\varepsilon) \right.$$

$$\left. + x_0^T \left(\sum_{i=1}^{k} \mu_i \mathcal{F}_i(\varepsilon, \mathcal{Y}, K_1, K_2) \right) x_0 + \sum_{i=1}^{k} \mu_i \mathcal{G}_i(\varepsilon, \mathcal{Y}) \right\} . \quad (33)$$

It is important to observe that

$$\sum_{i=1}^{k} \mu_i \mathcal{F}_i(\varepsilon, \mathcal{Y}, K_1, K_2) = -[A(\varepsilon) + B_1(\varepsilon)K_1 + B_2(\varepsilon)K_2]^T \sum_{i=1}^{k} \mu_i \mathcal{Y}_i$$

$$- \sum_{i=1}^{k} \mu_i \mathcal{Y}_i [A(\varepsilon) + B_1(\varepsilon)K_1 + B_2(\varepsilon)K_2]$$

$$- \mu_1 Q(\varepsilon) - \mu_1 K_1^T R_{11}(\varepsilon)K_1 + \mu_1 K_2^T R_{22}(\varepsilon)K_2$$

$$- \sum_{i=2}^{k} \mu_i \sum_{j=1}^{i-1} \frac{2i!}{j!(i-j)!} \mathcal{Y}_j G(\varepsilon) W G^T(\varepsilon) \mathcal{Y}_{i-j} ,$$

$$\sum_{i=1}^{k} \mu_i \mathcal{G}_i(\varepsilon, \mathcal{Y}) = - \sum_{i=1}^{k} \mu_i \operatorname{Tr} \left\{ \mathcal{Y}_i G(\varepsilon) W G^T(\varepsilon) \right\} .$$

Differentiating the expression within the bracket of (33) with respect to K_1 and K_2 yield the necessary conditions for an extremum of the performance index (25) on $[t_0, \varepsilon]$,

$$-2B_1^T(\varepsilon) \sum_{i=1}^{k} \mu_i \mathcal{Y}_i M_0 - 2\mu_1 R_{11}(\varepsilon) K_1 M_0 = 0 \ ,$$

$$-2B_2^T(\varepsilon) \sum_{i=1}^{k} \mu_i \mathcal{Y}_i M_0 + 2\mu_1 R_{22}(\varepsilon) K_2 M_0 = 0 \ .$$

Because M_0 is an arbitrary rank-one matrix, it must be true that

$$K_1(\varepsilon, \mathcal{Y}, \mathcal{Z}) = -R_{11}^{-1}(\varepsilon) B_1^T(\varepsilon) \sum_{r=1}^{k} \widehat{\mu}_r \mathcal{Y}_r \ , \tag{34}$$

$$K_2(\varepsilon, \mathcal{Y}, \mathcal{Z}) = R_{22}^{-1}(\varepsilon) B_2^T(\varepsilon) \sum_{r=1}^{k} \widehat{\mu}_r \mathcal{Y}_r \ , \tag{35}$$

where $\widehat{\mu}_r \triangleq \mu_i/\mu_1$ for $\mu_1 > 0$. Substituting the gain expressions (34) and (35) into the right member of the HJI equation (33) yields the value of the minimax

$$x_0^T \left[\sum_{i=1}^{k} \mu_i \frac{d}{d\varepsilon} \mathcal{E}_i(\varepsilon) - A^T(\varepsilon) \sum_{i=1}^{k} \mu_i \mathcal{Y}_i - \sum_{i=1}^{k} \mu_i \mathcal{Y}_i A(\varepsilon) \right.$$

$$- \mu_1 Q(\varepsilon) + \sum_{r=1}^{k} \widehat{\mu}_r \mathcal{Y}_r B_1(\varepsilon) R_{11}^{-1}(\varepsilon) B_1^T(\varepsilon) \sum_{i=1}^{k} \mu_i \mathcal{Y}_i$$

$$+ \sum_{i=1}^{k} \mu_i \mathcal{Y}_i B_1(\varepsilon) R_{11}^{-1}(\varepsilon) B_1^T(\varepsilon) \sum_{s=1}^{k} \widehat{\mu}_s \mathcal{Y}_s - \sum_{r=1}^{k} \widehat{\mu}_r \mathcal{Y}_r B_2(\varepsilon) R_{22}^{-1}(\varepsilon) B_2^T(\varepsilon) \sum_{i=1}^{k} \mu_i \mathcal{Y}_i$$

$$- \sum_{i=1}^{k} \mu_i \mathcal{Y}_i B_2(\varepsilon) R_{22}^{-1}(\varepsilon) B_2^T(\varepsilon) \sum_{s=1}^{k} \widehat{\mu}_s \mathcal{Y}_s - \mu_1 \sum_{r=1}^{k} \widehat{\mu}_r \mathcal{Y}_r B_1(\varepsilon) R_{11}^{-1}(\varepsilon) B_1^T(\varepsilon) \sum_{s=1}^{k} \widehat{\mu}_s \mathcal{Y}_s$$

$$+ \mu_1 \sum_{r=1}^{k} \widehat{\mu}_r \mathcal{Y}_r B_2(\varepsilon) R_{22}^{-1}(\varepsilon) B_2^T(\varepsilon) \sum_{s=1}^{k} \widehat{\mu}_s \mathcal{Y}_s$$

$$\left. - \sum_{i=2}^{k} \mu_i \sum_{j=1}^{i-1} \frac{2i!}{j!(i-j)!} \mathcal{Y}_j G(\varepsilon) W G^T(\varepsilon) \mathcal{Y}_{i-j} \right] x_0$$

$$+ \sum_{i=1}^{k} \mu_i \frac{d}{d\varepsilon} \mathcal{T}_i(\varepsilon) - \sum_{i=1}^{k} \mu_i \mathrm{Tr} \left\{ \mathcal{Y}_i G(\varepsilon) W G^T(\varepsilon) \right\} \ . \tag{36}$$

It is now necessary to exhibit time-dependent functions $\{\mathcal{E}_i(\cdot)\}_{i=1}^{k}$ and $\{\mathcal{T}_i(\cdot)\}_{i=1}^{k}$ which will render the left side of (36) equal to zero for $\varepsilon \in [t_0, t_f]$, when $\{\mathcal{Y}_i\}_{i=1}^{k}$ are evaluated along solution trajectories of the cumulant-generating equations.

Studying the expression (36) reveals that $\mathcal{E}_i(\cdot)$ and $\mathcal{T}_i(\cdot)$ for $1 \leq i \leq k$ satisfying the time-backward differential equations

$$\frac{d}{d\varepsilon}\mathcal{E}_1(\varepsilon) = A^T(\varepsilon)\mathcal{H}_1(\varepsilon) + \mathcal{H}_1(\varepsilon)A(\varepsilon) + Q(\varepsilon)$$

$$- \mathcal{H}_1(\varepsilon)B_1(\varepsilon)R_{11}^{-1}(\varepsilon)B_1^T(\varepsilon)\sum_{s=1}^{k}\widehat{\mu}_s\mathcal{H}_s(\varepsilon)$$

$$- \sum_{r=1}^{k}\widehat{\mu}_r\mathcal{H}_r(\varepsilon)B_1(\varepsilon)R_{11}^{-1}(\varepsilon)B_1^T(\varepsilon)\mathcal{H}_1(\varepsilon)$$

$$+ \mathcal{H}_1(\varepsilon)B_2(\varepsilon)R_{22}^{-1}(\varepsilon)B_2^T(\varepsilon)\sum_{s=1}^{k}\widehat{\mu}_s\mathcal{H}_s(\varepsilon)$$

$$+ \sum_{r=1}^{k}\widehat{\mu}_r\mathcal{H}_r(\varepsilon)B_2(\varepsilon)R_{22}^{-1}(\varepsilon)B_2^T(\varepsilon)\mathcal{H}_1(\varepsilon)$$

$$+ \sum_{r=1}^{k}\widehat{\mu}_r\mathcal{H}_r(\varepsilon)B_1(\varepsilon)R_{11}^{-1}(\varepsilon)B_1^T(\varepsilon)\sum_{s=1}^{k}\widehat{\mu}_s\mathcal{H}_s(\varepsilon)$$

$$- \sum_{r=1}^{k}\widehat{\mu}_r\mathcal{H}_r(\varepsilon)B_2(\varepsilon)R_{22}^{-1}(\varepsilon)B_2^T(\varepsilon)\sum_{s=1}^{k}\widehat{\mu}_s\mathcal{H}_s(\varepsilon) \ , \tag{37}$$

and, for $2 \leq i \leq k$

$$\frac{d}{d\varepsilon}\mathcal{E}_i(\varepsilon) = A^T(\varepsilon)\mathcal{H}_i(\varepsilon) + \mathcal{H}_i(\varepsilon)A(\varepsilon)$$

$$- \mathcal{H}_i(\varepsilon)B_1(\varepsilon)R_{11}^{-1}(\varepsilon)B_1^T(\varepsilon)\sum_{s=1}^{k}\widehat{\mu}_s\mathcal{H}_s(\varepsilon)$$

$$- \sum_{r=1}^{k}\widehat{\mu}_r\mathcal{H}_r(\varepsilon)B_1(\varepsilon)R_{11}^{-1}(\varepsilon)B_1^T(\varepsilon)\mathcal{H}_i(\varepsilon)$$

$$+ \mathcal{H}_i(\varepsilon)B_2(\varepsilon)R_{22}^{-1}(\varepsilon)B_2^T(\varepsilon)\sum_{s=1}^{k}\widehat{\mu}_s\mathcal{H}_s(\varepsilon)$$

$$+ \sum_{r=1}^{k}\widehat{\mu}_r\mathcal{H}_r(\varepsilon)B_2(\varepsilon)R_{22}^{-1}(\varepsilon)B_2^T(\varepsilon)\mathcal{H}_i(\varepsilon)$$

$$+ \sum_{j=1}^{i-1}\frac{2i!}{j!(i-j)!}\mathcal{H}_j(\varepsilon)G(\varepsilon)WG^T(\varepsilon)\mathcal{H}_{i-j}(\varepsilon) \ , \tag{38}$$

together with

$$\frac{d}{d\varepsilon}\mathcal{T}_i(\varepsilon) = \text{Tr}\left\{\mathcal{H}_i(\varepsilon)G(\varepsilon)WG^T(\varepsilon)\right\} \ , \qquad 1 \leq i \leq k \ , \tag{39}$$

will work. Furthermore, at the boundary condition, it is necessary to have $\mathcal{W}(t_0, \mathcal{H}_0, \mathcal{D}_0) = \phi_0(t_0, \mathcal{H}_0, \mathcal{D}_0)$. Or, equivalently, $x_0^T \sum_{i=1}^k \mu_i(\mathcal{H}_{i0} + \mathcal{E}_i(t_0)) x_0 + \sum_{i=1}^k \mu_i(\mathcal{D}_{i0} + \mathcal{T}_i(t_0)) = x_0^T \sum_{i=1}^k \mu_i \mathcal{H}_{i0} x_0 + \sum_{i=1}^k \mu_i \mathcal{D}_{i0}$. Thus, matching the boundary condition yields the corresponding initial value conditions $\mathcal{E}_i(t_0) = 0$ and $\mathcal{T}_i(t_0) = 0$ for (37)-(39). Applying the feedback gains specified in (34) and (35) along the solution trajectories of (23)-(24), these equations become Riccati-type

$$\frac{d}{d\varepsilon}\mathcal{H}_1(\varepsilon) = -A^T(\varepsilon)\mathcal{H}_1(\varepsilon) - \mathcal{H}_1(\varepsilon)A(\varepsilon) - Q(\varepsilon)$$

$$+ \mathcal{H}_1(\varepsilon)B_1(\varepsilon)R_{11}^{-1}(\varepsilon)B_1^T(\varepsilon)\sum_{s=1}^k \widehat{\mu}_s \mathcal{H}_s(\varepsilon)$$

$$+ \sum_{r=1}^k \widehat{\mu}_r \mathcal{H}_r(\varepsilon)B_1(\varepsilon)R_{11}^{-1}(\varepsilon)B_1^T(\varepsilon)\mathcal{H}_1(\varepsilon)$$

$$- \mathcal{H}_1(\varepsilon)B_2(\varepsilon)R_{22}^{-1}(\varepsilon)B_2^T(\varepsilon)\sum_{s=1}^k \widehat{\mu}_s \mathcal{H}_s(\varepsilon)$$

$$- \sum_{r=1}^k \widehat{\mu}_r \mathcal{H}_r(\varepsilon)B_2(\varepsilon)R_{22}^{-1}(\varepsilon)B_2^T(\varepsilon)\mathcal{H}_1(\varepsilon)$$

$$- \sum_{r=1}^k \widehat{\mu}_r \mathcal{H}_r(\varepsilon)B_1(\varepsilon)R_{11}^{-1}(\varepsilon)B_1^T(\varepsilon)\sum_{s=1}^k \widehat{\mu}_s \mathcal{H}_s(\varepsilon)$$

$$+ \sum_{r=1}^k \widehat{\mu}_r \mathcal{H}_r(\varepsilon)B_2(\varepsilon)R_{22}^{-1}(\varepsilon)B_2^T(\varepsilon)\sum_{s=1}^k \widehat{\mu}_s \mathcal{H}_s(\varepsilon) \ , \tag{40}$$

and, for $2 \le i \le k$

$$\frac{d}{d\varepsilon}\mathcal{H}_i(\varepsilon) = -A^T(\varepsilon)\mathcal{H}_i(\varepsilon) - \mathcal{H}_i(\varepsilon)A(\varepsilon)$$

$$+ \mathcal{H}_i(\varepsilon)B_1(\varepsilon)R_{11}^{-1}(\varepsilon)B_1^T(\varepsilon)\sum_{s=1}^k \widehat{\mu}_s \mathcal{H}_s(\varepsilon)$$

$$+ \sum_{r=1}^k \widehat{\mu}_r \mathcal{H}_r(\varepsilon)B_1(\varepsilon)R_{11}^{-1}(\varepsilon)B_1^T(\varepsilon)\mathcal{H}_i(\varepsilon)$$

$$- \mathcal{H}_i(\varepsilon)B_2(\varepsilon)R_{22}^{-1}(\varepsilon)B_2^T(\varepsilon)\sum_{s=1}^k \widehat{\mu}_s \mathcal{H}_s(\varepsilon)$$

$$- \sum_{r=1}^k \widehat{\mu}_r \mathcal{H}_r(\varepsilon)B_2(\varepsilon)R_{22}^{-1}(\varepsilon)B_2^T(\varepsilon)\mathcal{H}_i(\varepsilon)$$

$$- \sum_{j=1}^{i-1} \frac{2i!}{j!(i-j)!}\mathcal{H}_j(\varepsilon)G(\varepsilon)WG^T(\varepsilon)\mathcal{H}_{i-j}(\varepsilon) \ , \tag{41}$$

together, for $1 \leq i \leq k$

$$\frac{d}{d\varepsilon}\mathcal{D}_i(\varepsilon) = -\mathrm{Tr}\left\{\mathcal{H}_i(\varepsilon)G(\varepsilon)WG^T(\varepsilon)\right\} \tag{42}$$

where the terminal-valued conditions $\mathcal{H}_1(t_f) = Q_f$, $\mathcal{H}_i(t_f) = 0$ for $2 \leq i \leq k$ and $\mathcal{D}_i(t_f) = 0$ for $1 \leq i \leq k$. Thus, whenever these equations (40)-(42) admit solutions $\{\mathcal{H}_i(\cdot)\}_{i=1}^k$ and $\{\mathcal{D}_i(\cdot)\}_{i=1}^k$, then the existence of $\{\mathcal{E}_i(\cdot)\}_{i=1}^k$ and $\{\mathcal{T}_i(\cdot)\}_{i=1}^k$ satisfying (37)-(39) are assured. By comparing the equations (37)-(39) to those of (40)-(42), one may recognize that these sets of equations are related to one another by

$$\frac{d}{d\varepsilon}\mathcal{E}_i(\varepsilon) = -\frac{d}{d\varepsilon}\mathcal{H}_i(\varepsilon) \text{ and } \frac{d}{d\varepsilon}\mathcal{T}_i(\varepsilon) = -\frac{d}{d\varepsilon}\mathcal{D}_i(\varepsilon)$$

for $1 \leq i \leq k$. Enforcing the initial value conditions of $\mathcal{E}_i(t_0) = 0$ and $\mathcal{T}_i(t_0) = 0$ uniquely implies that

$$\mathcal{E}_i(\varepsilon) = \mathcal{H}_i(t_0) - \mathcal{H}_i(\varepsilon) \text{ and } \mathcal{T}_i(\varepsilon) = \mathcal{D}_i(t_0) - \mathcal{D}_i(\varepsilon)$$

for all $\varepsilon \in [t_0, t_f]$ and yields a value function

$$\mathcal{W}(\varepsilon, \mathcal{Y}, \mathcal{Z}) = \mathcal{V}(\varepsilon, \mathcal{Y}, \mathcal{Z})$$
$$= x_0^T \sum_{i=1}^k \mu_i \mathcal{H}_i(t_0) x_0 + \sum_{i=1}^k \mu_i \mathcal{D}_i(t_0) \ ,$$

for which the sufficient condition (29) of the verification theorem is satisfied. Therefore, the respective feedback gains (34) and (35) for Player 1 and Player 2 optimizing the performance index (25), become optimal

$$K_1^*(\varepsilon) = -R_{11}^{-1}(\varepsilon)B_1^T(\varepsilon)\sum_{r=1}^k \widehat{\mu}_r \mathcal{H}_r^*(\varepsilon) \ , \tag{43}$$

$$K_2^*(\varepsilon) = R_{22}^{-1}(\varepsilon)B_2^T(\varepsilon)\sum_{r=1}^k \widehat{\mu}_r \mathcal{H}_r^*(\varepsilon) \ . \tag{44}$$

Theorem 10. *Multi-Cumulant Saddle-Point Solution.*
Consider the linear-quadratic zero-sum stochastic differential game (5)-(6) in which the pairs (A, B_1) and (A, B_2) are uniformly stabilizable on $[t_0, t_f]$. Let $k \in \mathbb{Z}^+$ and the sequence $\mu = \{\mu_i \geq 0\}_{i=1}^k$ with $\mu_1 > 0$. Then, the optimal cost-cumulant control via state-feedback is achieved by the saddle-point gains

$$K_1^*(\alpha) = -R_{11}^{-1}(\alpha)B_1^T(\alpha)\sum_{r=1}^k \widehat{\mu}_r \mathcal{H}_r^*(\alpha) \ , \tag{45}$$

$$K_2^*(\alpha) = R_{22}^{-1}(\alpha)B_2^T(\alpha)\sum_{r=1}^k \widehat{\mu}_r \mathcal{H}_r^*(\alpha) \ , \tag{46}$$

where additional parametric design freedom $\widehat{\mu}_r \triangleq \mu_i/\mu_1$ mutually selected by Players 1 and 2 represent different levels of influence as they deem important to the global performance of the game and $\{\mathcal{H}_r^*(\alpha) \geq 0\}_{r=1}^k$ are the optimal solutions of the time-backward differential equations

$$\frac{d}{d\alpha}\mathcal{H}_1^*(\alpha) = - \left[A(\alpha) + B_1(\alpha)K_1^*(\alpha) + B_2(\alpha)K_2^*(\alpha)\right]^T \mathcal{H}_1^*(\alpha)$$
$$- \mathcal{H}_1^*(\alpha)\left[A(\alpha) + B_1(\alpha)K_1^*(\alpha) + B_2(\alpha)K_2^*(\alpha)\right]$$
$$- Q(\alpha) - K_1^{*T}(\alpha)R_{11}(\alpha)K_1^*(\alpha) + K_2^{*T}(\alpha)R_{22}(\alpha)K_2^*(\alpha) \ , \quad (47)$$

and, for $2 \leq r \leq k$

$$\frac{d}{d\alpha}\mathcal{H}_r^*(\alpha) = - \left[A(\alpha) + B_1(\alpha)K_1^*(\alpha) + B_2(\alpha)K_2^*(\alpha)\right]^T \mathcal{H}_r^*(\alpha)$$
$$- \mathcal{H}_r^*(\alpha)\left[A(\alpha) + B_1(\alpha)K_1^*(\alpha) + B_2(\alpha)K_2^*(\alpha)\right]$$
$$- \sum_{s=1}^{r-1}\frac{2r!}{s!(r-s)!}\mathcal{H}_s^*(\alpha)G(\alpha)WG^T(\alpha)\mathcal{H}_{r-s}^*(\alpha) \ , \quad (48)$$

with the terminal-boundary conditions $\mathcal{H}_1^*(t_f) = Q_f$, and $\mathcal{H}_r^*(t_f) = 0$ when $2 \leq r \leq k$.

4   Conclusions

This paper dealt with a class of two-player zero-sum differential games modeled in a stochastic environment for realistic conditions. Both players were assumed to have exact knowledge of the state, the payoff functional and the control capabilities of each. Matrix differential equations for generating statistics of the IQF random cost used in this game were derived. A more direct dynamic programming approach was used to solve for a saddle-point solution that can address both control strategy selection and performance analysis aspects. This saddle-point solution was computed by two multi-cumulant control gains within the class of linear memoryless-feedback strategies which then minimized a linear combination of first k cumulants of the IQF random cost of the game. Hopefully, these results will make some new theoretical contributions and performance analysis tools to differential game communities. Finally, this theoretical development provides framework and analyses to applications of boost phase missile interception whose the solution offers two optimal conflicting guidance laws: (1) a hit-to-kill homing guidance law for intercepting boosting ballistic missiles in minimum time and divert fuel and (2) an evasion strategy for a ballistic missile to achieve burnout before the kill vehicle arrives, and force the kill vehicle use maximum divert fuel. Future work will address the efficacy of the theoretical work herein via numerical simulation results.

References

1. Davis, M. H. A.: Linear Estimation and Stochastic Control. A Halsted Press. John Wiley & Sons. New York (1977)

2. Dieudonne, J.: Foundations of Modern Analysis. Academic Press. New York and London (1960)
3. Elliot, R. J.: The existence of value in stochastic differential games. SIAM Journal on Control and Optimization **14** (1976) 85–94
4. Fleming, W. H., Rishel, R. W.: Deterministic and Stochastic Optimal Control. New York: Springer-Verlag (1975)
5. Jacobson, D. H.: Optimal Stochastic Linear Systems with Exponential Performance Criteria and Their Relation to Deterministic Games. IEEE Transactions on Automatic Control **AC-18** (1973) 124–131
6. Pham, K. D., Liberty, S. R., Sain, M. K., Spencer, B. F., Jr.: Generalized Risk Sensitive Building Control: Protecting Civil Structures with Multiple Cost Cumulants. Proceedings of American Control Conference (1999) 500–504
7. Pham, K. D., Liberty, S. R., Sain, M. K.: Evaluating Cumulant Controllers on a Benchmark Structure Protection Problem in the Presence of Classic Earthquakes. Proceedings of 37th Annual Allerton Conference on Communication, Control, and Computing (1999) 617–626
8. Pham, K. D., Sain, M. K., Liberty, S. R., Spencer, B. F., Jr.: The Role and Use of Optimal Cost Cumulants for Protection of Civil Structures. The 14th ASCE Engineering Mechanics Conference. Proceeding CD-ROM (2000)
9. Pham, K. D., Liberty, S. R., Sain, M. K., Spencer, B. F., Jr.: First Generation Seismic-AMD Benchmark: Robust Structural Protection by the Cost Cumulant Control Paradigm. Proceedings of American Control Conference (2000) 1–5
10. Pham, K. D., Sain, M. K., Liberty, S. R., Spencer, B. F., Jr.: Optimum Multiple Cost Cumulants for Protection of Civil Structures. The 8th ASCE Specialty Conference on Probabilistic Mechanics and Structural Reliability Conference. Proceeding of CD-ROM (2000)
11. Pham, K. D., Sain, M. K., Liberty, S. R.: Robust Cost-Cumulants Based Algorithm for Second and Third Generation Structural Control Benchmarks. Proceedings of American Control Conference (2002) 3070–3075
12. Pham, K. D., Sain, M. K., Liberty, S. R.: Finite Horizon Full-State Feedback kCC Control in Civil Structures Protection. Stochastic Theory and Adaptive Control, Lecture Notes in Control and Information Sciences. Proceedings of a Workshop held in Lawrence, Kansas. Edited by B. Pasik-Duncan. Springer-Verlag, Berlin Heidelberg, Germany **280** (2002) 369–383
13. Pham, K. D., Sain, M. K., Liberty, S. R.: Cost Cumulant Control: State-Feedback, Finite-Horizon Paradigm with Application to Seismic Protection. Special Issue of Journal of Optimization Theory and Applications. Edited by A. Miele. Kluwer Academic/Plenum Publishers. New York **115** (2002) 685–710
14. Pham, K. D., Jin, G., Sain, M. K., Spencer, B. F., Jr., Liberty, S. R.: Generalized LQG Techniques for the Wind Benchmark Problem. Special Issue of ASCE Journal of Engineering Mechanics on the Structural Control Benchmark Problem **130** (2004) 466–470
15. Pham, K. D., Sain, M. K., Liberty, S. R.: Infinite Horizon Robustly Stable Seismic Protection of Cable-Stayed Bridges Using Cost Cumulants. Proceedings of American Control Conference (2004) 691–696
16. Pham, K. D., Sain, M. K., Liberty, S. R.: Statistical Control for Smart Base-Isolated Buildings via Cost Cumulants and Output Feedback Paradigm. Proceedings of American Control Conference (2005) 3090–3095
17. Pham, K. D.: Minimax Design of Statistics-Based Control with Noise Uncertainty for Highway Bridges. Proceedings of DETC 2005/2005 ASME 20th Biennial Conference on Mechanical Vibration and Noise (2005)

Decentralized Cooperative Optimization for Multi-criteria Decision Making

Cristinca Fulga

Academy of Economic Studies
Department of Mathematics
Bucharest, Romania
fulga@csie.ase.ro

Abstract. Motivated by recent research on multiobjective optimization, we focus on the problem of m interconnected systems characterized by multiple decision makers with limited centralized information. Two types of variables occur: local variables that appear in a single component and global variables which provide the connection between the m systems and appear in all of them. From the point of view of one system, the problem is seen as optimization of local costs using local control variables coupled with global variables, subject to local constraints. This is a decomposition of the general centralized vector optimization problem into a set of decentralized cooperative optimization problems with local mathematical models, coupled through constraints. In this chapter, we provide a method for solving this problem by using a decomposition technique combined with an exact penalty method.

1 Introduction

Modern industrial production systems demand methods for the automation and the optimization of their production processes, in order to improve their efficiency and adaptability to the flexibility and reactivity requested by the new global market. Generally, the management of multi-agent systems is difficult because their automation requires an approach to optimization that seeks to ensure global goals that are met by the collaboration of the agents which usually are only aware of their local surroundings. One approach is to solve this problem globally, but centralized algorithms scale very poorly with the large size of the problem because of the computational effort involved. This chapter presents a decentralized cooperative technique for multiobjective optimization which makes the initial problem more computationally manageable.

There are a lot of important contributors to this domain. Much of the current research on decentralized optimization uses a setup where each system solves a local problem and communicates the optimal solution to its neighbors (see Inalhan et al. [13]). The challenge in this case is how to achieve cooperation between all the systems.

Heiskanen [12] presents a decentralized method for computing Pareto-optimal solutions in multiparty negotiations over continuous issues. The method is called

M.J. Hirsch et al. (Eds.): Adv. in Cooper. Ctrl. & Optimization, LNCIS 369, pp. 65–80, 2007.
springerlink.com

decentralized because its use does not require decision makers to know each others' value functions and neither does it require anyone outside to know all value functions. The computation of Pareto-optimal solutions in a decentralized manner is interesting because of the negotiators' frequent failure to achieve efficient agreements in practice (see, for example Sebenius [22]) and their unwillingness to disclose private information due to strategic reasons.

A decentralized cooperative optimization method is proposed by Kuwata and How [14]. In this chapter, the approach is limited to a scalar global objective. The key difference is that when there are active coupling constraints, they are modified; therefore, this procedure could worsen the local performance, but increases the global performance.

For the particular case of a multi-agent system for which the objective function may be written as a sum of functions, the approach of decomposition technique received a great deal of attention, see the work of Benders [2], Geoffrion [10], Tammer [23], Braun [3], de Miguel [16].

In another approach for generating Pareto-optimal solutions in conflict situations, there is a mediator who works as a neutral coordinator and gathers information on decision makers' preferences during an interactive procedure (see Teich [24] and [25]).

Stochastic search algorithms imitating natural evolution have been developed and used for simulation and optimization. Traditionally, evolutionary algorithms have been used for scalar optimization. Especially during recent years, evolutionary algorithms have been applied to multiobjective programming (see Fonseca and Fleming [7], Hanne [11], Barrico and Antunes [1]).

In this chapter, we focus on the multiobjective optimization problem with objective functions and constraints depending on local variables and global variables. First, we seek the optimal solution for each of the m systems solved independently and we obtain the ideal point. Using the ideal point, the initial problem is transformed into an equivalent one which allows the decomposition into m subproblems. Each subproblem is then solved. The key difference is that the new approach recognizes that each system should consider sacrifices to its local performance if it is possible to reap a larger benefit to overall m systems performance. Indeed, after the optimization of each system, a master problem which gathers all the information from the optimal solutions is solved and the values of the global variables are improved by taking into account the ensemble of systems. The fact that each new iteration for the particular subproblem i, $i = \overline{1,m}$, begins with a starting point that was improved for increasing the performance of the ensemble, emphasizes the cooperative character of our method.

2   Preliminaries

Many optimization problems combine objective and constraint functions corresponding to a set of interconnected systems. In this chapter we consider m systems, each with their own independent cost function f_i, $i = \overline{1,m}$, and local constraints. We are concerned with those problems in which only a few of the variables, known as global variables, appear in all systems, while the remaining variables occur in only single systems.

Let $f : R^n \times R^p \to R^m$ be the vectorial cost function defined by $f(x,y) = (f_1(x_1,y), ..., f_i(x_i,y), ..., f_m(x_m,y))$, where $x_i \in R^{n_i}$ $\forall i = \overline{1,m}$, $\sum_{i=1}^{m} n_i = n$, $x = (x_1^T, ..., x_m^T)^T \in R^n$; x_i represents the vector of local variables of the ith system, $i = \overline{1,m}$, and $y \in R^p$ the vector of global variables.

The centralized optimization problem (CP) is defined as

$$\min_{(x,y) \in R^n \times R^p} f(x,y) = (f_1(x_1,y), ..., f_i(x_i,y), ..., f_m(x_m,y)) \tag{1}$$
$$s.t. \ g_i(x_i,y) \geq 0, \ i = \overline{1,m},$$

where $g_i = (g_{i1}, ..., g_{im_i}) : R^n \times R^p \to R^{m_i}$ is the vector function that formalizes the m_i constraints of the ith system, $i = \overline{1,m}$.

Let $X = \{(x,y) \in R^n \times R^p \,|\, g_i(x_i,y) \geq 0, \ i = \overline{1,m}\}$ be the set of feasible solutions of the problem (CP).

Definition 1. $(x^*, y^*) \in X$ *is a Pareto-optimal solution of the problem 1 if there is no* $(x,y) \in X$ *and* $j \in \{1, ..., m\}$ *such that* $f_i(x_i,y) \leq f_i(x_i^*, y^*), \forall i = \overline{1,m}$ *and* $f_j(x_j,y) < f_j(x_j^*, y^*)$.

Methodological approaches in multiobjective optimization are mostly based on the calculation of one or several, usually efficient solutions which can be interpreted as compromise solutions. Only for specific types of the general multiobjective optimization problem, (e.g, linear multiobjective optimization problems) have algorithms been developed for computing the complete efficient set. Often it is not only difficult but also not desirable to calculate the usually uncountable efficient set because the decision maker is interested in the selection of one alternative of his multicriteria decision problem; therefore, the proposed method does not look for the entire efficient set.

3 Decentralized Optimization Via Decomposition Technique

We consider the nonlinear programming problem (P_i) associated to the ith system

$$\min_{(x_i,y) \in R^{n_i} \times R^p} f_i(x_i,y) \tag{2}$$
$$s.t. \ g_i(x_i,y) \geq 0.$$

Let (x_i^*, y^*) the global optimal solution of the problem 2 and $f_i(x_i^*, y^*) = z_i^* \in R$ the corresponding optimal value. In this case, $z^* = (z_1^*, ..., z_m^*)$ is called the ideal point for 1 because, usually, $z^* = (z_1^*, ..., z_m^*) \in R^m$ is not attainable. Individually, the z_i^* are attainable but to find a point z^* which can simultaneously minimize each f_i, $i = \overline{1,m}$, is very difficult.

Now, given $(x, y) \in X \subset R^n \times R^p$, we define the regret function (Yu [26])

$$r(x, y) = (\|f(x, y) - z^*\|_2)^2 = \sum_{i=1}^{m} (f_i(x_i, y) - z_i^*)^2.$$

We say that $(\widetilde{x}, \widetilde{y}) \in X$ is a compromise solution of 1 with respect to the l_2-norm if it is the solution of the problem

$$\min_{(x,y) \in R^n \times R^p} \|f(x, y) - z^*\|_2^2 = \sum_{i=1}^{m} (f_i(x_i, y) - z_i^*)^2 \tag{3}$$

$$s.t. \ g_i(x_i, y) \geq 0, \ i = \overline{1, m}$$

Note that $r(x, y)$ treats each $(f_i(x_i, y) - z_i^*)^2$ as having the same importance in the regret function. In multiple-criteria problems, the criteria may have different degrees of importance; a weight vector $w = (w_1, ..., w_m) \geq 0$ may be assigned to signal the different degrees of importance. Therefore we define $r(x, y; w) =$

$$\left(\|f(x, y) - z^*\|_{w,2} \right)^2 = \sum_{i=1}^{m} w_i^2 (f_i(x_i, y) - z_i^*)^2 = \sum_{i=1}^{m} (w_i f_i(x_i, y) - w_i z_i^*)^2.$$

The concept of compromise solution has a natural extension to include the weighted regret function. In fact, the weight vector w changes the scale of each criterion. Once the scale is adjusted, the regret function is reduced to that of equal weight. Thus, without loss of generality, we will focus on the equal weight case 3.

In the sequel, we will use the notation $\widetilde{f}_i(x_i, y) = (f_i(x_i, y) - z_i^*)^2$. Therefore the problem 3 is equivalent with

$$\min_{(x,y) \in R^n \times R^p} \sum_{i=1}^{m} \widetilde{f}_i(x_i, y) \tag{4}$$

$$s.t. \quad g_i(x_i, y) \geq 0, \ i = \overline{1, m}$$

The interest for problems having the same structure as 4 and two types of variables that can be seen as local and global is justified by the wide applicability in the following practical example.

Example 1. A practical problem that can be formalized like a centralized optimization problem is presented in Escudero et al. [6]. In this paper, a modeling framework for the solution of hydroelectric power management problem with uncertainty in the values of the water inflows and outflows is presented. The problem under study is maximizing the hydropower generated along a time horizon by a multireservoir power system. We note that a multi-objective formulation could be very useful. Its formulation relies on a basic network, where the nodes represent the reservoirs and the arcs correspond to the sections that connect the reservoirs. Each node of the network can be considered as a component system within the optimization problem. The global variables are the amounts of water stored in the reservoirs at the end of each time period. Local variables include the amount of water released for generation at each hydropower plant and the energy generated using each of the thermal plants. Therefore, due to these features, this problem can by formalized as in 1 and solved using the algorithm presented in the next section.

Next, we allow the vector of global variables to take a different value $y_i \in R^p$ within each of the subproblems 4. The l_2-norm $\|y_i - u\|_2$ is used to force the global variables y_i to take the same value, equal to the so-called target variables $u \in R^p$, for all systems. Therefore, the resulting problem is

$$
\min_{u,x,y} \sum_{i=1}^m \widetilde{f}_i(x_i, y_i)
$$
$$
s.t. \ g_i(x_i, y_i) \geq 0, \ i = \overline{1, m} \tag{5}
$$
$$
\|y_i - u\|_2^2 \leq 0, \ i = \overline{1, m}
$$

where $y = \left(y_1^T, ..., y_m^T\right)^T \in R^{mp}$.

If we eliminate the global variables, problem 5 breaks into m independent subproblems:

$$
\min_{x_i, y_i} \widetilde{f}_i(x_i, y_i)
$$
$$
s.t. \ g_i(x_i, y_i) \geq 0, \tag{6}
$$
$$
\|y_i - u\|_2^2 \leq 0.
$$

where $g_i = (g_{i1}, ..., g_{im_i}) : R^n \times R^p \to R^{m_i}$. We consider the equivalent form of the problem 6:

$$
\min_{x_i, y_i} \widetilde{f}_i(x_i, y_i)
$$
$$
s.t. \ - g_{ij}(x_i, y_i) \leq 0, j = \overline{1, m_i}, \tag{7}
$$
$$
\|y_i - u\|_2^2 \leq 0.
$$

For the sake of clarity we use the notation $-g_{i,m_i+1}(x_i, y_i) = \|y_i - u\|_2^2 = \sum_{j=1}^p (y_{ij} - u_j)^2$. The problem formulation using the decomposition method is:

$$
\min_u \sum_{i=1}^m \widetilde{f}_i^*(u), \tag{8}
$$

where $\widetilde{f}_i^*(u)$ is the optimal value in 7 corresponding to the optimal solution (x_i^*, y_i^*):

$$
\widetilde{f}_i^*(u) = \min_{x_i, y_i} \widetilde{f}_i(x_i, y_i)
$$
$$
s.t. \ - g_{ij}(x_i, y_i) \leq 0, j = \overline{1, m_i + 1}. \tag{9}
$$

We present theoretical results that ensure the convergence of the algorithm in Section 5 based on an exact penalty method. We choose this approach for solving problem 9 because exact penalty methods have the ability to handle degenerate problems and inconsistent constraint linearizations.

4 Penalty Method

Penalty methods have undergone three stages of development. First, they were seen as vehicles for solving constrained optimization problems by means of unconstrained optimization techniques. In the second stage, the penalty problem is replaced by a sequence of linearly constrained subproblems. These formulations, which are related to the sequential quadratic approach, are much more effective than the unconstrained approach but they leave open the question of how to choose the penalty parameter. In the most recent stage of development, penalty methods adjust the penalty parameter at every iteration so as to achieve a prescribed level of linear feasibility. The choice of the penalty parameter then ceases to be a heuristic and becomes an integral part of the step computation, see Byrd et al. [4] and [5], Price [18], Fulga [8] and [9]. Therefore, we propose an algorithm for solving problem 7 that combines penalty methods so as to ensure balanced progress toward feasibility and optimality.

Throughout this chapter we assume that at each local minimizer of the nonlinear programming problem 9, an appropriate constraint qualification holds, thereby ensuring that any optimal point (x_i^*, y_i^*) satisfies the following Karush-Kuhn-Tucker conditions: there exists a vector of Lagrange multipliers $\lambda_i^* = \left(\lambda_{i1}^*, ..., \lambda_{i,\, m_i+1}^* \right) \in R^{m_i+1}$ such that

$$
\begin{cases}
-g_{ij}\left(x_i^*, y_i^*\right) \le 0 \; ; \quad j = \overline{1, m_i + 1} \\
\lambda_{ij}^* \ge 0 \; ; \quad j = \overline{1, m_i + 1} \\
\lambda_{ij}^* \cdot \left(-g_{ij}\left(x_i^*, y_i^*\right)\right) = 0 \; ; \quad j = \overline{1, m_i + 1} \\
\nabla \widetilde{f}_i\left(x_i^*, y_i^*\right) - \displaystyle\sum_{j=1}^{m_i+1} \lambda_{ij}^* \cdot \nabla g_{ij}\left(x_i^*, y_i^*\right) = 0.
\end{cases}
\tag{10}
$$

The nonlinear programming problem is not solved directly; instead a non-differentiable exact penalty function Φ_i is minimized, where the exact penalty function is constructed so that local minimizers of the nonlinear programming problem are also local minimizers of the penalty function Φ_i. The term exact refers to the fact that exact solutions are computed for finite values of the penalty parameters. Therefore, exact penalty functions avoid the ill-conditioning associated with large penalty parameter since they need no longer be driven to infinity. This advantage comes with an important difficulty: as a consequence of the use of exact penalty functions the problem optimal-value function becomes nonsmooth. The method that we propose handles this aspect quite well.

Following the approach in Price [18], we consider the penalty function

$$
\Phi_i\left(x_i, y_i\right) = \widetilde{f}_i\left(x_i, y_i\right) + \mu \cdot \theta\left(x_i, y_i\right) + \frac{1}{2}\nu \cdot \theta^2\left(x_i, y_i\right),
\tag{11}
$$

with $\mu > 0, \nu \ge 0$ and the degree of infeasibility, $\theta\left(x_i, y_i\right)$, is defined as

$$
\theta\left(x_i, y_i\right) = \max_{1 \le\, j \le m_i+1} \left\{ \left[-g_{ij}\left(x_i, y_i\right)\right]_+ \right\},
\tag{12}
$$

where $[y]_+ = \max\{0; y\}$. The penalty function Φ_i may be viewed as a hybrid of a quadratic penalty function based on the infinity norm and the single parameter exact penalty function of [15], [17] and [21]. Clearly θ is continuous $\forall (x_i, y_i) \in R^n \times R^p$, but it is usually not differentiable for some (x_i, y_i). However, the directional derivative

$$D_{(p,q)}\theta (x_i, y_i) = \lim_{\alpha \searrow 0} \frac{\theta (x_i + \alpha p, y_i + \alpha q) - \theta (x_i, y_i)}{\alpha}$$

exists for any $(x_i, y_i), (p, q) \in R^n \times R^p$. Definition 12 implies that for any $(x_i, y_i), (p, q) \in R^n \times R^p$ the directional derivative takes the form

$$D_{(p,q)}\theta (x_i, y_i) = \begin{cases} \max\limits_{i \in I(x_i, y_i)} \left(- \left(p^T, q^T\right) \nabla g_{ij} (x_i, y_i)\right), & \text{if} \begin{cases} \theta (x_i, y_i) > 0 \\ I (x_i, y_i) \neq \varnothing \end{cases}, \\ \max\limits_{i \in I(x_i, y_i)} \left[- \left(p^T, q^T\right) \nabla g_{ij} (x_i, y_i)\right]_+, & \text{if} \begin{cases} \theta (x_i, y_i) = 0 \\ I (x_i, y_i) \neq \varnothing \end{cases} \\ 0 \text{ if } I (x_i, y_i) = \varnothing \end{cases}$$

$$(13)$$

where $I (x_i, y_i) = \{j \,|-g_{ij} (x_i, y_i) = \theta (x_i, y_i)\}$.

Definition 2. *For fixed values of $\mu > 0$ and $\nu \geq 0$, a point (x_i^*, y_i^*) is a critical point of Φ_i if and only if for all $(p, q) \in R^n \times R^p$ the directional derivative $D_{(p,q)}\Phi_i (x_i^*, y_i^*)$ is non-negative.*

Given a suitable choice of the penalty parameters, problem 9 may be replaced by the problem

$$\begin{cases} \min \Phi_i (x_i, y_i) \\ (x_i, y_i) \in R^n \times R^p. \end{cases} \tag{14}$$

Definition 3. *The solution set of the penalty function problem 14 with fixed values for $\mu > 0, \nu \geq 0$ is defined as the set of critical points of Φ_i.*

Theorem 1. *Let (x_i^*, y_i^*) be an optimal solution of the nonlinear programming problem 9 at which Karush-Kuhn-Tucker conditions 10 hold and let $\lambda_i^* \in R^{m_i+1}$ be a vector of Lagrange multipliers satisfying these conditions for which $\|\lambda_i^*\|_1$ is minimal. If $\mu > \|\lambda_i^*\|_1$ then (x_i^*, y_i^*) is a critical point of Φ_i. Conversely, if (x_i^*, y_i^*) is both feasible and a critical point of Φ_i for some $\mu > 0, \nu \geq 0$, then (x_i^*, y_i^*) is a Karush-Kuhn-Tucker point of the nonlinear programming problem 9.*

Proof. The Karush-Kuhn-Tucker conditions 10 and definition 12 imply $\theta (x_i^*, y_i^*) = 0$ and $\lambda_{ij}^* = 0, \forall j \notin I (x_i^*, y_i^*)$. Therefore, combining 10 with 13, for any $(p, q) \in R^n \times R^p$ we have

$$D_{(p,q)}\Phi_i (x_i^*, y_i^*) = \left(p^T, q^T\right) \nabla \widetilde{f}_i (x_i^*, y_i^*) + \mu D_{(p,q)}\theta (x_i^*, y_i^*) =$$

$$= \left(p^T, q^T\right) \left[\sum_{j=1}^{m_i+1} \lambda_{ij}^* \nabla g_{ij} (x_i^*, y_i^*)\right] + \mu D_{(p,q)}\theta (x_i^*, y_i^*) \geq$$

$$\geq \left[- \sum_{j \in I(x_i^*, y_i^*)} \lambda_{ij}^* + \mu \right] D_{(p,q)} \theta \left(x_i^*, y_i^* \right) =$$

$$= (\mu - \|\lambda_i^*\|_1) \max_{i \in I(x_i, y_i)} \left[- \left(p^T, q^T \right) \nabla g \left(x_i^*, y_i^* \right) \right]_+ \geq 0.$$

We obtain $D_{(p,q)} \Phi_i \left(x_i^*, y_i^* \right) \geq 0, \forall (p,q) \in R^n \times R^p$ and thus, x^* is a critical point of Φ_i.

Conversely, if (x_i^*, y_i^*) is a critical point of Φ_i for some fixed $\mu > 0, \nu \geq 0$, then $D_{(p,q)} \Phi_i \left(x_i^*, y_i^* \right) \geq 0, \forall (p,q) \in R^n \times R^p$. For any (x_i, y_i) sufficiently close to (x_i^*, y_i^*) we have

$$\Phi_i (x_i, y_i) = \Phi_i \left(x_i^*, y_i^* \right) + D_{\left(x_i - x_i^*, y_i - y_i^* \right)} \Phi_i \left(x_i^*, y_i^* \right) + o \left\| (x_i - x_i^*, y_i - y_i^*) \right\|$$

$$\geq \Phi_i \left(x_i^*, y_i^* \right) + o \left\| (x_i - x_i^*, y_i - y_i^*) \right\|.$$

If (x_i, y_i) is a feasible point, $\Phi_i (x_i, y_i) = \widetilde{f}_i (x_i, y_i)$ and (x_i^*, y_i^*) satisfies Karush-Kuhn-Tucker conditions 10 and the theorem is proved.

Penalty problem 14 is solved by an iterative process. In order to determine a suitable descent direction at the k-th iterate, a continuous piecewise quadratic approximation to Φ_i near the current point is defined:

$$\psi_i^k (p, q) = \widetilde{f}_i \left(x_i^k, y_i^k \right) + \left(p^T, q^T \right) \nabla \widetilde{f}_i \left(x_i^k, y_i^k \right) +$$

$$+ \frac{1}{2} \left(p^T, q^T \right) H^k \left(p^T, q^T \right)^T + \mu^k \zeta (p, q) + \frac{1}{2} \nu^k \cdot \zeta^2 (p, q),$$

where

$$\zeta (p, q) = \max_{1 \leq j \leq m_i + 1} \left\{ \left[-g_{ij} \left(x_i^k, y_i^k \right) - \left(p^T, q^T \right) \nabla g_{ij} \left(x_i^k, y_i^k \right) \right]_+ \right\}$$

and H^k is positive definite. Clearly ψ_i^k is strictly convex in (p, q) and therefore ψ_i^k has an unique global minimizer (p^k, q^k) which also solves the quadratic programming problem (P_i^k):

$$\begin{cases} \min_{p,q,\zeta} \left(p^T, q^T \right) \nabla \widetilde{f}_i \left(x_i^k, y_i^k \right) + \frac{1}{2} \left(p^T, q^T \right) H^k \left(p^T, q^T \right)^T + \\ \qquad + \mu^k \zeta (p, q) + \frac{1}{2} \nu^k \cdot \zeta^2 (p, q) \\ g_{ij} \left(x_i^k, y_i^k \right) + \left(p^T, q^T \right) \nabla g_{ij} \left(x_i^k, y_i^k \right) \geq -\zeta, \quad j = \overline{1, m_i + 1} \\ \qquad \zeta \geq 0. \end{cases} \quad (15)$$

Theorem 2. *Let* $\left(p^k, q^k, \zeta^k \right)$ *be the unique solution of the quadratic programming problem* $\left(P_i^k \right)$ *, with* H^k *positive definite. Let* λ_i^k *denote an optimal Lagrange multiplier vector, which need not be unique, for which* $\left\| \lambda_i^k \right\|_1$ *is minimal. If* $\left(p^k, q^k \right) \neq (0, 0)$, $\zeta^k \leq \theta \left(x_i^k, y_i^k \right)$ *and* $\mu + \nu \theta \left(x_i^k, y_i^k \right) \geq \left\| \lambda_i^k \right\|_1$ *then* $\left(p^k, q^k \right)$ *is a descent direction for* Φ_i *at* $\left(x_i^k, y_i^k \right)$.

Proof. The Karush-Kuhn-Tucker conditions for problem $\left(P_i^k\right)$ are

$$
\begin{cases}
-g_{ij}\left(x_i^k, y_i^k\right) - \left(p^T, q^T\right)\nabla g_{ij}\left(x_i^k, y_i^k\right) - \zeta^k \le 0;\ \lambda_{ij}^k \ge 0 \\
\lambda_{ij}^k \left(-g_{ij}\left(x_i^k, y_i^k\right) - \left(p^T, q^T\right)\nabla g_{ij}\left(x_i^k, y_i^k\right) - \zeta^k\right) = 0 \\
\zeta^k \ge 0;\ \lambda_\zeta \le 0;\ \lambda_\zeta \zeta^k = 0
\end{cases}
\tag{16}
$$

and

$$
\begin{cases}
\nabla \widetilde{f}_i\left(x_i^k, y_i^k\right) + H^k\left(p^{k^{\,T}}, q^{k^T}\right)^T - \displaystyle\sum_{j=1}^{m_i+1}\lambda_{ij}^k \nabla g_{ij}\left(x_i^k, y_i^k\right) = 0 \\
\mu^k + \nu^k \zeta^k - \displaystyle\sum_{j=1}^{m_i+1}\lambda_{ij}^k + \lambda_\zeta = 0.
\end{cases}
\tag{17}
$$

Therefore, combining 16 and 17 we find

$$
\begin{aligned}
D_{(p^k, q^k)}\Phi_i\left(x_i^k, y_i^k\right) = &-\left(p^{k^{\,T}}, q^{k^T}\right)H^k\left(p^{k^{\,T}}, q^{k^T}\right)^T \\
&+ \sum_{j=1}^{m_i+1}\lambda_{ij}^k\left(-\nabla g_{ij}\left(x_i^k, y_i^k\right) - \zeta^k\right) \\
&+ \left(\mu^k + \nu^k \theta\left(x_i^k, y_i^k\right)\right)D_{(p^k, q^k)}\theta\left(x_i^k, y_i^k\right).
\end{aligned}
\tag{18}
$$

Since ζ is convex on $R^n \times R^p$, we have

$$
\zeta\left(p^k, q^k\right) - \theta\left(x_i^k, y_i^k\right) = \zeta\left(p^k, q^k\right) - \zeta(0,0) \ge D_{(p^k, q^k)}\zeta(0,0) = D_{(p^k, q^k)}\theta\left(x_i^k, y_i^k\right).
$$

Applying this result to 18 we obtain

$$
\begin{aligned}
D_{(p^k, q^k)}\Phi_i\left(x_i^k, y_i^k\right) \le &-\left(p^{k^{\,T}}, q^{k^T}\right)H^k\left(p^{k^{\,T}}, q^{k^T}\right)^T \\
&+ \left(-\left\|\lambda_i^k\right\|_1 + \mu^k + \nu^k \theta\left(x_i^k, y_i^k\right)\right)\left(\zeta^k - \theta\left(x_i^k, y_i^k\right)\right) \le 0.
\end{aligned}
$$

and thus, $\left(p^k, q^k\right)$ is a descent direction of Φ_i in $\left(x_i^k, y_i^k\right)$ and the theorem is proved.

The convergence properties of the algorithm are summarized in the following:

Theorem 3. *Assume that the sequence of iterates $\left\{\left(x_i^k, y_i^k\right)\right\}$ is bounded, the sequence of matrices $\left\{H^k\right\}$ generated is bounded in norm and the penalty parameters μ, ν are altered only a finite number of times. Then, every cluster point of the sequence of iterates $\left\{\left(x_i^k, y_i^k\right)\right\}$ generated by the algorithm is a critical point of $\Phi_i\left(x_i, y_i; \mu, \nu\right)$, where μ, ν are at their finite values.*

The following algorithm is based on the preceding theoretical results.

5   Penalty Algorithm

Step 1. Initialization
$k = 1$, $\mu^1 = 1$, $\nu^1 = 1$, $H^1 = I$, $\varepsilon = 10^{-5}$, $\rho = 0.02$, $\delta = 10^{-8}$, $\theta_{cross} = 1$,
$\theta_{cap} = 10$, $k_1 = 1.5$, $k_2 = 2$, $k_3 = 1.2$, $k_4 = 5$, $\delta = 10^{-8}$.

Step 2. Update H and the penalty parameters
This step is omitted from the first iteration. The matrix H is updated using
the Broyden-Fletcher-Goldfarb-Shanno update provided this maintains positive definiteness; otherwise H is not updated. The penalty parameters are
updated as follows:

(i)   If $\theta\left(x_i^k, y_i^k\right) \leq \theta_{cross}$ and $\mu^k < k_1 \left\|\lambda_i^k\right\|_1$, then $\mu^{k+1} = k_2 \left\|\lambda_i^k\right\|_1$ and

$\nu^{k+1} = \nu^k$.

(ii)   If $\theta\left(x_i^k, y_i^k\right) > \theta_{cross}$ and $\mu^k + \nu^k \theta\left(x_i^k, y_i^k\right) < k_3 \left\|\lambda_i^k\right\|_1$, then $\mu^{k+1} = \mu^k$

and $\nu^{k+1} = \frac{k_4 \left\|\lambda_i^k\right\|_1 - \mu^k}{\theta\left(x_i^k, y_i^k\right)}$.

Otherwise, the penalty parameters are not altered.

Step 3. Solve the $\left(P_i^k\right)$ problem

If $\theta\left(x_i^k, y_i^k\right) \leq \theta_{cap}$, then solve $\left(P_i^k\right)$; the solution will be denoted by $\left(p^k, q^k, \zeta^k\right)$
and the algorithm proceeds to Step 4.

If $\theta\left(x_i^k, y_i^k\right) > \theta_{cap}$, the capping constraint $\zeta \leq \theta\left(x_i^k, y_i^k\right)$ is also imposed in
$\left(P_i^k\right)$. Then, this problem is solved and the solution is denoted by $\left(p^k, q^k, \zeta^k\right)$.
If the capping constraint is not active at the $\left(P_i^k\right)$'s solution, the algorithm
proceeds directly to Step 4. Otherwise, the penalty parameters are updated as
described in Step 2, except that $\left\|\lambda_i^k\right\|_1$ is replaced by $\mu^k + \nu^k \theta\left(x_i^k, y_i^k\right) + |\xi|$, where
ξ is the Lagrange multiplier of the capping constraint. The $\left(P_i^k\right)$ problem is then
solved again.

Step 4. Attempt the proposed step
If (i) $\Phi_i\left(x_i^k, y_i^k\right) - \Phi_i\left(x_i^k + p^k, y_i^k + q^k\right) \geq \rho\left[\Psi_i^k(0,0) - \Psi_i^k\left(p^k, q^k\right)\right]$
(ii) $\theta\left(x_i^k + p^k, y_i^k + q^k\right) \leq \theta\left(x_i^k, y_i^k\right)$ are satisfied, then the proposed step $\left(p^k, q^k\right)$
is accepted and the algorithm proceeds to step 7. Otherwise, the execution
continues at the next step.

Step 5. Calculate the Maratos effect correction vector
Solve the following quadratic problem for the second order correction $\left(u^k, v^k\right)$:

$$\begin{cases} \min_{t \in R^n} \left\|(u, v)\right\|_2^2 \\ g_{ij}\left(x_i^k + p^k, y_i^k + q^k\right) + \left(u^T, v^T\right) \nabla g_{ij}\left(x_i^k, y_i^k\right) \leq 0, \ \forall j \in J_i^k \end{cases}$$

where J_i^k is the set of indices of the constraints active at the $\left(P_i^k\right)$'s solution in

Step 3 and $\left\|(u, v)\right\|_2 = \left(\sum_{i=1}^{n} |u_i|^2 + \sum_{i=1}^{p} |v_i|^2\right)^{1/2}$.

If $\left\|(u^k, v^k)\right\|_2 \geq \left\|(p^k, q^k)\right\|_2$ then set $\left(u^k, v^k\right) = (0, 0)$.

Step 6. Arc search

Consider successive values of the sequence $1, \frac{1}{2}, \frac{1}{4}, \frac{1}{8}, ...$ as trial values of α. The number of trial values for α is counted in N_α. If $(u^k, v^k) = (0, 0)$, then omit the first member of the sequence. Accept the first trial value which satisfies

(i) $\Phi_i\left(x_i^k, y_i^k\right) - \Phi_i\left(x_i^k + r^k(\alpha), y_i^k + s^k(\alpha)\right) \geq \rho\alpha\left[\Psi_i^k(0, 0) - \Psi_i^k(p^k, q^k)\right]$
where $r^k(\alpha) = \alpha p^k + \alpha^2 u^k$ and $s^k(\alpha) = \alpha q^k + \alpha^2 v^k$;
(ii) $\theta\left(x_i^k + r^k(\alpha), y_i^k + s^k(\alpha)\right) \leq \theta\left(x_i^k, y_i^k\right)$.
After a satisfactory value of α has been found, set
$\left(x_i^{k+1}, y_i^{k+1}\right) = \left(x_i^k + r^k(\alpha), y_i^k + s^k(\alpha)\right)$ and go to Step 7. If $N_\alpha > 10$ without finding a satisfactory value for α then, in order to get a feasible point, Rosen's method [19], [20] is employed; the new direction is $(p^k, q^k) = N\left(N^T N\right)^{-1} w$, where N is the matrix of normed column vectors of the linearly independent gradients of the violated constraints and w is the vector whose components are the absolute values of the constraint functions for the violated constraints . Then, go to Step 5.

Step 7. Check the stopping conditions

The algorithm halts if either the length of the previous step
$\left\|\left(x_i^k, y_i^k\right) - \left(x_i^{k-1}, y_i^{k-1}\right)\right\|_2 \leq \delta$ or both of the following conditions hold:

(i) $\theta\left(x_i^k, y_i^k\right) < \varepsilon$

(ii) $\left\|\nabla \widetilde{f_i}\left(x_i^k, y_i^k\right) - \sum_{j \in M} \lambda_{ij}^k \nabla g_{ij}\left(x_i^k, y_i^k\right)\right\|_2 < \varepsilon,$

where $M = \left\{ j \left| g_{ij}\left(x_i^k, y_i^k\right) \right| < \varepsilon \right\}$. Otherwise, k is incremented, and the algorithm proceeds to Step 2.

6   Decentralized Algorithm

The algorithm in this section coordinates the solution to the m subproblems to find the minimizer to the original problem. The coordination is carried out by the master problem 7, an optimization problem whose objective function is defined using information gathered from the subproblem solutions.

If we set the global variables to a fixed value, the problem breaks into m independent subproblems. At each iteration of the optimization algorithm solving the master problem, all of the m subproblems are solved and information is exchanged between the master problem and the subproblems. The master problem is used to find the optimal value of the global variables.

Step 1. Initialization

Initialize the target variables $u = u_0 \in R^p$, the Hessian approximation matrix $H_0 = I$, the maximal number of iterations N and the optimality tolerance ε.

Step 2. Solve (P_i), $i = \overline{1, m}$

Solve problem (P_i) using the Penalty Algorithm and let (x_i^*, y_i^*) be the optimal solution, $i = \overline{1, m}$. The optimal value of the objective function in (P_i) depends on the current iteration u_k, therefore we use the notation $\widetilde{f}_i^*(u_k) = \widetilde{f}_i(x_i^*, y_i^*)$, $i = \overline{1, m}$.

Step 3. Attempt the proposed target value

With the notation $F(u_k) = \sum_{i=1}^{m} \widetilde{f}_i^*(u_k)$ we check if u_k is the optimal solution of the master problem 8.

- If $\|\nabla F(u_k)\| < \varepsilon$ then stop.
- If $k = N$, then stop.
- If $\|u_k - u_{k-1}\| < \varepsilon$ then stop.

Otherwise, go to Step 4.

Step 4. BFGS update

H_k and u_k are updated by the BFGS formula.

a) Obtain s_k by solving $H_k s_k = -\nabla F(u_k)$.
b) Perform a line search to find the optimal α_k in the direction found, then update $u_{k+1} = u_k + \alpha_k s_k$.
c) Denote $y_k = \nabla F(u_{k+1}) - \nabla F(u_k)$.
d) $H_{k+1} = H_k + \frac{y_k y_k^T}{y_k^T s_k} - \frac{H_k s_k s_k^T H_k}{s_k^T H_k s_k}$.

Then k is incremented and algorithm proceeds to step 2.

Remark 1. Note that we don't use the actual Hessian matrix of F even if available, but instead, use the current approximation of it. The idea behind quasi-Newton methods is to start with a positive definite, symmetric approximation to H, usually the unit matrix, and build up the approximating H_k's in such a way that the matrix H_k remains positive definite and symmetric. Far from the minimum, this guarantees that we always move in a descent direction. Close to the minimum, the updating formula approaches the true Hessian and we have the convergence of Newton's method.

7   Computational Results

In this section we consider the initial multiobjective problem 1 already transformed into the equivalent nonlinear problem 4. We use a test-problem from de Miguel [16]:

$$\min_{x,y} k_1 \|y - a\|_2^2 + \frac{1}{2}k_2 \|x_{11} - y\|_2^2 + \frac{1}{2}k_2 \|x_{21} + y\|_2^2 + +\frac{1}{2} \|x_{12}\|_2^2 + \frac{1}{2} \|x_{22}\|_2^2$$

$$s.t. \begin{cases} e \le x_{11} + y \le 2e \\ -x_{11} + y \le e \\ e \le x_{21} - y \le 2e \\ -x_{21} - y \le e \end{cases} \tag{19}$$

where $y = (y_1, ..., y_m) \in R^m$ is the vector of global variables and $x_1 = (x_{11}, x_{12})$, $x_2 = (x_{21}, x_{22}) \in R^m \times R^r$ are the vectors of local variables of the first and second system respectively, and $e = (1, ..., 1)^T \in R^m$. In [16] problem 19 is solved by considering a different subproblem for each component of the m-dimensional vectors y, x_{11}, x_{21} using the fact that $x_{12} = x_{22} = 0 \in R^r$. The theoretical solution (see [16]) (x_1^*, x_2^*, y^*) is given by

$$\begin{pmatrix} x_{11i}^* \\ x_{21i}^* \\ y_i^* \end{pmatrix} = \begin{pmatrix} 1 - y_i^* \\ 1 + y_i^* \\ \frac{k_1}{k_1 + 4k_2} a \end{pmatrix}$$

and $x_{12\,j}^* = x_{22\,j}^* = 0, \forall i = \overline{1, m}$ and $j = \overline{1, r}$ if $0 \le a \le \frac{1}{2} + \frac{2k_2}{k_1}$.

Instead of solving m subproblems, our method requires solving only two subproblems (P_i), $i = \overline{1, 2}$, one for each system. We solve problem 19 using our decentralized algorithm for $a = 1.5$, $k_1 = k_2 = 1$ and $m = 3$, $r = 2$.

$$\min_{u,x,y} \frac{1}{2} \|y_1 - 1.5\|_2^2 + \frac{1}{2} \|x_{11} - y\|_2^2$$

$$s.t. \begin{cases} e \le x_{11} + y \le 2e \\ -x_{11} + y \le e \\ \|y_1 - u\| \le 0. \end{cases} \tag{20}$$

$$\min_{u,x,y} \frac{1}{2} \|y_2 - 1.5\|_2^2 + \frac{1}{2} \|x_{21} + y\|_2^2$$

$$s.t. \begin{cases} e \le x_{21} - y \le 2e \\ -x_{21} - y \le e \\ \|y_2 - u\| \le 0. \end{cases} \tag{21}$$

The theoretical solution (x_1^*, x_2^*, y^*), where $x_1^* = (x_{11}^*, x_{12}^*) = (x_{111}^*, x_{112}^*, x_{113}^*, x_{121}^*, x_{122}^*)$, $x_2^* = (x_{21}^*, x_{22}^*) = (x_{211}^*, x_{212}^*, x_{213}^*, x_{221}^*, x_{222}^*) \in R^3 \times R^2$, $y^* = (y_1^*, y_2^*, y_3^*) \in R^3$, is given by

$$\begin{pmatrix} x_{11i}^* \\ x_{21i}^* \\ y_i^* \end{pmatrix} = \begin{pmatrix} 0.7 \\ 1.3 \\ 0.3 \end{pmatrix}$$

and $x_{12\,j}^* = x_{22\,j}^* = 0, \forall i = \overline{1, 3}$ and $j = \overline{1, 2}$. The algorithm was tested and the results are listed in the table bellow. The optimal solution is marked with $(^*)$. Figure 5.1 depicts the variation of the objective function of the master problem, $F(u)$. The thin line represents the theoretical curve and the thick one the curve obtained using our algorithm.

Target variable	Theoretical solution			Algorithm solution		
u_i	y_i	x_{1j}	x_{2j}	y_i	x_{1j}	x_{2j}
0.05	0.05	0.95	1.05	0.055768	0.944232	1.044221
0.15	0.15	0.85	1.15	0.155774	0.844225	1.144226
0.25	0.25	0.75	1.25	0.255779	0.744221	1.244244
0.30*	0.30*	0.70*	1.30*	0.305774*	0.695266*	1.294247*
0.35	0.35	0.65	1.35	0.355772	0.644227	1.344244
0.45	0.45	0.55	1.45	0.455773	0.544226	1.444274
0.55	0.55	0.45	1.55	0.525773	0.485803	1.514153
0.65	0.65	0.35	1.65	0.655773	0.344229	1.644248
0.75	0.75	0.25	1.75	0.755773	0.244226	1.744223
0.85	0.85	0.15	1.85	0.865773	0.134223	1.844266
0.95	0.95	0.05	1.95	0.955773	0.04222	1.944162

Fig. 5.1. Master problem objective function

8   Conclusions

This chapter presents an efficient method for solving multiobjective optimization problems. The problem is the model of m interconnected systems. Each system has an objective function and constraints which depend on local and global variables. This centralized problem is decomposed in m subproblems and a master problem gathering informations from the solutions of the subproblems is constructed and then solved. Each subproblem is solved by a penalty-based algorithm; the convergence of the algorithm is established.

This decentralized algorithm offers significant reduction in computational time due to the possibility of parallel processing. Computational results show that the combined algorithm is effective in practice.

References

1. Barrico CMCS, Antunes CH, An Evolutionary Approach for assessing the degree of robustness of solutions to multi-objective models, Evolutionary Computation in dynamic and uncertain environments, Series on Studies in Computational Intelligence, Springer, 2006.
2. Benders, J. F., Partitioning procedures for solving mixed-variables programming problems, Numerische Mathematik, 4, 238-252, 1962.
3. Braun, R.D., Collaborative Optimization: An Architecture for Large-Scale Distributed Design. PhD thesis, Stanford University, 1996.
4. Byrd RH, Gould NIM, Nocedal J, Waltz RA, An algorithmfor nonlinear optimization using linear programming and equality constrained subproblems. Mathematical programming, Series B, 100(1), 27-48, 2004.

 5. Byrd RH, Nocedal J, Waltz RA, Steering exact penalty methods for optimization, Technical report, Optimization Technology center, Northwestern University, 2006.
 6. Escudero, L. F., de la Fuente, J. L., Garcia, C., Prieto, F. J.: Hydropower generation management under uncertainty via scenario analysis and parallel computation, IEEE Transactions on Power Systems, Vol. 11, No. 2, (1996).
 7. Fonseca CM, Fleming PJ, AN overview of evolutionary algorithms in multiobjective optimization, Evolutionary Computation, 3(1), 1-16, 1995.
 8. Fulga, C., Penalty Approach for Optimization in the Energy Sector, Proceedings of the 19th Mini-EURO Conference ORMMES 2006, Coimbra-Portugal.
 9. Fulga, C., A Sequential Quadratic Programming Technique with Two-Parameter Penalty Function, Bulletin Mathematique de la Societe des Sciences Mathematiques de Roumanie, Tome 49(97), No. 4, 2006.
10. Geoffrion, A. M., Generalized Benders Decomposition, Journal of Optimization Theory and Applications, 10(4), 237-260, 1972.
11. Hanne T, On the convergence of multiobjective evolutionary algorithms, European Journal of Operational Research, 117, 553-564, 1999.
12. Heiskanen P., Decentralized Method for Computing Pareto Solutions in Multiparty Negotiations, European Journal of Operational Research, 117 ,578-590, 1999.
13. Inalhan G, Stipanovic DM, Tomlin C, Decentralized optimization with application to multiple aircraft coordination, Proceedings of the 41st IEEE Conference on decision and control, Las Vegas, Nevada USA, 1147-1155, 2002.
14. Kuwata Y, How J, Decentralized cooperative trajectory optimization for UAVs with coupling constraints, Proceedings of the 45th IEEE Conference on decision and control, San Diego, CA, USA, 6820-6825, 2006.
15. D. Q. Mayne, E. Polak, *A Superlinearly Convergent Algorithm for Constrained Optimization Problems*, Mathematical Programming Study, **16**(1982), 45-61.
16. de Miguel, V., Two Decomposition Algorithms for Nonconvex Optimization Problems with Global Variables, PhD thesis, Stanford University, 2001.
17. Pantoja JFA, Mayne DQ, Exact penalty function algorithm with simple updating of the penalty parameter, Journal of optimization theory and applications, 69, 441-467, 1991.
18. Price CJ, Convergence properties of an algorithm for non-linear programmes which uses a two parameter exact penalty function, Research report no. 97, Department of mathematics and statistics, University of Canterbury, Christchurch, New Zealand, 1993.
19. Rosen JB, The gradient projection method for nonlinear programming. Part I. Linear constraints, Journal of the society for industrial and applied mathematics, 8, 181-217, 1960.
20. Rosen JB, The gradient projection method for nonlinear programming. Part II. Nonlinear constraints, Journal of the society for industrial and applied mathematics, 9, 514-532, 1961.
21. Sahba M, A globally convergent algorithm for nonlinearly constrained optimization problems, Journal of optimization theory and applications, 52, 291-307, 1987.
22. Sebenius JK, Negotiation Analysis: A characterization and review, Management Science 38(1), 18-38, 1992.
23. Tammer, K., The Application of Parametric Optimization and Imbedding to the Foundation and Realization of a Generalized Primal Decomposition Approach, Parametric Optimization and Related Topics, 376-386, Mathematical Research, 35, Akademie-Verlag, Berlin, 1987.

24. Teich JE, Wallenius H, Kuula M, Zionts S, A Decision support approach for negotiation with an application to agricultural income policy negotiations, European Journal of Operational Research, 81, 536-549, 1996.
25. Teich JE, Wallenius H, Kuula M, Zionts S, Identifying Pareto-optimal settlements for two-party resource allocation negotiations, European Journal of Operational Research, 93, 76-87, 1995.
26. Yu, Po-Lung: Multiple-Criteria Decision Making, Plenum Press, New York (1985).

Simultaneous Localization and Planning for Cooperative Air Munitions

Andrew J. Sinclair[1], Richard J. Prazenica[2], and David E. Jeffcoat[3]

[1] Auburn University, Auburn AL 36849, USA
[2] University of Florida, Shalimar FL 32579, USA
[3] Air Force Research Laboratory, Eglin Air Force Base FL, USA

Abstract. This chapter considers the cooperative control of aerial munitions during the attack phase of a mission against ground targets. It is assumed that sensor information from multiple munitions is available to refine an estimate of the target location. Based on models of the munition dynamics and sensor performance, munition trajectories are designed that enhance the ability to cooperatively estimate the target location. The problem is posed as an optimal control problem using a cost function based on the variances in the target-location estimate. These variances are computed by fusing the individual munition measurements in a weighted least squares estimate. Numerical solutions are found for several examples both with and without considering limitations on the munitions' field of view. These examples show large reductions in target-location uncertainty when these trajectories are used compared to other naively designed trajectories. This reduction in uncertainty could enable the attack of targets with greater precision using smaller, cheaper munitions.

1 Introduction

Research is in progress on the cooperative control of air armament designed to detect, identify, and attack ground targets. One class of this type of armament are wide-area search munitions, which can be deployed in an area of unknown targets. Current development is focused on possibilities of enhancing munition capabilities through cooperative control. This chapter presents a new concept for developing trajectories that enhance munitions' capability to cooperatively estimate target locations.

The tasks of intercepting a chosen target and estimating the target's location can represent competing requirements in the path planning of a munition. In a general sense, the problem posed here is to plan a path to a target while simultaneously estimating that target's location. This can be considered a simultaneous localization and planning (SLAP) problem. Whereas SLAP problems can be studied for a single agent, many interesting behaviors emerge when cooperative agents are considered.

Important work exists in the literature on the two related problems of cooperative search [1,2,3] and the design of optimal trajectories for single observers

M.J. Hirsch et al. (Eds.): Adv. in Cooper. Ctrl. & Optimization, LNCIS 369, pp. 81–93, 2007.
springerlink.com © Springer-Verlag Berlin Heidelberg 2007

[4,5,6,7,8,9,10]. Much of the work in optimal trajectories has focused on bearings-only measurements of a target, often focused on sonar applications. Fawcett investigated the impact of maneuvers on the Cramer-Rao lower bound for the target-state estimate [4]. Frew and Rock investigated a method to minimize a measure of the estimate error covariance [9]. Other works have studied optimal trajectories for cooperative observers [11,12]. These works have focused on reconnaissance of a target, relating the performance index to the quality of the target-location estimate at the end of the mission or a time interval.

Several related topics also capture aspects of both cooperative search and trajectory design. Dohner et al. used a Lyapunov approach to drive a vehicle swarm to an uncertain target location while simultaneously maintaining swarm spacing to ensure observability of the target [13]. Passino et al. developed a distributed cooperative search algorithm where decisions were made planning into the future to minimize a cost function representing several subgoals, such as covering areas in large uncertainty and minimizing overlap with other agents [14].

It is noteworthy that the problem considered in this chapter, trajectory design to enhance target-location estimation, is in some ways the dual of another problem that has received considerable attention, trajectory design to minimize detection by an enemy radar [15,16,17,18]. Pachter et al. have considered another related problem that used cooperative vehicles to project phantom tracks to an enemy radar [19,20].

This chapter extends the field of optimal observor trajectories to the cooperative-attack application. The methods presented in this chapter will be illustrated for a planar problem with two munitions and one target; the methods apply though to three-dimensional cases with general numbers of munitions and targets. In the following section, models for the munition motion and sensor performance are presented. Next, the SLAP trajectory design is posed as an optimal control problem. Several example numerical solutions are then presented. Finally, the performance of a target-location estimation algorithm is evaluated along the SLAP trajectories and compared to alternative trajectories.

2 Model Development

A scenario can be considered with the two-dimensional plane populated by n munitions and m fixed targets. The following developments will illustrate the method for two munitions and one target. The state of each munition is given by its position in two dimensional space, $\boldsymbol{x}_1 = [x_1 \ y_1]^\mathsf{T}$ and $\boldsymbol{x}_2 = [x_2 \ y_2]^\mathsf{T}$. A constant-speed kinematic model is used to describe the motion of the munitions. The heading angles of the munitions are ψ_1 and ψ_2, and the speed of each munition is v.

$$
\begin{aligned}
\dot{x}_1 &= v \cos \psi_1 \quad ; \quad & \dot{x}_2 &= v \cos \psi_2 \\
\dot{y}_1 &= v \sin \psi_1 \quad ; \quad & \dot{y}_2 &= v \sin \psi_2
\end{aligned}
\tag{1}
$$

$$
\dot{\boldsymbol{x}}_i = \boldsymbol{f}_i (\psi_i) , \quad i \in \{1, 2\}
\tag{2}
$$

Here, the heading angles are treated as control variables.

Additionally, each munition is considered to carry a sensor that is capable of measuring the target location in the xy plane. Again, the end goal will be to design trajectories that improve the estimation of the target location. Therefore, a model is needed of the sensor measurements and their uncertainties. The target has a position described by $\boldsymbol{x}_T = [x_T \; y_T]^\mathsf{T}$. The measurement of this target location by each munition, $\tilde{\boldsymbol{z}}_1 = [\tilde{x}_{T,1} \; \tilde{y}_{T,1}]^\mathsf{T}$ and $\tilde{\boldsymbol{z}}_2 = [\tilde{x}_{T,2} \; \tilde{y}_{T,2}]^\mathsf{T}$, is modeled as shown below.

$$\tilde{x}_{T,1} = x_T + w_{x,1}(0, \sigma_{x,1}) \quad ; \quad \tilde{x}_{T,2} = x_T + w_{x,2}(0, \sigma_{x,2})$$
$$\tilde{y}_{T,1} = y_T + w_{y,1}(0, \sigma_{y,1}) \quad ; \quad \tilde{y}_{T,2} = y_T + w_{y,2}(0, \sigma_{y,2}) \tag{3}$$

The measurement errors from each munition are assumed to be independent of the errors from the other munition. The x and y measurement errors from each individual munition, however, are treated as correlated Gaussian random variables with zero mean and standard deviations of $\sigma_{x,i}$ and $\sigma_{y,i}$, where $i \in \{1, 2\}$. It is these uncertainties that will drive the trajectory design, and they can be selected to model a particular sensor design.

The error in the target-location measurements from an individual munition is treated as following a zero-mean jointly-Gaussian distribution that is uncorrelated in the down-range and cross-range directions, relative to the true target and munition locations. The errors in these directions, $w_{d,i}(0, \sigma_{d,i})$ and $w_{c,i}(0, \sigma_{c,i})$, can therefore be treated as independent Gaussian random variables. The standard deviations in the down-range and cross-range directions are modeled as functions of the range from the munition to the target.

$$\sigma_{d,i} = 0.1 r_i \quad ; \quad \sigma_{c,i} = 0.01 r_i \tag{4}$$

This models a sensor that is more accurate when close to the target and more accurate in the transverse direction than in the radial direction. The uncertainty in the measurement of the target location by the ith munition is illustrated in Fig. 1.

From the down-range and cross-range variables, the errors and the covariance matrix in the x and y coordinates can be found.

$$\begin{bmatrix} w_{x,i} \\ w_{y,i} \end{bmatrix} = \begin{bmatrix} \cos\theta_i & \sin\theta_i \\ -\sin\theta_i & \cos\theta_i \end{bmatrix} \begin{bmatrix} w_{d,i} \\ w_{c,i} \end{bmatrix} \tag{5}$$

$$\boldsymbol{P}_i = \begin{bmatrix} \sigma_{x,i}^2 & \sigma_{xy,i} \\ \sigma_{xy,i} & \sigma_{y,i}^2 \end{bmatrix} = \begin{bmatrix} \cos\theta_i & \sin\theta_i \\ -\sin\theta_i & \cos\theta_i \end{bmatrix} \begin{bmatrix} \sigma_{d,i}^2 & 0 \\ 0 & \sigma_{c,i}^2 \end{bmatrix} \begin{bmatrix} \cos\theta_i & -\sin\theta_i \\ \sin\theta_i & \cos\theta_i \end{bmatrix} \tag{6}$$

Here, θ_i is the bearing angle of the target relative to the ith munition. The range and bearing angle for each target-munition pair are computed as shown below.

$$r_i = \sqrt{(x_T - x_i)^2 + (y_T - y_i)^2} \tag{7}$$

$$\theta_i = \tan^{-1}\left(\frac{y_T - y_i}{x_T - x_i}\right) \tag{8}$$

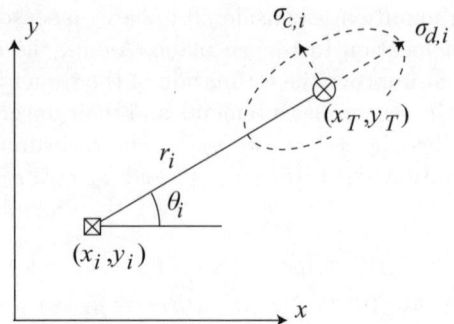

Fig. 1. Measurement of the target by the ith munition and the associated error probability ellipse

The significance of Eq. (6) is that it models the quality of the measurements from the ith munition based on its position relative to the target.

The measurements provided by both munitions can be fused into a single instantaneous estimate of the target location. This is done using a weighted least-squares estimator (WLSE) [21,22]. The measurements of the target location from each munition are grouped into a measurement vector $\tilde{z} = [\tilde{x}_{T,1}\ \tilde{y}_{T,1}\ \tilde{x}_{T,2}\ \tilde{y}_{T,2}]^{\mathsf{T}}$. This produces a linear measurement model in terms of the target location.

$$z = Hx_T + w \tag{9}$$

$$H = \begin{bmatrix} 1\,0\,1\,0 \\ 0\,1\,0\,1 \end{bmatrix}^{\mathsf{T}}; \qquad w = \begin{bmatrix} w_{x,1}\ w_{y,1}\ w_{x,2}\ w_{y,2} \end{bmatrix}^{\mathsf{T}} \tag{10}$$

Here, w is the vector of measurement errors. The covariance of this error vector is given by arranging the covariances from each munition.

$$R = \begin{bmatrix} P_1 & 0 \\ 0 & P_2 \end{bmatrix} \tag{11}$$

The instantaneous WLSE of the ith target location and the associated covariance are given by the following.

$$\hat{x}_T = \left(H^{\mathsf{T}}R^{-1}H\right)^{-1} H^{\mathsf{T}}R^{-1}\tilde{z} \tag{12}$$

$$P = \left(H^{\mathsf{T}}R^{-1}H\right)^{-1} \tag{13}$$

Considering the first of Eqs. (10), the WLSE reduces to the following.

$$\hat{x}_T = \begin{bmatrix} \hat{x}_T \\ \hat{y}_T \end{bmatrix} = \left(P_1^{-1} + P_2^{-1}\right)^{-1} \left(P_1^{-1}\tilde{z}_1 + P_2^{-1}\tilde{z}_2\right) \tag{14}$$

More importantly for the current purposes, the covariance of this combined estimate is related to the individual covariances of the measurements from each munition.

$$P = \begin{bmatrix} \sigma_x^2 & \sigma_{xy} \\ \sigma_{xy} & \sigma_y^2 \end{bmatrix} = \left(P_1^{-1} + P_2^{-1}\right)^{-1} \tag{15}$$

The covariance $\boldsymbol{P}$ now models the quality of the combined target-location estimate based on the positioning of the two munitions relative to the target. For cases with more than two munitions, similar expressions can be developed combining the measurements of each of the munitions. Additionally, for cases with multiple targets corresponding expressions can be used for the covariance of each target-location estimate.

3   Problem Formulation

The task of designing trajectories for the munitions in order to enhance the estimation performance can now be posed as the following optimal control problem. Consider the state vector $\boldsymbol{x} = [x_1 \; y_1 \; x_2 \; y_2]^{\mathsf{T}}$. The heading angles of the munitions can be organized into a control vector $\boldsymbol{u} = [\psi_1 \; \psi_2]^{\mathsf{T}}$. The state vector evolves according to the state equation found by grouping Eq. (2), $\dot{\boldsymbol{x}} = \boldsymbol{f}(\boldsymbol{u}) = [\boldsymbol{f}_1^{\mathsf{T}} \; \boldsymbol{f}_2^{\mathsf{T}}]^{\mathsf{T}}$. For boundary conditions, the initial positions of the munitions will be considered a given, and the final position of munition 1 is required to be the target location, $x_1(t_F) = x_T$ and $y_1(t_F) = y_T$. The final position of munition 2 is free.

The goal will be to find the trajectories that minimize the following cost function, which is based on the WLSE covariance.

$$J = \int_0^{t_F} \left(\sigma_x^2 + \sigma_y^2 \right) \mathrm{d}t \tag{16}$$

The variances of each target location are functions of the states describing the munition configuration. Clearly, this cost function emphasizes the uncertainty over the entire trajectory. Previous works have used performance indices related to the uncertainty at the end of the trajectory or a specified interval [11,12]. Compared to those alternative indices, the cost function used here encourages reduction in uncertainty earlier in the trajectory. It is also noted that other cost functions could be based on the determinant or other metrics of the covariance matrix.

Introducing the costates $\boldsymbol{\lambda}(t) = [\lambda_1 \; \lambda_2 \; \lambda_3 \; \lambda_4]^{\mathsf{T}}$, a time-varying vector of Lagrange multipliers, the Hamiltonian can be defined.

$$H = \sigma_x^2 + \sigma_y^2 + \boldsymbol{\lambda}^{\mathsf{T}} \boldsymbol{f}(\boldsymbol{u}) \tag{17}$$

From this, the first-order necessary conditions are derived [23].

$$\frac{\partial H}{\partial \boldsymbol{u}} = \left(\frac{\partial \boldsymbol{f}}{\partial \boldsymbol{u}} \right)^{\mathsf{T}} \boldsymbol{\lambda} = 0 \tag{18}$$

$$\dot{\boldsymbol{\lambda}} = -\frac{\partial H}{\partial \boldsymbol{x}} = -\frac{\partial}{\partial \boldsymbol{x}} \left(\sigma_x^2 + \sigma_y^2 \right) \tag{19}$$

From Eq. (18) the control law for the heading angles as a function of the costates can be found.

$$\frac{\partial H}{\partial \psi_1} = -\lambda_1 v \sin \psi_1 + \lambda_2 v \cos \psi_1 = 0 \tag{20}$$

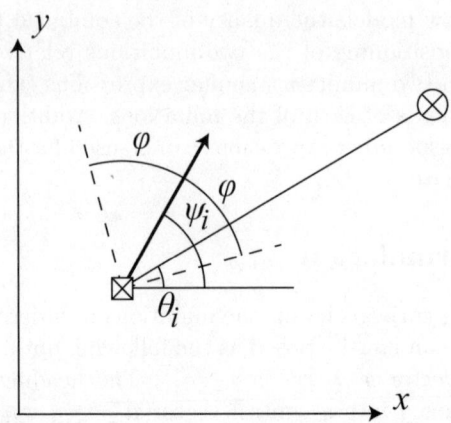

Fig. 2. Heading angle, field-of-view half angle, bearing angle of ith munition relative to the target

$$\frac{\partial H}{\partial \psi_2} = -\lambda_3 v \sin \psi_2 + \lambda_4 v \cos \psi_2 = 0 \qquad (21)$$

To find a minimum in the cost, the following curvature condition is additionally imposed.

$$\frac{\partial^2 H}{\partial \psi_1^2} = -\lambda_1 v \cos \psi_1 - \lambda_2 v \sin \psi_1 > 0 \qquad (22)$$

$$\frac{\partial^2 H}{\partial \psi_2^2} = -\lambda_3 v \cos \psi_2 - \lambda_4 v \sin \psi_2 > 0 \qquad (23)$$

This gives the optimal control as the following.

$$\psi_1 = \tan^{-1}\left(\frac{-\lambda_2}{-\lambda_1}\right) \quad ; \quad \psi_2 = \tan^{-1}\left(\frac{-\lambda_3}{-\lambda_4}\right) \qquad (24)$$

The costate equations, governing the evolution of $\boldsymbol{\lambda}$ are given by Eq. (19). These can be found by applying the chain rule to Eqs. (4), (6-8), and (15); however, they are rather extended and are not reproduced here. The terminal conditions for the problem are the specified conditions, $x_1(t_F) = x_T$ and $y_1(t_F) = y_T$, and the necessary conditions, $\lambda_3(t_F) = \lambda_4(t_F) = H(t_F) = 0$.

The above conditions have not accounted for limitations in the field of view of the vehicle sensors. This assumes either a sensor that has unlimited field of view or is gimbal mounted in order to view a target regardless of the vehicle orientation and heading. A sensor that is fixed mounted on the vehicle, though, may only offer a limited field of view relative to the vehicle heading. In this case a hard constraint can be enforced on the trajectory of the ith munition to keep the target in view. The field of view angle is labeled 2ϕ and is illustrated in Fig. 2.

Two inequality constraint functions can be enforced to keep the target in the field of view of the ith munition. For example, the following constraints keep the target in the field of view of munition 1.

$$c_1 = \psi_1 - \theta_1 - \phi \leq 0 \quad ; \quad c_2 = -\psi_1 + \theta_1 - \phi \leq 0 \tag{25}$$

Note that the bearing angles are functions of the states, and the heading angles are the controls. Arranging any desired constraints into a vector function $\boldsymbol{c}(\boldsymbol{x}, \boldsymbol{u}) \leq \boldsymbol{0}$ and introducing a second set of Lagrange multipliers $\boldsymbol{\mu}$, a revised Hamiltonian is developed [23].

$$H = \sigma_x^2 + \sigma_y^2 + \boldsymbol{\lambda}^\mathsf{T} \boldsymbol{f}(\boldsymbol{u}) + \boldsymbol{\mu}^\mathsf{T} \boldsymbol{c}(\boldsymbol{x}, \boldsymbol{u}) \tag{26}$$

During periods when one or more of these constraints are active, the target is kept on the edge of the field of view of the munition. The value of $\boldsymbol{\mu}$ is calculated from the revised stationary condition.

$$\frac{\partial H}{\partial \psi_1} = -\lambda_1 v \sin \psi_1 + \lambda_2 v \cos \psi_1 + \boldsymbol{\mu}^\mathsf{T} \frac{\partial \boldsymbol{c}}{\partial \psi_1} = 0$$

$$\frac{\partial H}{\partial \psi_2} = -\lambda_3 v \sin \psi_2 + \lambda_4 v \cos \psi_2 + \boldsymbol{\mu}^\mathsf{T} \frac{\partial \boldsymbol{c}}{\partial \psi_2} = 0 \tag{27}$$

The costate equations are revised as shown below.

$$\dot{\boldsymbol{\lambda}} = -\frac{\partial H}{\partial \boldsymbol{x}} = -\frac{\partial}{\partial \boldsymbol{x}} \left(\sigma_x^2 + \sigma_y^2 \right) + \left(\frac{\partial \boldsymbol{c}}{\partial \boldsymbol{x}} \right)^\mathsf{T} \boldsymbol{\mu} \tag{28}$$

The two-point boundary-value problem can now be posed to solve for $\boldsymbol{\lambda}(t_0)$ and t_F subject to the derived necessary conditions and the boundary conditions. When the field-of-view constraints are inactive or simply neglected, the necessary conditions are Eqs. (2), (19), and (24). When the field-of-view constraint is active, the necessary conditions are Eqs. (2), (27), and (28).

For cases with more than two munitions or more than one target, the terminal conditions could be specified by prechosen target-munition attack pairings. The final states for any munitions not assigned a target would be free. For multiple targets, the cost function could be augmented by summing the additional variances from their target-location estimates. For complex scenarios with many targets and munitions, difficulty may arise in the application of the field-of-view constraints. It may be desirable to let targets pass in and out of the field of view of some munitions.

For any scenario, the solution of the problem produces munition trajectories designed to reduce the uncertainty in the target-location estimates. These are referred to as the SLAP trajectories. Note that in a real-time application, the use of the true target positions as boundary conditions would not be possible. These must be estimated, which is the motivation behind finding the SLAP trajectories in the first place. Here, though, the true locations are used to illustrate the concept and potential benefit of these trajectories.

4   Sample SLAP Trajectories

The following are several example SLAP trajectories that have been found using a sequential quadratic programming algorithm to numerically search for the optimization parameters. Each of the examples considers the scenario of two munitions and one target. Munition 1 is assigned to attack the target, and munition 2 is free to assist in the estimation of the target location. Two different initial conditions are considered, and solutions are presented with and without the field-of-view constraint. Here, a munition speed of $v = 300$ ft/sec and a half field-of-view angle of $\phi = 45$ deg were used.

The first set of initial conditions are $x_1(0) = 0$ ft, $y_1(0) = -2000$ ft, $x_2(0) = 100$ ft, and $y_2(0) = -2000$ ft. The target is located at $x_T = y_T = 0$ ft. This problem was first solved neglecting any field-of-view constraints. The solution parameters found for this case are shown in Table 1 under problem 1.

Table 1. Solution parameters and cost for sample SLAP trajectories

solution parameters	problem 1	problem 2	problem 3	problem 4
$\lambda_1(0)$	29.495	32.226	30.550	32.453
$\lambda_2(0)$	-12.028	-12.745	-14.236	-15.170
$\lambda_3(0)$	-28.822	-49.167	30.551	49.532
$\lambda_4(0)$	-13.460	-40.658	14.258	40.001
t_F (sec)	8.0950	7.6325	8.2146	7.6769
J	1.59×10^4	1.70×10^4	1.89×10^4	2.02×10^4

The trajectories generated by these values are shown in Fig. 3(a). The marks along the trajectories in the figure indicate one-second intervals of flight time. In this case the SLAP trajectories are roughly symmetric about the y axis. Munition 1 intercepts the target at t_F as required by the boundary conditions, but munition 2 also approaches the target very closely. Intuitively this is because the measurement errors from either munition are reduced as the munition closes the range with the target. Instead of traveling directly to the target, however, near the initial time both munitions sweep out in the $\pm x$ directions. This gives the munitions differing perspectives on the target allowing them to compensate for the relatively large downrange errors in each other's measurements.

In the trajectories for problem 1, both munitions sweep out aggressively such that the target would be out of their fields of view during the initial stages of the trajectories. To correct for this, problem 2 is posed to enforce that both munitions keep the target within view. Problem 2 is identical to problem 1 in all other aspects. The solutions for this problem are shown in Table 1 and the corresponding trajectories are shown in Fig. 3(b).

In this case, the field-of-view constraint is active over the entire trajectory of munition 2. It is prevented from swinging wide during the initial periods, and instead munition 2 keeps the target on the edge of its field of view for the entire flight. The field-of-view constraint is also initially active for munition 1.

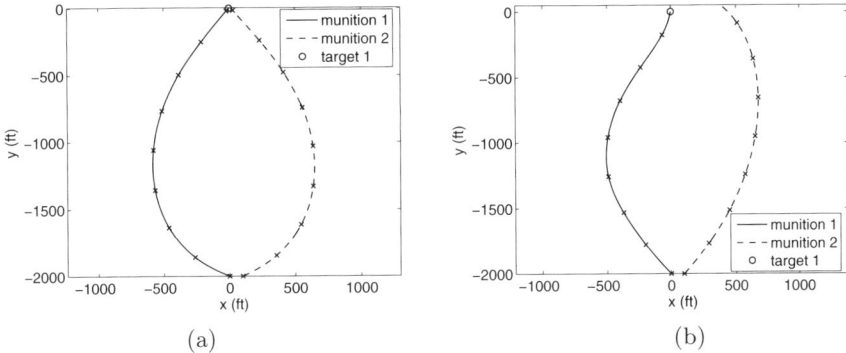

Fig. 3. SLAP trajectories for problems 1 and 2

After a short period, though, munition 1 turns to head more directly toward the target. Again, the intuitive behaviors of closing range to the target and achieving differing points of view are present in the SLAP trajectories for problem 2. The motions are restricted, however, by the field-of-view constraint.

Next, a different initial condition can be considered with munition 2 moved to an initial position $x_2(0) = 0$ ft, and $y_2(0) = 2000$ ft. Instead of starting nearby munition 1, munition 2 now starts on the opposite side of the target relative to munition 1. The solution for this case when neglecting the field-of-view constraint is shown as problem 3 in Table 1. The SLAP trajectories for this problem are shown in Fig. 4(a). The trajectories for the two munitions are nearly symmetric about the x axis. Similar to problem 1, the munitions sweep to the side to obtain differing viewpoints before closing in on the target.

The solution for the above initial conditions when applying the field-of-view constraint is listed as problem 4 in Table 1. The SLAP trajectories for this problem are shown in Fig. 4(b). The constraint is active for the early part of the trajectory of munition 1 and for the entire flight of munition 2.

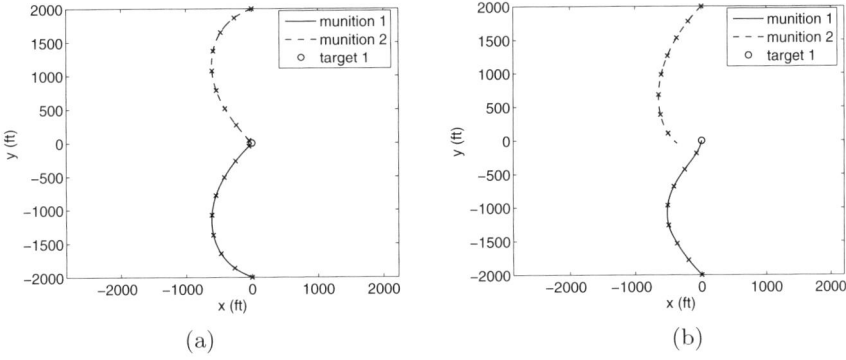

Fig. 4. SLAP trajectories for problems 3 and 4

5   Estimation Performance

The impact of the SLAP trajectories on the target-location estimation can now be evaluated. Although the trajectories were designed using a cost function based on the variances from a continuous WLSE algorithm, the estimation performance will be evaluated using a recursive weighted least squares estimation (RWLSE) algorithm with discrete measurement updates. First, the algorithm will be operated for a single munition following a trajectory from the initial condition straight to the target location (STT trajectory). Second, the estimation is performed for two munitions both following STT trajectories. Finally, the algorithm is implemented using two munitions following the field-of-view constrained SLAP trajectories. In each case, noisy measurements were simulated using the measurement model in Eq. (4).

The munition sensors were assumed to collect measurements of the target location at a rate of 10 Hz. The RWLSE algorithm operated as follows to determine the estimate and the uncertainty at the kth time step [21,22]. The current estimate is computed as follows.

$$K_k = P_{k-1}H^\top \left(HP_{k-1}H^\top + R\right)^{-1} \tag{29}$$

$$\hat{x}_k^{(T)} = \hat{x}_{k-1}^{(T)} + K_k \left(\tilde{z}_k - H\hat{x}_{k-1}^{(T)}\right) \tag{30}$$

The current covariance matrix is computed as shown.

$$P_k = \begin{bmatrix} \sigma_{x,k}^2 & \sigma_{xy,k} \\ \sigma_{xy,k} & \sigma_{y,k}^2 \end{bmatrix} = \left(P_{k-1}^{-1} + H_k^\top R_k^{-1} H_k\right)^{-1} \tag{31}$$

To compare the estimation performance along the different trajectories, the size of the one-sigma uncertainty ellipsoid in the target-location estimate can be used as a metric. At the kth time step, this is given by the product of π with the square root of the product of the eigenvalues of P_k. In particular, the ellipsoid size at $t_F - 2$ sec will be highlighted. Although t_F is different for each trajectory, at this point in time munition 1 is roughly 600 ft from the target.

Using the initial condition of $x_1(0) = 0$ ft, $y_1(0) = -2000$ ft the STT trajectory has a flight time given by $t_F = 6.67$ sec. Using a single munition on an STT trajectory, at $t_F - 2$ sec the one-sigma uncertainty ellipse has an area of 81.5 ft^2. For $x_2(0) = 100$ ft, and $y_2(0) = -2000$ ft, adding measurements from munition 2 on an STT trajectory reduces the uncertainty to 39.7 ft^2. When the two munitions follow the SLAP trajectory shown in Fig. 3b, however, the area is reduced to 9.1 ft^2.

The error histories for a sample simulation with noisy measurements and three-sigma error bounds ($\pm 3\sigma_{x,k}$ and $\pm 3\sigma_{y,k}$) generated by the RWLSE algorithm are shown in Fig. 5. Figure 5(a) shows the errors in the x and y estimates of the target location using the STT trajectories. Figure 5(b) show the errors using the SLAP trajectories. Clearly, both trajectories give similar good performance in estimating the x component of the target location, but the SLAP trajectories provide much better estimation of the y component.

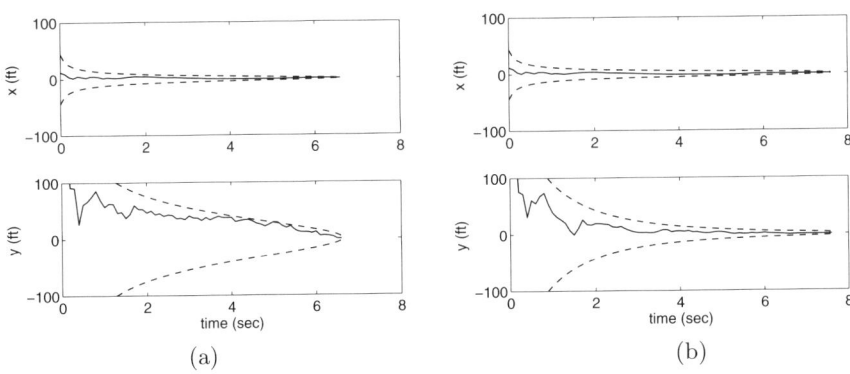

Fig. 5. Estimation errors using (a) STT and (b) SLAP trajectories with $x_2(0) = 100\,\text{ft}$, and $y_2(0) = -2000\,\text{ft}$

Moving munition 2 to the initial condition $x_2(0) = 0\,\text{ft}$, and $y_2(0) = 2000\,\text{ft}$ obviously does not change the results when only measurements from munition 1 are considered. For the cases with two munitions, however, the uncertainty areas at $t_F - 2\,\text{sec}$ are $40.8\,\text{ft}^2$ for the STT trajectories and $9.3\,\text{ft}^2$ for the SLAP trajectories. For these initial conditions, the error histories for a sample simulation with noisy measurements and three-sigma error bounds generated by the RWLSE algorithm are shown in Fig. 6.

These results give an indication of the impact of trajectory design on estimation performance. Significantly, for either initial condition, adding a second munition to help in the target-location estimation without paying attention to trajectory design improves performance to approximately half of the uncertainty achieved with a single munition. Careful use of the SLAP trajectories, however, further reduces the uncertainty to less than one quarter of what is achieved using

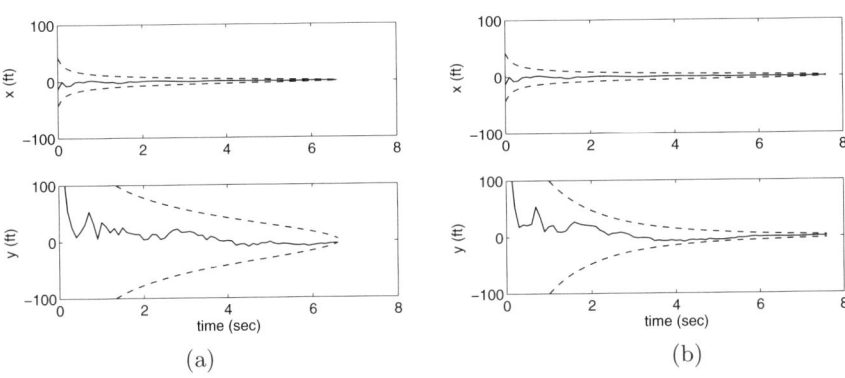

Fig. 6. Estimation errors using (a) STT and (b) SLAP trajectories with $x_2(0) = 0\,\text{ft}$, and $y_2(0) = 2000\,\text{ft}$

the STT trajectories. The SLAP trajectories benefit both from being delayed, which allows collection of more measurements, and their paths, which improve the quality of the measurements.

6 Conclusions

The results in the previous section demonstrate the impact that careful trajectory design can have on target-location estimation. Adding a second munition when following STT trajectories does significantly improve estimation performance. The use of the SLAP trajectories, however, reveals much greater further improvement. Furthermore, the complexity and cost of the second munition and communication between the two has already been accepted in taking the first step. The second step of following the SLAP trajectories only requires careful trajectory design.

Improvements in estimation performance like those demonstrated here could have significant impact on munition design and cost. More accurate target-location estimation could allow more accurate strike capability or the ability to attack targets that are difficult to detect. It is anticipated that the reduction in uncertainty early in the trajectory could be critical for the precision strike of these difficult targets; however, further work is needed to demonstrate the impact of these estimation enhancements on guidance and control performance. Combined, these effects could enable the use of smaller, cheaper munitions against targets in cluttered environments while limiting collateral damage.

The calculus-of-variations approach, used here, to solve for SLAP trajectories allowed for model-based trajectory design. This approach may not be the best approach, however, for real-time implementation. Future work for this application may require different solution approaches. The intuition gained from calculus-of-variations based sample solutions may allow the development of heuristic solutions that are better suited for real-time implementation.

References

1. Chandler, P.R., Pachter, M., Nygard, K.E., Swaroop, D.: Cooperative control for target classification. In Murphey, R., Pardalos, P.M., eds.: Cooperative Control and Optimization, Kluwer, Netherlands (2002) 1–19
2. Jeffcoat, D.E.: Coupled detection rates: An introduction. In Grundel, D., Murphey, R., Pardalos, P.M., eds.: Theory and Algorithms for Cooperative Systems, World Scientific, New Jersey (2004) 157–167
3. Frew, E., Lawrence, D.: Cooperative stand-off tracking of moving targets by a team of autonomous aircraft. In: AIAA Guidance, Navigation, and Control Conference, San Fancisco, California (2005) AIAA-2005-6363.
4. Fawcett, J.A.: Effect of course maneuvers on bearings-only range estimation. IEEE Transactions on Acoustics, Speech, and Signal Processing **36** (1988) 1193–1199
5. Hammel, S.E., Liu, P.T., Hilliard, E.J., Gong, K.F.: Optimal observor motion for localization with bearing measurements. Computers and Mathematics with Applications **18** (1989) 171–180

6. Logothetis, A., Isaksson, A., Evans, R.J.: Comparison of suboptimal strategies for optimal own-ship maneuvers in bearings-only tracking. In: American Control Conference, Phiadelphia, Pennsylvania (1998)
7. Passerieux, J.M., VanCappel, D.: Optimal observer maneuver for bearings-only tracking. IEEE Transactions on Aerospace and Electronic Systems **34** (1998) 777–788
8. Oshman, Y., Davidson, P.: Optimization of observer trajectories for bearings-only target localization. IEEE Transactions on Aerospace and Electronic Systems **35** (1999) 892–902
9. Frew, E.W., Rock, S.M.: Trajectory generation for constant velocity target motion estimation using monocular vision. In: IEEE International Conference on Robotics & Automation, Taipei, Taiwan (2003)
10. Watanabe, Y., Johnson, E.N., Calise, A.J.: Vision-based guidance design from sensor trajectory optimization. In: AIAA Guidance, Navigation, and Control Conference, Keystone, Colorado (2006) AIAA-2006-6607.
11. Grocholsky, B.: Information-Theoretic Control of Multiple Sensor Platforms. PhD thesis, University of Sydney, Sydney, Australia (2002)
12. Ousingsawat, J., Campbell, M.E.: Optimal cooperative reconnaissance using multiple vehicles. Journal of Guidance, Control, and Dynamics **30** (2007) 122–132
13. Dohner, J.L., Eisler, G.R., Driessen, B.J., Hurtado, J.: Cooperative control of vehicle swarms for acoustic target localization by energy flows. Journal of Dynamic Systems, Measurement, and Control **126** (2004) 891–895
14. Passino, K., Polycarpou, M., Jacques, D., Pachter, M., Liu, Y., Yang, Y., Flint, M., Baum, M.: Cooperative control for autonomous air vehicles. In Murphey, R., Pardalos, P.M., eds.: Cooperative Control and Optimization, Kluwer, Netherlands (2002) 233–271
15. Pachter, M., Hebert, J.: Cooperative aircraft control for minimum radar exposure. In Murphey, R., Pardalos, P.M., eds.: Cooperative Control and Optimization, Kluwer, Netherlands (2002) 199–211
16. Zabarankin, M., Uryasev, S., Pardalos, P.: Optimal risk path algorithms. In Murphey, R., Pardalos, P.M., eds.: Cooperative Control and Optimization, Kluwer, Netherlands (2002) 273–303
17. Murphey, R., Uryasev, S., Zabarankin, M.: Optimal path planning in a threat environment. In Butenko, S., Murphey, R., Pardalos, P., eds.: Recent Developments in Cooperative Control and Optimization, Kluwer, Netherlands (2004) 349–406
18. Kabamba, P.T., Meerkov, S.M., III, F.H.Z.: Optimal path planning for unmanned combat aerial vehicles to defeat radar tracking. Journal of Guidance, Control, and Dynamics **29** (2006) 279–288
19. Pachter, M., Chandler, P.R., Purvis, K.B., Waun, S.D., Larson, R.A.: Multiple radar phantom tracks from cooperating vehicles using range-delay deception. In Grundel, D., Murphey, R., Pardalos, P.M., eds.: Theory and Algorithms for Cooperative Systems, World Scientific, Singapore (2004) 367–390
20. Purvis, K.B., Chandler, P.R., Pachter, M.: Feasible flight paths for cooperative generation of a phantom radar track. Journal of Guidance, Control, and Dynamics **29** (2006) 653–661
21. Stengel, R.F.: Optimal Control and Estimation. Dover, New York (1986)
22. Crassidis, J.L., Junkins, J.L.: Optimal Estimation of Dynamic Systems. Chapman & Hall/CRC, Boca Raton, Florida (2004)
23. Bryson, A.E., Ho, Y.C.: Applied Optimal Control: Optimization, Estimation, and Control. Hemisphere Publishing Corporation, Washington, District of Columbia (1975)

Second-Order Cone Programming (SOCP) Techniques for Coordinating Large-Scale Robot Teams in Polygonal Environments

Jason C. Derenick and John R. Spletzer

Lehigh University, Bethlehem PA 18015, USA
{jcd6,spletzer}@cse.lehigh.edu

Abstract. In this paper, we present an online optimization approach for coordinating large-scale robot teams in both convex and non-convex polygonal environments. In the former, we investigate the problem of moving a team of m robots from an initial shape to an objective shape while minimizing the total distance the team must travel within the specified workspace. Employing SOCP techniques, we establish a theoretical complexity of $O(k^{1.5}m^{1.5})$ for this problem with $O(km)$ performance in practice – where k denotes the number of linear inequalities used to model the workspace. Regarding the latter, we present a multiphase hybrid optimization approach. In Phase I, an optimal path is generated over an appropriate tessellation of the workspace. In Phase II, model predictive control techniques are used to identify optimal formation trajectories over said path while guaranteeing against collisions with obstacles and workspace boundaries. Once again employing SOCP, we establish complementary complexity measures of $O(l^{3.5}m^{1.5})$ and $O(l^{1.5}m^{3.5})$ for this problem with $O(l^3m)$ and $O(lm^3)$ performance in practice – where l denotes the length of the optimization horizon.

1   Introduction

The robotics community has seen a tremendous increase in multi-agent systems research in recent years. This has been driven in part by the maturation of the underlying technology: advances in embedded computing, sensor and actuator technology, and perhaps most significantly pervasive wireless communication. However, the primary motivation is the diverse range of applications envisaged for large-scale robot teams, defined herein as formations ranging from tens to thousands of robots. These include support of first responders in search and rescue operations, autonomous surveillance and monitoring in support of military and homeland security operations, and environmental monitoring. Unfortunately, the effective coordination of a large-scale robot team in an arbitrary environment is a non-trivial problem – one that will need to be solved in order for such systems to find widespread use.

In this paper, we investigate an optimization approach to the coordination task. This is motivated by the realization that the effective operation of such a team is inherently a constrained resource allocation problem. A finite number of nodes are required to perform some task (*e.g.* area surveillance), perhaps with performance objectives (*e.g.* maximize coverage), while subjected to resources that are dictated by communication

M.J. Hirsch et al. (Eds.): Adv. in Cooper. Ctrl. & Optimization, LNCIS 369, pp. 95–108, 2007.
springerlink.com

and sensor ranges, motion constraints, environmental constraints, *etc.*. More precisely, we characterize the coordination task as an optimization problem geared towards minimizing the total distance traversed while transitioning the team to a new *shape* configuration subject to polygonal environmental constraints.

While the optimization construct has many advantages, its potential for use in multi-agent systems has never been fully realized due to scalability concerns. Complete solutions to problems of interest typically scale in super-linear time with the number of robots and/or the size of the environment. In this work, we leverage recent advances in convex optimization theory to develop motion planning strategies for effectively coordinating robot teams in both convex and non-convex polygonal work environments. In both cases, the proposed strategy in practice scales linearly with the number of team members. The result is a rich, optimization-based framework for coordinating a large-scale team of fully actuated robots in real-time.

2   Related Work

Control and coordination of mobile robots in polygonal environments has been extensively studied in the literature. Belta *et al.* proposed a computational framework for generating provably correct control laws for fully-actuated robots as well as unicycles in an arbitrary polygonal workspace [1]. In [15], Kloetzer and Belta define a computational framework for the deployment of robots in both 2D and 3D rectangular environments. In this work, obstacles were modeled as polytopes and robot motion was constrained to lie within polyhedral sets. Lindemann and LaValle also considered robot control in polygonal spaces [16]. In particular, they focused upon "car-like" vehicles with bounded path curvature constraints. In their work they partition the polygonal environment into a collection of convex cells before developing safe control laws that obey specified smoothness constraints. Conner *et al.* considered global control laws based upon the utility of local potential functions [6]. They partition the environment into discrete cells and then associate each with control laws which they model as vector fields.

Formations of robot teams have also been extensively studied. As a complete survey is beyond the scope of this paper, we instead focus on those where the notion of *shape* – defined differently under different contexts – was of significant relevance to the research. Das *et al.* described a vision-based formation control framework [8]. This focused on achieving and maintaining a given formation shape using a leader-follower framework. Control of formations using Jacobi shape coordinates was addressed by Zhang *et al* [22]. The approach was applied to a formation of a small number of robots which are modeled as point masses. Abstraction based control was used by Belta and Kumar as a mechanism to coordinate a large number of fully actuated mobile robots moving in formation [2]. The main idea was to map the configuration space of the robots Q to a lower dimensional manifold $\mathcal{A}$. The concept of *shape* refers to the area spanned by the robots. A local controller was designed based on the state of the robot and the state on the manifold $\mathcal{A}$.

There has also been significant interest in applying optimization based techniques to coordinate robot teams and deploy sensor networks. Contributions in this area include the work by Cortes *et al* [7]. Here the focus is on autonomous vehicles performing distributed sensing tasks. Recently Feddema *et al.* applied decentralized optimization based control to solve a variety of multi-robot problems [12]. Optimal motion planning was considered by Belta and Kumar [3]. In this work, the authors generate a family of optimal smooth trajectories for a set of fully actuated mobile robots.

3   Defining the Coordination Problem

In developing our strategies, we first consider the problem of transitioning a robot team, constrained to lie within a convex polygonal space, to a new shape formation while minimizing the total distance that the team must travel. As the operating environment is assumed both convex and polygonal, we define it as the affine set:

$$E_c = \{x \in \mathbb{R}^2 : A_c x \le b_c\} \tag{1}$$

where $A_c \in \mathbb{R}^{k \times 2}$ with k denoting the the finite number of linear inequalities used to model the team workspace.

Since the coordination problem is defined as a function of *shape*, it is imperative to first solidify what is precisely meant by this term, as it is often defined differently in different contexts. For our purposes, we adopt the traditional definition of shape that is often employed in statistical shape analysis [11]:

Definition 1. *The shape of a formation is the geometrical information that remains when location, scale, and rotational effects are removed.*

Thus, formation shape is invariant under the Euclidean similarity transformations of translation, rotation, and scale [11].

Given this definition, we can now provide a formal statement of the coordination problem. We begin by letting $Q = [q_1, \ldots, q_m]^T \in \mathbb{R}^{m \times 2}$ denote the concatenated coordinates of the objective shape formation with respect to some world frame $\mathcal{W}$ and by letting $S = [s_1, \ldots, s_m]^T \in \mathbb{R}^{m \times 2}$ denote an instance of our objective shape with respect to some local frame $\mathcal{F}$. Given our convex polygonal workspace E_c, we see that solving the coordination problem reduces to identifying the optimal similarity transformation that when applied to $S \subset \mathcal{F}$ yields an equivalent shape $Q \subset E_C \subset \mathcal{W}$ such that our total distance objective is minimized with respect to Q and the initial robot positions $P = [p_1, p_2, \ldots, p_m]^T$. In other words, we must identify the optimal transformation parameters $[\alpha, \theta, t^x, t^y]^T$ such that $q_i = \alpha R(\theta)s_i + [t^x, t^y]^T \in E_c$ for $i = 1, \ldots, m$ where $\alpha \in R_+, R(\theta) \in SO(2)$ and $t^x, t^y \in \mathbb{R}$. Given these observations, the coordination problem can be formulated as the following constrained non-linear optimization problem:

$$
\begin{aligned}
\min \ & f(q) = \sum_{i=1}^{m} \| q_i - p_i \|_2 \\
\text{s.t.} \ & q_i = \alpha R(\theta)s_i + [t^x, t^y]^T , i = 1, \ldots, m \\
& q_i \in E_c, i = 1, \ldots, m \\
& \alpha > 0, \theta \in [0, 2\pi)
\end{aligned} \tag{2}
$$

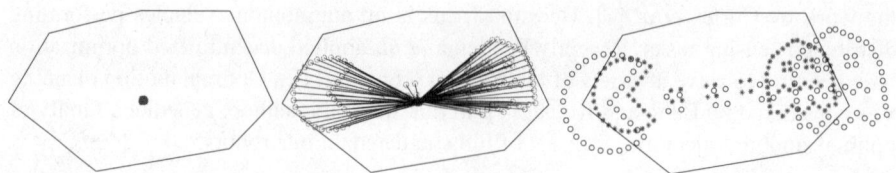

Fig. 1. (Left) The initial formation pose for 101 nodes living in a convex heptagonal (7-sided) environment. (Center) The final formation trajectories to achieve the desired shape configuration while ensuring maximal sensor network coverage in E_c. (Right) The final formation pose (red) after achieving the optimal configuration overlaid with the optimal solution (blue) obtained when the environmental model is ignored. The respecitve optimal parameters were $\alpha = 59.85$ with $[t^x, t^y]^T = [5.585, 5.169]^T$ and $\alpha = 80$ with $[t^x, t^y]^T = [9.893, 4.530]^T$. In this example, θ was fixed at $7.5°$, and scale was constrained to $\alpha \in [10, 80]$.

Unfortunately, this formulation is non-convex due to the $2m$ non-linear constraints used to capture the full set of similarity transformations (as a function of $[\alpha, \theta, t^x, t^y]^T$) that characterize the desired shape geometry S. To remedy this, we employ our results from [9]. In this work, we showed that the optimization variables $[\alpha, \theta, t^x, t^y]^T$ can be implicitly rewritten as a function of the optimal shape configuration Q. More precisely, we can supplant the non-linear equalities in (2) with the following linear (homogenous) constraints (while retaining all original problem information):

$$\left. \begin{array}{l} \| s_2 \|_2 \, (q_i^x - q_1^x) - (s_i^x, -s_i^y)^T (q_2 - q_1) = 0 \\ \| s_2 \|_2 \, (q_i^y - q_1^y) - (s_i^y, s_i^x)^T (q_2 - q_1) = 0 \end{array} \right\} i = 3, \ldots, m \qquad (3)$$

by defining without loss of generality $\left[\alpha, \; \theta, \; t^x, \; t^y \right]^T \triangleq \left[\frac{\|q_2 - q_1\|_2}{\|s_2\|_2}, \; \arctan \frac{q_2^y - q_1^y}{q_2^x - q_1^x}, \; q_1^x, \; q_1^y \right]^T$.

Given this constraint set, we can now write (2) in convex form; however, doing so would be premature as the objective is non-smooth due to the Euclidean norms inherent in its definition. To handle this, we simply introduce m auxiliary variables $[t_1, t_2, \ldots, t_m]^T$. Doing so allows us to rewrite our non-smooth objective function as a sum of upper bounds on the given Euclidean measures. In other words, the introduction of these variables induces m second-order cone constraints.

Making these adjustments, we can now formally state the coordination problem as the following SOCP in standard form:

$$\begin{aligned} \min_q \; & f(t) = 1_m^T t \\ \text{s.t.} \; & \| q_i - p_i \|_2 \leq t_i, \; i = 1, \ldots, m \\ & \begin{bmatrix} A_w \; I \\ A_s \; 0 \end{bmatrix} \begin{bmatrix} q \\ r \end{bmatrix} = \begin{bmatrix} b \\ 0 \end{bmatrix} \\ & r \geq 0 \end{aligned} \qquad (4)$$

where $A_s \in \mathbb{R}^{2(m-1) \times 2m}$ corresponds to the coefficient structure for the linear equalities given in (3) and $A_w \in \mathbb{R}^{km \times 2m}$ denotes the structure for the linear inequalities used to model E_c. We also introduce km non-negative slack variables $r = [r_1, \ldots, r_{km}]^T$.

As SOCPs are convex programs, a local minimum corresponds to a global minimum. This allows optimal solutions to be obtained through a variety of ways such as descent techniques [13] or (more efficiently) by interior point methods (IPMs) [4,13].

Figure 1 illustrates a simple application of our framework for a team of 101 robots charged with maximizing sensor network coverage within a convex heptagonal (7-sided) space. In this case, θ was fixed at $7.5°$, and scale was constrained to $\alpha \in [10, 80]$.

3.1   On Complexity

In this section, we solve (4) by adapting the logarithmic penalty-barrier approach outlined in [4]. In so doing, we establish a theoretical complexity of $O(k^{1.5}m^{1.5})$, where k once again denotes the number of linear inequalities used to model E_c.

Like other IPMs, the total complexity of the penalty-barrier approach is largely defined by solving a linear system of equations. In this case, Equality-constrained Newton's method (ENM) is used for internal minimization and the linear system is in KKT form. As solving this system provides a solution to the Newton step sub-problem, we accordingly refer to it as the "Newton KKT system." We show that by reformulating (4), we can band the coefficient matrix to solve the system in $O(km)$ time via algorithms that exploit knowledge of matrix bandwidth.

Reformulating the Coordination Problem. Problem (4) can be restated in a relaxed form suitable for solving via the barrier approach by simply augmenting the objective function with log-barrier terms corresponding to both the problem's conic inequalities as well as the inequalities used to ensure the non-negativity of the associated slack variables. Doing so yields the following equivalent problem statement:

$$\min f(q,t,r) = \tau_l 1_m^T t - \sum_{i=1}^{m} \log\left(t_i^2 - (q_i - p_i)^T(q_i - p_i)\right) - \sum_{i=1}^{km} \log r_i$$

$$\text{s. t. } \begin{bmatrix} A_w & I \\ A_s & 0 \end{bmatrix} \begin{bmatrix} q \\ r \end{bmatrix} = \begin{bmatrix} b \\ 0 \end{bmatrix} \tag{5}$$

where τ_l is the inverse log-barrier scaler for the lth iteration. Essentially, solving our SOCPs reduces to solving a sequence of convex optimization problems of this form, where after each iteration τ_{l+1} is chosen such that $\tau_{l+1} > \tau_l$ [4].

Banding the Newton KKT System. During each iteration of the log-barrier approach, we aim to minimize the second-order Taylor approximation of our objective function as a function of the Newton step, $\delta x = [\delta q, \delta r]^T$, subject to $A\delta x = 0$. As a result, obtaining δx is equivalent to analytically solving the KKT conditions associated with this equality-constrained sub-problem. In other words, we must solve the following linear system of equations [4]:

$$\begin{bmatrix} H & \hat{A}^T \\ \hat{A} & 0 \end{bmatrix} \begin{bmatrix} \delta x \\ v \end{bmatrix} = \begin{bmatrix} -g \\ 0 \end{bmatrix} \tag{6}$$

where H and g respectively denote the evaluated Hessian and gradient of the objective function given in (5) at x, v is the corresponding dual variable for δx, and $\hat{A} = \begin{bmatrix} A_w & I \\ A_s & 0 \end{bmatrix}$. Undoubtedly, solving (6) is the bottleneck of the algorithm, requiring $O(k^3 m^3)$ basic operations in a naive implementation; however, we will show that it can be solved very efficiently by simply reposing the problem given in (5).

Noting that the coefficient matrix of (6) is symmetric indefinite, we employ Gaussian elimination with non-symmetric partial pivoting. The performance of this technique suffers significantly when the linear system in question features dense rows and/or columns due to fill-in [21]. In particular, the algorithm could yield a worst-case performance of $O(k^3 m^3)$ when solving an instance of (6) associated with the nominal problem formulation given in (5). To illustrate this point, we include Figure 2 (Left) which shows the corresponding non-zero sparsity structure of the Newton KKT system. As the rows of system are permuted during reduction, the dense rows and columns respectively located in the upper-right and lower-left quadrants of (6) could introduce a solid sub-block of order $km \times km$, which itself would require $O(k^3 m^3)$ basic operations to reduce. Such a workload is highly impractical, especially when considering large-scale configurations that inherently feature 1000's of decision variables.

To address this issue, we present the following auxiliary formulation of (5) that facilitates transforming the Newton KKT system into a mono-banded form:

$$\min \; f(q,t,r) = \tau_l 1_m^T t - \sum_{i=1}^{m} \log \left(t_i^2 - (q_i - p_i)^T (q_i - p_i)\right) - \sum_{i=1}^{km} \log r_i$$

s. t.

$$\| s_2 \|_2 \, (q_i^x - c_{ik+1}^x) - (s_i^x, -s_i^y)^T (d_{(i-1)k+1} - c_{ik+1}) = 0, \; i = 3, \ldots, m$$
$$\| s_2 \|_2 \, (q_i^y - c_{ik+1}^y) - (s_i^y, s_i^x)^T (d_{(i-1)k+1} - c_{ik+1}) = 0, \; i = 3, \ldots, m$$
$$a_j^T w_{(i-1)k+j} + r_{(i-1)k+j} = b_{(i-1)k+j}, \; i = 1, \ldots, m, j = 1, \ldots, k$$
$$w_{(i-1)k+j} = w_{(i-1)k+j+1}, \; i = 1, \ldots, m, j = 1, \ldots, k-1 \qquad (7)$$
$$w_{(i-1)k+1} = q_i, \; i = 1, \ldots, m$$
$$c_i = c_{i+1}, \; i = 1, \ldots, km - 1$$
$$d_i = d_{i+1}, \; i = 1, \ldots, km - k - 1$$
$$c_1 = q_1$$
$$d_1 = q_2$$

In this formulation, the shape constraints are given by the first two sets of equalities while the environmental bounds are given by the third. Essentially, the additional c and d variables allow us to "chain" the values of q_1 and q_2 respectively through the corresponding Newton KKT system, which eliminates the dense row and column features that would otherwise be present. Similarly, as the k linear inequalities defining E_c bound the final objective position of each node (*i.e.* q_i), we introduce k auxiliary variables (*i.e.* w) for each node in order to locally chain q_i. Doing so ensures a bandwidth that will ultimately remain independent of both k and m.

Given this augmented formulation, our claim is that the system can be made mono-banded. To show this, we begin by defining the nominal solution vector for the coefficient structure of (6) as follows:

$$\left[\delta \eta^T, \delta \kappa_1^T, \ldots, \delta \kappa_{m-2}^T, \delta \zeta^T, \mu^T \right]^T \qquad (8)$$

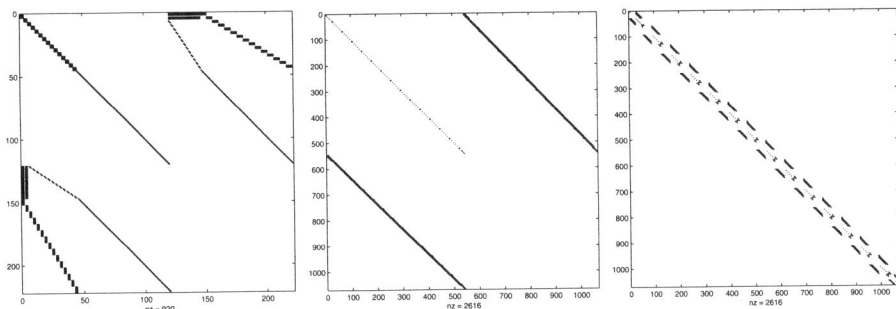

Fig. 2. (Left) The nominal Newton KKT system structure for a team of 15 robots constrained to a pentagonal workspace. (Center) The augmented Newton KKT system for the same configuration. This system is derived from (8) and (9). (Right) The banded system with lower and upper bandwidths of 37. The bandwidth is independent of both team size and the number of constraints used to model E_c. In this form, the system is now solvable in $O(km)$.

$$\delta\eta = \begin{bmatrix} \delta q_1 \\ \delta t_1 \\ \delta w_1 \\ \delta r_1 \\ \delta c_1 \\ \vdots \\ \delta w_k \\ \delta r_k \\ \delta c_k \end{bmatrix} \quad \delta\kappa_i = \begin{bmatrix} \delta q_{i+1} \\ \delta t_{i+1} \\ \delta c_{ik+1} \\ \delta d_{(i-1)k+1} \\ \delta w_{ik+1} \\ \delta r_{ik+1} \\ \vdots \\ \delta c_{(i+1)k} \\ \delta d_{ik} \\ \delta w_{(i+1)k} \\ \delta r_{(i+1)k} \end{bmatrix} \quad \delta\zeta = \begin{bmatrix} \delta q_m \\ \delta t_m \\ \delta c_{(m-1)k+1} \\ \delta d_{(m-2)k+1} \\ \delta w_{(m-1)k+1} \\ \delta r_{(m-1)k+1} \\ \vdots \\ \delta w_{mk} \\ \delta r_{mk} \end{bmatrix} \quad \mu = \begin{bmatrix} v_1 \\ \vdots \\ v_{(7m-6)k+2m} \end{bmatrix}$$

where the δ variables correspond to the primal Newton step components associated with each of the respective system variables.

In order to yield the mono-banded form, we begin by stating the constraint/row permutation for A that yields the tri-banded system appearing Figure 2 (Center). We assume that A is already arbitrarily constructed with random row and column permutations. For the sake of clarity, we group constraints by associating them with the respective nodes that introduce them into the system. In doing so, we employ a slight abuse of notation by allowing the variable q_i to also denote the ith robot in the configuration. That stated, we can now define the constraints associated with q_1:

$$\left. \begin{array}{l} q_1^x = c_1^x \\ q_1^y = c_1^y \\ q_1^x = w_1^x \\ q_1^y = w_1^y \\ a_1^T w_1 + r_1 = b_1 \end{array} \right\} \triangleq \varrho_1 \qquad \left. \begin{array}{l} c_{j-1}^x = c_j^x \\ c_{j-1}^y = c_j^y \\ w_{j-1}^y = w_j^y \\ w_{j-1}^y = w_j^y \\ a_j^T w_j + r_j = b_j \end{array} \right\} \triangleq \varrho_j, \; j = 2, \ldots, k$$

Similarly for q_2, we associate

$$
\left.\begin{array}{r}
c_k^x = c_{k+1}^x \\
c_k^y = c_{k+1}^y \\
q_2^x = d_1^x \\
q_2^y = d_1^y \\
q_2^x = w_{k+1}^x \\
q_2^y = w_{k+1}^y \\
a_{k+1}^T w_{k+1} + r_{k+1} = b_{k+1}
\end{array}\right\} \triangleq \varrho_{k+1}
\qquad
\left.\begin{array}{r}
c_{k+j-1}^x = c_{k+j}^x \\
c_{k+j-1}^y = c_{k+j}^y \\
d_{j-1}^x = d_j^x \\
d_{j-1}^y = d_j^y \\
w_{k+j-1}^y = w_{k+j}^y \\
w_{k+j-1}^y = w_{k+j}^y \\
a_{k+j}^T w_{k+j} + r_{k+j} = b_{k+j}
\end{array}\right\} \triangleq \varrho_{k+j}, \; j = 2, \ldots, k
$$

For $3 \leq i \leq (m-1)$, we define the constraints associated with q_i as:

$$
\left.\begin{array}{r}
c_{(i-1)k}^x = c_{(i-1)k+1}^x \\
c_{(i-1)k}^y = c_{(i-1)k+1}^y \\
d_{(i-2)k}^x = d_{(i-2)k+1}^x \\
d_{(i-2)k}^y = d_{(i-2)k+1}^y \\
\| s_2 \|_2 \, (q_i^x - c_{(i-1)k+1}^x) = (s_i^x, -s_i^y)^T (d_{(i-2)k+1} - c_{(i-1)k+1}) \\
\| s_2 \|_2 \, (q_i^y - c_{(i-1)k+1}^y) = (s_i^y, s_i^x)^T (d_{(i-2)k+1} - c_{(i-1)k+1}) \\
q_i^x = w_{(i-1)k+1}^x \\
q_i^y = w_{(i-1)k+1}^y \\
a_{(i-1)k+1}^T w_{(i-1)k+1} + r_{(i-1)k+1} = b_{(i-1)k+1}
\end{array}\right\} \triangleq \varrho_{(i-1)k+1}
$$

$$
\left.\begin{array}{r}
c_{(i-1)k+j-1}^x = c_{(i-1)k+j}^x \\
c_{(i-1)k+j-1}^y = c_{(i-1)k+j}^y \\
d_{(i-2)k+j-1}^x = d_{(i-2)k+j}^x \\
d_{(i-2)k+j-1}^y = d_{(i-2)k+j}^y \\
w_{(i-1)k+j-1}^y = w_{(i-1)k+j}^y \\
w_{(i-1)k+j-1}^y = w_{(i-1)k+j}^y \\
a_{(i-1)k+j}^T w_{(i-1)k+j} + r_{(i-1)k+j} = b_{(i-1)k+j}
\end{array}\right\} \triangleq \varrho_{(i-1)k+j}, \; j = 2, \ldots, k
$$

Finally, we associate the remaining constraints with q_m:

$$
\left.\begin{array}{r}
c_{(m-1)k}^x = c_{(m-1)k+1}^x \\
c_{(m-1)k}^y = c_{(m-1)k+1}^y \\
d_{(m-2)k}^x = d_{(m-2)k+1}^x \\
d_{(m-2)k}^y = d_{(m-2)k+1}^y \\
\| s_2 \|_2 \, (q_m^x - c_{(m-1)k+1}^x) = (s_m^x, -s_m^y)^T (d_{(m-2)k+1} - c_{(m-1)k+1}) \\
\| s_2 \|_2 \, (q_m^y - c_{(m-1)k+1}^y) = (s_m^y, s_m^x)^T (d_{(m-2)k+1} - c_{(m-1)k+1}) \\
q_m^x = w_{(m-1)k+1}^x \\
q_m^y = w_{(m-1)k+1}^y \\
a_{(m-1)k+1}^T w_{(m-1)k+1} + r_{(m-1)k+1} = b_{(m-1)k+1}
\end{array}\right\} \triangleq \varrho_{(m-1)k+1}
$$

$$
\left.\begin{array}{r}
w_{(m-1)k+j-1}^y = w_{(m-1)k+j}^y \\
w_{(m-1)k+j-1}^y = w_{(m-1)k+j}^y \\
a_{(m-1)k+j}^T w_{(m-1)k+j} + r_{(m-1)k+j} = b_{(m-1)k+j}
\end{array}\right\} \triangleq \varrho_{(m-1)k+j}, \; j = 2, \ldots, k
$$

Again we employ a slight abuse of notation by letting each ϱ_j also denote the initial row indices of the constraints with which it is associated. Preserving the relative ordering of the constraints as they appear in the respective definition of each ϱ_j, we provide the following row permutation for A. This ordering yields the tri-banded form as it appears in 2 (Center):

$$\left[\varrho_1^T, \varrho_2^T, \ldots, \varrho_{mk}^T\right]^T \tag{9}$$

Given this definition of A as well as (8), the mono-banded form of (6) can be constructed. Symmetrically applying the permutation that yields the following Newton KKT system solution vector ordering:

$$\left[\lambda^T, \gamma^T, \xi_1^T, \ldots, \xi_{(m-3)}^T, \chi^T\right]^T \tag{10}$$

$$\lambda = \begin{bmatrix} \delta q_1 \\ \delta l_1 \\ v_1 \\ \vdots \\ v_{5k} \\ \delta w_1 \\ \delta r_1 \\ \delta c_1 \\ \vdots \\ \delta w_k \\ \delta r_k \\ \delta c_k \end{bmatrix} \quad \gamma = \begin{bmatrix} \delta q_2 \\ \delta t_2 \\ v_{5k+1} \\ \vdots \\ v_{12k} \\ \delta c_{k+1} \\ \delta d_1 \\ \delta w_{k+1} \\ \delta r_{k+1} \\ \vdots \\ \delta c_{2k} \\ \delta d_k \\ \delta w_{2k} \\ \delta r_{2k} \end{bmatrix} \quad \xi_i = \begin{bmatrix} \delta q_{i+2} \\ \delta t_{i+2} \\ v_{(7i+5)k+2i-1} \\ \vdots \\ v_{(7i+12)k+2i} \\ \delta c_{(i+1)k+1} \\ \delta d_{ik+1} \\ \delta w_{(i+1)k+1} \\ \delta r_{(i+1)k+1} \\ \vdots \\ \delta c_{(i+2)k} \\ \delta d_{(i+1)k} \\ \delta w_{(i+2)k} \\ \delta r_{(i+2)k} \end{bmatrix} \quad \chi = \begin{bmatrix} \delta q_m \\ \delta t_m \\ v_{(7m-9)k+2m-5} \\ \vdots \\ v_{(7m-6)k+2m} \\ \delta c_{(m-1)k+1} \\ \delta d_{(m-2)k+1} \\ \delta w_{(m-1)k+1} \\ \delta r_{(m-1)k+1} \\ \vdots \\ \delta w_{mk} \\ \delta r_{mk} \end{bmatrix}$$

produces a mono-banded coefficient structure having a respective upper and lower bandwidths of 37.

Figure 2 illustrates the process of transforming the KKT system via our approach. The "augmented" Newton KKT system derived from the permutations given in (8) and (9) is shown in Figure 2 (Center). Taking the coefficient structure of (6) in this form and symmetrically permuting its rows and columns according to (10) yields the mono-banded system appearing in Figure 2 (Right). It can now be solved in $O(km)$ using a band-diagonal LU-based solver [20].

Applying these alterations effectively reduces the per-iteration complexity of the penalty-barrier method to $O(km)$ for the coordination problem. As the iteration complexity of the barrier approach scales as $O(\sqrt{km})$, we see that the total complexity is $O(k^{1.5}m^{1.5})$ in theory. However, it should be emphasized that this bound is highly conservative as it is well-known that iteration complexity scales as $O(1)$ in practice [4]. As such, solving the coordination problem will require a number of basic operations that grows more like $O(km)$. In other words, the computational workload scales linearly with the number environmental constraints and the configuration size.

3.2 Simulation Results

The results presented thus far correspond to an application of a simple penalty-barrier approach. Although effective, such an IPM is not typically used in practice as more

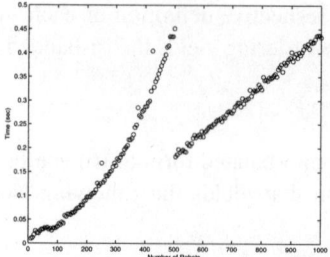 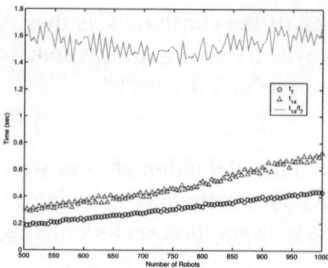

Fig. 3. (Left) MOSEK CPU utilization time for teams operating in a heptagonal environment. When problem structure is fully exploited ($m \gtrsim 500$), the trend becomes highly linear with $r^2 = 0.9822$. (Right) The highly linear trends for teams operating in heptagonal and tetradecagonal (14-sided) environments. Regarding the latter, we have $r^2 = 0.9645$.

sophisticated and robust solvers exist [17,19]. As such, we carried-out a sequence of trials whereby the coordination problem was solved using the MOSEK industrial solver package, which utilizes a homogenous self-dual IPM [19]. For our trials, we varied the given team size m from 10 to 1000 at intervals of 5 with the mean CPU time being recorded over a sample size of 10 trials for each value. All problems were solved using a standard desktop PC having a 3.0 GHz Pentium 4 processor and 2.0 GB of RAM.

In Figure 3 (Left) the CPU utilization trend is provided for a team confined to operations in a heptagonal (7-sided) environment. Notice that below ≈ 500 nodes, the complexity scales cubicly ($r^2 = 0.9933$). This appears to be the result of the solver not fully exploiting problem structure in obtaining its solution. Beyond 500 this is not the case as performance is highly linear with linear regression analysis revealing $r^2 = 0.9822$. Perhaps more importantly, we see that solutions for configurations having up to 1000 nodes are obtainable in less than 0.45 seconds.

Figure 3 (Right) shows the highly-linear performance trends for robot teams operating respectively in heptagonal and tetradecagonal (14-sided) environments. Moreover, the linear growth of the complexity as a function of k is evident by considering the comparative performance ratio $\frac{t_{14}}{t_7}$ which remains essentially constant as $m \to 1000$. Together, these results highlight the efficacy of our approach.

4   Coordination in Non-convex Polygonal Environments

We now compose our previous results into a more general instance of the coordination task. Specifically, we consider motion planning in an arbitrary polygonal environment with obstacles. As the space of feasible robot positions is no-longer convex by assumption, solving this problem directly would require more general and less-efficient nonlinear programming techniques that guarantee only convergence to local minima. Thus, in an effort to obtain a similar complexity results as those seen in Section 3.1, we propose a hybrid multi-phase optimization approach over a discrete convex tessellation of the work environment.

4.1   Generalizing the Coordination Problem

As noted, we assume that the configuration space $\mathcal{C}$ for the robot team is a polygonal environment with obstacle subspace $\mathcal{O}$ and free space $\mathcal{C}_{free}$ such that $\mathcal{C}_{free} = \mathcal{C} - \mathcal{O}$. Using exact cell decomposition methods (*e.g.* triangulation, trapezoidal decomposition, *etc.*), $\mathcal{C}_{free}$ can be tessellated into convex polygonal cells $C_1, \ldots, C_z$, where $\mathcal{C}_{free} = \bigcup_{i=1}^{z} C_i$ [5]. The resulting partition induces an undirected graph $G = (V, E)$, where vertex $v_i \in V$ corresponds to cell C_i, and edge $e_{ij} \in E$ implies that there exists a common edge between C_i and C_j. Paths between cells can then be efficiently computed using traditional graph optimization algorithms (*e.g.* [10]). The coordination problem can then be reposed as transitioning the formation from cell to cell along the specified path. In the sequel, we assume a triangulation partition of $\mathcal{C}_{free}$. We also assume that the union of adjacent cells $C_{ij} = C_i \bigcup C_j \ \forall \ (C_i, C_j) \in E$ is convex. This is hardly restrictive as it is straightforward to refine any pair of adjacent triangles to three such triangles where both of the resulting adjacent pairs meet this constraint.

Remark 1. Given two adjacent cells $(C_i, C_j) \in E$, where $C_{ij} = C_i \bigcup C_j$ is convex, if node $x_i \in C_i$ and $x_j \in C_j$, then by convexity $\lambda x_i + (1 - \lambda) x_j \in C_{ij}$, $\lambda \in [0, 1]$. This implies that for a formation of m nodes with initial pose $X_i = (x_{i1}, \ldots, x_{im})^T \in \mathbb{R}^{2m}$ in triangle C_i, and final pose X_j in triangle C_j, the paths of each node will remain entirely in $C_{ij} \subseteq \mathcal{C}_{free}$. This guarantees against collisions with obstacles.

Let us assume that such a path $C_p = \{C_1, \ldots, C_l\} \subseteq \mathcal{C}_{free}$ has been specified by a higher level planner. The coordination problem can then be written as follows:

Problem 1. Given a path specification $C_p = \{C_1, \ldots, C_l\}$, a corresponding shape specification $S = \{S_1, \ldots, S_l\}$, and an initial formation pose X_0, find a motion sequence $X = \{X_1, \ldots, X_l\}$ for the formation such that

1. $X_i \sim S_i$, $i = 1, \ldots, l$
2. $X_i \in C_i$, $i = 1, \ldots, l$
3. The distance traveled by the formation is minimized in accordance with the criteria from Problem 4.

In solving Problem 1, we employ optimization techniques from model predictive control [14,18]. In this context however, the length of the horizon is not defined by time, but rather the length of the path over which the optimization problem is solved.

To constrain the pose of the formation during each step of the horizon, each triangle can be modeled as a set of three linear inequality constraints on the position of each robot

$$c_{ik}^T x_{ij} \leq 0, \ i = 1, \ldots, l, \ j = 1, \ldots, m, \ k = 1, 2, 3 \tag{11}$$

In a slight abuse of notation, we also let $C_i = (c_{11}, \ldots, c_{m3})^T \in \mathbb{R}^{3m \times 2m}$ denote the set of linear constraints on the formation pose such that $X_i \in C_i$ We can now write the solution to Problem 1 for our total distance metric as

$$\min_{X} \sum_{i=1}^{l} \sum_{j=1}^{m} t_{ij}, \ i = 1, \ldots, l, \ j = 1, \ldots, m$$

$$\text{s.t. } \| x_{ij} - x_{i-1,j} \|_2 \leq t_{ij} \tag{12}$$

$$A_i X_i = 0$$

$$C_i X_i \leq 0$$

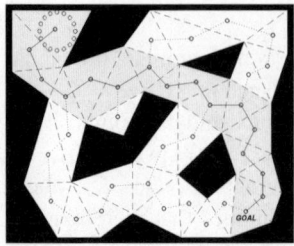 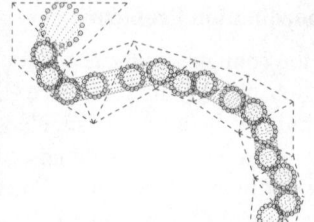 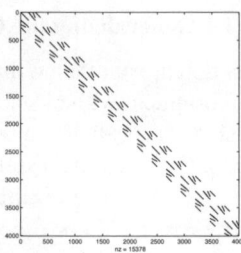

Fig. 4. (Left) $\mathcal{C}_{path}$ as specified by the higher level planner. (Center) The corresponding motion sequence obtained from solving the associated SOCP. The formation is guaranteed to follow $\mathcal{C}_{path}$ while minimizing the total distance traveled, avoiding obstacles, and maintaining the desired formation shape. In this example, both the orientation and minimum scale for the formation were constrained. (Right) The associated linear system remains mono-banded, and in this case, the bandwidth is defined as a function of the configuration size m.

where A_i are the constraints associated with shape S_i as defined in Section 3. By now, we can readily recognize the form of this problem as a SOCP. More significantly perhaps, the corresponding KKT matrix corresponds to the chaining of l instances of our single step problem. As a result, the associated linear system will remain mono-banded; however, in this case the bandwidth will grow as either a function of m or l depending upon the selected permutation of the augmented KKT system. As such, we conclude the theoretical complexity is $O(l^{1.5}m^{3.5})$ or $O(l^{3.5}m^{1.5})$ – once again depending upon the chosen ordering. In cases where the problem demands $l \gg m$ – $i.e.$ the horizon length far exceeds the team size – a permutation yielding a bandwidth as a function of m is best. Similarly, when the problem requires $m \gg l$, the bandwidth is best defined as a function of horizon length as that yields the best performance bound.

Once again, these theoretical results are highly conservative as iteration complexity scales as $O(1)$ in practice [4]. Thus, solving the generalized coordination problem will require a number of basic operations that scales more like $O(l^3m)$ (or $O(lm^3)$). In the former case, complexity scales linearly with configuration size making it well-suited for coordinating a large-scale robot team.

4.2 Simulation Results

A sample simulation trial for a formation of 16 robots is shown in Figure 4. The path of the formation is specified by a higher level planner after a discrete optimization phase on the corresponding graph G (Left). The formation then solves the continuous optimization problem specified in (12). The resulting path of the formation is shown in Figure 4 (Right). In this example, the optimization was over the entire path length ($l = 16$), the shape was held constant, and the minimum scale of the formation was specified as a premise for inter-robot collision avoidance.

5   Discussion and Future Work

In this paper, we developed strategies for coordinating large-scale robot teams in both convex and non-convex polygonal environments. We began by formulating the coordination problem as a constrained optimization problem in which the objective was to minimize the distance a team, living in a convex polygonal workspace, must travel while transitioning to a new objective shape configuration. We showed that this problem can be formulated as a convex mathematical program. Solving with a log-barrier IPM, we also showed that its solvable in $O(k^{1.5}m^{1.5})$ time in theory with $O(km)$ performance in practice – where k denotes the number of affine constraints used to model the convex workspace and m denotes the configuration size.

After establishing these results, we then extended them to solve the coordination problem in a non-convex polygonal workspace. By using an appropriate tessellation of the environment along with model predictive control techniques, we showed that a large-scale team of robots can obtain an objective position while successfully avoiding collisions with both workspace boundaries and static obstacles. This problem is also presented in convex form, and we showed that complexity scales as $O(l^{3.5}m^{1.5})$ with $O(l^3m)$ performance in practice – when the bandwidth of the IPM's core linear system is defined as a function of the optimization horizon length l. In the case where bandwidth is defined in terms of configuration size m, the theoretical complexity is then $O(l^{1.5}m^{3.5})$ with $O(lm^3)$ performance in practice.

We are currently extending these results to a more general multi-objective framework for large-scale coordination in $SE(2)$. Such an extension is invaluable as many applications require teams of robots to perform well with respect to multiple goals. Additionally, we are exploring the possibility of extending the framework to $SE(3)$. However, such an extension is not obvious as a direct formulation of the coordination problem in this higher dimensional space introduces imaginary terms. As a result, alternate approaches and possible relaxations are being evaluated to achieve this end.

References

1. C. Belta, V. Isler, and G. Pappas. Discrete abstractions for robot motion planning and control in polygonal environments. *IEEE Transactions on Robotics*, 17(6):864–875, 2005.
2. C. Belta and V. Kumar. Abstraction and control for groups of robots. IEEE Trans. on Robotics and Automation, 2004.
3. C. Belta and V. Kumar. Optimal motion generation for groups of robots: a geometric approach. *ASME Journal of Mechanical Design*, 126, 2004.
4. S. Boyd and L. Vandenberghe. *Convex Optimization*. Cambridge Unviersity Press, 2004.
5. H. Choset, K. Lynch, S. Hutchinson, G. Kantor, W. Burgard, L. Kavraki, and S. Thrun. *Principles of Robot Motion Planning*. MIT Press, 2005.
6. D. C. Conner, A. A. Rizzi, and H. Choset. Composition of local potential functions for global robot control and navigation. In *IEEE/RSJ Int. Conf. on Intelligent Robots and Systems*, Las Vegas, Nevada, USA, October 2003.
7. J. Cortés, S. Martínez, T. Karatas, and F. Bullo. Coverage control for mobile sensing networks. *IEEE Trans. on Robotics and Automation*, 20(2):243–255, April 2004.
8. A. K. Das, R. Fierro, V. Kumar, J. P. Ostrowski, J. Spletzer, and C. J. Taylor. A vision-based formation control framework. *IEEE Trans. on Robotics and Automation*, 18(5):813–825, October 2002.
9. J. Derenick, C. Mansley, and J.Spletzer. Efficient motion planning strategies for large-scale sensor networks. In *Proceedings of the Seventh International Workshop on the Algorithmic Foundations of Robotics (WAFR 2006)*, New York, NY, USA, July 2006.

10. E. Dijkstra. A note on two problems in connexion with graphs. *Numerische Mathematik*, 1:269–272, 1959.
11. I. L. Dryden and K. V. Mardia. *Statistical Shape Analysis*. John Wiley and Sons, 1998.
12. J. T. Feddema, R. D. Robinett, and R. H. Byrne. An optimization approach to distributed controls of multiple robot vehicles. In *Workshop on Control and Cooperation of Intelligent Miniature Robots, IEEE/RSJ IROS*, Las Vegas, Nevada, October 31 2003.
13. R. Horst, P. M. Pardalos, and N. V. Thoai. *Introduction to Global Optimization*. Springer, 2nd edition, 2000.
14. A. Jadbabaie, J. Yu, and J. Hauser. Unconstrained receding-horizon control of nonlinear systems. *IEEE Trans. on Automatic Control*, 46(5):776–783, May 2001.
15. M. Kloetzer and C. Belta. A framework for automatic deployment of robots in 2d and 3d environments. In *IEEE/RSJ Int. Conf. on Intelligent Robots and Systems*, pages 953–958, Beijing, China, October 2006.
16. S. Lindemann and S. LaValle. Smooth feedback for car-like vehicles in polygonal environments. In *Proc. IEEE Int. Conf. Robot. Automat.*, Roma, Italy, April 2007.
17. M. Lobo, L. Vandenberghe, S. Boyd, and H. Lebret. Applications of second-order cone programming. *Linear Algebra and Applications, Special Issue on Linear Algebra in Control, Signals and Image Processing*, 1998.
18. D. Mayne, J. Rawings, C. Rao, and P. Scokaert. Constrained model predictive control: Stability and optimality. *Automatics*, 36(6):789–814, June 2000.
19. MOSEK ApS. *The MOSEK Optimization Tools Version 3.2 (Revision 8) User's Manual and Reference*. http://www.mosek.com.
20. W. Press et al. *Numerical Recipes in C*. Cambridge University Press, 1993.
21. Y. Saad. *Iterative Methods for Sparse Linear Systems*. Society for Industrial and Applied Mathematics, Philadelphia, PA, USA, 2003.
22. F. Zhang, M. Goldgeier, and P. S. Krishnaprasad. Control of small formations using shape coordinates. In *Proc. IEEE Int. Conf. Robot. Automat.*, volume 2, Taipei, Sep 2003.

UAV Splay State Configuration for Moving Targets in Wind

Derek Kingston and Randal Beard

Brigham Young University

Abstract. Cooperative surveillance problems require members of a team to spread out in some fashion to maximize coverage. In the case of single target surveillance, a team of UAVs angularly spaced (i.e. in the splay state configuration) provides the best coverage of the target in a wide variety of circumstances. In this chapter we propose a decentralized algorithm to achieve the splay state configuration for a team of UAVs tracking a moving target. We derive the allowable bounds on target velocity to generate a feasible solution as well as show that, near equilibrium, the overall system is exponentially stable. Monte Carlo simulations indicate that the surveillance algorithm is asymptotically stable for arbitrary initial conditions. We conclude with high fidelity simulation tests to show the applicability of the splay state controller to actual unmanned air systems.

1 Introduction

A primary use of unmanned air vehicle (UAV) systems is in surveillance and reconnaissance missions [1] [2]. We investigate the use of a team of multiple UAVs orbiting a target with application to target tracking and convoy support.

The payload of choice for most small UAVs is a camera. The objective of our work is to develop a cooperative guidance strategy to distribute UAV agents around an orbit spaced equally in angle. The equal angle spacing allows the team to cooperatively overcome possible line-of-sight occlusions, i.e. equal spacing gives the team the best chance to track a target in the presence of occlusions. We note that for two UAVs carrying radar sensors, line-of-sight angles separated by 90 degrees provide better statistical performance in the tracking problem [3] and when the team size is greater than two, equal spacing has good performance. In a general surveillance mission, the equal spacing of the sensors provides the best overall coverage of a target and its surroundings.

The design of a spacing controller is strongly influenced by the capabilities of the UAVs on the team. For instance, helicopters can hover at a specific location and thereby maintain persistant coverage of a ground based target, however fixed-wing aircraft must fly above the stall velocity, and may therefore not be able to maintain persistent coverage. Furthermore, fixed-wing aircraft fly most efficiently at a fixed, nominal airspeed. One approach to equal spacing is to adjust the local velocity of the agents along the desired orbit. However, for small allowable velocity bounds, the convergence to the equilibrium configuration may

M.J. Hirsch et al. (Eds.): Adv. in Cooper. Ctrl. & Optimization, LNCIS 369, pp. 109–128, 2007.
springerlink.com

be sluggish. Additionally, maintaining fixed-wing aircraft at their constant fuel efficient velocity is desirable from a mission duration standpoint. In this chapter we develop a spacing controller that steers the UAVs to the desired configuration while holding a constant airspeed.

Other researchers have studied the problem of spacing fixed-speed UAVs around a possibly moving target. Paley et al. introduce the notion of the splay state configuration and give an elegant control solution for fixed target problems [4]. Their approach relies on invariant set arguments to show that the splay state configuration is the stable equilibrium of the system. The main drawback of their work is the inability to specify the orbit center. The splay state configuration is shown to be stable around the collective center of mass *not* a specific target location which makes tracking a moving target infeasible without modifications. Additionally, the control signal exhibits slow transient response for large initial errors.

Paley's splay state configuration work is extended by Klein and Morgansen in [5] to moving targets. By choosing a control signal that preserves the invariant sets introduced by Paley, they are able to design an algorithm to track a moving target in the splay state configuration with 3 UAVs. Unfortunately, the method does not currently extend to team sizes other than $N = 3$.

Frew and Lawrence [1] use vector field notions to steer a team of two UAVs to an orbit centered on a moving target. A limit cycle is designed as the equilibrium of the vector field dynamics and is modified to account for spacing errors. No formal proof is offered in their method and only team sizes of $N = 2$ are considered.

The unique features of our approach are the ability to include an arbitrary number of team members in a moving target scenario and the determination of bounds on target velocity for which the algorithm satisfies the UAV's kinematic constraints. Additionally, the transient response is qualitatively better than other approaches. Of note is that our algorithm is completely decentralized where agents base their actions only on communication from immediate team members. This allows for dynamic changes to the team to be accounted for without global communication or replanning. A drawback to our approach is that global stability is not conclusively shown, although Monte-Carlo simulations indicate that the splay state configuration is the globally stable equilibrium of the system.

The aim of this chapter is to present a stable, decentralized spacing controller for fixed velocity UAVs tracking moving targets in the presence of wind. Section 2 formally defines the notion of equal spacing and describes the mathematical model that we use for the UAVs. Section 3 establishes the heading design for a group of UAVs monitoring a stationary target. In Section 4, we analyze the stability of the system for the stationary target case. These results are extended to the moving target/wind case in Section 5 and we conclude with simulation results in Section 6. Concluding remarks are offered in Section 7.

2   Problem Description

In a variety of applications the ability for a team of UAVs to spread out in some manner increases the efficiency of the team as a whole. For single target surveillance, a team of UAVs spaced equally around an orbit centered on the target gives the best line-of-sight coverage in the presence of occlusions. This chapter focuses on constructing a desired heading for each UAV in the team to achieve equal spacing. The desired heading is calculated based on the distance away from the desired orbit and the spacing error from the splay state configuration.

Definition 1 (Splay State Configuration). *A set of agents $\mathcal{I}$, all of which are following the same periodic trajectory, is said to have reached the* splay state configuration *if for each agent i, the time difference of arrival to a specific point on the trajectory between agent i and its two immediate neighbors is constant for all $i \in \mathcal{I}$.*

Definition 1 describes the splay state configuration as equally spaced in time along a periodic trajectory. When agents pass a reference point (arbitrarily chosen) on the trajectory at equal time intervals, the team has reached the splay state configuration. For simple circular trajectories, the splay state configuration is achieved when agents are equally spaced in angle around the circle perimeter. Note that equal angular spacing matches the definition of the splay state configuration in [4]. Definition 1 extends the splay state notion to non-circular trajectories which occur when the center of the desired orbit is changing in time due to wind or target motion.

Consider a circular trajectory with all agents traveling at constant speed V. The time difference of arrival corresponds to the angle separation between neighbors. When the angle between all agents is the same then the splay state configuration has been reached, i.e. the agents are equally spaced in angle around the circle. Now consider the trajectory shown in Figure 1, which is an example of a UAV orbiting a moving target. Note that as the target speed increases, the ability for the UAV to maintain an orbit around the target depends on its ability to make increasingly sharp turns. Constraints on the turning radius of the UAV will lead to a threshold value of target speed where feasible tracking is no longer possible (see Section 5). In a moving reference frame (with the target in the

Fig. 1. For a UAV orbiting a moving target, the trajectory exhibits loops corresponding to the times when the UAV and the target are moving in opposite directions and long arcs when both are moving in the same direction

center) the motion of the UAV traces out a circle, but the splay state configuration does *not* correspond to equal spacing in angle around that circle. Since the target is moving, a much greater amount of time is spent on the part of the trajectory where the UAV and the target are moving in the same direction. When the target and UAV are moving in opposite directions, the UAV quickly travels around a large portion of the circle. Figure 2 shows the splay state configuration for 5 UAVs when the target is moving at 75% of V in zero wind conditions.

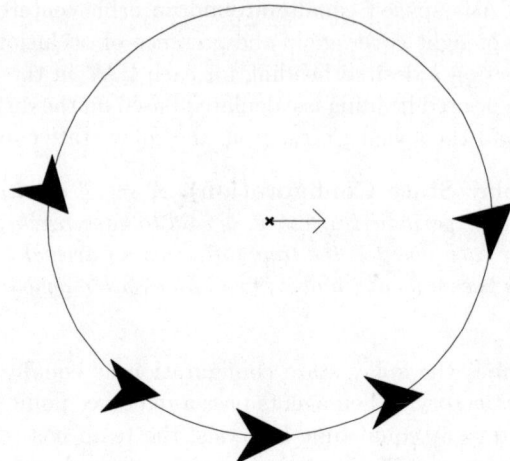

Fig. 2. A target moving at 75% of UAV speed has a splay state configuration with 5 vehicles that corresponds to the spacing in this figure. Note that at the bottom of the orbit, the target and the UAV are moving in the same direction, so the UAV slowly turns the corner. However, at the top of the orbit, the UAV and the target are moving in opposite directions, so the UAV quickly moves around the arc.

2.1 UAV Modeling

To maximize fuel efficiency each UAV maintains a constant airspeed. Additionally, we assume that all UAVs fly at a fixed altitude. A kinematic model for a constant airspeed, constant altitude UAV in wind, is given by

$$
\begin{aligned}
\dot{p}_N &= V_a \cos \psi + V_w \cos \psi_w \\
\dot{p}_E &= V_a \sin \psi + V_w \sin \psi_w \\
\dot{\psi} &= \frac{g}{V_a} \tan \phi \\
\dot{\phi} &= u
\end{aligned}
\tag{1}
$$

where (p_N, p_E) are the (North, East) coordinates of the UAV in a flat earth model, ψ is the heading of the UAV (with the $\dot{\psi}$ equation given by the coordinated turn assumption), ϕ is the roll angle, V_a is the constant airspeed of the vehicle, V_w is the magnitude of the wind vector and ψ_w is the heading of the wind vector (note that this is not the meteorological definition of wind heading, i.e. ψ_w is the direction the wind is blowing *to* as opposed to the direction the wind is blowing *from*). In addition to these dynamics, a constraint on roll angle $-\phi_{\max} \leq \phi \leq \phi_{\max}$ is enforced that stall conditions are avoided.

We consider the motion of the UAV relative to a target position. Let

$$
\begin{aligned}
x &= p_N - q_N \\
y &= p_E - q_E
\end{aligned}
\tag{2}
$$

where $(q_N,\ q_E)$ is the position of the target. The dynamics of (1) become

$$\dot{x} = V_a \cos \psi + W_x$$
$$\dot{y} = V_a \sin \psi + W_y$$
$$\dot{\psi} = \frac{g}{V_a} \tan \phi \tag{3}$$
$$\dot{\phi} = u$$

where $W_x = V_w \cos \psi_w - \dot{q}_N$ and $W_y = V_w \sin \psi_w - \dot{q}_E$. Target velocity and wind are indistinguishable with respect to the relative motion of the UAV to the target. This allows the control design to maintain constant airspeed and account for wind disturbances and target motion with only regard to (W_x, W_y).

Model (3) can be reduced further by letting

$$u = \frac{g V_a \dot{\omega}}{g^2 + V_a^2 \omega^2}$$

where ω is the heading rate of the UAV, i.e. $\omega = \frac{g}{V_a} \tan \phi$. Model (3) then becomes the kinematic unicycle model

$$\dot{x} = V_a \cos \psi + W_x$$
$$\dot{y} = V_a \sin \psi + W_y \tag{4}$$
$$\dot{\psi} = \omega$$

where we constrain $|\omega| \leq \frac{g}{V_a} \tan(\phi_{\max})$ to ensure that $|\phi| \leq \phi_{\max}$. The constraint on ω can be thought of as a curvature constraint on the system kinematics from which it follows that the UAV can be considered a Dubins-type vehicle. This model has shown great value for design of UAV systems as it captures the essential navigational kinematics of UAV motion while at the same time being of low enough order to allow tractable analysis [2] [6] [7].

The heading design and analysis is performed at a level of abstraction greater than the unicycle level by computing a desired heading ψ^d and using it as a feed-forward term to the model (4). Feedback is then introduced at the control signal ω while maintaining the saturation constraints on ω. Let

$$\omega = \dot{\psi}^d + \nu \tag{5}$$

where ν is the feedback term driving ψ to ψ^d. This chapter shows that ψ^d can be chosen so that a team of UAVs with individual dynamics

$$\dot{x} = V_a \cos \psi^d$$
$$\dot{y} = V_a \sin \psi^d \tag{6}$$

can reach the splay state configuration. Control gains in the calculation of ψ^d can then be chosen to allow the saturation constraints on ω to be satisfied. Note that ψ^d can be considered a sliding surface along which the specifications of the mission are satisfied. If ψ reaches ψ^d in finite time via the feedback term ν, then the overall system can be guaranteed to converge to the splay state configuration. Theoretically, a sliding mode controller of the form

$$\nu = \beta \text{sign}(\psi - \psi^d)$$

ensures that ψ reaches ψ^d in finite time, however in practice, a control law of the form

$$\nu = \beta \text{sat}\left(\frac{\psi - \psi^d}{\epsilon}\right)$$

is used, where β is a positive control gain. We do not show the overall system stability with this control strategy, but refer the reader to [8] where this choice of ν is shown to ensure path convergence for an arbitrary path in the single UAV case.

2.2   Orbit Dynamics

We will be concerned with the behavior of UAV teams while orbiting a target at a fixed radius R_{nom}. To analyze the stability of the orbit system, we make a change of variables by letting

$$R = \sqrt{x^2 + y^2}$$
$$\theta = \tan^{-1}\left(\frac{y}{x}\right) \tag{7}$$

where R is the distance of the UAV from the target and θ is the "clock angle" of the UAV around the orbit.

In the static target, no wind case (i.e. $W_x = W_y = 0$), the dynamics of R and θ can be calculated as follows. Let

$$\chi \triangleq \psi - \psi^p \tag{8}$$

be the difference between the actual heading, ψ, and the heading of the tangent vector to the orbit, i.e. $\psi^p = \theta + \pi/2$. Therefore $\dot{R}$ can be calculated as

$$\dot{R} = \frac{d}{dt}\sqrt{x^2 + y^2}$$
$$= \frac{x\dot{x} + y\dot{y}}{\sqrt{x^2 + y^2}}$$
$$= \frac{V_a}{R}[x\cos\psi + y\sin\psi] \ .$$

Since $\psi = \chi + \theta + \pi/2$, we obtain

$$\dot{R} = \frac{V_a}{R}[-x\sin(\chi + \theta) + y\cos(\chi + \theta)] \ .$$

Using the relations $\frac{x}{R} = \cos\theta$ and $\frac{y}{R} = \sin\theta$ we get that

$$\dot{R} = -V_a[\cos\theta\sin(\chi + \theta) - \sin\theta\cos(\chi + \theta)]$$
$$= -V_a\{\sin\chi\cos^2\theta + \cos\chi\sin\theta\cos\theta - \cos\chi\sin\theta\cos\theta + \sin\chi\sin^2\theta\}$$
$$\Rightarrow \dot{R} = -V_a\sin\chi \ .$$

Similar arguments are used to derive the equation of motion for θ resulting in

$$\dot{R} = -V_a\sin\chi$$
$$\dot{\theta} = \frac{V_a}{R}\cos\chi \ . \tag{9}$$

In the case of a moving target and/or wind, the motion is abstracted by calculating the path heading ψ^p, i.e. the heading which the UAV should be traveling if directly on the path. By accounting for target motion and wind via the ψ^p term, the radial orbit dynamics remain identical to those in (9) [8]. We show in Section 5 the calculation of ψ^p for moving targets.

To accommodate the multiple UAV splay state configuration, a spacing term is defined. For the static target, no wind scenario, the separation of the ith agent from the angular mean of its neighbors is

$$\delta\theta_i = \frac{1}{2}\left((\theta_i - \theta_{i-1}) - (\theta_{i+1} - \theta_i)\right) \tag{10}$$

where a ring topology is assumed (i.e. addition is defined modulo N). The term $\delta\theta_i$ captures how far away agent i is from being equally spaced between its two immediate neighbors on the ring. When all agents are on the nominal radius with spacing terms $\delta\theta_i$ equal to zero, then the team has achieved the splay state configuration. Although the calculation of $\delta\theta_i$ is more complicated in the moving target case, the principle is the same: $\delta\theta_i$ captures how far away from the splay state configuration agent i is with regards to its immediate neighbors along the ring.

A visual representation of the notation used to describe the desired heading calculation is shown in Figure 3 where d_i is the radial error from the nominal radius, i.e. $d_i \triangleq R_i - R_{\text{nom}}$.

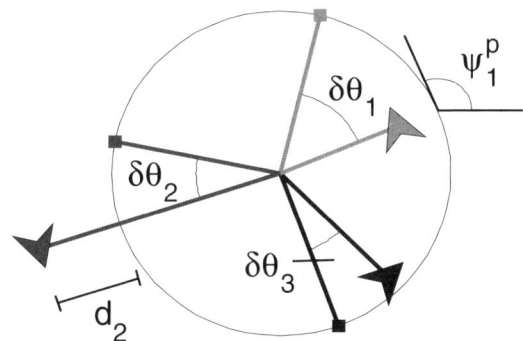

Fig. 3. Spacing error and radial error are combined to construct a desired heading for each UAV. Radial error is determined by the distance from the desired orbit (d_i) and spacing error is the distance from the angular center of an agent's two immediate neighbors ($\delta\theta_i$).

3   Heading Calculation for Non-moving Targets

This section details the construction of a desired heading to achieve the splay state configuration in the case of zero wind and a non-moving target. The basis of the splay state configuration controller is the calculation of an appropriate heading command that steers the agents to the proper steady state behavior. By creating a desired heading for the UAV, a reliable, robust heading controller can be used to track the heading commands. For a single UAV, a desired heading of the form

$$\psi^d = \psi^p + \tan^{-1}(kd) \tag{11}$$

will draw the agent onto the path, where d is the distance from the path and ψ^p is the heading along the path at $d = 0$ [8]. Using definition (8) equation (11) can be reduced to

$$\chi = \tan^{-1}(kd) \ . \tag{12}$$

Note that when d is large, the commanded heading is almost perpendicular to the heading along the path, effectively steering the UAV toward the path before beginning to follow it. For a simple orbit maneuver, ψ^p is selected to be tangent to the circle of interest along the ray connecting the agent and the target position. The radial distance of the agent from the nominal orbit constitutes d and a heading field constructed via (11) is shown in Figure 4. The gain k determines how aggressive the field is in steering the agent to the desired path.

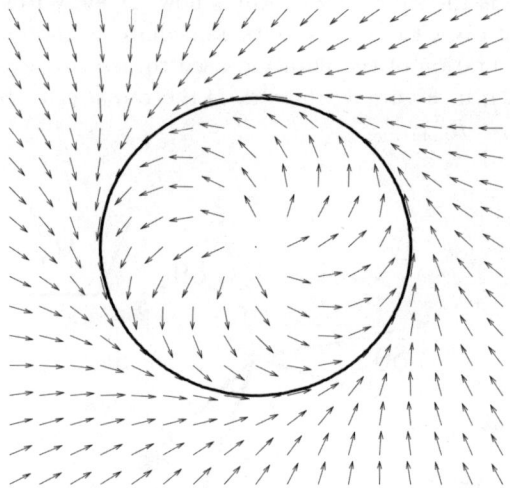

Fig. 4. A single UAV orbiting a stationary target has a commanded heading computed at each point given by (11). Note that when the agent is far from the orbit, the heading steers it toward the target. As it gets near the desired trajectory, the desired heading transitions to tangent to the nominal circular motion.

The constraint on ω is satisfied when

$$\max |\omega| = \max |\dot{\psi}^d| + \beta \le \omega_{\max}$$

where $\omega_{\max} = \frac{g}{V_a} \tan(\phi_{\max})$ and β is the maximum control allowed for the feedback control term (see Equation (5)). Due to the relationship in Equation (11), the term $\max |\dot{\psi}^d|$ can be bounded by

$$\max |\dot{\psi}^d| < \max |\dot{\psi}^p| + \max |\dot{\chi}| \ .$$

The term $\max |\dot{\psi}^p|$ can be determined using *a priori* knowledge or an estimate of the path to be tracked (e.g. moving orbit, straight line, etc.); for stationary orbits, $|\dot{\psi}^p| = V_a/R_{\mathrm{nom}}$. The term $\max |\dot{\chi}|$ directly depends on the strength of the field through the gain k. Recalling that $\chi = \tan^{-1}(kd)$ gives

$$|\dot{\chi}| = \left| \frac{k\dot{d}}{1 + (kd)^2} \right| = \left| \frac{-kV_a \sin \chi}{1 + (kd)^2} \right| \leq kV_a$$

which when coupled with knowledge of $\dot{\psi}^p$, the gain k can be chosen so as not to violate the UAV turn rate/roll angle constraints.

For a single UAV, a commanded heading of the form $\chi = \tan^{-1}(kd)$ guarantees asymptotic convergence to an orbit at radius R_{nom} about the target. A simple Lyapunov argument supports this assertion. Letting $W = \frac{1}{2}\chi^2$ and using (9) gives

$$\dot{W} = \chi\dot{\chi} = \frac{-kV_a\chi \sin \chi}{1 + (kd)^2} \quad . \tag{13}$$

Since $\chi \in (-\pi/2, \pi/2)$ (χ is the output of an inverse tangent), the term $\chi \sin \chi$ is always greater than zero for nonzero χ. Therefore, $\dot{W} < 0$ and $\chi \to 0$ asymptotically. By LaSalle's invariance principle [9], it follows that $d \to 0$. Again we note that a complete proof for system (4) requires a sliding mode controller to guarantee that ψ reaches ψ^d in finite time, however, this can be relaxed as in [8]. Qualitatively, the commanded heading simply points the UAV directly toward the target if d is large and transitions to tangent to the orbit when near R_{nom}.

To account for spacing, the single agent heading command (11) is augmented as

$$\psi_i^d = \psi_i^p + \tan^{-1}(kd_i - \gamma\delta\theta_i) \tag{14}$$

where γ is a control gain weighting the value of spacing the UAVs to the value of converging to R_{nom}. The spacing term effectively increases the radius of the orbit when a UAV is too close to the agent in front of it and decreases the radius of the orbit if it is behind. This allows agents to "catch up" when the spacing is not at the desired state. An example of the heading field for an agent when $\delta\theta = \pi/2$ is shown in Figure 5. Notice the agent is drawn away from the nominal radius to allow the agent in front to increase its angular separation.

By constructing $\delta\theta_i$ to be only a function of its immediate neighbors, the error signal (heading field calculation) is local to each agent in the system. This allows the implementation to be completely decentralized. The advantage to decentralization is that the overall system will scale to any number of agents and be robust to insertion and deletion of team members. When agents are tasked to leave the formation for high priority assignments, the rest of the group can adjust to a new configuration without any centralized planning. Similarly, if a new agent is added (e.g. returns from a high priority task) the group will adjust through local interaction without any global communication.

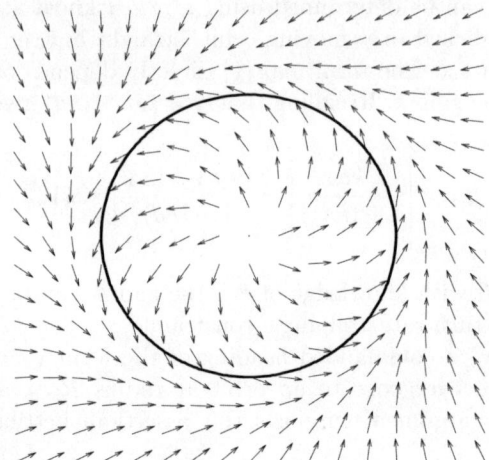

Fig. 5. A single UAV orbiting a stationary target with spacing error $\pi/2$ has desired heading given by (14). Note that a positive spacing error will cause the agent to effectively increase its radius, allowing the neighbor in front to gain distance and increase their relative spacing.

4   Stability Analysis

In the static target, no wind case, the splay state configuration coincides with the team members being equally spaced around an orbit. This section investigates the stability of the entire system when each agent follows the heading defined by (14). Figure 6 shows the behavior exhibited by a team of three UAVs.

A complete Lyapunov argument (or other method) may be used to determine the stability of the system to the splay state configuration. We have been unable to find a Lyapunov function that shows the stability of the entire system. For this reason, the convergence of the team of UAVs using (14) to the splay state configuration is argued as follows. We first show that the radial error is bounded by a function of the control gains k and γ. Near equilibrium, the overall system is shown to be exponentially stable. Finally, Monte-Carlo simulations are used to investigate system stability for initial conditions lying in the bounded region.

4.1   Ultimately Bounded

Lemma 1. *The system of agents described by (6) when following heading (14) is ultimately bounded in radial error d_i, i.e.*

$$|d_i| \leq R_\delta \tag{15}$$

where $R_\delta \triangleq \gamma\pi/k$ is less than R_{nom}.

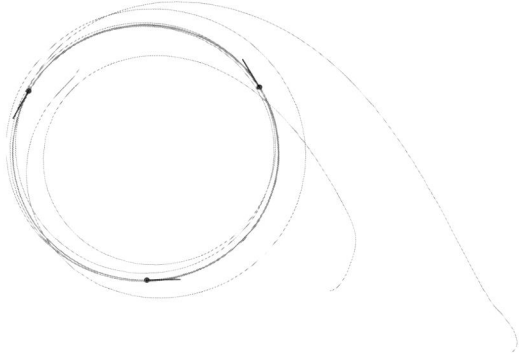

Fig. 6. Three UAVs following the heading defined by (14) converge to the splay state configuration along a non-moving orbit

Proof. For any agent, $\delta\theta_i$ is constrained to the region $(-\pi, \pi)$, i.e. agent cannot have an angular spacing error greater than π radians. If $|d_i| > R_\delta$, then

$$\text{sign}(k_i d_i - \gamma\delta\theta_i) = \text{sign}(d_i)$$
$$\Rightarrow \text{sign}(\chi_i) = \text{sign}(d_i)$$
$$\Rightarrow \text{sign}(\sin\chi_i) = \text{sign}(d_i)$$
$$\Rightarrow \text{sign}(-V_a \sin\chi_i) = \text{sign}(-d_i)$$
$$\Rightarrow \text{sign}(\dot{d}_i) = \text{sign}(-d_i)$$
$$\Rightarrow d_i\dot{d}_i < 0 \ .$$

Therefore, the Lyapunov function $W = d_i^2$ has a negative definite derivative whenever d_i is outside the bound (15). When $|d_i| > \gamma\pi/k$, the kd_i terms dominates the $\gamma\delta\theta_i$ term in (14) effectively steering the UAV to reduce radial error regardless of spacing error. Therefore, $|d|$ is decreasing when $|d| > \gamma\pi/k$ and so all d_i are ultimately bounded to the region $(-R_\delta, R_\delta)$.

4.2   Local Stability

The splay state configuration in the no wind, non-moving target case corresponds to all the UAVs traveling on the orbit equally spaced, i.e. $d_i = 0$ and $\delta\theta_i = 0$ for all agents on the team. The change of variables introduced in Section 2.2 allows analysis of the system dynamics where each UAV has equations of motion determined by (9). Rewriting (9) using the definition of $\delta\theta_i$ in (10) to evaluate the error signals for each agent, we obtain

$$\begin{aligned}\dot{d}_i &= -V_a \sin\chi_i \\ \dot{\delta\theta}_i &= \frac{V_a}{R_i}\cos\chi_i - \frac{1}{2}\left[\frac{V_a}{R_{i+1}}\cos\chi_{i+1} + \frac{V_a}{R_{i-1}}\cos\chi_{i-1}\right]\end{aligned} \qquad (16)$$

In the calculation of the linearization of (16), it is helpful to compute the partial derivatives of χ_i with respect to the system state variables d_i and $\delta\theta_i$. Since

$\chi_i = \tan^{-1}(kd_i - \gamma\delta\theta_i)$, the partial derivatives evaluated at the equilibrium point $d_i = 0$, $\delta\theta_i = 0$ are calculated as

$$
\begin{aligned}
\frac{\partial \chi_i}{\partial d_i} &= k \\
\frac{\partial \chi_i}{\partial d_{\neg i}} &= 0 \\
\frac{\partial \chi_i}{\partial \delta\theta_i} &= -\gamma \\
\frac{\partial \chi_i}{\partial \delta\theta_{\neg i}} &= 0
\end{aligned}
\tag{17}
$$

where $\neg i$ represents any value in $\mathcal{I}$ not equal to i. The partial derivative of $\dot{d}_i$ can be calculated as

$$
\frac{\partial}{\partial *}\left(\dot{d}_i\right) = \frac{\partial}{\partial *}\left(-V_a \sin \chi_i\right) = -V_a \cos \chi_i \left(\frac{\partial}{\partial *}\chi_i\right) . \tag{18}
$$

The matrix composing the partial derivatives of the system dynamics (16) has the structure

$$
F = \begin{bmatrix} A & B \\ C & D \end{bmatrix} \triangleq \left[\begin{array}{c|c} \frac{\partial}{\partial d_i}(\dot{d}_i) & \frac{\partial}{\partial \delta\theta_i}(\dot{d}_i) \\ \hline \frac{\partial}{\partial d_i}(\dot{\delta\theta}_i) & \frac{\partial}{\partial \delta\theta_i}(\dot{\delta\theta}_i) \end{array}\right] . \tag{19}
$$

Combining (18) with (17), the matrices A and B are calculated as $A = -kV_a I_N$ and $B = \gamma V_a I_N$ where I_N is the $N \times N$ identity matrix.

The linearization of the $\delta\theta$ dynamics reveals the ring structure inherent in the spacing calculation used to construct the desired heading. The function $\dot{\delta\theta}_i$ is composed of terms

$$
\frac{V_a}{R_i} \cos \chi_i
$$

which when linearized become

$$
\frac{V_a}{R_i^2}\left(\frac{\partial}{\partial *}R_i\right)\cos \chi_i - \frac{V_a}{R_i}\sin \chi_i \left(\frac{\partial}{\partial *}\chi_i\right) .
$$

At the equilibrium, the only term that does not become zero is the term containing $\partial R_i/\partial d_i$. Note that since R_i does not depend on $\delta\theta_i$, the partial derivative with respect to $\delta\theta_i$ will be zero. The linearized dynamics of $\delta\theta_i$ become

$$
\begin{aligned}
\frac{\partial}{\partial d_i}\left(\dot{\delta\theta}_i\right) &= \frac{-V_a}{R_{\mathrm{nom}}^2} \\
\frac{\partial}{\partial d_{i\pm 1}}\left(\dot{\delta\theta}_i\right) &= \tfrac{1}{2}\frac{V_a}{R_{\mathrm{nom}}^2} \\
\frac{\partial}{\partial \delta\theta_i}\left(\dot{\delta\theta}_{i,\neg i}\right) &= 0 .
\end{aligned}
\tag{20}
$$

We conclude that the matrix D in (19) is simply the zero matrix of size $N \times N$ and matrix C is a circulant matrix

$$
C = \frac{1}{2}\frac{V_a}{R_{\mathrm{nom}}^2}\begin{bmatrix} -2 & 1 & 0 & \cdots & 0 & 1 \\ 1 & -2 & 1 & \cdots & 0 & 0 \\ \vdots & \vdots & \vdots & \ddots & \vdots & \vdots \\ 1 & 0 & 0 & \cdots & 1 & -2 \end{bmatrix} . \tag{21}
$$

Of particular note is the structure of C

$$C = \frac{1}{2} \frac{V_a}{R_{\text{nom}}^2} \left(-2I_N + \mathcal{C}_N\right) \qquad (22)$$

where

$$\mathcal{C}_N = \begin{bmatrix} 0 & 1 & 0 & \cdots & 0 & 1 \\ 1 & 0 & 1 & \cdots & 0 & 0 \\ \vdots & \vdots & \vdots & \ddots & \vdots & \vdots \\ 1 & 0 & 0 & \cdots & 1 & 0 \end{bmatrix} \qquad (23)$$

is the adjacency matrix corresponding to the ring graph of size N. The eigenvalues of F can be formulated in terms of the eigenvalues of C which are known using results from algebraic graph theory [10].

Lemma 2. *Consider the matrix*

$$F = \begin{bmatrix} -kV_a I_N & \gamma V_a I_N \\ C & 0_N \end{bmatrix} \qquad (24)$$

where C is given by (21), I_N is the $N \times N$ identity matrix and 0_N is an $N \times N$ matrix of zeros. The eigenvalues of F are given by

$$\lambda_j = -\frac{1}{2}kV_a \pm \sqrt{\left(\frac{1}{2}kV_a\right)^2 + \gamma V_a \mu_j} \quad \text{for } j = 1 \ldots N \qquad (25)$$

where

$$\mu_j = \frac{1}{2} \frac{V_a}{R_{\text{nom}}^2} \left(2\cos\left(\frac{2\pi}{N}(j-1)\right) - 2\right) \qquad (26)$$

is an eigenvalue of C.

Proof. We begin by showing that the eigenvalues of C are given by (26). From (22) we conclude that

$$\mu_j = \frac{1}{2} \frac{V_a}{R_{\text{nom}}^2} \left(-2 + \gamma_j\right)$$

where γ_j is an eigenvalue of $\mathcal{C}_N$. Results from algebraic graph theory show that the eigenvalues of $\mathcal{C}_N$ are

$$\gamma_j = 2\cos\left(\frac{2\pi}{N}(j-1)\right) \quad \text{for } j = 1 \ldots N \ .$$

Let λ be an eigenvalue of F and x its corresponding eigenvector. Partition x into blocks corresponding with the blocks of F, i.e. $x = \begin{bmatrix} x_d^T & x_{\delta\theta}^T \end{bmatrix}^T$ where both x_d and $x_{\delta\theta}$ are of length N. The eigenvector relationship $Fx = \lambda x$ can be written

$$-kV_a x_d + \gamma V_a x_{\delta\theta} = \lambda x_d \Rightarrow \gamma V_a x_{\delta\theta} = (\lambda + kV_a) x_d \qquad (27)$$

$$Cx_d = \lambda x_{\delta\theta} \ . \qquad (28)$$

From (27) we see that

$$x_{\delta\theta} = \frac{\lambda + kV_a}{\gamma V_a} x_d \tag{29}$$

which when applied to (28) yeilds

$$Cx_d = \left(\frac{\lambda(\lambda + kV_a)}{\gamma V_a} \right) x_d \ .$$

Note that this is exactly the eigenvector relationship for the matrix C where $Cx = \mu x$ for

$$\mu = \left(\frac{\lambda(\lambda + kV_a)}{\gamma V_a} \right) \ .$$

Solving this for λ yields Equation (25).

Theorem 1. *Consider the matrix F as defined in (24). All eigenvalues except for $\lambda = 0$ of F are located in the open left half plane. Additionally, the eigenvectors associated with $\lambda = 0$ and $\lambda = -kV_a$ span a subspace of $\mathbb{R}^{2N}$ orthogonal to the remaining $2N - 2$ eigenvectors of F.*

Proof. Equation (25) gives the relationship of the eigenvalues of F to the eigenvalues of C. Only a single eigenvalue of C is equal to zero, all other $N - 1$ values are strictly less than zero. The zero eigenvalue in C maps to the eigenvalues $\lambda = -kV_a$ and $\lambda = 0$ in F. The remaining eigenvalues of C (all strictly less than zero) have discriminant strictly less than $(\frac{1}{2}kV_a)^2$ thus ensuring that each λ has real part in the open left half plane.

The proof of Lemma 2 gives the relationship between the eigenvectors of C and those of F via (29) where x_d is the eigenvector of C corresponding to eigenvalue

$$\mu = \left(\frac{\lambda(\lambda + kV_a)}{\gamma V_a} \right) \ .$$

Since C is a symmetric matrix, its eigenvectors form an orthonormal basis of $\mathbb{R}^N$. Note that C has constant row sums of zero, so the eigenvector associated with the zero eigenvalue of C is the vector of all ones, **1**. Due to the orthogonality of the eigenvectors of C, $\mathbf{1}^T u_j = 0$ for all eigenvectors of C, $u_j \neq \mathbf{1}$. Using (29), the eigenvectors for $\lambda = 0$ and $\lambda = -kV_a$ are

$$x_0 = \begin{bmatrix} \mathbf{1} \\ \frac{k}{\gamma}\mathbf{1} \end{bmatrix}, \quad x_{-kV_a} = \begin{bmatrix} \mathbf{1} \\ \mathbf{0} \end{bmatrix} \ . \tag{30}$$

The inner product of these eigenvectors with all other eigenvectors of F can be written as

$$\begin{bmatrix} \mathbf{1}^T & \frac{k}{\gamma}\mathbf{1}^T \end{bmatrix} \begin{bmatrix} u_j \\ \frac{\lambda + kV_a}{\gamma V_a} u_j \end{bmatrix} = 0 \quad \text{and} \quad \begin{bmatrix} \mathbf{1}^T & \mathbf{0}^T \end{bmatrix} \begin{bmatrix} u_j \\ \frac{\lambda + kV_a}{\gamma V_a} u_j \end{bmatrix} = 0 \ .$$

Corollary 1. *The linearization of system (16) is exponentially stable.*

Proof. Linearization of (16) yields the state equation $\dot{x} = Fx$ where F is given in equation (24), and whose solution is $x(t) = e^{Ft}x_0$. By Theorem 1 all but one eigenvalue is in the open left half plane, so any part of the initial condition x_0 that lies in the span of the eigenvectors associated with those eigenvalues exponentially decays to zero. By definition of $\delta\theta_i$, the constraint

$$\sum_{i=1}^{N} \delta\theta_i = 0 \tag{31}$$

must hold for any state vector associated with the original system. The eigenvectors associated with $\lambda = 0$ and $\lambda = -kV_a$ are given in (30). These eigenvectors form a subspace orthogonal to all other eigenvectors in the linearized system. To lie in the subspace spanned by the eigenvectors (30), all $\delta\theta_i$ must be equal. However, the only vector $\boldsymbol{\delta\theta}$ that satisfies the constraint (31) and is in this subspace is $\boldsymbol{\delta\theta} = \mathbf{0}$, which is either along the eigenvector associated with $\lambda = -kV_a$ or in the subspace spanned by the remaining eigenvectors of the system. In other words, it is impossible to have an initial condition in the subspace spanned by the eigenvector associated with $\lambda = 0$. Therefore, the initial condition x_0 lies in the space spanned by eigenvectors whose eigenvalues are in the open left half plane and the linearized system is exponentially stable.

4.3   Global Stability

The system (16) is ultimately bounded to $d_i \in (-R_\delta,\ R_\delta)$, $\delta\theta_i \in (-\pi,\ \pi)$ and locally asymptotically stable. Monte-Carlo simulations are used to infer the stability of the system in the remaining region between the ultimate bound and the equilibrium path.

The Monte-Carlo simulations use the model (4) with desired heading given by (14). For team sizes $N = 2, 3, 4, 5,$ and 6, a set of 10,000 simulations with random initial conditions in d_i and $\delta\theta_i$ were run to verify the stability of the system. An error metric

$$e(t) = \sqrt{\sum_{i=1}^{N} d_i(t)^2 + \delta\theta_i(t)^2}$$

captures the error from the splay state configuration at time t. The largest error at $t = 100$ seconds over all 50,000 simulations was $2e^{-4}$ indicating that the actual region of convergence is likely to be global.

5   Extension to Moving Targets

The ability for a UAV to orbit a target in the presence of wind or target motion is crucial. Modifications to the static target, no wind case can be made to allow UAVs to track moving targets.

To extend the approach of (14) to moving targets, the path heading term ψ^p must be calculated to allow a UAV to remain on a moving orbit. Essentially, the steady state behavior of a UAV on the orbit is determined by ψ^p: while following ψ^p at $d = 0$, a UAV should remain on the moving orbit.

Consider the behavior of a particle orbiting a constant speed target at fixed radius R_n then

$$
\begin{aligned}
x^p(t) &= R_n \cos(\theta(t)) + W_x t \\
y^p(t) &= R_n \sin(\theta(t)) + W_y t
\end{aligned}
\tag{32}
$$

where W_x and W_y are the velocity of the orbit center. Differentiating (32) results in the expression

$$
\begin{aligned}
\dot{x}^p &= -R_n \dot{\theta} \sin\theta + W_x \\
\dot{y}^p &= R_n \dot{\theta} \cos\theta + W_y \ .
\end{aligned}
\tag{33}
$$

The path heading is chosen as

$$
\psi^p = \tan^{-1}\left(\frac{\dot{y}^p}{\dot{x}^p}\right)
\tag{34}
$$

which is the direction of the vector that is tangent to the moving orbit. To ensure that the UAV maintains constant airspeed, the magnitude of the tangent vector must equal V. This constraint allows the calculation of $\dot{\theta}$ from (33) as

$$
\begin{aligned}
V_a^2 = (\dot{x}^p)^2 + (\dot{y}^p)^2 &= \left(-R_n \dot{\theta} \sin\theta + W_x\right)^2 + \left(R_n \dot{\theta} \cos\theta + W_y\right)^2 \\
\Rightarrow \dot{\theta}^2 \left(R_n^2\right) + \dot{\theta}\left(2R_n W_y \cos\theta - 2R_n W_x \sin\theta\right) &+ \left(W_x^2 + W_y^2 - V_a^2\right) = 0
\end{aligned}
$$

$$
\begin{aligned}
\Rightarrow \dot{\theta} = -\tfrac{1}{R_n}\left(W_y \cos\theta - W_x \sin\theta\right) \pm \\
\tfrac{1}{R_n}\sqrt{\left(W_y \cos\theta - W_x \sin\theta\right)^2 - \left(W_x^2 + W_y^2 - V_a^2\right)} \ .
\end{aligned}
\tag{35}
$$

The discriminant in (35) shows that when the magnitude of the velocity of the target is greater than the speed of the UAV, a real solution does not exist. In practical terms, this means that for the agent to properly maintain its orbit around the target, the speed of the wind plus the speed of the target cannot exceed the speed of the UAV.

The turn rate constraint of the UAV must also be accounted for in determining the allowable magnitude of motion that can be feasibly tracked. Disregarding the other components of heading rate,

$$
\left|\dot{\psi}^p\right| \le \frac{g}{V_a} \tan(\phi_{\max})
\tag{36}
$$

ensures that the path satisfies the turn rate constraints. The maximum value of $\dot{\psi}^p$ depends on V_w, the magnitude of the motion in the system (note $V_w^2 = W_x^2 + W_y^2$). To ensure that the orbit can feasibly be followed with regard to the turn constraints of the UAV, V_w must satisfy

$$
\frac{(2V_w + V_a)(V_w + V_a)^2}{R_n V_a^2} \le \frac{g}{V_a} \tan(\phi_{\max}) \ .
\tag{37}
$$

Intuitively, a UAV can follow a moving target in wind if the magnitude of the wind and target velocity are not too great to violate the velocity or turn rate constraints of the UAV. For example, a UAV with maximum bank angle of 35 degrees, airspeed of 15 meters per second and desired orbit of 100 meters can track a target with speed less than 5.17 m/s.

With ψ^p determined by (34), a desired heading of (11) can be used for a single UAV to follow a moving target in the presence of wind given that the turn rate constraint of the UAV is satisfied. For multiple UAVs, the definition of the splay state configuration is used to develop a spacing error term. Note that achieving equal angle spacing around a moving orbit is impossible when the velocity of the UAVs is held constant. For this reason, the actual time along the steady-state orbit between neighbors is used to compute the error from the splay state configuration. Similar to the static target case, the timing error is computed by assuming that all UAVs are on the desired orbit (i.e. $d_i = 0$). Consider two agents on the orbit with clock angles θ_i and θ_j. The time difference from agent i to agent j is given by $T_{i \rightarrow j} = t - t_0$ such that $\theta(t) = \theta_j$ where $\theta(t)$ is determined by solving the initial value problem

$$\dot{\theta} = -\tfrac{1}{R_n}\left(W_y \cos\theta - W_x \sin\theta\right) \pm$$
$$\tfrac{1}{R_n}\sqrt{\left(W_y \cos\theta - W_x \sin\theta\right)^2 - \left(W_x^2 + W_y^2 - V_a^2\right)} \qquad (38)$$
$$\theta(t_0) = \theta_i \ .$$

The timing error for a specific agent i can then be defined as

$$\delta t_i = \frac{1}{2}\left(T_{(i-1)\rightarrow i} - T_{i \rightarrow (i+1)}\right) \ . \qquad (39)$$

The δt term is used in exactly the same manner as the $\delta\theta$ term in the static target case, i.e. a desired heading is calculated as

$$\psi_i^d = \psi_i^p + \tan^{-1}(kd_i - \gamma\delta t_i) \ . \qquad (40)$$

Many of the stability notions from the non-moving target case carry over to the moving target case. A maximum δt exists since agents can only be of finite angle apart. Therefore, for large errors in radial distance d, the kd_i term will dominate the heading calculation and force the system to be ultimately bounded. A linearization of the system dynamics for the moving target case also shows many similarities to static case. In particular the upper two blocks of the state matrix are identical to the blocks in the static target linearization. We postulate that the lower blocks are identical up to a positive scale factor, i.e. the circulant structure of the lower left block is preserved which allows us to conclude linear stability via the same arguments as in the static target case. Additionally, Monte-Carlo simulations are used to indicate that the system converges to the splay state configuration in the moving target case. For team sizes $N = 2, 3$, and 4, a set of 1,000 simulations with random initial conditions in d_i, δt_i and V_w were run to verify the stability of the system. An error metric

$$e(t) = \sqrt{\sum_{i=1}^{N} \delta t_i(t)^2}$$

captures the error from the splay state configuration at time t. The largest error at $t = 100$ seconds over 3,000 simulations was 0.5 indicating that control (40) leads to convergence to the splay state configuration. Figure 7 shows typical behavior of 4 UAVs orbiting a moving target. The timing error from the splay state configuration for this scenario is shown in Figure 8.

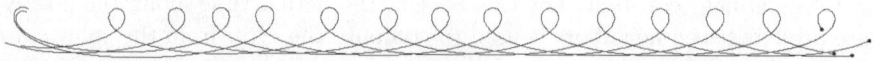

Fig. 7. Trajectories of 4 UAVs orbiting a moving target trace out routes similar to those in this figure

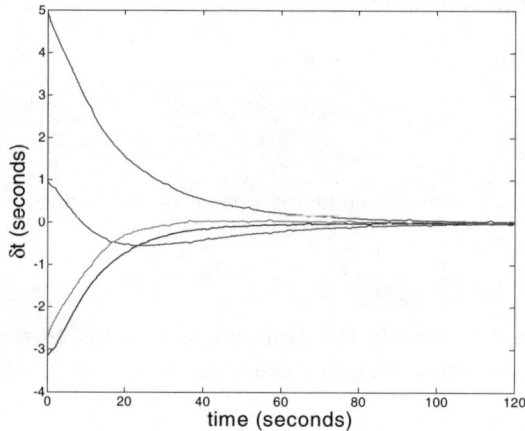

Fig. 8. Error from the splay state configuration for 4 UAVs tracking a moving target is driven to zero using (40)

6   Simulation Results

The splay state controller is based upon choosing a heading that draws the UAVs to the splay state configuration. The design of the heading command is accomplished by assuming a simple kinematic model given by (4). To validate the design, the splay state controller is tested in high fidelity simulation. Each UAV is simulated with full 6 degree of freedom dynamics model with aerodynamic parameters that match the small UAVs flown at BYU [11]. Additionally, the human interface and autopilot code are emulated to match actual flight conditions as closely as possible.

Trajectories of 3 UAVs that loiter at fixed locations and are then commanded to reach the splay state configuration are shown in Figure 9. The radial error of a UAV is approximately one meter and the spacing error about 3 degrees. These errors are due mainly to the update rate of the team - each UAV only communicates to its neighbors when a new GPS packet is received at approximately 1 Hertz.

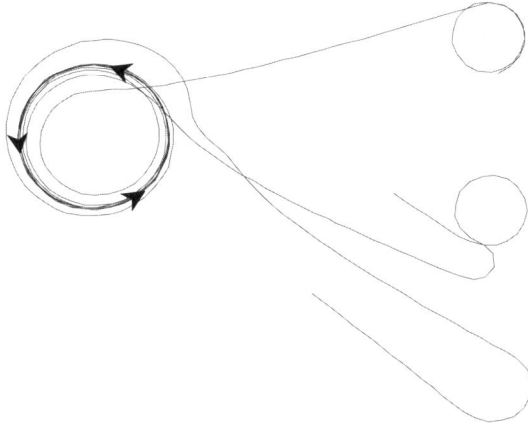

Fig. 9. High fidelity simulation results of the splay state controller indicate that the method can be effective in actual implementation

Despite design of the splay state controller in a low-order environment, application of the control in high fidelity simulation shows that the splay state controller may be effective in hardware implementation.

7   Conclusions and Future Work

This chapter has developed a decentralized splay state controller for a team of UAVs monitoring a target. In the static case (i.e. non-moving target and no wind), the controller spaces UAVs equally around an orbit centered on the target. The decentralized nature of the control strategy allows the the team to be robust to insertion, deletion and re-assignment of team members. The controller is shown to be linearly stable in the static target case and Monte-Carlo simulations indicate global stability in all cases. By defining an appropriate measure of spacing around the orbit, the splay state configuration can be reached for moving targets in the presence of wind. High fidelity simulation results show that the controller may be practical in actual hardware implementation.

There are still many open questions in regards to the convergence of a team of UAVs to the splay state configuration. Monte-Carlo simulations indicate that the region between the ultimate bound and the equilibrium is stable, but a formal proof of this assertion remains an open problem. Additionally, the design of the commanded heading is based on a low-order UAV model. Extending the analysis to the model (1) and finding an appropriate control u, rather than relying on a sliding mode inner-loop control, is also an important extension.

Acknowledgements

This work is partially funded by the National Science Foundation under Information Technology Research Grant CCR-0313056 and by the Air Force Office of Scientific Research award No. FA9550-04-0209.

References

1. Frew, E.W., Lawrence, D.A.: Cooperative stand-off tracking of moving targets by a team of autonomous aircraft. In: Proceedings of the AIAA Guidance, Navigation, and Control Conference. (2005)
2. Tang, Z., Ozguner, U.: Motion planning for multitarget surveillance with mobile sensor agents. IEEE Transactions on Robotics (October 2005)
3. Gu, G., Chandler, P.R., Schumacher, C., Sparks, A., Pachter, M.: Optimal cooperative sensing using a team of UAVs. IEEE Transactions on Aerospace and Electronic Systems (October 2006)
4. Sepulchre, R., Paley, D.A., Leonard, N.E.: Stabilization of planar collective motion: All-to-all communication. IEEE Transactions on Automatic Control (June 2007)
5. Klein, D.J., Morgansen, K.A.: Controlled collective motion for trajectory tracking. In: Proceedings of the IEEE American Control Conference. (2006)
6. Yang, G., Kapila, V.: Optimal path planning for unmanned air vehicles with kinematic and tactical constraints. In: Proceedings of the IEEE Conference on Decision and Control. (2002)
7. Chandler, P.R., Pachter, M., Rasmussen, S.: UAV cooperative control. In: Proceedings of the IEEE American Control Conference. (2001)
8. Griffiths, S.R.: Vector field approach for curved path following for miniature aerial vehicles. In: Proceedings of the AIAA Guidance, Navigation, and Control Conference. (2006)
9. Khalil, H.K.: Nonlinear Systems. Prentice Hall (1996) Theorem 4.4, p. 128.
10. Godsil, C., Royle, G.: Algebraic Graph Theory. Springer-Verlag New York, Inc. (2001)
11. Beard, R., Kingston, D., Quigley, M., Snyder, D., Christiansen, R., Johnson, W., Mclain, T., Goodrich, M.: Autonomous vehicle technologies for small fixed wing UAVs. AIAA Journal of Aerospace Computing, Information, and Communication (2005)

A Risk-Based Approach to Sensor Resource Management

Dimitri Papageorgiou and Maxim Raykin

The Raytheon Company
Integrated Defense Systems
Woburn, MA 01801
{Dimitri_J_Papageorgiou,Maxim_Raykin}@raytheon.com

Abstract. We investigate the benefits of employing a suitable risk-based metric to determine in real-time the high level actions that an agile sensor should execute during a mission. Faced with a barrage of competing goals, a sensor resource manager must optimize system performance while simultaneously meeting all requirements. Numerous authors advocate the use of information-theoretic measures for driving sensor tasking algorithms, wherein the relative value of different sensing actions is calculated in terms of the expected gain in information. In this chapter, motivated by the sensor resource allocation problem in missile defense, we deviate from the information-based trend and propose an approach for determining sensor tasking decisions based on risk, or expected loss of defended assets. We present results of a missile defense simulation that illustrate the advantages of our risk-based objective function over its information-theoretic and rule-based counterparts.

Keywords: Sensor Resource Management, Risk, Missile Defense[1].

1   Introduction

This chapter addresses what we refer to as the sensor resource allocation problem, the problem of tasking a multi-modal sensor to perform high level sensing actions, e.g., search, track maintenance, and discrimination, over the course of a mission. A multi-modal sensor can collect data on objects and areas of interest using a variety of sensing modalities. This flexibility has led to marked improvements in detection of targets, kinematic estimation, and classification capabilities. At the same time, this large palette of sensing actions has also introduced challenges concerning the timely and efficient use of limited sensing resources. This chapter focuses on a specific sensor resource management problem that appears in the context of ballistic missile defense (BMD).

To date, a majority of sensor systems employ a prioritization scheme to determine which actions should be taken during a data collection interval. In this approach, a slew of Boolean conditions are quickly checked and then actions are

---

[1] The United States Missile Defense Agency approved this work for public release (07-MDA-2387).

M.J. Hirsch et al. (Eds.): Adv. in Cooper. Ctrl. & Optimization, LNCIS 369, pp. 129–144, 2007.
springerlink.com   © Springer-Verlag Berlin Heidelberg 2007

chosen based on some pre-specified action plan. While this method has its advantages – requirements can be plainly stated, Boolean conditions are typically easy to verify in real-time, and the decision chain is traceable – it suffers in several notable ways. First, an approach based on a fixed set of rules cannot be completely adaptable to the evolving battlespace environment and can therefore be far from optimal. Second, adjustment of priorities is a delicate, time-consuming procedure. As new requirements are introduced into the system, it is difficult to ensure that the conditions are appropriately agreed upon so that a new requirement's priority is properly set. Third, glitches and irregularities in algorithm behavior may be difficult to diagnose due to tacit assumptions and makeshift implementation choices.

To circumvent these deficiencies, numerous suggestions have been propounded that provide a single metric able to automatically and simultaneously capture the complex tradeoffs involved when choosing between sensor allocations. A metric that has received considerable attention is entropy, which attempts to measure the uncertainty associated with random variables of interest. Within this information-theoretic framework, authors typically focus on Shannon entropy [6,12,15], Kullback-Leibler divergence [8,11,17,18], and Rényi divergence [9,10]. In this approach, sensor tasking decisions are made based on the principle that actions should be chosen to maximize the information expected to be extracted from the scene of interest. Within a Bayesian estimation framework, a good measure of the quality of a sensing action is the reduction in entropy of the posterior distribution that is expected to be induced by a measurement. For instance, when evaluating the benefits of a track update (or propagation without an update), these algorithms use the logarithm of the ratio of the determinants of the a priori and a posteriori covariance matrices as a measure of sensor effectiveness.

Although the use of entropy for judging a sensor's performance can be justified for a conventional battlefield situation, we believe that it is not the most suitable metric for BMD. Indeed, for traditional military applications, e.g., surveillance of enemy troops, ground target tracking, etc., the battlespace has infinite variability and the ultimate objective often cannot be stated precisely. The goal of a sensor or system of sensors in this situation may only be to maximize the amount of information collected for subsequent use in decision making. In contrast, the situation arising in BMD can be stated in precise terms (i.e., we have finite number of objects, each with finite degrees of freedom) and the underlying goal of BMD is clear and always the same – to minimize our losses from an enemy's missile attack. We will use the term *risk* for the expected value of this loss and consider risk reduction as a driver for a sensor's action, by which its performance should be judged. We briefly note that other authors [3,4,19] have considered a similar metric, but they may define it in different terms or apply it in different contexts.

The chapter is organized as follows. In the next section, we introduce discrimination risk and discuss the meaning of cost coefficients. The expressions for risk reduction are derived in Section 3. In Section 4, we describe a myopic approach to scheduling based on risk reduction and a heuristic approach to non-myopic scheduling. In Section 5, we present the results of a missile defense simulation that compares our risk-based approach with other competing methods. Conclusions are presented in Section 6.

2   Discrimination Risk and the Cost Coefficients

Consider the following situation: an object is to be classified into one of two classes, C_1 (threat) or C_2 (nonthreat). Let p_i be the current probability that the object belongs to class C_i, $i = 1, 2$. An object that is classified as a threat will be fired upon and destroyed by an interceptor, while an object that is classified as a nonthreat will be left unscathed. Let c_{12} denote the cost of an interceptor and c_{21} the cost of leakage, i.e., the cost of misclassifying a threat as a nonthreat. Then, the risk of declaring an object as belonging to class i, for $i = 1, 2$, is given by

$$R_1 = c_{12} \, ,$$
$$R_2 = c_{21} p_1 \, . \tag{2.1}$$

Note that while R_2 depends on the probability p_1, R_1 does not depend on a probability because once an interceptor is launched, its cost is incurred regardless of whether or not the object was a threat.

The decision rule for class selection minimizes the risk R; that is, the object is declared to belong to the class C_i with the smallest R_i:

$$R = \min_{i=1,2} R_i = \min \left(c_{12}, \, c_{21} p_1 \right). \tag{2.2}$$

We assume that $c_{12} < c_{21}$, i.e., the cost of an interceptor is less than the cost of leakage, otherwise the decision is always made in favor of C_2 and the problem becomes trivial. Observe that while p_1 grows from zero to the "critical value" c_{12}/c_{21}, the decision is made in favor of C_2 (nonthreat) and the risk of this decision grows linearly from zero to c_{12}. Similarly, while p_1 grows from c_{12}/c_{21} to 1, the decision is made in favor of C_1 (threat), and its risk remains constant at c_{12}, representing the loss of an interceptor.

Regarding the origin and value of the cost coefficients c_{12} and c_{21}, it is a common misconception that the cost of an interceptor is just the monetary price of its production and is, therefore, negligible with respect to the potential loss of defended assets (quantified by a cost of leakage). We argue, however, that this line of reasoning is incorrect. Indeed, interceptors are our last defense against a missile attack. Moreover, at any given moment, we have a limited supply of them, which cannot be increased instantaneously. Expending interceptors now depletes their availability for future defense. Consequently, the cost of interceptors should regulate their use and reflect the balance between the demand for them now (or in the near future, before new interceptors can be produced) and their current supply. As such, the cost of interceptors has nothing or very little to do with the price of their production; rather, this cost is just a parameter, which should be selected by a commanding entity in such a way that expending interceptors at their current cost would be optimum with respect to the current military and political situation. Guidelines for how cost coefficients should be set are suggested in [14].

3 Risk Reduction

In this section, we derive expressions for the risk reduction due to two critical sensor actions, discrimination and tracking, and show that the expected value of the discrimination part of risk reduction is always nonnegative. In the first subsection we consider discrimination risk when the target can be classified into one of two classes, e.g., lethal and nonlethal. We then proceed by incorporating the risk due to the uncertainty in a target's kinematic state, which we call *track risk*, and conclude by considering the combined influence of classification and kinematic uncertainties on risk estimation in a general case of n classes.

3.1 Discrimination Risk in the Case of Two Classes

Suppose a sensor is trying to classify an object into one of two possible classes. If the sensor has an opportunity to collect an additional measurement on this object before making a classification decision, the risk associated with this object may be reduced. We now derive an expression for the expected value of this risk reduction. Let x be the feature which we measure and let $p(x|i)$, $i = 1, 2$ be the corresponding class-conditional probability density functions (PDFs). We assume that a sufficient amount of time has passed from the previous measurement of x so that the new measurement can be considered independent from the previous one. Then after the new value of x is measured, the probabilities are updated according to Bayes' rule and the new probabilities become

$$p_i' = \frac{p(x|i)p_i}{p(x)}, \quad i = 1, 2, \tag{3.1}$$

where $p(x) = \sum_{i=1}^{2} p(x|i)p_i$ is the PDF of the feature x. The updated risk of a classification decision, which is based on probabilities p_i', is

$$R' = \min\left(c_{12}, c_{21}p_1'\right) = \min\left[c_{12}, c_{21}\frac{p(x|1)\, p_1}{p(x)}\right], \tag{3.2}$$

and its expected value is

$$\langle R' \rangle = \int R' p(x)\, dx = \int \min\left[c_{12}p(x), c_{21}p_1 p(x|1)\right] dx. \tag{3.3}$$

Using the normalization of $p(x|i)$, we have from Equation (3.3)

$$\langle R' \rangle \leq \min\left[\int c_{12}p(x)\, dx, \int c_{21}p_1 p(x|1)\, dx\right] = \min\left(c_{12}, c_{21}p_1\right) = R, \tag{3.4}$$

which means that the expected value of the new risk after an additional measurement is never larger than the old risk. This is a desirable mathematical property as we never anticipate, in expectation, to increase risk by collecting more information. Note that risk itself (as opposed to its expected value) can increase after an additional measurement due to an atypical result of the measurement.

3.2 Track Risk

Imperfect knowledge of a target's kinematic state may lead to an additional risk, which we term *track risk*. As before, we assume that the object may be either a threat (C_1) or a nonthreat (C_2). The case of several classes may be considered in a similar fashion (see Section 3.3). Since in the case of a nonthreat decision we will not shoot at the target, the risk of this decision remains the same as in Equation (2.1). The risk of a threat decision, however, will change. Namely, if we make this decision and shoot at the target, there is still some probability p_{miss} that the interceptor will miss, in which case, with probability p_1, we will suffer a loss of c_{21}. Correspondingly, the term $c_{21}p_1p_{miss}$ should be added to the risk of a threat decision, where we assume that only one interceptor is fired at the target. As a result, with track risk taken into account, instead of Equation (2.1), we will have

$$R_1 = c_{12} + c_{21}p_1p_{miss},$$
$$R_2 = c_{21}p_1.$$
(3.5)

Apparently, $R_1 = c_{12} + p_{miss}R_2$. Therefore, if p_{miss} is sufficiently large, R_1 might become larger than R_2 even when p_1 is large (e.g., even when $p_1 = 1$). In particular, this will always be the case when $p_{miss} = 1$. In this situation, a nonthreat decision should be made regardless of the value of p_1. Thus, as one would expect, we should not shoot (and waste) an interceptor if the interceptor is guaranteed to miss the target in the first place.

The probability p_{miss} depends on, among other factors, a state estimation error covariance matrix Σ. With each successive track measurement, Σ changes as described by Kalman filtering equations, and so p_{miss} and R_1 will change accordingly. This change will measure the risk reduction utility of a track measurement. Namely, the corresponding risk reduction is $\Delta R = R(\Sigma) - R(\Sigma')$, where Σ' is a state error covariance matrix after the measurement and $R(\Sigma) = \min[c_{21}p_1, c_{12} + c_{21}p_1p_{miss}(\Sigma)]$.

3.3 Modifications to Discrimination Risk Due to the Presence of Track Risk

Here we consider the situation when an object is to be classified into one of n classes $C_1, \ldots, C_n$. We denote the current probabilities as p_k, $k = 1, \ldots, n$, and introduce the set of nonnegative costs c_{kl} of declaring an object a member of class k when in fact it belongs to class l. Consequently, $R_k = \sum_{l=1}^{n} c_{kl}p_l$ is the risk of declaring an object a member of class C_k. In keeping with the convention that C_1 represents the class of lethal objects, we will set $c_{1l} = c_{int}$ for all $l = 1, \ldots, n$, where c_{int} is the cost of an interceptor. The risk of a threat decision R_1 will then be the same as in Equation (2.1).

Taking track risk into account implies corrections to our expressions for expected risk after an additional discrimination measurement. Indeed, during the time interval between discrimination measurements, the error covariance matrix evolves from its current value Σ to some new value Σ', as is typically described by Kalman filtering equations *without* a track measurement. The current risk of declaring the object as a member of class k is

$$R_k = \sum_{l=1}^{n} c_{kl}p_l + \delta_{1k}c_{leak}p_{miss}(\Sigma)p_1 \,, \tag{3.6}$$

where δ_{1k} is the Kronecker delta, equal to 1 for $k = 1$ and 0 otherwise, and c_{leak} is the cost of leakage. Correspondingly, the current risk is

$$R(\Sigma,p) = \min_k R_k = \min_k \left[\sum_{l=1}^{n} c_{kl}p_l + \delta_{1k}c_{leak}p_{miss}(\Sigma)p_1 \right] . \tag{3.7}$$

After a discrimination measurement, class probabilities get updated and Σ gets propagated. The new risk becomes

$$\begin{aligned} R(\Sigma',p') &= \min_k \left[\sum_{l=1}^{n} c_{kl}p'_l + \delta_{1k}c_{leak}p_{miss}(\Sigma')p'_1 \right] \\ &= \min_k \left[\sum_{l=1}^{n} c_{kl} \frac{p(x|l)\,p_l}{p(x)} + \delta_{1k}c_{leak}p_{miss}(\Sigma') \frac{p(x|1)\,p_1}{p(x)} \right] , \end{aligned} \tag{3.8}$$

and its expected value is

$$\begin{aligned} \langle R(\Sigma',p') \rangle &= \int R(\Sigma',p')\,p(x)\,dx \\ &= \int \min_k \left[\sum_{l=1}^{n} c_{kl}p(x|l)p_l + \delta_{1k}c_{leak}p_{miss}(\Sigma')p(x|1)p_1 \right] dx \,, \end{aligned} \tag{3.9}$$

while the expected risk reduction is $\langle \Delta R \rangle = R(\Sigma,p) - \langle R(\Sigma',p') \rangle$. Following the same logic as in the derivation of Equation (3.4), one can show that the expected value of the discrimination part of the decision risk [represented by the first term in Equation (3.9)] never grows as a result of a discrimination measurement.

4   Sensor Resource Management Algorithms

Having derived expressions for risk and risk reduction associated with kinematic estimation and classification, we now incorporate these calculations into various sensor resource management (SRM) algorithms, wherein a resource manager tasks a sensor to perform actions in an effort to minimize expected risk. After describing myopic and far-sighted SRM algorithms, we outline how a far-sighted risk-based approach can be extended to facilitate hierarchical control in a multisensor system.

4.1   Myopic Sensor Resource Management

Using the expressions for risk reduction derived in the previous section, it is straightforward to suggest a myopic resource management algorithm for a single sensor which strives to achieve the fastest possible rate of risk reduction (RRR) over the next data collection interval. Prior to every data collection interval, we assume the sensor has the choice of applying one of several waveforms to any target. If there are n_w available waveforms and n_t targets, then there are a total of $n_t n_w$ action-object pairs from which to choose. For each pair we can calculate the fraction $f_{ij} = \text{ERR}_{ij}/d_i$, where ERR_{ij} is the expected risk reduction due to the application of waveform i to target j, and d_i is the amount of sensor resources or duty required to perform action i. Obviously, f_{ij} represents the rate at which the risk is expected to decrease due to resources spent. Being myopic, we would like to maximize this rate, and so, the algorithm selects the action-object (here, the waveform-object) pair that maximizes f_{ij}.

4.2   Far-Sighted Sensor Resource Management

The myopic algorithm just described minimizes expected risk after the next sensor action is taken. If that were the time when a final decision had to be made, then this algorithm would be optimal. This, however, is rarely the case, as the information collected now is usually used (much) later. An ideal planner would, instead, have a far-sighted planning horizon and be able to enumerate all possible action-object pairs up to some future deadline for all threats. For each threat, it would compute the expected risk resulting from a sequence of actions taken up to that deadline. Finally, based on the risk associated with the various action sequences, it would then task the sensor with the best possible action-object pair for next planning interval, allow the system to evolve, and then repeat the process. A standard approach to tackling such a problem is to formulate it as a finite-horizon Markov Decision Process, also known as Stochastic Dynamic Program, although some authors reserve the latter name to characterize solution methods for this class of problems. Classic references include [1,2,13,20]. Although we have investigated a number of approximate dynamic programming approaches, our formulations and solution methods lie outside the scope of this discussion. Instead, we briefly describe a heuristic approach for far-sighted SRM, akin to the "critical ratio" algorithm given in Feinberg et al. [5], which will also set the stage for our discussion of hierarchical control in multisensor resource management.

Our heuristic approach is based on the observation that the myopic algorithm leads to the appearance of an expected residual loss, or residual risk, i.e., a risk which is impossible or very difficult to eliminate once it has been incurred. For example, residual risk appears if the sensor fails to detect a new target before it leaves a search volume, or fails to collect enough information about a target which is due to be intercepted. Obviously, resource management should be done in such a way as to avoid the appearance of residual risk. The reason it appears in the myopic approach is that the sensor fails to accomplish some goals by their corresponding deadlines. We therefore conclude that for critical tasks similar to those just mentioned, goals and corresponding deadlines should be imposed on the resource manager in addition to the objective of maximizing RRR.

Let N be the total number of tasks the sensor is executing, where by task we mean a particular sensor activity, such as tracking or discrimination, performed on a particular object. For every task i, let d_i be the remaining time until the deadline by which this task should be accomplished, i.e., its goal should be achieved. Goals and deadlines are set by a so-called battle manager. We will assume that given the goal i for task i and the current state of our knowledge, we have some predictive capability to determine a conservative estimate of the expected time t_i the sensor needs to spend on the corresponding task in order to accomplish it. On every iteration, the algorithm orders tasks according to their deadlines, so that $d_1 \leq d_2 \leq \cdots \leq d_N$, and for every $k = 1, \ldots, N$, it checks if there is enough time left to accomplish the first k tasks with some safety margin. In other words, the algorithm verifies if $\alpha \sum_{i=1}^{k} t_i < d_k$, $k = 1, \ldots, N$, where α is a "safety factor" which should be greater than 1. If this inequality holds for all k, then there is no need to worry about deadlines, and the algorithm follows the original RRR logic. If, however, for some k the inequality is violated, then the algorithm finds the "most critical" task index $\hat{k}$ such that

$$\hat{k} = \arg\max_{k} \left(\alpha \sum_{i=1}^{k} t_i - d_k \right) \tag{4.1}$$

and schedules a measurement required by the task $\hat{k}$. If this measurement takes time τ to be executed, then after this measurement both $t_{\hat{k}}$ and $d_{\hat{k}}$ become smaller by τ, and $\alpha \sum_{i=1}^{\hat{k}} t_i - d_{\hat{k}}$ becomes smaller by $(\alpha - 1)\tau$. Since $\alpha > 1$, the task is now less critical than it was before the measurement. In the event it is impossible to satisfy all deadlines, the algorithm first sacrifices the task with the smallest residual risk.

4.3   A Hierarchical Multisensor Control Architecture

Thus far, we have limited our discussion to risk-based resource management algorithms for a single sensor in missile defense. We assumed that for each task, a sensor has a corresponding goal and deadline, which can be incorporated into a risk-based approach for optimizing the set of actions taken in the subsequent data collection interval. In this section, we describe a natural extension of our risk-based approach to a hierarchical decision-making architecture for multisensor resource management. Such hierarchical approaches have gained increasing attention over the past decade in the reinforcement learning domain [16], and are well suited for the missile defense problem in which a distributed architecture is already in place. This hierarchical architecture facilitates solution of the (intractable) global problem of assigning all sensors a set of actions to perform over the course of a mission by decomposing the larger long-term problem into smaller short-term problems. In this way, a hierarchical architecture exploits a "divide-and-conquer" mentality for solving complex, large-scale problems.

In the current missile defense decision-making architecture, a Battle Manager (BM) acts as a commanding entity that tasks participants (e.g., sensors, platforms, and interceptors) throughout a given mission to collect information or execute some plan. A BM maintains an integrated picture of the battlespace, or in dynamic programming terminology, the state of the system, including (1) target-related information like track accuracy, classification, predicted impact point, and estimated time to impact, (2) sensor-related information, including sensor capabilities and current tasking, and (3) weapon system-related information, including the number of available interceptors and interceptor capabilities.

The ability of our approach towards sensor resource management to accommodate goals and deadlines allows us to naturally insert it into a general hierarchical framework of system management for coordinating multiple sensors. Given all available information about existing missile complexes, which consist of one or more targets spawned from the same object, the BM interacts with individual sensors with some periodicity, known as a Battle Manager Planning Interval (BMPI), collecting information obtained during the previous data collection interval, and giving assignments for the next interval. Based on known trajectories of the missile complexes relative to the positions of the sensors and known performance characteristics of the sensors, the BM creates a battle plan. For each BMPI and for each missile complex, the plan specifies which sensor or sensors will observe this missile complex, and with which task (tracking and/or discrimination). Search can be considered on the same grounds as a sensor activity related to a potential additional threat. Included in the plan, therefore, is the expected improvement of our knowledge of this missile complex's tracking and classification characteristics (or of the presence of a threat in a search volume). The plan is designed in such a way as to provide the smallest expected loss of defended assets and gets updated every BMPI according to the evolving situation. The generation of a battle plan is a separate (and complex) problem, which is not considered here. We do, however, consider the interaction of the BM with individual sensors assuming this problem has been solved.

In our hierarchical structure this interaction is organized as follows. At the beginning of each BMPI, after computing its own long-term plan, the BM assigns each sensor a set of targets and search volumes along with the associated cost coefficients and search/track/discrimination goals corresponding to the expected improvements mentioned above. The natural deadline for these goals is the end of this BMPI, although in certain situations the deadline could vary. Since it is possible for a sensor to achieve all of its goals by the corresponding deadlines and still have some remaining resources, the BM also informs each sensor about other targets, not assigned to it, and the value of their cost coefficients. Now each sensor finds itself in a situation described in the previous section: it faces a number of targets with associated cost coefficients, goals, and deadlines. Accordingly, each sensor acts as described above, without regard to the presence of other sensors, by attempting to determine an optimal plan over a shorter time horizon (a BMPI) that simultaneously meets all goals and deadlines while minimizing risk.

The results are reported to the BM at the end of a BMPI and will be used for updating the battle plan and creating an assignment list for the next BMPI.

5 Simulation Results

A low fidelity simulation environment was constructed in MATLAB to test and compare myopic sensor resource management algorithms that assign track and discrimination actions to multiple objects based on a heuristic policy or the maximization of a single objective function. The application of far-sighted approaches, which incorporate search management, is not considered here, but is discussed in [14].

We assume a single sensor has just begun tracking N objects. There is only one lethal target, known as a re-entry vehicle (RV), amongst the targets. At each time step, an action-object pair is selected depending on the algorithm used, and that action is then performed on that object. Each action takes the same amount of time to complete. Each object is assumed to belong to one of three classes (lethal objects belong to class 1), and classification is based on the measurement of three independent features. We assume that class-conditional PDFs are known and are Gaussian for each feature (see Figure 1). The sensor can make an observation on exactly one feature at a time when performing a discrimination action. This assumption could easily be relaxed. If the manager decides to measure a particular feature of an object, then the result of this measurement is generated as a random variable whose distribution corresponds to this feature's distribution for the true class of the observed object. An object's posterior probability of belonging to any class is then computed using Bayes' rule, where we have made the simplifying assumption that observations are independent from one time step to the next.

We assume a simple tracking model of a target moving with a constant velocity without a process noise. We model the probability that an interceptor successfully "kills" a target (*probability of kill* for short) as a function of a track's error covariance matrix. In particular, we used a sigmoidal function of the form

$$p_{kill}(x) = \frac{1 + \exp(-m/s)}{1 + \exp((x-m)/s)}$$

to determine the probability of kill, where x denotes the Euclidean norm of a track's position error, m defines the "midpoint," i.e., the point at which $p_{kill}(x) \approx 1/2$, and s represents the "spread" of the curve. Note that small values of s result in near step functions where p_{kill} is either close to one or zero. Such a function could easily be extended to incorporate additional factors beyond just the position error of track (e.g., velocity errors, classification information, etc.). In fact, Kalandros and Pao [7] give several examples of why more information may be necessary.

As described in Section 2, cost coefficients are needed to compute the risk of making a particular decision. We set the cost of incorrectly declaring a threat a nonthreat to 6, the cost of incorrectly declaring a nonthreat as a threat to 1, and

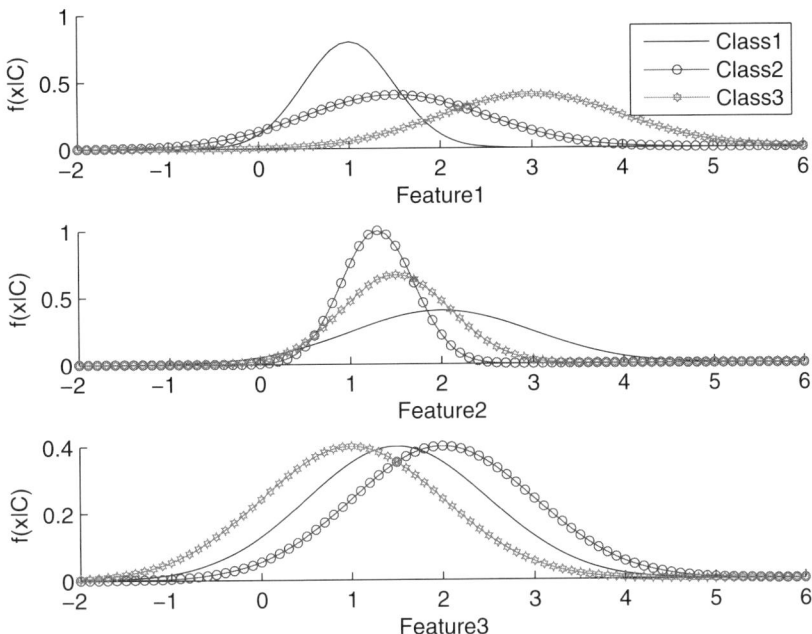

Fig. 1. Class-conditional PDFs used in simulations

the cost of incorrectly classifying a nonthreat as a different type of nonthreat to 0.2. We assume there are three objects, and, in truth, object i belongs to class i, $i = 1, 2, 3$. (Initial results with more than three objects demonstrated that our conclusions remain the same.)

For general class-conditional PDFs, the expected risk reduction from an additional measurement cannot be computed analytically. Thus, we turned to numerical integration techniques in our computations.

There are six different planners (or resource management algorithms) that we tested for comparison. At each planning interval (each time step), the planner assigns the sensor to perform a single action on a specific object during the subsequent time interval. The planners (and their symbols used in the figure legends) are:

1. Risk Reduction Planner (maxRRR): Enumerates all action-object pairs and determines which action-object pair will yield the largest expected reduction in risk.
2. Information Gain Planner (maxInfoG): Enumerates all action-object pairs and determines which action-object pair will yield the largest expected information gain.
3. Improved Information Gain Planner (maxIInfoG): Operates exactly like the Information Gain Planner except there is no information gain for performing

a track update on an object whose track error is below a pre-defined threshold.

4. Highest Probability of RV Planner (maxPRV): Identifies the object with the largest current probability of lethality (i.e., of being a re-entry vehicle) and randomly generates an action to be performed on this object. The purpose of this planner is to dispel the oft-held belief that spending the most resources on the most threatening object is an optimal use of resources.

5. Highest Risk Object Planner (maxRiskObject): Identifies the object with the largest current risk and randomly generates an action to be performed on this object.

6. Round Robin Planner (RoundRobin): First performs action 1 on all objects, then action 2, and so on.

The different resource management algorithms were compared with respect to five different metrics: (1) average loss; (2) average probability of correct classification of the lethal object; (3) average probability of correct classification of all nonlethal objects; (4) average track quality of the lethal object; and (5) average track quality of all nonlethal objects. In general, we found that all planners maintain a very high track quality on the object it believes to be lethal and a sufficient track quality on all remaining objects. Results with respect to the first three metrics are described below.

As one would expect, the risk reduction planner, which strives to reduce risk as quickly as possible, outperforms all other planners in the average loss category (see Figure 2). What is interesting is that the planner that attempts to maximize pure information gain (maxInfoG) over the course of the mission dedicates the majority of its resources to performing track maintenance actions. Under the assumptions of this simulation, this corroborates our initial statement that metrics based on pure information gain may not be well suited in the context of missile defense.

To give a more mathematical explanation as to why a purely information-based approach may yield inferior results, consider the following classification problem involving an object that can belong to one of n possible classes $C_1, \ldots, C_n$, where C_1 is the class of lethal objects and all other classes represent various nonlethal objects. The object's class can be represented as a discrete random variable X, which must take on one of the values $x_1, \ldots, x_n$ with probabilities $p_1, \ldots, p_n$, respectively. It is well known that the (Shannon) entropy of the object's class, $H(X) = -\sum_{i=1}^{n} p_i \log_2 p_i$, is maximized when all of the p_i are equal because the object is equally likely to belong any of the n classes. In a similar way, suppose that an object has been perfectly classified as a nonlethal object, i.e., $p_1 = 0$, but that the exact type of nonlethal object is completely unknown, i.e., $p_2 = \ldots = p_n = 1/(n-1)$. Then, the entropy associated with this object is still relatively large. However, from the standpoint of risk, or expected loss, this object is of little concern. One could then argue that it would be an inappropriate use of scarce resources to determine precisely what class of nonlethal object it is, when its associated risk has already been determined to be zero.

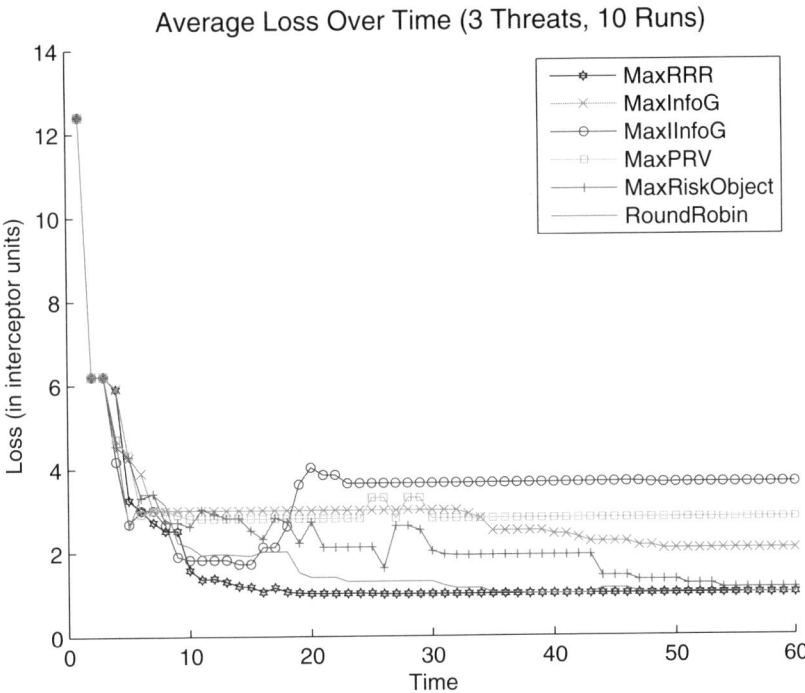

Fig. 2. Performance comparison of different resource managers

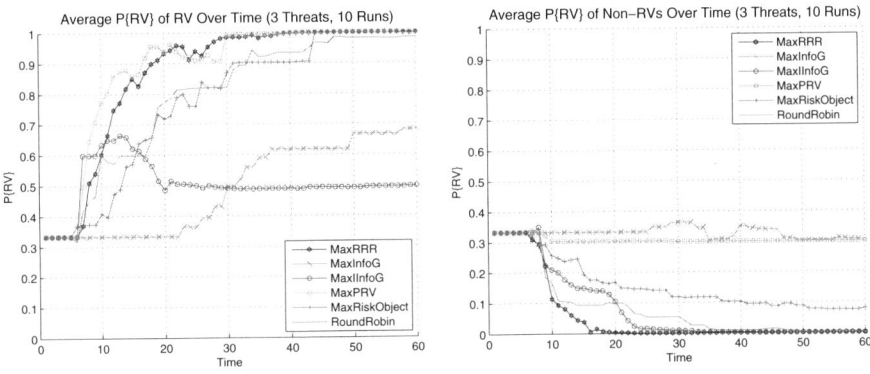

Fig. 3. Discrimination performance of different resource managers. P{RV} is the probability that an object is lethal.

Besides average loss, it is illustrative to compare the various planners based on two important questions related to classification: (1) How well were the targets classified? (2) How long did it take the sensor to classify the targets? Focusing solely on classification, one would expect an ideal sensor to quickly classify threatening objects as threatening and identify nonthreatening objects as nonthreatening. The sensor could then provide higher quality information in less time to an interceptor whose goal is to prosecute all threatening targets and reduce expected loss. It turns out that this goal is achieved as a byproduct of the risk reduction planner, and is illustrated in Figure 3. To understand this, recall that a nonzero probability of lethality directly contributes to the risk of a nonthreat decision. Consequently, if the cost of leakage is relatively high and several objects have a probability of lethality well above zero, then it is beneficial to perform additional classification measurements in order to reduce this probability. Reducing this probability is one way to possibly decrease total risk. Thus, a natural consequence of the risk reduction planner is to reduce the probability of lethality on all nonthreatening objects by performing additional discrimination actions.

6   Conclusions and Future Work

This chapter advocates the use of a risk-based objective function for sensor resource management in the context of missile defense. After presenting a formal description of the equations and update formulas needed to compute risk and risk reduction quantities, we outlined a risk-based approach to single-sensor resource management as well as a hierarchical approach to multisensor control. We performed a comparative analysis of various myopic approaches for tasking a sensor to track and discriminate targets (without the presence of deadlines) and found that maximizing the expected rate of risk reduction produced superior results.

Although not presented in this work, we have conducted an investigation of a modified rate of risk reduction method when presented with Battle Manager goals and deadlines [14]. Future research includes refinement of a mathematical solution methodology as needed to solve a finite-horizon dynamic program. Likewise, we are currently working to incorporate the risk due to possible misassociation of closely-spaced targets, which adds yet another layer of complexity into our formulation. It can be shown that our risk-based approach is also applicable in a situation when some contextual information is available for discrimination. Although the calculations become more involved, the results will be very similar. Finally, we are continuing to develop and test our proposed hierarchical resource management approach and the associated dynamic programming techniques involved.

Acknowledgements

The authors are grateful to Fred Daum, Herb Landau, and Manny Silvia for many valuable discussions, which stimulated a number of new ideas.

References

1. D.P. Bertsekas, *Dynamic Programming and Optimal Control*, Athena Scientific, Belmont, MA, 1995.
2. D.P. Bertsekas and J.N. Tsitsiklis, *Neuro-Dynamic Programming*, Athena Scientific, Belmont, MA, 1996.
3. F. Bolderheij and P. van Genderen, "Mission Driven Sensor Management," Proceedings of the 7th International Conference on Information Fusion, pp. 799-804, Stockholm, Sweden, 2004.
4. F. Bolderheij, F.G.J. Absil, and P. van Genderen, "A Risk-Based Object-Oriented Approach to Sensor Management," Proceedings of the 7th International Conference on Information Fusion, vol. 1, pp. 598-605, July 2005.
5. E.A. Feinberg, M.A. Bender, M.T. Curry, D. Huang, T. Koutsoudis, and J.L. Bernstein, "Sensor Resource Management for an Airborne Early Warning Radar," In O.E. Drummond, editor, Signal and Data Processing of Small Targets, Proceedings of SPIE vol. 4728, pp. 145-156, 2002.
6. K.J. Hintz, "A Measure of the Information Gain Attributable to Cueing," IEEE Transactions on Systems, Man, and Cybernetics, vol. 21, no. 2, pp. 237-244, 1991.
7. M. Kalandros and L.Y. Pao, "Covariance Control for Multisensor Systems," IEEE Transactions on Aerospace and Electronic Systems, vol. 38, No. 4, pp. 1138-1157, 2002.
8. K. Kastella, "Discrimination Gain to Optimize Detection and Classification," IEEE Transactions on Systems, Man, and Cybernetics, Part A: Systems and Humans, 27:1, pp. 112-116, 1997.
9. C. Kreucher, A.O. Hero, and K. Kastella, "A Comparison of Task Driven and Information Driven Sensor Management for Target Tracking," Proceedings of the 44th IEEE Conf. on Decision and Control, Seville, Spain, pp. 4004-4009, 2005.
10. C. Kreucher, "An Information-Based Approach to Sensor Resource Allocation," Ph.D. dissertation, The University of Michigan, 2005.
11. R. Mahler, "Global Optimal Sensor Allocation," in Proceedings of the Ninth National Symposium on Sensor Fusion, vol. I, pp. 167-172, Monterey, CA, 1996.
12. G.A. McIntyre and K.J. Hintz, "An Information-Theoretic Approach to Sensor Scheduling," Proceedings of SPIE, 2755, pp. 304-312, 1996.
13. M.L. Puterman, *Markov Decision Processes: Discrete Stochastic Dynamic Programming*. John Wiley & Sons, Inc., 1994.
14. M. Raykin and D. Papageorgiou, "A Risk-Based Approach to Sensor Resource Management," Raytheon Technical Report HQ0006-06-D-0001, 2007.
15. M. Rudary, D. Khosla, J. Guillochon, P.A. Dow, and B. Blyth, "A Sparse Sampling Planner for Sensor Resource Management," In I. Kadar, editor, Proceedings of SPIE vol. 6235, Signal Processing, Sensor Fusion, and Target Recognition XV, pp. 62350A1-62350A9, 2006.
16. M. Ryan. Hierarchical Decision Making. In J. Si, A.G. Barto, W.B. Powell, and D. Wunsch, editors, "Handbook of Learning and Approximate Dynamic Programming," IEEE Press Series on Computational Intelligence, 2004.
17. W.W. Schmaedeke, "Information-based Sensor Management," Proceedings of SPIE vol. 1955, Signal Processing, Sensor Fusion, and Target Recognition II, pp. 156-164, April 1993.

18. W.W. Schmaedeke and K. Kastella, "Information-based Sensor Management and IMMKF," Proceedings of the SPIE Conference on Signal Processing of Small Targets, vol. 3373, Orlando, FL, 1998.
19. M.K. Schneider and C. Chong, "A Rollout Algorithm to Coordinate Multiple Sensor Resources to Track and Discriminate Targets," In I. Kadar, editor, Proceedings of SPIE vol. 6235, Signal Processing, Sensor Fusion, and Target Recognition XV, pp. 62350E1-62350E10, 2006.
20. R.S. Sutton and A.G. Barto, *Reinforcement Learning: An Introduction.* The MIT Press, Cambridge, MA, 1998.

Constructing Optimal Cyclic Tours for Planar Exploration and Obstacle Avoidance : A Graph Theory Approach

Abhishek Tiwari, Harish Chandra, Jacob Yadegar, and Junxian Wang

Utopia Compression Corporation
11150 West Olympic Boulevard, Suite 1020,
Los Angeles, CA-90064
{abhishek, harish, jacob, junxian}@utopiacompression.com

Abstract. We propose a graph theory approach for planar autonomous exploration. We first partition the planar space using Peano-Cesaro triangular tiling and then construct an edge adjacency dual graph of the tiling pattern. The dual graph of the Peano Cesaro triangulation is obtained by defining a vertex for each triangular tile and drawing an edge between two tiles that share an edge. In the presence of obstacles we analyze the subgraph induced by the non-obstacle tiles in the dual graph. We prove the existence of Hamiltonian cycles in this induced subgraph for a certain class of obstacles. We also prove the non-existence of Hamiltonian cycles for certain other obstacle configurations. We present heuristic based algorithms and compare their results for the cases where we have a definitive answer to the existence of Hamiltonian cycles. Examples with figures are included to illustrate the concept.

1   Introduction

Exploratory path planning with obstacle avoidance finds application in areas involving autonomous search/exploration tasks like mine sweeping, rescue operations, locating survivors in a disaster struck area, ocean exploration, monitoring coast lines, protecting borders etc. Some other exciting commercial applications of interest involve autonomous coverage applications for lawn mowing and vacuum cleaning robots.

Autonomous path planning and obstacle avoidance has been studied by numerous researchers and over the years quite a few interesting approaches have been proposed. [7] and [10] present a comprehensive treatment of these approaches. In this chapter, we bring together concepts from the fields of space filling curves, graph theory and path planning and merge them to gain insights into the optimal solutions of a problem of considerable practical interest. We propose a graph theory approach to the exploratory path planning problem. The existing approaches are based mainly on heuristic and there exist no provable guarantees. Graph theory has been applied for path planning problems earlier in [9] by Jun and D'Andrea, the difference is that the authors use hexagonal cells

M.J. Hirsch et al. (Eds.): Adv. in Cooper. Ctrl. & Optimization, LNCIS 369, pp. 145–165, 2007.
springerlink.com   © Springer-Verlag Berlin Heidelberg 2007

which are not well suited for multi-resolution decomposition and it is not clear whether an irregular decomposition of the exploration space is possible. In [14] and [15], the Savla et al. tile the exploration space with bead tiles. The bead tiles are constructed assuming a Dubin's vehicle model for the UAVs (with a maximum allowable radius of curvature of the planned paths). The bead tiles even though they encode the mobility constraints satisfactorily, but again there is no mention of irregular decomposition. We believe our approach of Peano-Cesaro tiling is more flexible in terms of representation of the exploration space. However, it remains to be seen how the novel ideas proposed in this chapter compare with the existing exploratory path-planning approaches in terms of actual implementation on a real system. The concept of Peano-Cesaro tiling used in this paper for path planning application, has been successfully used for pattern based image compression in our earlier work [5].

In this chapter, we use Peano-Cesaro sweep [13] to partition the territory map (including obstacles, if any) into triangular tiles and use the associated Sierpinski tour for exploration. The Peano-Cesaro sweep is a space filling heuristic. Space filling heuristics have been successfully used to solve the Traveling Salesman Problem in a time efficient manner [3], [12]. Even though the solutions obtained are suboptimal, the savings in computation time are immense. The authors in [3] have proved a bounded distance from the global optimum. The Sierpinski space filling curve has been proven to be optimal in terms of the overall tour length when compared to other space filling curves like the Hilbert curve [1].

We represent the mobility constraint of the autonomous vehicle by allowing moves only between tiles that share a side in one time step. We overlay the Peano-Cesaro tiling pattern with a graph, where the tiles are represented by vertices and the edges represent allowed moves. This graph is the same as the edge adjacency dual graph of the Peano-Cesaro triangulation. If the planar region has obstacles, in order to disallow movement to the obstacle tiles, we remove the obstacle vertices and edges incident on them from the dual graph. The resulting graph is a subgraph induced by the non-obstacle vertices of the dual graph. Our objective then is to find a cyclic tour of the induced subgraph such that it includes all the non-obstacle tiles. In practical implementation, this is equivalent to exploring a planar region with obstacles in the most efficient manner, so that it is possible to visit all the non-obstacle regions and also save fuel, time and energy.

A no repetition cyclic tour of the vertices of a graph is known as a Hamiltonian cycle. Thus, an optimal exploration tour, in the presence of obstacles, is a Hamiltonian cycle in the subgraph induced by the non-obstacle tiles in the dual graph. In general, determining whether a graph is Hamiltonian is an NP-complete problem. But, in this paper we use the special properties of the induced dual subgraph to prove existence of Hamiltonian cycles for a certain class of obstacles. Similar work for Triangulated Irregular Networks has been done in [2], which has immense application in computer graphics.

The contribution of this chapter can be summarized under three specific headings: Firstly we have devised an algorithm called ESSENTIAL-CHAINS, which starts with an assumption that a given dual subgraph is Hamiltonian and then proceeds by logical reasoning collecting all components of the graph which provably are essential components of the Hamiltonian cycle, if one exists. For certain obstacle configurations the algorithm comes up with a counter example thus proving that the dual subgraph is non-Hamiltonian. Secondly we prove the existence of Hamiltonian cycles for a special class of obstacles. Lastly we have devised a few heuristics based algorithms to find the minimum repetition tours for a given obstacle configuration. We compare the result of the heuristics based algorithms, based on the attainable optimum established using our existence results.

In section 2, we introduce the ideas of Peano-Cesaro sweep and the associated Sierpinski bucketing tour. In section 3 we use the concepts from section 2 and graph theory to pose a combinatorial optimization problem for exploratory path planning in the presence of obstacles. In section 4 we prove existence and non-existence results of Hamiltonian cycles for a certain class of obstacle configurations. In section 5 we present algorithms to find long cycles in our graph of interest. Finally in section 6, we sum up the contributions of this paper.

2   Peano-Cesaro Tiling and the Associated Sierpinski Bucketing Tour

2.1   Peano-Cesaro Fractal Sweep

The central idea behind a fractal sweep is to find a recursive mapping that takes the unit interval into the plane. The key concepts in such constructions are initiator, generator, sweep and rules of arms placement. An example of a fractal is the Peano-Cesaro fractal sweep as illustrated in figure 1(a). If production of the fractal, as in Peano-Cesaro fractal sweep, proceeds indefinitely, a trace that is everywhere continuous but nowhere differentiable would be obtained. The generator of a fractal consists of arms, each treated as scaled down initiators for the next stage of construction. Patterns that replicate the generator as recursion deepens are called self-similar [11] and admit to the concept of dimensionality $D = \log N / \log(1/r)$ where, N is the number of arms of the generator and r is the similarity ratio defined as the length of an arm of the generator to that of the initiator. Thus, the Peano-Cesaro curve, in which the generator has two arms ($N = 2$), each ($r = 1/\sqrt{2}$) factor of the length of the initiator, has dimensionality $D = \log 2 / \log \sqrt{2} = 2$. Fractals defined over the plane with D close to 1 are smoother and better behaved than those with D close to 2, which are more plane-filling. Fractals with $D > 2$ exhibit chaotic behavior by multiply crossing regions trapped by the fractal sweep. Brownian motion is a primary example of sweep patterns with $D > 2$, while Koch curve [11] has $1 < D < 2$.

A *sweep* is a walk from the start to the end of the initiator along a defined path and this path defines the fractal. Taking one of the diagonals of the unit square $\mathcal{U}$ on the plane as the initiator, the Peano-Cesaro fractal sweep can be developed using the generator shown in figure 1(a), and the following rule of

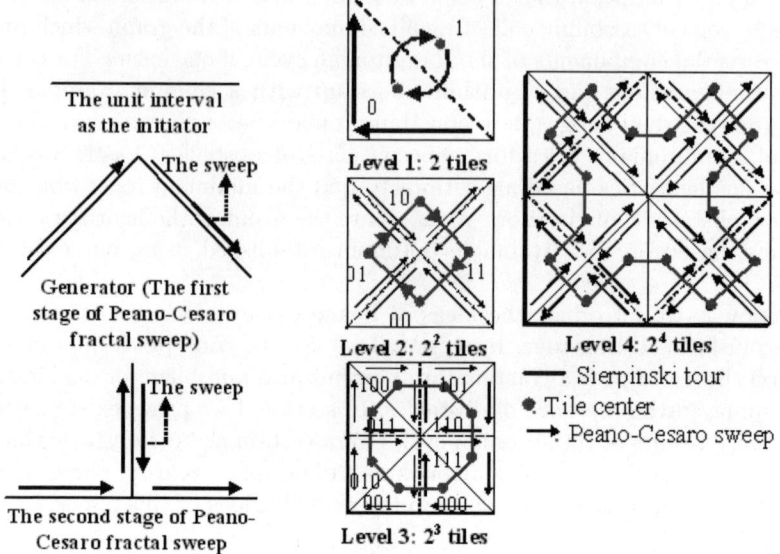

(a) Peano-Cesaro fractal sweep.

(b) Peano-Cesaro tiling and the associated Sierpinski bucketing tour.

Fig. 1. Peano-Cesaro fractal sweep and tiles

placement: **At every even (odd) stage $L = 1$, walk along $(L-1)$th sweep and place the generator to the right (left) of each and every arm.** Four stages of the Peano-Cesaro fractal sweep in figure 1(b) are indicated by directed paths all beginning and ending at a corner of the unit square $\mathcal{U}$. Note that the Peano-Cesaro sweep as illustrated in figure 1(b) shows no region crossing behavior. To be more accurate, the sweep is linear-wise degenerate, as it visits the vertices of the tiles generated multiple times - for instance the center of $\mathcal{U}$ in stage 2 of the decomposition is visited four times, which in fact can be proven to be the maximum degeneracy for all recursion levels. However, the sweep is planar-wise non-degenerate (non-crossing), precisely because $D = 2$.

The Peano-Cesaro sweep tiles the unit square $\mathcal{U}$ with right-angled isosceles triangles. A tile is composed of the current initiator forming the hypotenuse (denoted by $\mathbf{B}$), and the first and second arms of the corresponding generator as the other two sides (denoted by $\mathbf{F}$ and $\mathbf{S}$ respectively.) The number of tiles is doubled each time the sweep is advanced by one recursion stage (also referred to as recursion level), yielding 2^L triangular tiles at the Lth stage ($L \geq 1$). The domain $\mathcal{U}$ is decomposable to any desired degree and tiling is always regular and isotropic such that at every level of decomposition each tile is visited by the sweep (in the sense that it traverses the two ($\mathbf{F}$ and $\mathbf{S}$) sides of the tile defined by the arms of the generator). These are the properties that make the Peano-Cesaro sweep amenable to analysis of the neighborhood of a triangular tile and, as we shall discuss in this chapter, highly suitable for autonomous

Unmanned Vehicle (UV) path planning, object avoidance and also cooperative strategies for exploration and goal-seeking missions amongst a team of UVs. The decomposition procedure highlighted above can be represented by a binary tree structure wherein nodes on left (right) branches are recursively assigned the values 0(1). This, in a natural way, maps each tile with a binary code sequence indicating a tree descent inheritance from the root (representing the unit square $\mathcal{U}$) to tiles/nodes at various levels of the tree. Denoting a code sequence by C, a tile at the Lth recursion level may be expressed by $C = c_1 \cdots c_L$, where $c_i \in \{0, 1\}$, $i = 1 \cdots L$. Tiles at stages 1, 2 and 3 in figure 1(b) carry their code sequences. Later we use the decimal equivalents of these code sequences to distinguish tiles, we will refer to this decimal indexing as the *Sierpinski ordering*. This concept of distinguishing regions in space using indexing of the subdivisions has also been analyzed in [1], where it was proved that the Sierpinski ordering is the best in terms of preserving spatial adjacency information.

The Peano-Cesaro sweep in figure 1(b) shows four regular stages of tiling and as mentioned the isotropic and self-similarity properties of the tiles ensure that all regions of U at any recursion level have equal probability of being visited, where the probability measure is proportional to the area of the tile at the specific recursion level. These properties, which turn out to be important for single or cooperative autonomous path planning, are not met in most of the other fractal patterns - the reader is invited to consult [11] for a number of examples such as the snowflake, the monkeys tree and the dragon sweep formations. The regular tiling pattern in figure 1(b), though extremely efficient and simple, is not mandatory. Figure 2 shows two (non-homogenous) deviations of the regular decomposition. Figure 2(a) depicts an example of variable tile size Peano-Cesaro decomposition along with its associated Sierpinski sweep. The deviation in figure 2(b) is more drastic, where tile splitting is no longer constrained to the mid-point of B side, though the topological structure of the binary decomposition is left invariant.

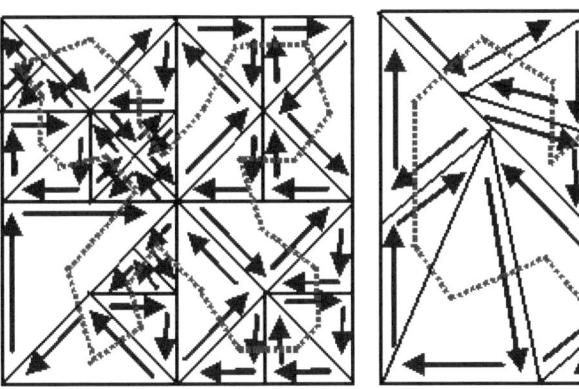

(a) Variable size decomposition.   (b) Irregular adaptive decomposition.

Fig. 2. Non-homogenous Peano-Cesaro tiling

2.2   Sierpinski Bucketing Tour

The Peano-Cesaro sweep in figure 1(b) induces what is referred to as a Sierpinski Bucketing tour according to the following rule [8]:

Whenever a tile is swept by the Peano-Cesaro fractal sweep, connect the center of the tile to the center of the preceding tile visited by the Peano-Cesaro sweep.

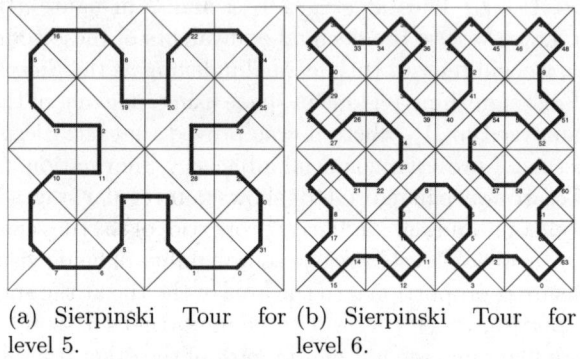

(a) Sierpinski Tour for level 5.

(b) Sierpinski Tour for level 6.

Fig. 3. Sierpinski Tour for decomposition levels 5 and 6

Application of the above rule to regular (and irregular) sweeps yields the regular (and irregular) Sierpinski tours. Figure 3 illustrates the Sierpinski tour for level 5 and 6. Sierpinski tour generation algorithm forms the basis for autonomous agent path planning whether on an exploratory or surveillance missions starting and ending at the same location or a goal oriented mission starting from a source location to a destination. The hierarchical (multi-resolution) nature of the Peano-Cesaro tiling, represented by sparse binary tree structures, entirely carries over to the Sierpinski tour. This hierarchical behavior is highly desirable for path planning missions in situations where the autonomous agent is required to probe more carefully and search finer grain territories while cruising at distance in open spaces (see figure 2(a)) for a scenario. Thus, if the granularity of tiles in the 3rd stage of Sierpinski tour in figure 1(b) is not sufficient, the autonomous agent can easily penetrate one (or more) level(s) deeper into the tree and search the region according to the 4th (or n^{th} in general) stage of the Sierpinski tour where granularity shrinks at a rate of two per level. Due to the fractal dimension of the Peano-Cesaro sweep $D = 2$, the Sierpinski tour drains U to any level given a sufficiently large n, where n is the recursion level. This property guarantees an exploratory tour from a point A and back to A, or a Hamiltonian chain from a point A to a point B, if such a path exists.

3   Exploring a Planar Region with Minimum Number of Repetitions in the Presence of Obstacles

In section 2 we proposed Peano-Cesaro tiling and the associated Sierpinski bucketing tour as tools to generate way points for a planar autonomous mobile agent. If the agent follows the Sierpinski tour, it visits each tile in the planar region and returns to its origin. The problem of generating exploration way points in an obstacle strewn space is of extreme practical interest. Consider for example, an autonomous patrol boat, to be used for exploration of a coastline which is full of small islands. For the purpose of this paper we assume that the territory map of the planar region is available to the autonomous agent. Application of the proposed tools to path-planning and obstacle avoidance with local sensors and in the presence of mobile obstacles is an avenue of current and future research. It is often the case that the capability of an autonomous agent is limited by it's battery life. For an exploration application it is desirable that the autonomous agent visits every tile in the planar space with obstacles. This mission should be accomplished with minimum repetitions of the tiles to extend battery life. In this section we pose a combinatorial optimization problem to find a cyclic tour that visits all the non-obstacle tiles, with minimum number of repetitions.

Let L be the level of decomposition of the planar region. This will result in the decomposition of the planar region into 2^L tiles. Let the tiles be numbered from 0 to $2^L - 1$ under the regular Sierpinski ordering, as defined in section 2. We will call the set of all tiles as $\mathcal{T}$. We now state our main assumption for this section.

Assumption 1 (Obstacle set). *Any obstacle in the planar region can be represented as a union of tiles. Let $\mathcal{O}$ be the set of tiles marked as obstacles. We call $\mathcal{O}$ the obstacle set.*

We need the above assumption to retain the mathematical structure of the problem. For exploration applications this assumption does not create any limitations as it is always possible (as shown in section 2) to set the decomposition level L sufficiently high such that the obstacle can be closely approximated by a set of contiguous tiles.

Constraint 1 (Allowed moves). *If the autonomous vehicle is at tile i at time instant t, then at time $t + 1$, it can only move to the side neighbors of tile i.*

This assumption conforms to the mobility constraints of unmanned surface vehicles. We now state the problem objective:

Given a set of covering tiles $\mathcal{T}$ and the obstacle set $\mathcal{O}$. Under constraint 1, find a cyclic tour that covers all the tiles in the set $\mathcal{T} - \mathcal{O}$ at least once, with minimum number of revisits of tiles.

In order to analyze the performance of any algorithm that tries to find a cyclic tour with minimum number of tile revisits, it is important that the minimum number of repetitions needed for a given obstacle configuration be known. In the following section we present and prove a few existence results.

4   Results on Existence and Non-existence of Hamiltonian Cycles

We analyze the optimization problem from section 3 using results from graph theory. We first create a edge adjacency dual graph. For each triangular tile we have a vertex in the dual graph and an undirected edge exists between two vertices if and only if it is possible to move between the corresponding tiles under Constraint 1. We refer to the set of all tiles in the plane as $\mathcal{T}$.

The following definition puts these concepts in a mathematically precise formulation.

Definition 1. *The* edge adjacency dual graph *is an undirected graph* $\mathcal{S} = (V_{\mathcal{S}}, E_{\mathcal{S}})$*, such that for every tile in* $\mathcal{T}$*, there is a vertex in* $V_{\mathcal{S}}$ *and for a pair of vertices* $x, y \in V_{\mathcal{S}}$*,* $xy \in E_{\mathcal{S}}$ *if and only if* x *and* y *are reachable from each other in one time step under Constraint 1.*

Note that the dual graph is a 2-connected graph and the vertices can have maximum degree 3. Since vertices in the dual graph represent tiles, in this section we will use the terms tile and vertex interchangeably. We now define Hamiltonian cycles and graphs.

Definition 2. *For a graph with more than two vertices, a* **Hamiltonian cycle** *is a cycle that contains each vertex of the graph exactly once. If a graph has a Hamiltonian cycle, it is called Hamiltonian.*

A cyclic tour of the planar region that visits all the tiles with no revisits is a Hamiltonian cycle in the associated dual graph.

The Sierpinski tour, as defined in section 2, is a Hamiltonian cycle in $\mathcal{S}$. Thus the dual graph, $\mathcal{S}$, is a Hamiltonian graph. Now since we know that $\mathcal{T} - \mathcal{O} \subseteq V_{\mathcal{S}}$, $\mathcal{T} - \mathcal{O}$ induces a subgraph $\mathcal{S}'$ in $\mathcal{S}$. This can be represented as $\mathcal{S}' = \mathcal{S}[\mathcal{T} - \mathcal{O}]$ following the notation of [6].

The existence of Hamiltonian cycle in $\mathcal{S}'$, implies that a cyclic exploration tour exists that visits all the tiles in the obstacle strewn planar region with no revisits.

The problem of verifying whether a graph contains a Hamiltonian cycle has been studied for over a hundred years. It is well known that the Hamiltonian-cycle problem is NP complete [4]. We prove existence and non-existence of Hamiltonian cycles for different obstacle configurations using special properties of the dual graph. The optimum of the combinatorial optimization problem described earlier, is zero repetitions in the case a Hamiltonian cycle exists. Once we know that a Hamiltonian cycle exists, we can evaluate the distance from optimum for heuristic based algorithms.

Definition 3. *A* **simple cycle** *is a cycle without any chords. Here, a chord is defined as an edge that joins two vertices of a cycle but is not itself an edge of that cycle.*

As evident in figures 4(a) and 4(b), simple cycles in the dual graph either have 4 or 8 nodes. Such simple cycles look like squares or octagons respectively.

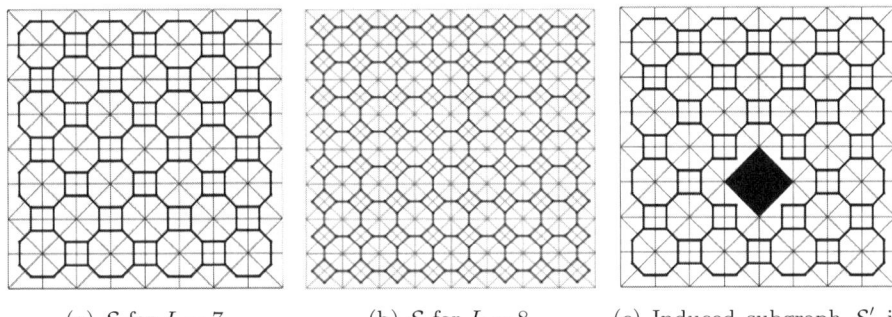

(a) $\mathcal{S}$ for $L = 7$.   (b) $\mathcal{S}$ for $L = 8$.   (c) Induced subgraph $\mathcal{S}'$ in the presence of an obstacle.

Fig. 4. $\mathcal{S}$ and $\mathcal{S}'$

Definition 4. *All tiles that share an edge or a vertex with the boundary of the square region will be referred to as the* **periphery tiles**. *We will call all the other tiles as* **interior tiles**.

Our first result establishes non-existence of Hamiltonian cycles when the number of tiles in the obstacle set is odd.

Proposition 1 (Properties)

1. *If $L > 1$, $\mathcal{S}$ is 2-connected.*
2. *$\mathcal{S}$ and $\mathcal{S}'$ are bipartite graphs.*
3. *$|\mathcal{O}|$ is odd then $\mathcal{S}'$ is non-Hamiltonian*

Proof

1. $L > 1$ implies $|\mathcal{T}| = |V_{\mathcal{S}}| > 2$. One needs to remove at least two vertices from $V_{\mathcal{S}}$, such that the subgraph induced by the remaining vertices on $\mathcal{S}$ is disconnected. Hence $\mathcal{S}$ is 2-connected.
2. All cycles in $\mathcal{S}$ have even number of vertices, therefore by proposition 1.6.1 in [6] we know $\mathcal{S}$ is a bipartite graph. $\mathcal{S}'$ being the induced subgraph retains the bipartite property.
3. If $\mathcal{S}'$ is disconnected or 1-connected, then the results hold trivially. If $\mathcal{S}'$ is 2-connected, then because $|\mathcal{O}|$ is odd, $|V_{\mathcal{S}'}| = |\mathcal{T} - \mathcal{O}|$ is also odd. Now if a Hamiltonian cycle exists it will have an odd number of vertices in it, which is a contradiction because $\mathcal{S}'$ is bipartite. Hence $\mathcal{S}'$ is non-Hamiltonian.

The above result proves non-existence of Hamiltonian cycles. We now develop an algorithm which can prove non-existence of Hamiltonian cycles for a larger class of obstacles by logical reasoning. We refer to this algorithm as ESSENTIAL-CHAINS. We need to first define what we mean by a chain.

Definition 5

1. *A* **tree** *is a connected graph that does not have any cycles.*

2. A **chain** *is a tree, in which all vertices have degree less than or equal to 2. There are exactly 2 vertices with degree 1 in a chain, which we refer to as the* **terminal vertices** *of a chain.*

The ESSENTIAL-CHAINS algorithm starts with the assumption that S' is Hamiltonian. It then finds chains in S' that should form parts of any existing Hamiltonian cycle. For some obstacle scenarios, the algorithm ends up with a contradiction, hence proving S' is non-Hamiltonian. To develop concepts for this algorithm, we begin with the following definition:

Definition 6

1. **Chain edge**: *An edge $xy \in E_{S'}$ is a chain edge, if either x or y or both have degree 2. We refer to the set of all chain edges as $E_{S'}^c$.*
2. **Chain vertex**: *A vertex $x \in V_{S'}$ is a chain vertex, if one of the edges incident on x is a chain edge. We refer to the set of all chain vertices as $V_{S'}^c$.*

Consider the following simple observations:

Proposition 2

1. *If S' is Hamiltonian, then any Hamiltonian cycle of S' will contain all the edges in $E_{S'}^c$.*
2. *If S' is Hamiltonian and $|E_{S'}^c| = |V_{S'}|$, then there exists a unique Hamiltonian cycle in S'*
3. *All degree 2, degree 1 or disconnected vertices in S' are either periphery tiles or they share an edge with an obstacle tile.*
4. *If S' is Hamiltonian, a Hamiltonian cycle of S' must contain exactly two of the edges, incident on the every vertex in $V_{S'}$.*

Proof

1. By definition of Hamiltonian cycle must visit all vertices in S'. Every edge in $E_{S'}^c$ is incident to a degree 2 vertex. Therefore, in order to visit the degree 2 vertices, the Hamiltonian cycle must traverse through all edges in $E_{S'}^c$.
2. Any Hamiltonian cycle in S' has $|V_{S'}|$ edges, and we already know all members of $E_{S'}^c$ should be part of every Hamiltonian cycle that exists. Hence there is a unique choice for a Hamiltonian cycle.
3. Follows by observation.
4. Follows from the definition of Hamiltonian cycle.

If S' has any disconnected or degree 1 vertex then it is trivially non-Hamiltonian. From proposition 1, we know if $|\mathcal{O}|$ is odd, then S' is non-Hamiltonian. Thus the added utility of ESSENTIAL-CHAINS algorithm is evident when the obstacle set has an even number of tiles and S' is 2-connected. For the rest of the discussion on ESSENTIAL-CHAINS algorithm we will assume $|\mathcal{O}|$ is even and S' is 2-connected.

4.1 ESSENTIAL-CHAINS Algorithm

The ESSENTIAL-CHAINS algorithm begins by assuming, $\mathcal{S}'$ is Hamiltonian. The input to ESSENTIAL-CHAINS is the graph $\mathcal{S}'$. There are three possible outcomes of ESSENTIAL-CHAINS:

1. Either the algorithm comes up with a contradiction and exits abruptly, thus proving $\mathcal{S}'$ is non-Hamiltonian. Or,
2. it finds a graph K whose components are chains. These chains are essential components of any Hamiltonian cycle of $\mathcal{S}'$. Or,
3. it comes up with a K, such that $V_K = V_{\mathcal{S}'}$, thus finding a unique Hamiltonian cycle K for $\mathcal{S}'$.

We refer to the neighborhood set of x in $\mathcal{S}'$ as $n_{\mathcal{S}'}(x)$ and the neighborhood set in K as $n_K(x)$. Using the same notation, for an $x \in K$, $d_K(x)$ and $d_{\mathcal{S}'}(x)$ refer to the degree of the vertex x in K and $\mathcal{S}'$ respectively.

As a first step of ESSENTIAL-CHAINS, we add all chain vertices of $\mathcal{S}'$ and all chain edges incident on them, into the vertex and edge set of K respectively.

$K = (V_K, E_K)$
FIRST-STEP$(\mathcal{S}')$
```
1    Initialize V_K = {} ,  E_K = {}
2    for  all x, such that x ∈ V_S', s.t. d_S'(x) = 2
3      do  Let y, z ∈ n_S'(x)
4          if ∃ a path between y and z in K
5            then S' is non-Hamiltonian  EXIT
6            else  V_K ← V_K ∪ x ∪ y ∪ z
7                  E_K ← E_K ∪ xy ∪ xz
```

Note that after the execution of FIRST-STEP, $V_K = V_{\mathcal{S}'}^c$ and $E_K = E_{\mathcal{S}'}^c$. The next module is central to the ESSENTIAL-CHAINS algorithm. This module loops over all vertices which are in V_K, and are degree 3 in $\mathcal{S}'$. We update a candidate set $\mathcal{X}_c$, after execution of this module.

ESSENTIAL-CHAINS$()$
```
 1    Initialize X_c = {x : x ∈ V_K, d_S'(x) = 3}
 2    while X_c ≠ {}
 3      do
 4          Pick an x ∈ X_c Let n_S'(x) = {y, z, w}
 5          switch
 6            case d_K(x) = 1 : Let y ∈ n_K(x) and z, w ∉ n_K(x)
 7              switch
 8                case ∃ a path in K between z and w :
 9                  switch
10                    case d_K(z) = 2 and d_K(w) = 2 :
11                      S' is non-Hamiltonian  EXIT
12                    case d_K(z) = 2 and d_K(w) = 1 :
13                      V_K ← V_K ∪ w
```

14                                     $E_K \leftarrow E_K \cup xw$
15   **if** K has changed
16   **then** UPDATE $\mathcal{X}_c([w, z], x)$
17   **case** $\exists$ a path in K between z and x :
18   **if** $d_K(w) \neq 2$
19   **then if** path-length in K between $z, x < |V_{\mathcal{S}'}|$
20   **then** $V_K \leftarrow V_K \cup w$
21 $E_K \leftarrow E_K \cup xw$
22   **if** K has changed
23   **then** UPDATE $\mathcal{X}_c([w, z], x)$
24   **else** $\mathcal{S}'$ is Hamiltonian and K
25   is the unique Hamiltonian
26   cycle EXIT
27   **else** $\mathcal{S}'$ is non-Hamiltonian EXIT
28   **case** $d_K(x) = 2$:
29   Let $y, z \in n_K(x)$ and $w \notin n_K(x)$
30   **switch**
31   **case** $d_{\mathcal{S}'}(w) = 2$: $\mathcal{S}'$ is non-Hamiltonian EXIT
32   **case** $d_{\mathcal{S}'}(w) = 3$: Let $n_{\mathcal{S}'}(w) = \{a, b, x\}$
33                       $V_K \leftarrow V_K \cup w \cup a \cup b$
34                       $E_K \leftarrow E_K \cup aw \cup bw$
35   **if** K has changed
36   **then** UPDATE $\mathcal{X}_c([a, b, w, y, z], x)$
37   **case** $d_K(x) = 3$:
38                 $\mathcal{S}'$ is non Hamiltonian EXIT
39     $\mathcal{X}_c \leftarrow \mathcal{X}_c - x$

The function UPDATE $\mathcal{X}_c(Y, x)$, first selects all vertices in Y with degree 3 in $\mathcal{S}'$, lets call this set of vertices $Y_3 \subset Y$. The function deletes all previous occurrences of the elements of Y_3 from $\mathcal{X}_c$ and then adds the elements of Y_3 right after the occurrence of x. The function then deletes x from $\mathcal{X}_c$ before returning.

We now illustrate the execution of the ESSENTIAL-CHAINS algorithm using an example

Consider $L = 5$ and $\mathcal{O} = \{3, 4\}$, as shown in figure 5. Figure 5(a), shows K after the execution of the FIRST-STEP. The candidate set $\mathcal{X}_c$ after FIRST-STEP is initialized to $\{10, 12, 13, 18, 21, 26, 27, 29\}$ Now we present a step by step execution of the ESSENTIAL-CHAINS algorithm

1. Pick $x = 10$ from $\mathcal{X}_c$, $d_K(x = 10) = 2, \{y, z, w\} = \{9, 11, 13\}$. The control goes to the case in line 32, finds $\{a, b\} = 14, 12$ and adds edge $12 - 13$ to K, figure 5(b). $Y = \{13, 9, 11, 12, 14\}$, $Y_3 = \{13, 12\}$ UPDATE $\mathcal{X}_c$ returns $\{12, 13, 18, 21, 26, 27, 29\}$.

2. Pick $x = 12$, $d_K(x = 12) = 2, \{y, z, w\} = \{11, 13, 19\}$. Again in line 32, the algorithm finds $\{a, b\} = 18, 20$, adds edges $18 - 19$, $19 - 20$ and vertices $\{19, 20\}$ to K, figure 5(c). $Y = \{19, 11, 13, 18, 20\}$, $Y_3 = \{19, 13, 18, 20\}$ UPDATE $\mathcal{X}_c$ returns $\{20, 18, 13, 19, 21, 26, 27, 29\}$.

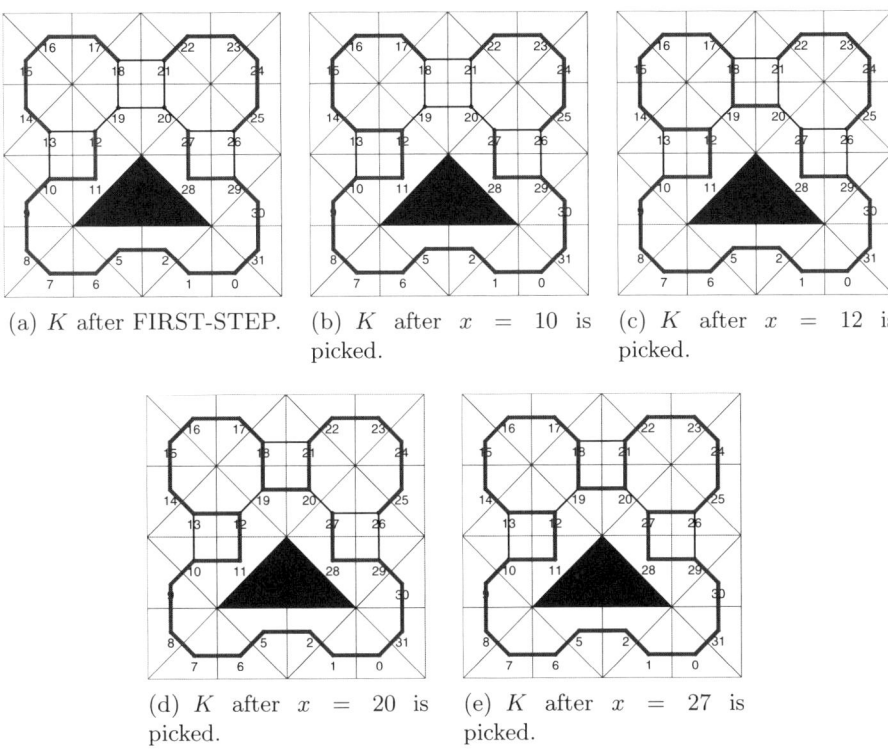

(a) K after FIRST-STEP.

(b) K after $x = 10$ is picked.

(c) K after $x = 12$ is picked.

(d) K after $x = 20$ is picked.

(e) K after $x = 27$ is picked.

Fig. 5. Illustration of ESSENTIAL-CHAINS algorithm. Graph K is shown using thick edges.

3. Pick $x = 20$, $d_K(x = 20) = 1$, $\{y, z, w\} = \{19, 27, 21\}$. There exists a path in K between $x = 20$ and $z = 27$, and its length is less than $|V_{\mathcal{S}'}|$. Therefore, control goes to the case in line 20, and adds edge $20 - 21$ to K, figure 5(d). $Y = Y_3 = \{21, 27\}$ and UPDATE $\mathcal{X}_c$ returns $\{27, 21, 18, 13, 19, 26, 29\}$.

4. Pick $x = 27$, $d_K(x = 27) = 1$, $\{y, z, w\} = \{28, 20, 26\}$. There exists a path in K between $z = 20$ and $w = 26$, $d_K(z = 20) = 2$ and $d_K(w = 26) = 1$. Therefore, control goes to the case in line 12, and adds edge $27 - 26$ to K, figure 5(e). $Y = Y_3 = \{20, 26\}$ and UPDATE $\mathcal{X}_c$ returns $\{26, 21, 18, 13, 19, 29\}$.

5. For the remaining elements in $\mathcal{X}_c$, K does not change. Therefore, there are no further additions to $\mathcal{X}_c$. The algorithm eventually terminates when $\mathcal{X}_c$ becomes empty. In this case, for the given obstacle configuration, the algorithm finds the unique Hamiltonian cycle.

In figure 6(a) the ESSENTIAL-CHAINS algorithm finds a contradiction. When vertex 37 is picked from the candidate set $\mathcal{X}_c$, the set $n_{\mathcal{S}'}(37) - n_K(37) = \{36, 34\}$. Both 34 and 36 have degree 2 in K, thus the algorithm concludes that $\mathcal{S}'$ is non-Hamiltonian. This can be seen in the following manner: under the assumption that a Hamiltonian cycle exists, K should contain exactly two edges incident on every node. One edge incident on 37 is already a member of K. For the other edge, one has a choice between $37 - 34$ and $37 - 36$. Now since two edges incident on each of the vertices 36 and 34 are already in K, for both choices $37 - 34$ and $37 - 36$, a degree 3 vertex will result. This is a contradiction.

In figure 6(b), ESSENTIAL-CHAINS is unable to determine whether $\mathcal{S}'$ is Hamiltonian or not. The thick edges are essential in a Hamiltonian cycle if one exists. Our main result of this paper, proves that $\mathcal{S}'$ Hamiltonian for obstacle configuration shown in figure 6(b).

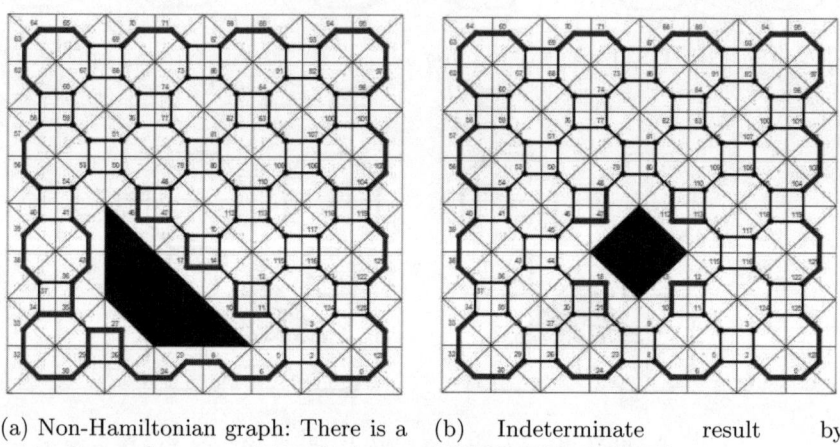

(a) Non-Hamiltonian graph: There is a contradiction for vertex 37 in the lower left corner.

(b) Indeterminate result by ESSENTIAL-CHAINS.

Fig. 6. Outcomes of ESSENTIAL-CHAINS

4.2 Main Result

The main result of the chapter is a positive result. But before we present our main result we need the following Lemma to prove it.

Lemma 1. *Let L be odd and $V \subset V_S$ such that the induced subgraph $\mathcal{S}[V]$ is a rectangular lattice with octagons on the four corners. Then $\mathcal{S}[V]$ is Hamiltonian.*

Proof. We prove this lemma by mathematical induction on the dimension of the lattice. First step verification: the lattice is 1×1. In other words, if V is such that $\mathcal{S}[V]$ is an octagon, then $\mathcal{S}[V]$ is trivially Hamiltonian.

All chains of octagons with interleaving squares can be proved to be Hamiltonian using induction as shown in figure 7(a). Now we know that all chains of m octagons with interleaving squares are Hamiltonian. As our new induction hypothesis, we assume that a rectangular lattice $m \times n$ of octagons with interleaving squares is Hamiltonian. Now as shown in figure 7(b), we apply the induction step again to prove that the rectangular lattice of dimension $m+1 \times n$ bounded by octagons on the four sides is Hamiltonian.

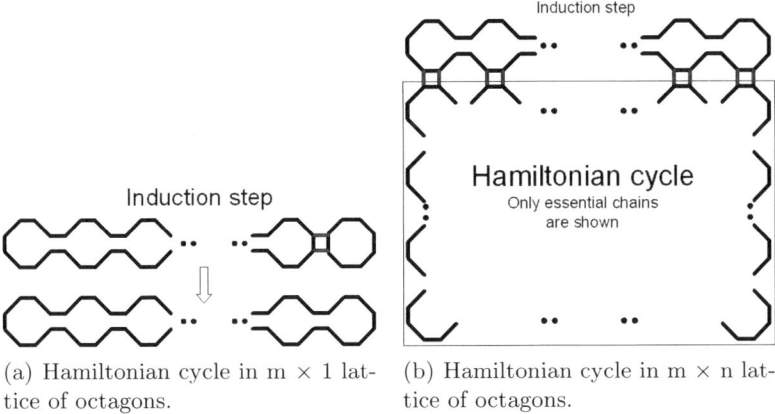

(a) Hamiltonian cycle in m × 1 lattice of octagons.

(b) Hamiltonian cycle in m × n lattice of octagons.

Fig. 7. Odd level dual graph for any rectangular region is Hamiltonian

Theorem 1. *If the induced graph $\mathcal{S}[\mathcal{O}]$ is a simple cycle, and ESSENTIAL-CHAINS does not come up with a contradiction, then $\mathcal{S}'$ is Hamiltonian.*

Proof. We prove this result by explicitly showing that a Hamiltonian cycle can be constructed in $\mathcal{S}'$. Here we give a sketch of the proof for an odd level of decomposition $L = 2k+1$. Since $\mathcal{S}[\mathcal{O}]$ is a simple cycle, it can only be a square or an octagon. If $\mathcal{S}[\mathcal{O}]$ is an octagon, it partitions the entire graph into rectangular regions, both when it is an interior octagon (figure 8(a)) and a periphery octagon (figure 8(b)). Now we know by Lemma 1 that the rectangular partitions of the graph are Hamiltonian. The Hamiltonian cycles of each of these rectangular partitions can be stitched together to form one Hamiltonian cycle for $\mathcal{S}'$, as shown in figure 8(e). If $\mathcal{S}[\mathcal{O}]$ is a square, then for all the squares that exist on the periphery of the graph, the ESSENTIAL-CHAINS algorithm finds a counter example, thus proving the non-existence of a Hamiltonian cycle. However, for other squares, ESSENTIAL-CHAINS does not come up with a contradiction and for all such cases, a bounding cycle can be found (figures 8(c) and 8(d)). Again, as in the case of an octagon, the rectangular partitions created are Hamiltonian, by Lemma 1. Stitching together Hamiltonian cycles as shown in figure 8(f) gives us a Hamiltonian cycle for $\mathcal{S}'$.

The above result holds true for even values of L and the proof is very similar to the one above.

Theorem 1 presents a lot of interesting and intriguing research questions. The idea of finding a bounding box and connecting Hamiltonian cycles can be used when the planar region to be explored has multiple disconnected obstacles. This idea can also be utilized for collaborative exploration by multiple vehicles, where each vehicle is assigned a rectangular partition and it executes the optimum cyclic tour computed for that particular partition. Collaborating vehicles can also exploit the multiple connecting edges their optimal cycles to those of their partners and switch tours. This may be helpful when the collaborating agents

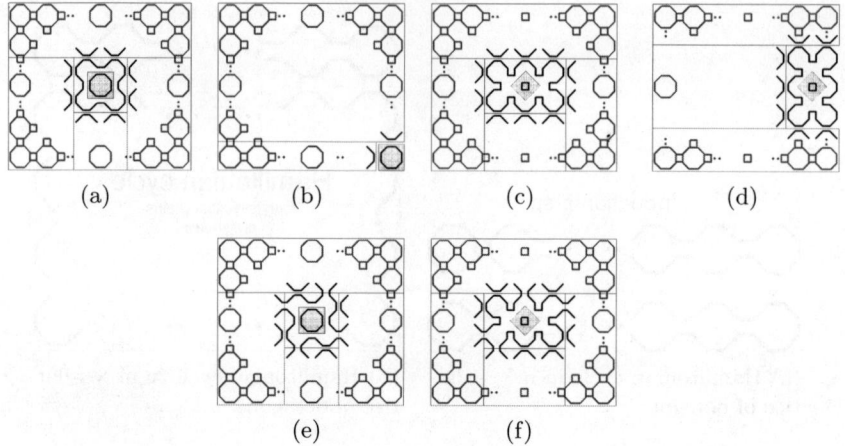

Fig. 8. Illustration of proof of the main result

are heterogenous and a vehicle with a certain kind of capability is needed at a certain location. A stronger and more useful result will be: to find the largest dimension of a bounding box such that if the obstacle configuration results in a non-Hamiltonian $\mathcal{S}'$ for this bounding box, the obstacle configuration will result in non-Hamiltonian $\mathcal{S}'$ for any bounding box, no matter how large it is. We are currently investigating the properties of the dual graph to gain further insight into the problem. The eventual aim of this research is to characterize the minimum number of repetitions of tiles for all obstacle configurations and to find provable algorithms to find the optimal cyclic tours. We now present a few heuristics based algorithms. Using results from this section we evaluate the performance of the algorithms for obstacle configurations where we know the optimum.

5   Algorithm Development and Comparison

In section 4.1, we described the ESSENTIAL-CHAINS algorithm that determines the set of chains K (as in figure 6(a)) that are essential in a Hamiltonian cycle, if one exists. As we observed in figure 6(b), the algorithm can terminate without a conclusive answer to the existence of a Hamiltonian cycle. In this section, we describe a heuristic based cycle maximization algorithm that tries to find the longest cycle in the induced dual subgraph S'. Later in this section, we show that the cycle maximization algorithm does not necessarily find the optimal solution. We also show that by including a few conditions, we can make the cycle maximization step always output a Hamiltonian cycle for the class of obstacles considered in Theorem 1.

5.1   Pre-cycle Computation

In this section, we present the first stage of the cycle maximization algorithm. We refer to this as the pre-cycle computation step. In the presence of obstacles, the Sierpinski tour gets partitioned into one or more chains as shown in figure 9(a). The idea is to use these chains to find a Hamiltonian cycle in $\mathcal{S}'$. We describe the algorithm in more detail below.

Group the obstacle tiles into k subsets $\mathcal{O}_1, \mathcal{O}_2, \cdots \mathcal{O}_k$, where each subset $\mathcal{O}_i$ is the largest set such that all its members have continuous Sierpinski ordering. For example (figure 9(a)), for the Sierpinski decomposition level $L = 7$, let $\mathcal{O}$ be $\{47, 80, 81, 82\}$. It turns out, for this example, $k = 2$, $\mathcal{O}_1 = \{47\}$ and $\mathcal{O}_2 = \{80, 81, 82\}$.

The k subsets, $\mathcal{O}_1, \mathcal{O}_2, \cdots, \mathcal{O}_k$, generate an equal number of chain partitions $\mathcal{P}_1, \mathcal{P}_2, \cdots, \mathcal{P}_k$ as shown in figure 9(a). In our example, the two subsets $\mathcal{O}_1$ and $\mathcal{O}_2$ create two chain partitions $\mathcal{P}_1 = \{48, 49, \cdots 78, 79\}$ and $\mathcal{P}_2 = \{83, 84, \cdots 126, 127, 0, 1, 2, \cdots, 45, 46\}$ respectively.

After finding the chain partitions, we find the candidate bridges. A candidate bridge is essentially a pair of vertices (x, y) in the induced subgraph $\mathcal{S}'$ such that the edge $xy \in E_{\mathcal{S}'}$ (figure 9(a)). In other words, tiles that correspond to x and y in the original Peano-Cesaro triangulation are adjacent. To find the candidate bridges for a subset, say, $\mathcal{O}_i = \{w_{m+1}, w_{m+2}, \cdots, w_{m+r}\}$, we first identify the pair of vertices (u, v) such that u is the largest tile index less than $(m + 1)$ and v is the smallest tile index greater than $(m + r)$. In our example, $(u, v) = (46, 48)$ corresponding to the subset $\mathcal{O}_1$ and $(79, 83)$ corresponding to the subset $\mathcal{O}_2$.

Now, for each subset $\mathcal{O}_i$, starting from the pair (u, v), we search (tile indices lesser than u for x and tile indices greater than v for y) for the first occurrence of a pair of vertices $(x, y) \in \mathcal{S}'$, where, x and y are defined in the previous paragraph. Geometrically interpreting the figure 9(a), we search (outwards and starting from the pair (u, v)) for the pair of vertices $(x, y) \in \mathcal{S}'$ that minimizes the distance d_c between them. Here, the distance d_c between two vertices u and v is expressed as:

$$d_c = min(abs(u - v) - 1, 2^L - abs(u - v) - 1) \qquad (1)$$

where L is the Peano-Cesaro decomposition level, $abs()$ is the absolute value function and $min()$ returns the minimum value. In the above example, the two candidate bridge pairs (x, y) determined are $(46, 49)$ and $(73, 86)$ for $\mathcal{O}_1$ and $\mathcal{O}_2$ respectively. The pre-cycle computation outputs a cycle (figure 9(b)) which is the input to the cycle maximization step discussed in the next section.

5.2   Cycle Maximization

Denote the cycle found in section 5.1 by $\mathcal{C}$ (figure 9(b)). The cycle $\mathcal{C}$ is the input to the *cycle maximization* step. Denote the set of vertices in the cycle $\mathcal{C}$ by $V_{\mathcal{C}}$. We call the graph $\mathcal{S}[V_{\mathcal{S}'} - V_{\mathcal{C}}]$ as $\mathcal{G}$. Now if $\mathcal{S}'$ is 2-connected, there exists at least 2 vertices in $\mathcal{C}$ whose neighborhood set $n_{\mathcal{S}'}()$ in $\mathcal{S}'$ has vertices in $\mathcal{G}$. Let us denote two such vertices in $\mathcal{C}$ by $x_{\mathcal{C}}$ and $y_{\mathcal{C}}$. Let their neighbors in $\mathcal{G}$ be $x_{\mathcal{G}}$

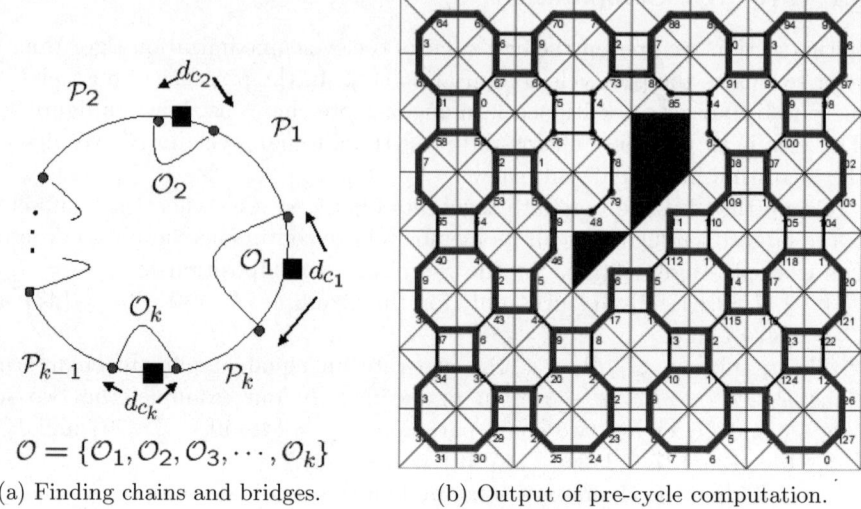

(a) Finding chains and bridges.   (b) Output of pre-cycle computation.

Fig. 9. Illustration for finding partition chains, finding bridges and computing pre-cycle

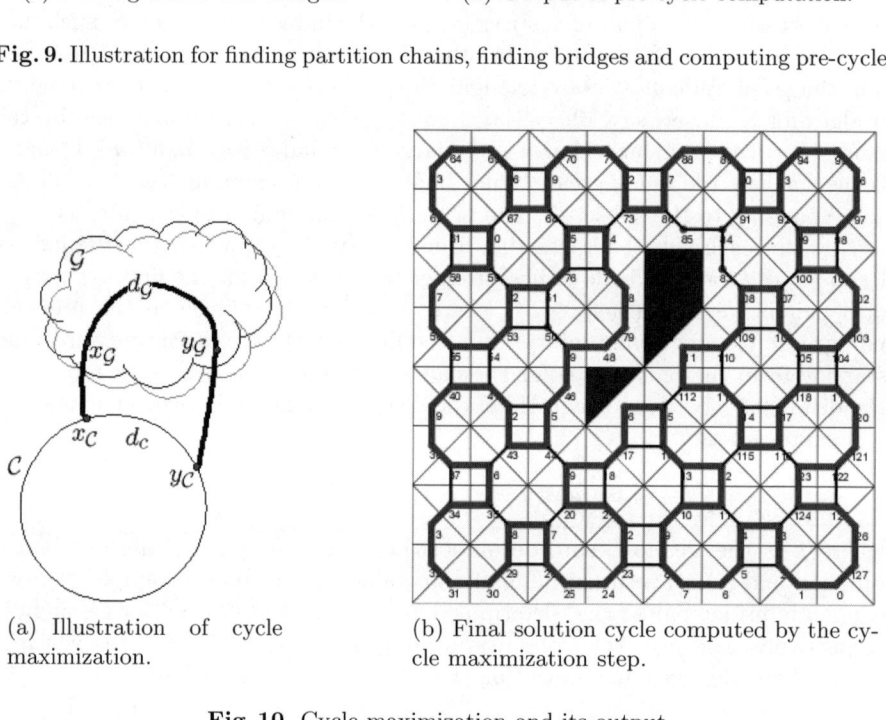

(a) Illustration of cycle maximization.

(b) Final solution cycle computed by the cycle maximization step.

Fig. 10. Cycle maximization and its output

and $y_{\mathcal{G}}$ respectively (figure 10(a)). The algorithm now finds the shortest path in $\mathcal{G}$ between $x_{\mathcal{G}}$ and $y_{\mathcal{G}}$. Let us denote the distance of this shortest path as $d_{\mathcal{G}}$. Therefore the distance between $x_{\mathcal{C}}$ and $y_{\mathcal{C}}$ in $\mathcal{G}$ is $d_{\mathcal{G}} + 2$. Now if $d_{\mathcal{G}} + 2 > d_{\mathcal{C}}$, where $d_{\mathcal{C}}$ is the shortest path distance between $x_{\mathcal{C}}$ and $y_{\mathcal{C}}$ in $\mathcal{C}$, then it is possible to increase the number of nodes in the initial cycle $\mathcal{C}$. This can be done by

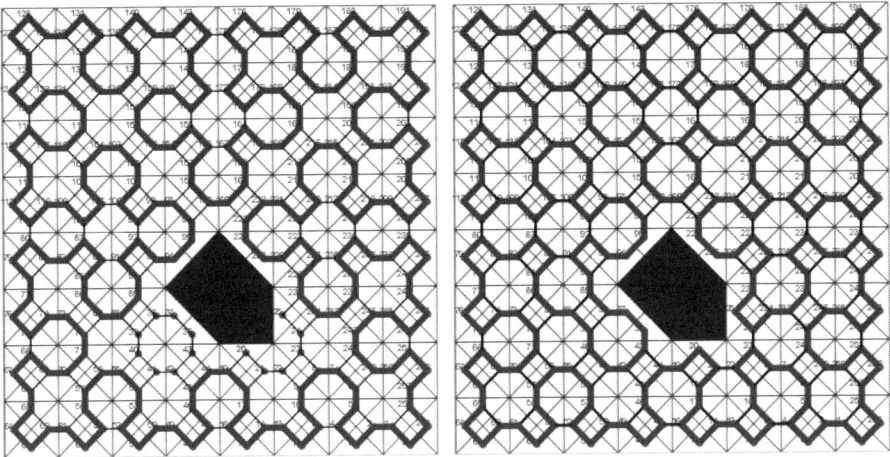

(a) Initial cycle $\mathcal{C}$ (*thick lines*) determined by the pre-cycle computation step.

(b) A Hamiltonian cycle *(thick lines)* determined as the solution cycle using cycle maximization step.

Fig. 11. Illustration of cycle maximization. In this example, cycle maximization comes up with a Hamiltonian cycle (an optimal solution in this case).

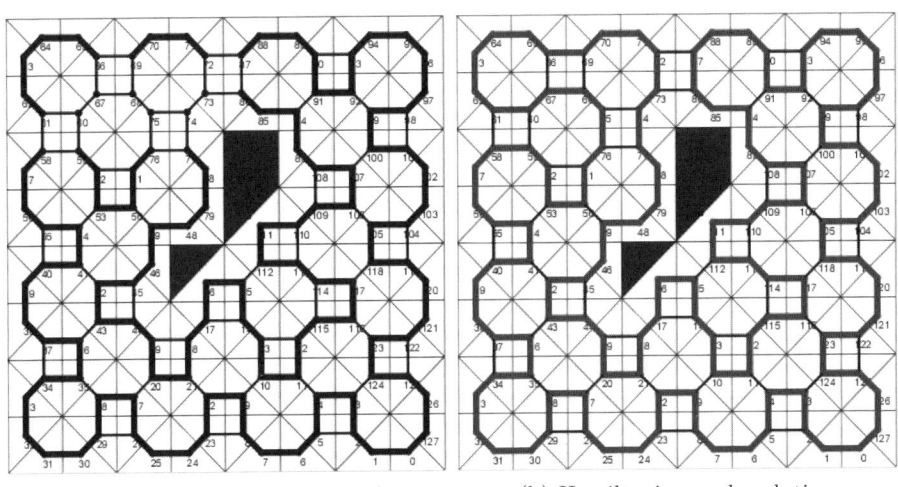

(a) Essential Chains generated.

(b) Hamiltonian cycle solution.

Fig. 12. Performance of ESSENTIAL-CHAINS algorithm

replacing the path in $\mathcal{C}$ between $x_\mathcal{C}$ and $y_\mathcal{C}$ by the path between them in $\mathcal{G}$. The algorithm terminates when no such increase in the cycle length is possible. This concept is illustrated in figure 10(a).

After all the possible replacements are done, the resulting cycle $\mathcal{C}$ is the final solution cycle. Figure 10(b) illustrates a solution cycle determined using the cycle maximization step. For certain cases, cycle maximization finds the optimal tour as illustrated in figures 11(a) and 11(b).

In order to assess the performance of our heuristic based cycle maximization algorithm, we need to determine the minimum number of repetitions attainable. For this, we ran our *ESSENTIAL- CHAINS* algorithm and the computed set of essential chains, K, is illustrated in figure 12(a). Now by observation, we join the computed essential chains to find a Hamiltonian cycle for this obstacle configuration as shown in figure 12(b). Thus, for this example, we observe that our heuristic based cycle maximization algorithm is four vertices away from the attainable optimal solution.

We also observe that the vertices missed $(50, 83, 84, 85)$ by the cycle maximization step (figure 10(b)) were a part of the essential chains set, K (figure 12(a)). Therefore, it may be possible that by forcing the cycle maximization not to drop vertices that are essential, the optimal solution can always be obtained. We are currently working on conditions, which, if included in the cycle maximization step, can always output the optimal tour for the class of obstacles described in Theorem 1.

6   Conclusion

In this chapter we use Peano-Cesaro tiling to divide the exploration region into triangular tiles, with some tiles marked as obstacles. The problem of finding the minimum repetition cyclic tour of the non-obstacle tiles, under mobility constraints has been posed using results from graph theory, where vertices and edges represent tiles and allowed moves respectively. The resulting graph is referred to as S'. We have devised an algorithm that collects all the essential components of a Hamiltonian cycle (assuming one exists). The algorithm determines whether S' is Hamiltonian or not in some cases. In cases where the number of obstacle tiles is sufficiently low, and the obstacles are far away from the boundary of the region, the algorithm does not provide a deterministic answer. However, it does output chains that are essential components of a Hamiltonian cycle, if one exists. The main result is: S' *is Hamiltonian if the obstacle tiles form a simple cycle*. We also provide a heuristic algorithm and compare its performance for the obstacle scenarios where the minimum number of repetitions is known.

Acknowledgements

The authors would like to acknowledge the help provided by Dr. Junxian Wang, UtopiaCompression Corporation for the preliminary part of this work. This work was supported under NASA phase II SBIR grant number NNM06AA08C.

References

1. J. J. Bartholdi and P. Goldsman. Continuous Indexing of Hierarchical Subdivisions of the Globe. *International Journal of Geographical Information Science*, 15(6):489–522, 2001.

2. J. J. Bartholdi and P. Goldsman. The Vertex-Adjacency Dual of a Triangulated Irregular Network has a Hamiltonian Cycle. *Operations Research Letters*, 32:304–308, 2004.

3. J. J. Bartholdi and L. K. Platzman. Heuristics based on spacefilling curves for combinatorial problems in the plane. *Management Science*, 34(3):291–305, 1988.

4. T. H. Cormen, C. E. Leiserson, R. L. Rivest, and C. Stein. *Introduction to Algorithms*. The MIT Press, second edition.

5. UtopiaCompression Corporation. Advanced, Disruptive Intelligent Based Image Compression Technologies. *National Institute of Standards and Technology: Advanced Technology Program (Final report)*, Nov 2006.

6. R. Diestel. *Graph Theory*. Springer, second edition.

7. S. Hutchinson G. Kantor W. Burgard L. Kavraki S. Thrun H.Choset, K. Lynch. *Principles of Robot Motion: Theory, Algorithms, and Implementation*. MIT Press, 2005.

8. H. Imai. Worst-case Analysis for Planar Matching and Tour Heuristics with Bucketing Techniques and Spacefilling Curves. *Journal of the Operations Research Society of Japan*, 29(1):43–68, 1986.

9. M. Jun and R. D'Andrea. *Path Planning for Unmanned Aerial Vehicles in Uncertain and Adversarial Environments*, chapter 6, pages 95–111. Cooperative Control: Models, Applications and Algorithms. Kluwer Academic Publisher, 2002.

10. S. M. LaValle. *Planning Algorithms*. Cambridge University Press, 2006.

11. B. Mandelbrot. *The Fractal Geometry of Nature*. W. H. Freedman and Co., 1983.

12. L. K. Platzman and J. J. Bartholdi. Spacefilling curves and the planar travelling salesman problem. *Journal of the Association for Computing Machinery*, 36(4):719–737, 1989.

13. H. Sagan. *Space filling curves*. Springe-Verlag, 1994.

14. K. Savla, F. Bullo, and E. Frazolli. On Traveling Salesperson Problems for Dubin's Vehicle: Stochastic and Dynamic Environments. *Proceedings of the IEEE Conference on Decision and Control*, 2005.

15. K. Savla, F. Bullo, and E. Frazolli. Traveling Salesperson Problems for the Dubin's Vehicle. *IEEE Transactions on Automatic Control*, June 2006.

An Analysis and Solution of the Sensor Scheduling Problem

Mesut Yavuz[1] and David Jeffcoat[2]

[1] Research and Engineering Education Facility, University of Florida, Shalimar, FL
yavuz@reef.ufl.edu
[2] Air Force Research Laboratory, Munitions Directorate, Eglin AFB, FL
david.jeffcoat@eglin.af.mil

Abstract. This chapter addresses the scheduling problem of a sensor that constantly collects information from multiple sites. In the existing literature, the problem is solved by probabilistic approaches, potentially generating schedules in which a site is not visited for a long time. To overcome this deficiency, this chapter presents a deterministic approach formulated as an integer linear program. Upon showing that the problem is NP-Hard, the chapter develops valid lower and upper bounds and proposes two constructive heuristic methods. Tested via an extensive computational study, the heuristic methods are proven efficient and effective in solving the problem.

1 Introduction

This chapter is concerned with scheduling a single sensor to maintain an estimate of a dynamic physical attribute (e.g., position) of multiple targets. The research builds on previous work by Tiwari et al. [9], Yerrick et al. [11] and Yerrick et al. [12]. Tiwari et al. [9] present a feasibility criterion for a single sensor to maintain a bounded estimate of an attribute at multiple locations. Yerrick et al. [11] demonstrate by simulation the feasibility criterion presented in [9] and develop a heuristic to find a good sensor motion model given the dynamics of the system under observation. Yerrick et al. [12] provide an optimal sensor coverage solution for two sensor motion models given a model of the observed system's dynamics. All three papers consider probabilistic strategies for the motion of the single sensor among the sites. A similar model in the literature is known as the traveling inspector model [4, 5]. In this chapter, we focus on deterministic methods to schedule the sensor's motion. A deterministic approach overcomes one disadvantage of probabilistic motion: with any random motion strategy, there is nonzero probability that a particular site will not be visited at all in any finite time horizon.

Figure 1 provides an illustration for a three-site scenario. At the time instant pictured, the sensor is focused on site three. In its current position, the sensor can observe the characteristics of site three, but cannot observe sites one or two. In the next discrete time step, we assume that the sensor can move (or refocus) from site three to either of the other two sites, or can maintain its

M.J. Hirsch et al. (Eds.): Adv. in Cooper. Ctrl. & Optimization, LNCIS 369, pp. 167–177, 2007.

current position. At each time step, since the sensor focuses on exactly one of the sites, the sensor's processor can update its information for only one site and must estimate the rest. Therefore, information loss at a site increases with the number of time steps that the site has not been visited. Sites may have different rates of change, and, hence, the criticality of information loss may vary among the sites. A successful sequence is one that balances the visit frequencies of the sites to minimize overall information loss. The goal of this chapter is to develop methods to construct such sequences.

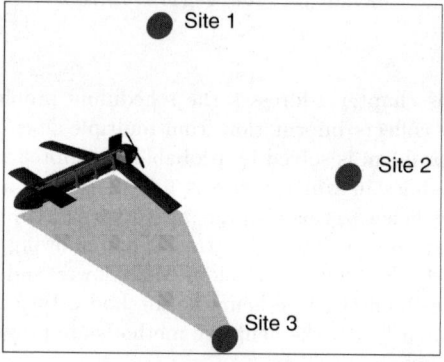

Fig. 1. Three site example

The remainder of this chapter is organized as follows. In Section 2 we formulate the problem as an integer linear programming problem and in Section 3 we analyze its properties. In Section 4 we develop a lower and an upper bound on the objective function of our model. In Section 5 we propose two heuristic solution procedures and in Section 6 we evaluate their performance. In Section 7, we conclude by summarizing our contribution and discussing possible future research directions.

2   A Mathematical Model

Let $x_{i,t}$ be the binary decision variable denoting whether the sensor is scheduled to visit site i at time t, and $y_{i,t}$ denote the last time site i was visited as of the end of time t. Note that $y_{i,t} = t$ happens only in time intervals in which the sensor visits site i. When the sensor is focused on site i, it updates the status of the site. In other words, we have perfect information of the site in that time step. Since the sensor cannot focus on more than one site at the same time, focusing on one site means losing information about the current states of the other sites. The extent of the information loss depends on the activity rate of a site. We can afford to ignore less active sites for a large number of time steps, whereas more active sites must be visited frequently. In this chapter, we assume that there is no cost for movement or observation; our whole concern is the cost of lost information.

We associate with each site i a fixed cost a_i and a variable cost b_i of information loss. More specifically, a fixed penalty of not visiting a certain site is incurred for each time step in which the sensor is away from the site. In addition, a variable cost is incurred for each time unit that has passed since the sensor's last visit to that site, providing ever-increasing motivation for the sensor to return to a neglected site. (The cost parameters implicitly model the activity level at a site or the importance of a site.) The objective function in our model minimizes the maximum penalty incurred for a sensor schedule defined over a finite time horizon. If we are given a planning horizon consisting of T periods, then the following integer linear program can be formulated.

$$\text{Minimize} \qquad C \tag{1}$$

Subject to

$$C + a_i x_{i,t} + b_i y_{i,t} \geq a_i + b_i t, \qquad \forall i = 1, .., n; \forall t = 1, .., T \tag{2}$$

$$\sum_{i=1}^{n} x_{i,t} = 1, \qquad \forall t = 1, .., T \tag{3}$$

$$y_{i,t} - y_{i,t-1} \leq t x_{i,t}, \qquad \forall i = 1, .., n; \forall t = 1, .., T \tag{4}$$

$$y_{i,t} \leq t, \qquad \forall i = 1, .., n; \forall t = 1, .., T \tag{5}$$

$$y_{i,0} = 0, \qquad \forall i = 1, .., n \tag{6}$$

$$C > 0, \tag{7}$$

$$x_{i,t} \in \{0, 1\}, \qquad \forall i = 1, .., n; \forall t = 1, .., T \tag{8}$$

$$y_{i,t} \in \{0\} \cup \mathbb{Z}^+, \qquad \forall i = 1, .., n; \forall t = 1, .., T \tag{9}$$

The model is built as a fully linear model, that is, the objective function and constraints are all linear functions of the decision variables. Note that defining C as a variable and defining it in a constraint is critical for the linearity of the formulation. The objective function of the model (1) simply aims to minimize the maximum cost defined by the first constraint (2). More specifically, $C \geq a_i(1 - x_{i,t}) + b_i(t - y_{i,t})$, for all i and t. Constraint (3) assures that the sensor visits exactly one site in each stage. Constraints (4) and (5) together assure that $y_{i,t}$ is updated only when the sensor is on site i, and remains constant at other times. Constraint (6) initializes variable y. Finally, constraints (7-9) define the decision variables C, x and y as nonnegative, binary and nonnegative-integer variables, respectively.

3   Structural Properties of the Problem

The optimization model of the previous section is built upon a given sequence length, T. However, in practice, we may not be given such a length but asked to find infinitely long sequences. This property, regardless of the computational complexity of the formulated integer programming model, makes the problem a challenging one. This property motivates us to study the problem from a different perspective, that is, periodic scheduling. If an infinite sequence can be constructed such that the objective function value is C, every site i must be visited by a period p_i where $a_i + (p_i - 1)b_i \leq C$ and $a_i + p_i b_i > C$. Various periodic scheduling problems arise in the context of computer and telecommunications systems, and have received significant academic interest. Consider a satellite with finite memory capacity that needs to download its memory periodically. If we have multiple satellites each serviced by a single download facility, then construction of a download sequence would constitute a periodic scheduling problem. The first result we use from the periodic scheduling literature proves the computational complexity of the sensor scheduling problem as follows.

Theorem 1. *The sensor scheduling problem is NP-hard.*

Proof. Bar-Noy et al. [1] show that the periodic scheduling problem is NP-hard, with a reduction from the graph coloring problem. Here, we reduce the periodic scheduling problem to our problem. An instance of the periodic scheduling problem is given as follows.

Given m machines and service intervals $p_1, p_2, .., p_m$ such that $\rho = \sum_{i=1}^{m} 1/p_i$ ≤ 1, does there exist an infinite maintenance service schedule of these machines in which consecutive maintenance times for machine i are exactly p_i time-slots apart and no more than one machine is serviced in a single time-slot?

For a given instance of the periodic scheduling problem, we first create m sites with $a_i = 0$ and $b_i = C/p_i$, where C is an arbitrarily selected constant. Next, we find the smallest positive integers c and d such that $c/d = 1 - \sum_{i=1}^{m} 1/p_i$ (if $\sum_{i=1}^{m} 1/p_i = 1$, then we assign $c = d = 0$). Then we create c additional sites each with $a_i = 0$ and $b_i = C/d$. Note that $d = 0$ is only possible when $c = 0$, in which case no additional sites are created. If we can find a solution to this problem with $n = m + c$ sites such that maximum cost is at most C, then in that solution the first m sites will be visited exactly every p_i time-slots, since the density (ρ) is now 1.   □

Theorem 2. *There exists an optimal solution in which no site is visited in two consecutive stages.*

Proof. It is clear that when $n > 1$, the minimum cost for site $i = 1, .., n$ will be at least $a_i + b_i$, since at least one of the other sites must be visited between two consecutive visits to site i. Therefore, at any stage in the sequence, staying at the same site results in a zero cost for site i, whereas it increases the variable cost for all other sites $i' \neq i$ by $b_{i'}$. Hence, staying at the same site can only increase the maximum cost with respect to the other sites and it can never decrease the maximum cost factor at that site. Using this property, any optimal solution to the problem can be converted to another optimal solution in which no site is visited in two consecutive stages.   □

Corollary 1. *Instances with two sites ($n = 2$) are trivial.*

Proof. This result directly follows from Theorem 2: if an optimal solution exists such that the sensor never stays at the same site in two consecutive stages and there are only two sites, then in each stage there is exactly one site that the sensor can focus on.   □

Corollary 2. $\max_i(a_i + b_i)$ *is a lower bound for* C.

Proof. From Theorem 2 we know that an optimal solution to the sensor's schedule can be found by constantly moving between the sites. Therefore, there will be at least one time-slot in which a site (i) is not visited, thus the cost incurred at site i will be at least $a_i + b_i$. Since the objective function C is greater than or equal to those cost factors, it must be at least as large as the largest of them, which completes the proof.   □

4   Lower and Upper Bounds on the Objective Function

Corollary 2 provides a loose lower bound on C. Before obtaining a tight lower bound, we first elaborate our discussion on periodic scheduling. A special version of the periodic scheduling class of problems is known as *pinwheel scheduling*, see [2, 3] for further reading. In the pinwheel scheduling problem, a number (n) of tasks each with a possibly distinct period (p_i) are aimed to be scheduled such that two consecutive executions of task i are not separated by more than p_i time steps. The sensor scheduling problem reduces to the pinwheel scheduling problem for a given C, and, hence, is a general case thereof. An instance of the pinwheel scheduling problem is characterized by its density $\rho = \sum_{i=1}^{n} 1/p_i$. It is well known that instances with $\rho > 1$ cannot be scheduled. Instances with $\rho \leq 1$ may or may not be scheduled. A widely believed conjecture is that all instances with $\rho \leq 5/6$ are schedulable. However, no one to date has been able to prove or disprove this conjecture.

   We use the properties of the pinwheel scheduling problem to develop a tight lower bound and conjecture an upper bound. Both bounds are obtained using the search procedure, i.e., Algorithm `Search_on_C`$(n,\mathbf{a},\mathbf{b},\rho^U)$, depicted in Figure 2. The algorithm first uses Corollary 2 to find the minimum C value that is possible, and then performs an increasing search on C until a pinwheel instance with a density less than or equal to the designated threshold is obtained. In a basic setting, if $\mathbf{a}$ and $\mathbf{b}$ are integer vectors, the search can be performed by increasing C by one. Our algorithm performs the search intelligently in that it calculates the smallest candidate for the increased C value that will change at least one p_i value. Therefore, it is guaranteed that every time C is increased, a pinwheel instance with a lower density is obtained. We also denote the optimal solution of the sensor scheduling problem by C^*.

Theorem 3. $C^L = $ `Search_on_C` $(n,\mathbf{a},\mathbf{b},1)$ *is a lower bound for* C^*.

Proof. The proof is based upon the following two simple observations: ρ is non-increasing in C and there is no feasible schedule with $\rho > 1$. Thus, terminating the search when $\rho \leq 1$ assures that the minimum C value that may be schedulable is returned.   □

Proposition 1. $C^C = $ `Search_on_C`$(n,\mathbf{a},\mathbf{b},5/6)$ *is always schedulable, and, hence, is an upper bound for* C^*.

Algorithm Search_on_$C(n,\text{a},\text{b},\rho^U)$

BEGIN

1. Set $p_i = 2, i = 1, 2, .., n$.
2. Set $C_i = a_i + b_i, i = 1, 2, .., n$.
3. Set $C = \max_i C_i$.
4. Set $\rho = \sum_{i=1}^{n} \frac{1}{p_i}$.
5. While $\rho > \rho^U$

   BEGIN

   6. Set $p_i = \left\lfloor \frac{C - a_i}{b_i} \right\rfloor + 1, i = 1, 2, .., n$.
   7. Set $C'_i = a_i + p_i b_i, i = 1, 2, .., n$.
   8. Set $C' = \min_i C'_i$.
   9. Update $\rho = \sum_{i=1}^{n} \frac{1}{p_i}$.
   10. If $\rho > \rho^U$, then update $C = C'$.

   END.

END.

Fig. 2. Pseudo-code for Algorithm Search_on_$C(n,\text{a},\text{b},\rho^U)$

This is a direct extension of the conjecture on the schedulability of pinwheel instances. Therefore, its proof does not exist in the literature and is out of the scope of this chapter.

At this point, we focus on obtaining a valid upper bound on C^*, based on a special type of periodic scheduling problem in the context of just-in-time (JIT) manufacturing. An ultimate goal of the JIT philosophy is to manufacture products at the exact time of demand, and, thus, minimize the costs associated with carrying inventories as well as backlogging or losing orders. Since the exact time of demand cannot be known in advance, demand is assumed uniformly distributed over the planning horizon. Accordingly, an ideal manufacturing schedule would produce each product in the exact rate of its demand. For example, if demand is expected to be 10 units for a given product in a 30-day horizon, then we should produce one unit every three days. For more on the JIT scheduling problem, we refer the reader to [6, 7, 10]

Steiner and Yeomans [8] address the JIT scheduling problem and prove that there always exists a sequence in which the ideal and actual cumulative production quantities of a product differ by at most one. Here, each product has a demand d_i in the planning horizon. The total demand $D = \sum_i d_i$ defines the length of the sequence, and, hence, the length of the planning horizon. Ideal cumulative production quantity up to stage k is defined by kd_i/D. Actual cumulative production quantity is the number of units of a product sequenced in the first k stages.

The JIT scheduling problem is similar to the sensor and pinwheel scheduling problems in structure, that is, the goal of evenly spacing products/sites/tasks over the sequence is common to all. Building on this point, we define a period $p_i = D/d_i$ for the production of i. Steiner and Yeomans's result [8] shows that there always exists a sequence that produces exactly one unit of product i in stages $(r - 1)p_i + 1, .., rp_i$, for all $r = 1, 2, .., d_i$. Revisiting the above example, this result means that exactly one unit is produced in stages 1-3, one in 4-6, and so forth. Here, note that sequencing the product in stages $1 - 6 - 7 - 12 - \ldots$ is possible; we relate this result to the sensor scheduling problem as follows.

Lemma 1. *For an instance of the sensor scheduling problem, if $\mathbf{p}$ is the vector of periods obtained using C^L; then a sequence always exists such that two consecutive visits to site i are at most $2p_i - 1$ time steps apart.*

Proof. Let LCM be the least common multiplier of $p_1, p_2, .., p_n$. We can create an instance of the JIT scheduling problem by creating n products each with a demand of $d_i = LCM/p_i$. The summation of the demands may be less than LCM, in which case the gap should be filled by creating dummy products with a demand of 1 so that the dummies do not have an effect on the sequence. Through JIT scheduling, one can obtain a sequence where product i is produced exactly once in stages $(r-1)p_i + 1, .., rp_i$ for all $r = 1, 2, .., d_i$ and $i = 1, 2, .., n$. Also note that both d_i and p_i are integers. The largest possible distance between the positions r and $(r+1)$th copies of product i is $(r+1)p_i - ((r-1)p_i + 1) = 2p_i - 1$, for $r = 1, 2, .., d_i - 1$. □

Theorem 4. $C^U = 2C^L - \min_i a_i$ *is a valid upper bound on C^*.*

Proof. Given a C^L, we obtain the periods for each site with $p_i = \left\lfloor \frac{C^L - a_i}{b_i} \right\rfloor + 1$. From Lemma 1, we know that we can always find a sequence in which two consecutive visits to site i are at most $2 \left\lfloor \frac{C^L - a_i}{b_i} \right\rfloor + 1$ time steps apart. Therefore the cost of information loss for site i is $C_i = a_i + (2 \left\lfloor \frac{C^L - a_i}{b_i} \right\rfloor)b_i \leq a_i + 2 \lfloor C^L - a_i \rfloor = 2C^L - a_i$. □

5   Heuristic Solution Approaches

The sensor scheduling problem is NP-Hard as shown earlier in this chapter. Furthermore, infinitely long sequences are sought as complete solutions to the problem. These two facts render exact solution methods impractical. Therefore, developing time-efficient constructive heuristic procedures is beneficial.

A constructive heuristic starts with a null solution, which is an empty sequence in our case. Recalling decision variable $y_{i,t}$ of our optimization model, we assume $y_{i,0} = 0$ for all $i = 1, 2, .., n$. In other words, it is assumed that at the beginning, we have perfect information about all sites. In each stage, exactly one site is visited, and, hence, there is exactly one i satisfying $y_{i,t} = t$ $(t = 1, 2, \ldots)$. For the $n - 1$ sites not visited in stage t, we have $y_{i,t} = y_{i,t-1}$. Now we define a time since the last visit to site i by stage t: $z_i(t) = t - y_{i,t}$. Note that in each stage exactly one $z_i(t) = 0$ and the remaining $n - 1$ are positive integers. Moreover, $z_i(t)$ increases in t until site i is visited.

As discussed earlier in the chapter, for a given C, we can derive visit periods p_i for each site and reduce the problem to an instance of the pinwheel scheduling problem. If this instance is schedulable, then $z_i(t) \in \{0, 1, .., p_i\}$, for all $i = 1, 2, .., n$ and $t = 1, 2, \ldots$ Therefore, the number of different values the vector

$\mathbf{z}(t)$ can take is finite. This result implies that after a finite number of steps, the $\mathbf{z}(t)$ vector will repeat itself. That is, if we can identify such a stage, we can build a cyclic sequence by repeating the stages between consecutive occurrences of the same $\mathbf{z}(t)$. This result constitutes a main principle used in both our heuristic procedures. Another common principle is based upon Theorem 2, prohibiting the sensor from staying focused on the same site in two consecutive stages. In other words, our constructive heuristics evaluate all sites but the one that has been just visited for the next move, in all stages.

Our first constructive heuristic is a greedy procedure. It starts with the initial visit history as described above ($\mathbf{z}(0) = \mathbf{0}$). It evaluates all n sites that can be visited in the first stage. The selection of the site to visit is made to minimize the penalty of information loss, i.e., penalty of not visiting a site. After the selection the visit history is updated. Note that in the later stages the method evaluates $n-1$ sites for its next visit. Repeating this simple selection and update operations until a repetition in the visit history is observed constitutes the framework of our greedy heuristic. We improve its performance by adding a look-ahead feature. The heuristic still evaluates $n-1$ possible sites to visit in each stage, but makes the decision based on the cost observed in the next ℓ stages. Larger ℓ values are expected to yield better (lower cost) solutions on the average, however it is not guaranteed. On the other hand, the number of operations to perform increases with ℓ, rendering small ℓ values more computationally efficient. We call our first heuristic *greedy with look-ahead* (GLA).

Our second heuristic is an alternative greedy approach that dynamically sets a deadline to visit each site and then selects the site with the earliest deadline for the sensor's next move. More specifically, it starts with a small C and calculates periods p_i for each site to achieve that C value. For each $i = 1, 2, .., n$ and $t = 1, 2, ..$, the deadline for the next visit to site i is set to $y_{i,t-1} + p_i$. Then, the site with the earliest deadline is selected for the next visit (ties can be broken arbitrarily). However, if the C value at hand is too small, then in some stage the method will unavoidably have more than one site that must be visited in that stage. Since this is infeasible, the method increases C until at most one site must be visited in that stage. The termination again is based on observing a repetition in the visit history. We call this heuristic *dynamic deadlines* (DD).

6   Computational Study

We consider four different numbers of sites: $n \in \{4, 6, 8, 10\}$. The number of sites can also be considered the problem size. For each problem size, we pseudo-randomly create 100 test instances with $b_i \in \{1, .., 10\}$ and $a_i \in \{11, .., 100\}$. Therefore, we have a total of 400 test instances.

In this study, for each instance, we first obtain C^L, C^U and C^C. Then, we run the two heuristics. From our preliminary experiments we have observed that $\ell = n$ works best. Therefore, we run the GLA method with $\ell = n$ only.

We know that C^U is always greater than C^L and less than $2C^L$. However, we cannot make such clear inferences about C^C. Therefore, we are interested in the relative position of C^C to C^L and C^U. We calculate the relative position with $(C^C - C^L)/(C^U - C^L)$. Similarly, we calculate the relative positions of the solutions obtained by the GLA and DD heuristics, as well. For example, if the relative position of the GLA method's solution is calculated as 0.25, we understand it is located at 25% of the distance from the lower bound to the upper bound. In other words, smaller values represent better solutions. C^L, C^U and C^C are computed almost instantly, thus we are not concerned about their time consumption. The heuristic methods, on the other hand, can take a significant amount of computation time depending on the problem size. The results are summarized in Figures 3 and 4.

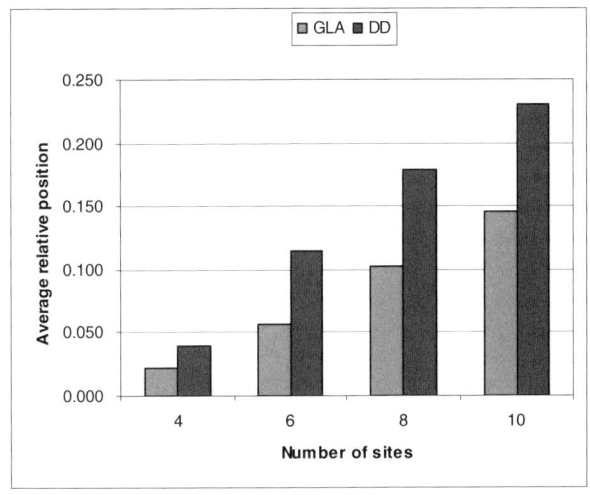

Fig. 3. Relative positions

Our two heuristics perform well in general as their relative position is always closer to the lower bound than the upper bound. Hence, we state that our methods are effective in solving the sensor scheduling problem. When the two heuristics are compared, we see that GLA outperforms DD on all problem sizes. Computation time of both methods increases significantly with problem size, DD taking longer than GLA. Thus, we state that GLA heuristic is superior to DD. Even so, the results show that both methods are computationally efficient in that they solve the problem in seconds.

The conjectured upper bound is found to be tighter than the valid upper bound developed in this chapter. Furthermore, the conjectured upper bound seems to work better than the heuristic methods on larger problem sizes. However, a sequencing procedure is not known in the existing literature to support the conjecture. Therefore, with their negligible computational burden and high performance, the heuristics proposed in this chapter can be used to solve the sensor scheduling problem in practice.

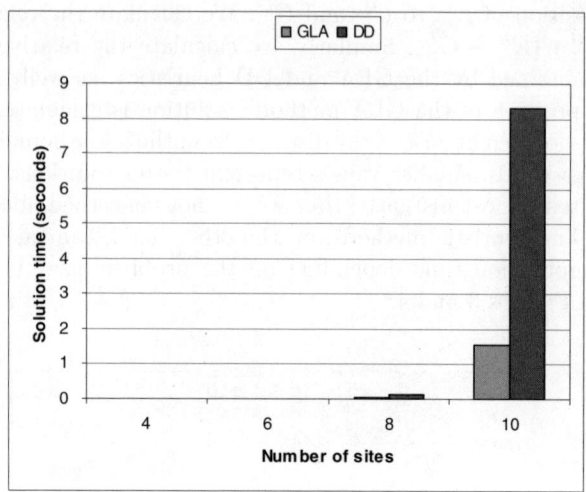

Fig. 4. Computation times

7 Conclusions

This chapter addresses the scheduling problem of a sensor that constantly collects information from multiple sites. In the earlier work, the problem is solved by probabilistic approaches, potentially generating schedules in which a site is not visited for a long time. To overcome this deficiency, this chapter presents a deterministic approach formulated as an integer linear program. Upon showing that the problem is NP-Hard, the chapter develops valid lower and upper bounds and proposes two constructive heuristic methods. Tested via an extensive computational study, the heuristic methods are efficient and effective in solving the problem.

The results also pinpoint the need to prove the widely believed "5/6" conjecture of the pinwheel scheduling literature and to develop efficient algorithms to solve the pinwheel scheduling problem. In the existing literature, algorithms developed for the pinwheel problem either require a small number (2-3) of distinct periods or have low density guarantees. A comprehensive scheduling method for the pinwheel is therefore critical for the solution of the sensor scheduling problem in the general case.

The problem studied in this chapter belongs to a rich and relatively unexplored area. Promising future research directions in the area include multiple sensors in a cooperative framework and non-unit switch-over/observation times between the sites, with a combination of the two being the ultimate goal. Also, investigation of the problem under time-variant site dynamics, and comparison of the deterministic heuristic procedures with probabilistic approaches are possible future research directions.

Bibliography

[1] A. Bar-Noy, R. Bhatia, J.S. Naor, and B. Schieber. Minimizing service and operations costs of periodic scheduling. *Mathematics of Operations Research*, 27: 518–544, 2002.

[2] D. Chen and A. Mok. *The pinwheel: A real-time scheduling problem*, chapter 27. Chapman & Hall/CRC, 2004.

[3] E.A. Feinberg and M.T. Curry. Generalized pinwheel problem. *Mathematical Methods of Operations Research*, 62:99–122, 2005.

[4] J. Filar. Player aggregation in the traveling inspector model. *IEEE Transactions on Automatic Control*, 30:723–729, 1985.

[5] J.A. Filar and T.A. Schultz. The traveling inspector model. *OR Spectrum*, 8: 33–36, 1986.

[6] J. Miltenburg. Level schedules for mixed-model assembly lines in just-in-time production systems. *Management Science*, 35(2):192–207, Feb. 1989.

[7] Y. Monden. *Toyota Production System: An Integrated Approach to Just-In-Time*. Engineering & Management Press, third edition, 1998.

[8] G. Steiner and S. Yeomans. Level schedules for mixed-model, just-in-time processes. *Management Science*, 39(6):728–735, June 1993.

[9] A. Tiwari, M. Jun, D. Jeffcoat, and R. Murray. The dynamic sensor coverage problem. In *Proceedings of the 16th International Federation of Automatic Control (IFAC) World Congress*, Prague, Czech Republic, July 2005.

[10] M. Yavuz and E. Akcali. Production smoothing in just-in-time manufacturing systems: A review of the models and solution approaches. *International Journal of Production Research*, 2007. Forthcoming.

[11] N. Yerrick, A. Tiwari, and D. Jeffcoat. An investigation of a dynamic sensor motion strategy. In *Proceedings of the 6th Cooperative Control and Optimization Conference*, Gainesville, FL, February 2006. World Scientific.

[12] N. Yerrick, M. Yavuz, and D. Jeffcoat. Two sensor motion models for the dynamic sensor coverage problem. *Military Operations Research*, 2007. Forthcoming.

Cooperative Vision Based Estimation and Tracking Using Multiple UAVs

Brett Bethke, Mario Valenti, and Jonathan How

Massachusetts Institute of Technology, Cambridge, MA
{bbethke,valenti,jhow}@mit.edu

Abstract. Unmanned aerial vehicles (UAVs) are excellent platforms for detecting and tracking objects of interest on or near the ground due to their vantage point and freedom of movement. This paper presents a cooperative vision-based estimation and tracking system that can be used in such situations. The method is shown to give better results than could be achieved with a single UAV, while being robust to failures. In addition, this method can be used to detect, estimate and track the location and velocity of objects in three dimensions. This real-time, vision-based estimation and tracking algorithm is computationally efficient and can be naturally distributed among multiple UAVs. This chapter includes the derivation of this algorithm and presents flight results from several real-time estimation and tracking experiments conducted on MIT's Real-time indoor Autonomous Vehicle test ENvironment (RAVEN).

Keywords: Cooperative multi UAV vision tracking estimation.

1 Introduction

Unmanned Aerial Vehicles (UAVs) have attracted significant interest in recent years. Due to improvements in embedded computing, communications, and sensing technologies, UAVs have become increasingly capable of carrying out sophisticated tasks. Because UAVs lack a human occupant and are generally simpler and less expensive than their manned counterparts, they are well suited to perform a wide range of "dull, dirty and/or dangerous" missions. Examples of such missions include traffic monitoring in urban areas, search and rescue operations, military surveillance, and border patrol [5].

For many mission scenarios, the deployment of video cameras onboard UAVs is of particular interest due to the richness of information and real-time situational assessment capabilities that can be provided by the video stream. Researcher have used onboard cameras for remote detection of forest fires [3,6,7]. In addition, a number of researchers have used vision-based techniques for object detection, tracking and surveillance [1,2,10]. The measurements of the target location are inherently nonlinear in the single-vehicle case because the observed state variables are measured angles to the target. As such, numerous researchers have investigated using nonlinear estimators, such as Extended Kalman Filters and Unscented Kalman Filters [4,9], to determine the target state. In addition, observations from multiple vantage points are required to provide depth

M.J. Hirsch et al. (Eds.): Adv. in Cooper. Ctrl. & Optimization, LNCIS 369, pp. 179–189, 2007.
springerlink.com
© Springer-Verlag Berlin Heidelberg 2007

perception and to obtain a good estimate of the target's position and orientation. Geometrically, using only a single observation (via a single camera), the UAV can only determine a ray along which the target lies. Therefore, the UAV must be maneuvered around the target to provide multiple vantage points to gain a more accurate position estimate of the vehicle. However, if the object of interest is moving, the UAV may not be able to complete the necessary maneuvers to gain a more accurate estimate.

In this chapter, we present a vision-based estimation and tracking algorithm that exploits cooperation between multiple UAVs in order to provide accurate target state estimation and allow good tracking of the target without the need for a single vehicle to execute maneuvers to gain better vantage points. The method uses an optimization technique to combine the instantaneous observations of all UAVs, allowing for very rapid estimation. Furthermore, the algorithm can be naturally distributed among all participating UAVs with very modest communication bandwidth requirements and is computationally efficient, making it well suited to implementation on real-time applications. The vision processing is done in a manner designed to be robust to noise in the video stream, which is often present, especially in applications where the video signal is wirelessly transmitted. Flight results from several estimation and tracking experiments conducted on MIT's Real-time indoor Autonomous Vehicle test ENvironment (RAVEN) are presented [8].

2 Vision Based Tracking and Estimation: A Cooperative Approach

Using multiple UAVs in the real-time, vision-based detection, estimation and tracking problem is advantageous for a number of reasons. First, multiple UAVs provide redundancy, allowing for continued tracking even when individual vehicles experience failures. Second, the presence of obstructions in the environment may temporarily block the field of view of a UAV as it attempts to observe the target. Using multiple UAVs with different lines of sight increases the probability that the target will remain observable to the group of UAVs even when individual vehicles' lines of sight are blocked. Third, because more observations are available at a given time, multiple UAVs working together can estimate the target's state more accurately than a single UAV could.

In addition, the cooperative UAV vision tracking problem can be reformulated as a linear estimation problem. Using the observed bearings of the target from each UAV, an estimate of the absolute target position can be obtained by minimizing the errors in distance from the estimate to each measurement. This estimate is then used as a measurement input to a simple linear Kalman filter that uses the target location as the state variables $\{x,y,z\}$.

The statement of the multi-UAV vision-based detection, estimation and tracking problem is as follows. Assume that there are n UAVs, each equipped with a camera. The location of each UAV is given by

$$\mathbf{x}_i = \hat{\mathbf{x}}_i + \delta\mathbf{x}_i, \quad \mathbf{x}_i \in \mathbb{R}^3 \tag{1}$$

Fig. 1. Five vehicle coordinated flight test on MIT's RAVEN [8]

where $\hat{\mathbf{x}}_i$ is the estimated position of the UAV (as given by the UAV's onboard navigation sensors), and $\delta\mathbf{x}_i$ is a random variable that captures the uncertainty in the UAV's position. The distribution of $\delta\mathbf{x}_i$ is assumed to be known.

From each UAV, there is a director vector to the target

$$\mathbf{d}_i = \hat{\mathbf{d}}_i + \delta\mathbf{d}_i, \quad \mathbf{d}_i \in \mathbb{R}^3 \tag{2}$$

where again $\hat{\mathbf{d}}_i$ is the estimated (unit-length) direction vector generated by the vision system, described below, and $\delta\mathbf{d}_i$ represents uncertainty in the direction vector (i.e., uncertainty in the precise direction in which the camera is pointing). Assume that

$$\delta\mathbf{d}_i^T \hat{\mathbf{d}}_i = 0 \tag{3}$$

This assumption is reasonable given that $\hat{\mathbf{d}}_i$ is most naturally characterized by uncertainty in the angles from the camera to the target, so that $\mathbf{d}_i^T \mathbf{d}_i \approx 1$. Again, assume that the distribution of $\delta\mathbf{d}_i$ is known. Finally, a weight w_i is associated with each UAV's estimate. This weight may be used to account for differences in the quality of each UAV's estimate (i.e, differences in video quality). Note that given $\mathbf{x}_i$ and $\mathbf{d}_i$, the target must lie along the ray

$$\mathbf{l}_i(\lambda_i) = \mathbf{x}_i + \lambda_i\mathbf{d}_i, \quad \lambda_i \geq 0 \tag{4}$$

In order to solve this problem, the true position of the object, $\mathbf{q}$, must be estimated given the set of all measurements $\{\hat{\mathbf{x}}_i, \hat{\mathbf{d}}_i : i = 1, \ldots, n\}$. This estimate $\mathbf{q}$ should minimize the error

$$E(\mathbf{q}) = \sum_{i=1}^{n} w_i h_i(\mathbf{q}) \tag{5}$$

where $h_i(\mathbf{q})$ is the square of the minimum distance from $\mathbf{q}$ to the ray $\mathbf{l}_i(\lambda_i)$:

$$h_i(\mathbf{q}) = \min_{\lambda_i} ||\mathbf{q} - \mathbf{l}_i(\lambda_i)||^2 = \min_{\lambda_i} ||\mathbf{q} - (\mathbf{x}_i + \lambda_i\mathbf{d}_i)||^2 \tag{6}$$

Minimizing $h_i(\mathbf{q})$ with respect to λ_i yields the result

$$h_i(\mathbf{q}) = \mathbf{q}^T\mathbf{q} - 2\mathbf{q}^T\mathbf{x}_i + \mathbf{x}_i^T\mathbf{x}_i - (\mathbf{d}_i^T\mathbf{q} - \mathbf{d}_i^T\mathbf{x}_i)^2 \tag{7}$$

Substituting this result into Eq. 5 and minimizing $E(\mathbf{q})$ with respect to $\mathbf{q}$ yields the equation that the optimal estimate must satisfy:

$$\mathcal{A}\mathbf{q}^\star = \mathbf{b} \tag{8}$$

where

$$\mathcal{A} = \sum_{i=1}^{n} w_i \left(I - \mathbf{d}_i\mathbf{d}_i^T \right) \tag{9}$$

$$\mathbf{b} = \sum_{i=1}^{n} w_i \left(\mathbf{x}_i - (\mathbf{x}_i^T\mathbf{d}_i)\mathbf{d}_i \right) \tag{10}$$

However, $\mathcal{A}$ and $\mathbf{b}$ cannot be calculated by the algorithm directly because only the the noisy measurements $\hat{\mathbf{x}}_i$ and $\hat{\mathbf{d}}_i$ are known. To compensate for these errors, $\mathcal{A}$ and $\mathbf{b}$ are expanded by substituting Eqs. 1 and 2 into Eqs. 9 and 10. After dropping second-order terms and grouping the known and unknown terms, the equations become

$$\mathcal{A} = \hat{\mathcal{A}} + \delta\mathcal{A} \tag{11}$$
$$\mathbf{b} = \hat{\mathbf{b}} + \delta\mathbf{b} \tag{12}$$

where

$$\hat{\mathcal{A}} = \sum_{i=1}^{n} w_i \left(I - \hat{\mathbf{d}}_i\hat{\mathbf{d}}_i^T \right) \tag{13}$$

$$\delta\mathcal{A} = -\sum_{i=1}^{n} w_i(\delta\mathbf{d}_i\hat{\mathbf{d}}_i^T + \hat{\mathbf{d}}_i^T\delta\mathbf{d}_i) \tag{14}$$

$$\hat{\mathbf{b}} = \sum_{i=1}^{n} w_i \left(\hat{\mathbf{x}}_i - (\hat{\mathbf{x}}_i^T\hat{\mathbf{d}}_i)\hat{\mathbf{d}}_i \right) \tag{15}$$

$$\delta\mathbf{b} = \sum_{i=1}^{n} w_i(\delta\mathbf{x}_i - (\hat{\mathbf{x}}_i^T\delta\mathbf{d}_i)\hat{\mathbf{d}}_i - (\delta\mathbf{x}_i^T\hat{\mathbf{d}}_i)\hat{\mathbf{d}}_i - (\hat{\mathbf{x}}_i^T\hat{\mathbf{d}}_i)\delta\mathbf{d}_i) \tag{16}$$

Note that $\hat{\mathcal{A}}$ and $\hat{\mathbf{b}}$ are known terms, because they involve only quantities that are measured directly. $\delta\mathcal{A}$ and $\delta\mathbf{b}$ are random variables because they involve the uncertain quantities $\delta\mathbf{x}_i$ and $\delta\mathbf{d}_i$. The optimal estimate can now be written as

$$\mathbf{q}^\star = \mathcal{A}^{-1}\mathbf{b} = (\hat{\mathcal{A}} + \delta\mathcal{A})^{-1}(\hat{\mathbf{b}} + \delta\mathbf{b}) \tag{17}$$

We assume that the error terms are small ($\delta\mathcal{A} \ll \hat{\mathcal{A}}$). Expanding the matrix inverse function in a Taylor series around $\hat{\mathcal{A}}$ gives

$$(\hat{\mathcal{A}} + \delta\mathcal{A})^{-1} \approx \hat{\mathcal{A}}^{-1} - \hat{\mathcal{A}}^{-1}\delta\mathcal{A}\hat{\mathcal{A}}^{-1} \tag{18}$$

Thus, Eq. 17 becomes

$$\mathbf{q}^\star = \hat{\mathcal{A}}^{-1}\hat{\mathbf{b}} + \hat{\mathcal{A}}^{-1}\delta\mathbf{b} - \hat{\mathcal{A}}^{-1}\delta\mathcal{A}\hat{\mathcal{A}}^{-1}\hat{\mathbf{b}} - \hat{\mathcal{A}}^{-1}\delta\mathcal{A}\hat{\mathcal{A}}^{-1}\delta\mathbf{b} \qquad (19)$$

$$\approx \hat{\mathbf{q}}^\star + \hat{\mathcal{A}}^{-1}\delta\mathbf{b} - \hat{\mathcal{A}}^{-1}\delta\mathcal{A}\hat{\mathcal{A}}^{-1}\hat{\mathbf{b}} \qquad (20)$$

where

$$\hat{\mathbf{q}}^\star = \hat{\mathcal{A}}^{-1}\hat{\mathbf{b}} \qquad (21)$$

is the optimal estimate which can be calculated from the measurements. The error $\delta\mathbf{q}^\star$ in the estimate is

$$\delta\mathbf{q}^\star = \hat{\mathcal{A}}^{-1}\delta\mathbf{b} - \hat{\mathcal{A}}^{-1}\delta\mathcal{A}\hat{\mathcal{A}}^{-1}\hat{\mathbf{b}} \qquad (22)$$

Since the probability distributions of the random variables $\delta\mathbf{x}_i$ and $\delta\mathbf{d}_i$ are known, the covariance of $\delta\mathbf{q}^\star$ can be calculated. This covariance is needed in order to implement the Kalman filter, discussed below.

Eq. 21 demonstrates that the optimal estimate $\hat{\mathbf{q}}^\star$ can be computed in time that is *linear* in the number of measurements to the object, n. $\hat{\mathcal{A}}$ and $\hat{\mathbf{b}}$ can be constructed in linear time since they are sums over all rays. Once $\hat{\mathcal{A}}$ and $\hat{\mathbf{b}}$ are known, Eq. 21 can be solved in constant time by inverting the 3 x 3 matrix $\hat{\mathcal{A}}$. Since the entire process runs in linear time with respect to n, this method is very computationally efficient. Note that if there is only a single vehicle, $n = 1$, the matrix $\hat{\mathcal{A}}$ is singular and Eq. 21 cannot be solved. In this case, a single vehicle would have to make an additional assumption about the location of the vehicle, such that it was located on the ground ($z = 0$), in order to calculate a solution. In all other cases, however, $\hat{\mathcal{A}}$ is invertible as long as the observed direction vectors $\hat{\mathbf{d}}_i$ are not all parallel to each other. As long as the observation points $\hat{\mathbf{x}}_i$ are not the same, which cannot happen since the UAVs cannot occupy the same physical point in space, a solution can always be found.

Once the estimate $\hat{\mathbf{q}}^\star$ is known, it can be used as the measurement into a simple linear Kalman filter based on the assumed dynamics of the target vehicle [1]. This paper uses a system model with state vector

$$\mathbf{X} = [x, y, z, \dot{x}, \dot{y}, \dot{z}]^T \qquad (23)$$

The discrete time system dynamics are then given by

$$\mathbf{X}_{k+1} = A\mathbf{X}_k + \mathbf{v}_k \qquad (24)$$

$$\mathbf{Y}_k = \hat{\mathbf{q}}^\star = C\mathbf{X}_k + \delta\mathbf{q}^\star \qquad (25)$$

where $\mathbf{v}_k$ is the process noise and $\delta\mathbf{q}^\star$ is the measurement noise. The process noise covariance is assumed to be known, and the covariance of $\delta\mathbf{q}^\star$ is found as discussed above. A and C are given by

$$A = \begin{pmatrix} 1 & 0 & 0 & \Delta t & 0 & 0 \\ 0 & 1 & 0 & 0 & \Delta t & 0 \\ 0 & 0 & 1 & 0 & 0 & \Delta t \\ 0 & 0 & 0 & 1 & 0 & 0 \\ 0 & 0 & 0 & 0 & 1 & 0 \\ 0 & 0 & 0 & 0 & 0 & 1 \end{pmatrix} \qquad C = \begin{pmatrix} 1 & 0 & 0 & 0 & 0 & 0 \\ 0 & 1 & 0 & 0 & 0 & 0 \\ 0 & 0 & 1 & 0 & 0 & 0 \end{pmatrix} \qquad (26)$$

where Δt is the sampling rate of the filter. Using these dynamics, a linear Kalman filter can be easily designed and implemented. This filter can be run on each UAV; the only requirement is that the UAVs communicate their locations $\mathbf{x}_i$ and estimation directions $\mathbf{d}_i$ to each other. Since each of these quantities is a three dimensional vector, this method requires only six numbers to be transmitted by each UAV, making it well suited for environments where communication bandwidth is limited.

3   Hardware and Software Setup

A real-time, vision-based tracking system was implemented on the quadrotor platform. The motivation behind implementing the vision system was to allow the quadrotors to carry their own sensor payloads and make decisions based on their own sensor data instead of relying upon artificially synthesized sensor data, allowing for a more realistic overall hardware platform.

The hardware setup consists of two Draganfly quadrotors outfitted with a Draganfly SAVS wireless camera. The SAVS camera broadcasts to a diversity receiver on the ground. The receiver is connected to a PC with a LifeView FlyVideo 3000FM video capture card. The Intel OpenCV video capture and processing library is used to interface with the video card and provide a software API for accessing and processing the images.

3.1   Persistent Object Filter

Given the low-power nature of the wireless camera's transmitter, the presence of RF noise, and other impediments to receiving a clear video signal, it is important that the vision tracking system be able to function even when the received video stream is noisy. Experimental data from several wireless cameras shows that noise is often present in the processed images from the camera (as shown in Figure 2).

An expected but important characteristic of the noise seen in the video stream is that it is highly uncorrelated from one frame to the next (as shown in Figure 2). Based on this observation, a filtering algorithm was designed to extract the features of the images that are persistent across multiple frames while rejecting the transient noise components. The key component of the algorithm is a dynamic list P of objects which have been seen in previous frames. By comparing the objects in P with the objects that are in the current frame, denoted by C, the algorithm can decide which objects in C have appeared before, and these objects are given higher weight in the filter.

A detailed description of the algorithm follows.

```
1    set C = ∅; (The set of objects in the current frame)
2    set P = ∅; (The set of persistent objects)
3    while true do:
4        for p in P do:
5            p.f = false; (Mark p as "not found")
```

Fig. 2. Two images taken within one second of each other showing noise in the video stream

```
6        end for;
7        F = getCurrentFrame(); (Get the current image frame)
8    G=preprocessFrame(F); (Downsample image and convert to binary image)
9        C = findObjectsInFrame(G); (Find objects in the current frame)
10       for c in C do:
11           f = false;
11           for p in P do:
12               if (||c.x − p.x|| < ε_x) and (|c.A − p.A| < ε_A) do:
                     (Determine whether c is similar to p in terms of location in the
                     image x and area a)
13                   p.f = true; (Mark p as "found")
14                   p.n = p.n + 1; (Add 1 to the running frame count of p)
15                   f = true;
16                   break; (Since p is determined, there is no need to continue
                     examining the other elements of P)
17           end for; (Ends for p in P)
18           if f == false do:
19               P.push(c); (If c is not found anywhere in P, append c to P)
20           end if;
21       end for; (Ends for c in C)
22       for p in P do:
23           if p.f == false do: (If p was not found in this frame...)
24               p.n=p.n − 1; (...subtract 1 from the running frame count of p)
25           end if;
26           if p.n == 0 do: (If the frame count of p is zero...)
27               P.pop(p); (...remove p from P)
28           end if;
29       end for;
30   end while;
```

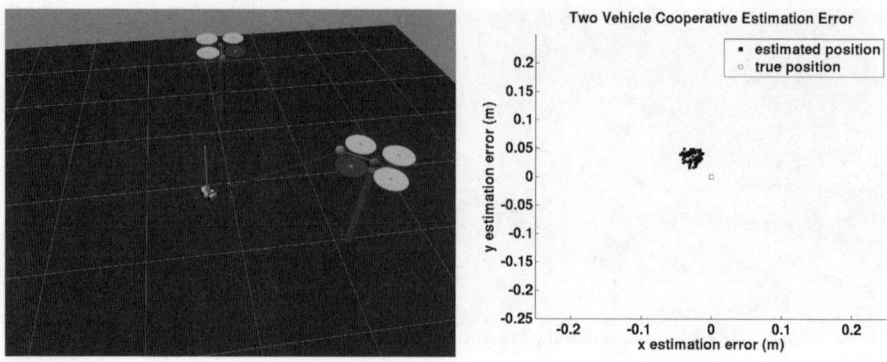

Fig. 3. Left: First experiment configuration. Right: First experiment results. Mean estimation error: x = -0.0265 m, y = 0.0368 m. Standard deviation: σ_x = 0.0082 m, σ_y = 0.0083 m

The current implementation detects objects based on image intensity, although other schemes for object recognition are possible as well. Continuous dark areas ("blobs") in the image are assumed to be objects of interest. The preprocessing step in Line 7 of the algorithm downsamples the image (this is to allow faster processing of the image) and applies a threshold filter in order to convert it to a binary image. This binary image is then passed to a function in Line 8 that detects the centers of each object in the image.

Once the objects are found, the filtering algorithm is applied. The output of the filter is a list of persistent objects in the video stream. The positions of these objects can then be estimated.

4   Results

A number of multi-vehicle vision tracking experiments were conducted to verify the performance of the vision estimation and tracking system. Two vehicles equipped with Draganfly SAVS camera systems were used as the test platform. A small, radio-controlled truck (shown in Figure 2) was used as a target.

In the first experiment, the goal of the UAVs was to hover with the target in the field of view and cooperatively estimate its position. Figure 3 shows the configuration of the experiment using a screenshot from a real-time 3D data visualization tool that was developed for use with the system. The visualization tool shows the locations of the UAVs and target, as well as the rays from each UAV's camera (red lines) and the cooperatively estimated position of the target (yellow sphere with vertical yellow line). Data is displayed in real-time as the system is running, and can also be logged and played back later for analysis purposes.

Results of the first experiment are shown in Figure 3. The scatter plot shows the estimated X-Y positions of the target over a period of about 60 secs of flight. Note that during this time, the UAVs were subject to disturbances in

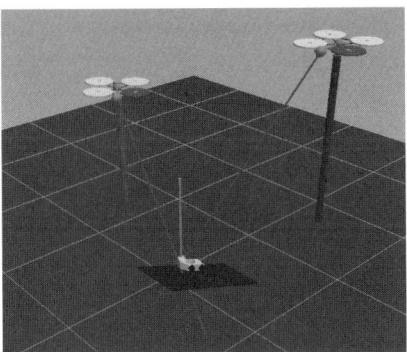

Fig. 4. Second experiment configuration

the form of wind (from ambient air currents as well as propeller wash from the neighboring UAV) which caused them to execute small corrective pitch and roll maneuvers to maintain their assigned hover locations. In the presence of these disturbances, the estimated position of the target remained within 0.04 m of its true location, an excellent result considering that the target itself is over 0.2 m in length. Note that the estimator also gives the Z position of the target (not plotted in Figure 3), and the estimated Z position was also within 0.04m of the true location.

The second experiment shows the advantage of cooperation in the vision estimation problem. In this experiment, two UAVs hovered near a target vehicle which drove up a ramp at constant speed. The goal of the vehicles was to estimate the position and velocity of the target as it moved up the ramp, using both the cooperative estimation method and the noncooperative, single vehicle estimation method. Figure 4 shows the experiment setup, including the ramp and ground vehicle. Note that the arrow protruding from the estimated position of the target shows the estimated velocity, which is clearly aligned with the slope of the ramp.

Figure 5 shows that as the target vehicle moves up the ramp, the estimated error for the noncooperative estimation technique grows due to the fact that the target vehicle is farther from the ground. Without another UAV's perspective, the single UAV is unable to accurately determine the position of the target. Furthermore, the single UAV is unable to determine the velocity of the target well, since it is unable to estimate the ground object's z-axis velocity. Meanwhile, the cooperative estimates for target position and velocity remain very accurate in all three dimensions.

The third experiment incorporated active tracking into the detection and estimation problem. Two UAVs were commanded to detect a ground vehicle and estimate its position. The estimated position was then passed to a tasking system that generated waypoints for the UAVs to follow. The waypoints were chosen in a way that was designed to keep the ground vehicle in the field of view of the UAVs, thus enabling them to continue tracking. In this case, the tasking

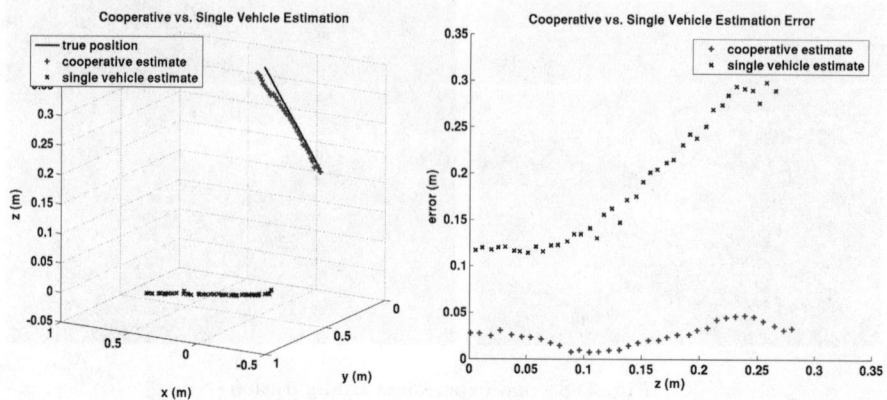

Fig. 5. Second experiment results: cooperative and noncooperative estimated trajectory

system kept one UAV two meters south of the ground vehicle and the other UAV two meters west of the ground vehicle.

Results of the third experiment are shown in Figure 6. The results show that the UAVs were able to estimate the position of the ground vehicle well (within about 5cm) even while they were moving cooperatively in order to keep the vehicle in the field of view of both UAVs. Note that in this experiment, the ground vehicle moved enough that it would have been outside the field of view of both UAVs at times had the UAVs not moved along with it.

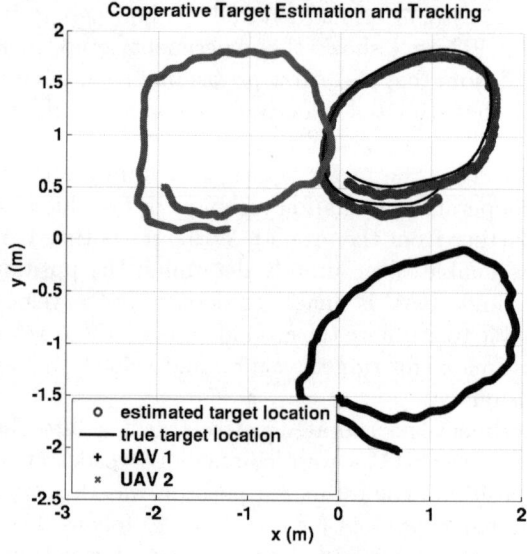

Fig. 6. Results of cooperative estimation and tracking

5   Conclusions

This paper has demonstrated a cooperative target estimation algorithm that is well-suited for real-time implementation and can be distributed among the cooperating vehicles. This approach to target tracking may be advantageous in terms of performance and robustness when multiple vehicles are available to perform the tracking. Flight results indicate that the algorithm performs well in tracking both stationary and maneuvering targets.

Acknowledgments. The authors would like to thank Daniel Dale, Adrian Frank, Jim McGrew, and Spencer Ahrens for their assistance in the project. Brett Bethke would like to thank the Hertz Foundation and the American Society for Engineering Education for their support of this research. This research has been supported in part by the Boeing Company (Dr. John Vian at the Boeing Phantom Works, Seattle) and by AFOSR grant FA9550-04-1-0458.

References

1. N. Gordon B. Ristic, S. Arulampalam. *Beyond the Kalman Filter: Particle Filters for Tracking Applications.* Artech House, Boston, MA, 2004.
2. C. Sharp, O. Shakernia, and S. Sastry. A Vision System for Landing an Unmanned Aerial Vehicle. In *Proceedings of the 2001 IEEE International Conference on Robotics and Automation*, volume 2, pages 1720–1727, 2001.
3. D. Casbeer, S. Li, R. Beard, R. Mehra, T. McLain. Forest Fire Monitoring With Multiple Small UAVs. Porland, OR, April 2005.
4. E. Wan, R. Van Der Merwe. The unscented Kalman filter for nonlinear estimation. In *Adaptive Systems for Signal Processing, Communications, and Control Symposium*, Alta, Canada, October 2000.
5. G. Goebel. In the Public Domain: Unmanned Aerial Vehicles. http://www.vectorsite.net/twuav.html, April 2006.
6. L. Merino, F. Caballero, J. R. Martinez de Dios, A. Ollero. Cooperative Fire Detection using Unmanned Aerial Vehicles. In *Proceedings of the 2005 IEEE International Conference on Robotics and Automation*, Barcelona, Spain, April 2005.
7. L. Merino, F. Caballero, J. R. Martinez de Dios, J. Ferruz, A. Ollero. A Cooperative Perception System for Multiple UAVs: Application to Automatic Detection of Forest Fires. *Journal of Field Robotics*, 23:165–184, 2006.
8. M. Valenti, B. Bethke, G. Fiore, J. How, and E. Feron. Indoor multi-vehicle flight testbed for fault detection, isolation, and recovery. In *Proceedings of the AIAA Guidance, Navigation, and Control Conference and Exhibit*, Keystone, CO, August 2006.
9. S. Julier, J. Uhlmann. A new extension of the Kalman filter to nonlinear systems. In *Proceedings of the 11th International Symposium on Aerospace/Defense Sensing, Simulation and Controls*, 1997.
10. T. McGee, R. Sengupta, K. Hedrick. Obstacle Detection for Small Autonomous Aircraft Using Sky Segmentation. In *Proceedings of the 2005 IEEE International Conference on Robotics and Automation*, Barcelona, Spain, April 2005.

Waypoint Selection in Constrained Domains (for Cooperative Systems)

Yongjie Zhu, Yongling Zheng, and Ümit Özgüner*

Department of Electrical and Computer Engineering
The Ohio State University
2015 Neil Ave. Columbus, OH 43210, USA
ozguner.1@osu.edu

Abstract. This chapter presents a new framework for multiple vehicle systems modeling and control, emphasizing team behavior in a multi-level, multi-resolution way. To set the common reference trajectory for team vehicles, a waypoint selection strategy is proposed taking into account the dimensions of the free space and practical aspects of motion generation. The multi-vehicle cooperative parking strategy is proposed so that a "class" of problems can be solved by formation reconfiguration. The study focuses on several cases corresponding to different scenarios.

1 Introduction

Coordination of multi-vehicle systems to fulfill a mission will yield more benefits than single vehicles performing solo missions. Recent years have seen much research work on this field [[7], [8], [11], [9]]. One motivation for cooperative autonomous vehicle systems is to follow global trajectories and accomplish task as has been done by single vehicles so that the input trajectory and path following strategy are mainly designed for the leader. Team members will share specific information and achieve the final goal by formation reconfiguration [10].

Usually, the reconfiguration mode will be set corresponding to different type of cooperative control problems. In some of them, rigid formation is kept since the common reference state will be assigned to individual vehicles. In other problems, each vehicle will access the information from its neighbor and the team behavior is determined by recurring local inter-vehicle communication. Also, the combination of the above two kinds of problems exists. That is, the team members have similar missions. They keep rigid formation in some of the scenario and reconfigure their formation for a new situation. Under some circumstances in this process, the temporal/sequenced formation is required.

Our problem falls into the last scope. We are interested in teams of vehicles "going somewhere(transition)", and ultimately, "doing something(ultimate task)". A new model for multiple vehicle systems during transition phase has

---

* The research was supported by the Collaborative Center of Control Science(CCCS) under Contract F33615-01-2-3154 issued by the AFRL/AFOSR.

M.J. Hirsch et al. (Eds.): Adv. in Cooper. Ctrl. & Optimization, LNCIS 369, pp. 191–202, 2007.
springerlink.com

been proposed in [11], emphasizing team behavior in a multi-level, multi-resolution way. This framework integrates issues like team formation and path following, so that tasks can easily be allocated to individual and teams of vehicles. The movement of the leader is modeled as a discrete state system within a cellular map, and the movement of the follower is modeled as a hybrid system, including the leader-follower interface.

To set the common reference trajectory, we present a waypoint selection strategy taking into account the dimensions of the free space and practical aspects of motion generation. A mainline approach and a number of special cases are investigated. The maneuvering task is finished by approaching the target cell and stabilizing to the final parking position. A multi-vehicle cooperative parking framework is proposed based on the above model and the waypoint selection strategy.

This chapter is organized as follows: in Section 2, we give the models of leading vehicle, leader-follower interface and following vehicles. A Waypoint selection strategy in constrained domain and a multi-vehicle cooperative parking framework are introduced in Section 3. After that, we provide simulations with a Dubin's car model in Section 4. Conclusions are drawn in Section 5.

2   A Model During Transition Phase

For autonomous multi-vehicle navigation, coordination of multi-vehicle systems to fulfill a mission will yield more benefits than single vehicles performing solo missions. In this field, how to simplify vehicle dynamics by systems modeling is important for efficiently realizing the transition among different formation modes.

The model framework proposed in [11] embodies the idea of decomposition and synthesis for large scale systems. We repeat the key concepts here for completeness. In the original paper [11] we considered convoy type driving along roadways. Here, we shall consider motion in larger open areas, hence the need for path planning. The hierarchical layered structure is shown in Figure 1. A graph can be used to represent the map where vehicles are moving on, with vertices representing crossings and edges representing roads.

Consider a planar digraph $G = (V, E)$, where V is the vertex set and E is the edge set, and $e^{ij} = (v^i, v^j) \in V \times V$. If at one moment $e_k = e^i$, and at the next moment $e_{k+1} = e^j$, the adjacency matrix A is defined as

$$[A]_{ji} = \begin{cases} 1, & \text{if } e_k = e^i \text{ and } e_{k+1} = e^j \\ 0, & \text{otherwise} \end{cases}$$

Where A_{ji} means j−th row, i−th column element of A. Consider a physical road map located in a coordinated system Ω_0 corresponding to the digraph G. Each road can be divided into segments that are connected as a chain, in an approximate sense. We call these segment cells and name these cells along the edge direction as $s_m^{ij} = s_m(e^{ij})$ with ascending subscripts $m \in \{1, 2, \dots, N_{e^{ij}}\}$, where $N_{e^{ij}} \in \mathbb{N}$ is the number of cells on edge e^{ij}.

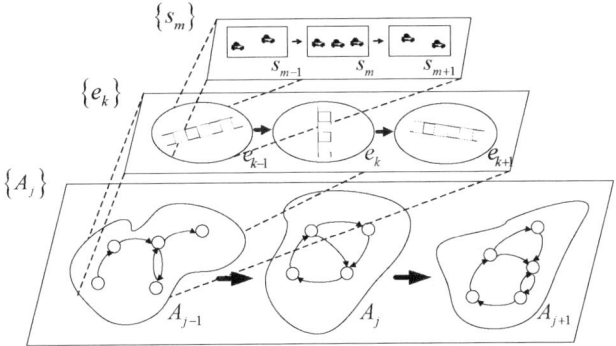

Fig. 1. Hierarchical layered model structure

2.1 Leader Model

With cellular structure modeling, the dynamics of the leader is modeled as a discrete system in a slow time scale. Between "jumps" from one cell to the next, there is a *continuous* movement of the follower in a *fast time scale*.

The cellular movement of the leader is described as

$$x_{k+1} = \varsigma(x_k, u_k) = x_k \oplus u_k$$
$$= \begin{cases} (s_k + u_k, e_j), & s_k + u_k \leq N_{e_j} \\ (s_k + u_k - N_{e_j}, e_{j+1}), & \text{otherwise.} \end{cases} \tag{1}$$

and control variable u_k takes "quantized" values such as

$$u_k = \begin{cases} 0, & \text{stop,} \\ 1, & 1^{st} \text{ speed level,} \\ \dots & \dots \\ u^{\max}, & \text{max speed level.} \end{cases} \tag{2}$$

The evolution of road states is as follows.

$$e_{j+1} = A_j e_j \tag{3}$$

2.2 Leader-Follower Interface

An interface is usually needed to connect two types of systems. This operation defines the rotation of each cell and consequently builds a one-to-one mapping for any position between two cells spatially.

Position Mapping. To better describe the movement of the leader and the followers in a common framework, the following notation system is introduced in [4].

Let $T = [t_i, t_f] \subset \mathbb{R}$ be the time zone of interest. Introduce time index

$$\mathcal{T} = \{[\tau_0', \tau_1], [\tau_1', \tau_2], \ldots, [\tau_{n-1}', \tau_n]\} \tag{4}$$

where $\tau_i \in T$ for all i, $\tau_0' = t_i$, $\tau_n = t_f$, and $\tau_i = \tau_i' \le \tau_{i+1}$ for all $i = 1, 2, \ldots, n-1$.

Position mapping for $t \in [\tau_i', \tau_{i+1})$ corresponds to in-cell dynamics while that for $t = \tau_i$ corresponds to inter-cell dynamics.

Coordinate Rotation. A coordinate system $\tilde{\Omega}_m^{ij}$ will be built within each cell s_m^{ij}, with the origin set at the center g_m^{ij}. One of the axes $\mathbf{n}_m^{ij}$ can be chosen as the normal of the arc passing through the cell center, and the other axis, therefore, can be chosen as $\mathbf{n}_m^{ij\perp}$ which rotates $\mathbf{n}_m^{ij}$ by $\pi/2$ counterclockwise, as the tangent of the arc.

The rotation of $\tilde{\Omega}_m^{ij}$ with respect to Ω_0 is recorded in matrix

$$R_m^{ij} = \begin{bmatrix} \cos\phi_m^{ij} & -\sin\phi_m^{ij} \\ \sin\phi_m^{ij} & \cos\phi_m^{ij} \end{bmatrix} \tag{5}$$

where $\phi_m^{ij} = \angle\mathbf{n}_m^{ij}$. As a result, $\hat{\Omega}_m^{ij}$ is the corresponding *upright* coordinate rotated by $\tilde{\Omega}_m^{ij}$ and they are constrained by the relationship, $\tilde{\Omega}_m^{ij} = R_m^{ij}\hat{\Omega}_m^{ij}$. An

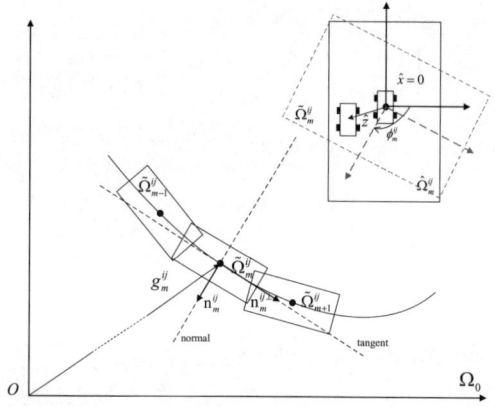

Fig. 2. Illustration for coordinate systems in cells

illustration for the above concept is provided in Figure 2.

Continuous and Discrete Evolution. During the in-cell phase, in the individual up-right coordinate in Figure 2, for $t \in [\tau_i', \tau_{i+1}]$, $x(t) \equiv x(\tau_i')$, $g(t) \equiv g(\tau_i')$, $R^{-1}(\tau_i') = (R_m^{ij})^{-1}$. Assume the leading vehicle stays in cell s_m^{ij}. The leader's position in Ω_0 is $g(t) = g_m^{ij}$ and in $\hat{\Omega}_m^{ij}$ is

$$\hat{x}(t) \equiv 0, \forall t \in [\tau_i', \tau_{i+1}] \tag{6}$$

Let $\hat{z}(t) = (R_m^{ij})^{-1}(z(t) - g(t))$ represent the position of the follower in $\hat{\Omega}_m^{ij}$ with respect to the leader. The movement of followers is simplified from $\dot{z}(t) = \zeta(z(t), g(t))$ to

$$\dot{\hat{z}}(t) = \hat{\zeta}(\hat{z}(t), 0) = \hat{\zeta}(\hat{z}(t)), \forall t \in [\tau_i', \tau_{i+1}] \tag{7}$$

In the global rotated coordinate

$$g(t) \equiv g(\tau_i'), \forall t \in [\tau_i', \tau_{i+1}] \tag{8}$$

$$z(t) = g(t) + R(\tau_i')\hat{\zeta}(R^{-1}(\tau_i')(z(t) - g(t)), 0), \tag{9}$$
$$\forall t \in [\tau_i', \tau_{i+1}]$$

3   Waypoint Selection for Parking Maneuver

Maximum curvature and space limit are the two most important factors when generating parking trajectories autonomously. Considering the nonholonomic constraint of rolling without slipping, the common reference trajectory for team vehicles is not allowed to make sharp turns. Especially when the operation area is a constrained domain, waypoints should be selected to satisfy both the vehicle dynamics and the area constraints. The problem is related to many others, from a so-called SOFA problem, to path planning algorithms like A-star, to potential field approaches. It has a number of new applications, on the ground with autonomous cars, in the air, with UAV's flying around buildings.

We will consider parking maneuvers in a constrained parking zone. In this zone, upper bound for trajectory curvature is required. The vehicle's pose (position and orientation) should be adjusted to a suitable one before entering the parking bay. We prefer forward maneuvers unless straight reverse is necessary in several special cases. Therefore, the waypoint selection strategy should take into account the dimensions of the free space and practical aspects of motion generation. The strategy is designed in a hybrid framework.

3.1   High Level Decomposition of Configuration Space

Different from the other parking maneuver cases [[5],[6],[2]], which usually concentrate on how to enter the parking bay by robotic behavior from a relatively friendly initial posture, we care how to make use of the open space to adjust a vehicle's posture so that it can enter the parking bay smoothly.

In the higher level, first we divide the whole parking spot into a few cells around the goal parking position and specify the area near the parking bay as the target cell. This cell is a highly constrained area since as long as the vehicle entered this cell the "pose" must meet some requirements so that parking maneuver can be easily finished within this small cell. Secondly, the planar cell structure is extended to three dimensional space in which the orientation is expressed by the third dimension. And now, given a vehicle pose, we can index it by (x, y, z) information of each cubic cell. Finally, for each cell that has a different pose, a corresponding waypoint selection method is developed in the lower level considering vehicle dynamic motion constrains.

The configuration space decomposition is described in Figure 3. We use parking bay fixed coordinate system. The directed arrow denotes the vehicle which has a certain length and direction. The origin (x_p, y_p) is the parking bay position. The constant R is the minimum turning radius for the vehicle. The two gray cells constitute the highly constrained area. In this area, the vehicle's position should not be inside the two quarter circles no matter what its direction is. It should enter the gray rectangular from the upper side.

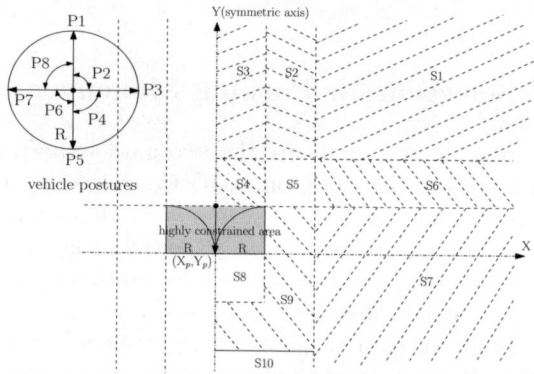

Fig. 3. Configuration space decomposition

Since (x, y) only gives the position information, the third dimension z can denote the heading direction. In the upper left of Figure 3, the z axis is divided to eight levels. That is

$$P1 = \tfrac{\pi}{2}, \qquad P2 \in (0, \tfrac{\pi}{2}), \qquad P3 = 0, \qquad P4 \in (-\tfrac{\pi}{2}, 0),$$
$$P5 = -\tfrac{\pi}{2}, \qquad P6 \in (-\pi, -\tfrac{\pi}{2}), \qquad P7 = \pi, \qquad P8 \in (\tfrac{\pi}{2}, \pi).$$

Thus for each section in the planar plane, it has eight levels in the 3 dimensional space.

It is worth noting that by this decomposition method, only half plane solution need to be given since the whole open space is symmetry by y axis. We will only analyze the right half plane. In this part, S1 is a friendly area. As long as the vehicle is in this section, it will be easy to find a maneuver adjusting "pose". Our basic idea is to find solution for S1 and drive the vehicle to this section first when the initial position is in other sections. To simplify the algorithm, we try to merge the sections having the same waypoint selection method. In total, ten sections in each z level are needed.

Now the problem can be rephrased as follows. Given a vehicle pose (indexed by x_0, y_0, z_0) and the parking bay (the origin of the configuration space), find a waypoint selection strategy so that the vehicle system will start from (x_0, y_0, z_0) and converge to $(0, 0, 0)$.

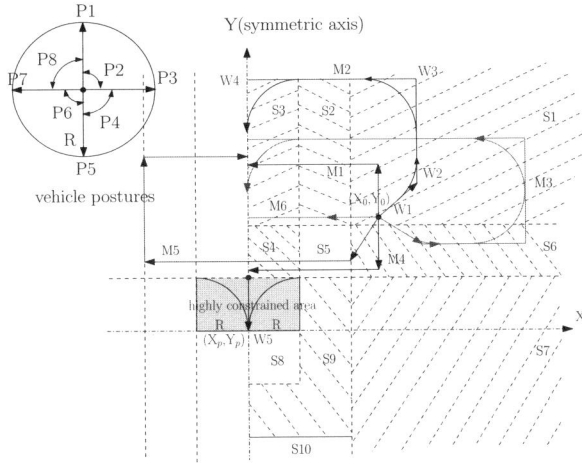

Fig. 4. Six typical maneuvers

3.2 Low Level Waypoint Selection

In the low level, both the vehicle desired (parking) position and its initial position is defined as two points in the configuration space. The maneuvering task is finished by approaching the target cell and stabilizing to the final parking position. A mainline approach and a number of special cases are investigated.

Considering the initial position in S1, we mentioned that this section is a friendly section and the vehicle can drive to the parking bay easily by some maneuver. For different z levels, we define corresponding waypoints and they could be connected by "real-time dynamic trajectory smoothing algorithm[1]" so that a feasible trajectory will be provided to the vehicle. Typical lower level waypoint selection strategy is shown in Table 1. For different initial posture, different selected waypoints are given. The length of the line segment connecting the waypoints is $n * R$ where n is an integer and R is the minimum turning radius, a parameter of the vehicle. The waypoints should be selected so that an obtuse angle or right angle is formed by adjacent line segments connecting waypoints. This method could ensure an arc tangent to both adjacent line segments exists.

A few cubic cells such as S4:P4/P5/P6, S5:P6, S8:P1 and so on, need straight reverse maneuver. Otherwise, there will be no solution.

This method is a resolution-complete algorithm. A solution is guaranteed if it exists at a given resolution when modeling the search space by grids [3]. The completeness of the geometric planner assures the completeness of the algorithm. The resulting trajectory could be tracked by the leader of a team of vehicles.

Table 1. Six typical maneuvers in S1. Note: (x_p, y_p) is the target parking position and θ_0 is the initial yaw angle.

Method	Initial posture	Selected waypoints
M1	P1 $(x_0, y_0, \frac{\pi}{2})$	$(x_0, y_0) \rightarrow (x_0, y_0 + r) \rightarrow (x_p, y_0 + r) \rightarrow (x_p, y_p)$
M2	P2$(x_0, y_0, (0, \frac{\pi}{2}))$ P3$(x_0, y_0, 0)$	$(x_0, y_0) \rightarrow (x_0 + r\cos\theta_0, y_0 + r\sin\theta_0)$ $\rightarrow (x_0 + r\cos\theta_0, y_0 + r\sin\theta_0 + 2r) \rightarrow$ $(x_p, y_0 + r\sin\theta_0 + 2r) \rightarrow (x_p, y_p)$
M3	P4 $(x_0, y_0, (0, -\frac{\pi}{2}))$	$(x_0, y_0) \rightarrow (x_0 + r\cos\theta_0, y_0 + r\sin\theta_0)$ $\rightarrow (x_0 + r\cos\theta_0 + 2r, y_0 + r\sin\theta_0)$ $\rightarrow (x_0 + r\cos\theta_0 + 2r, y_0 + r\sin\theta_0 + 2r)$ $\rightarrow (x_p, y_0 + r\sin\theta_0 + 2r) \rightarrow (x_p, y_p)$
M4	P5 $(x_0, y_0, -\frac{\pi}{2})$	$(x_0, y_0) \rightarrow (x_0 + r\cos\theta_0, y_0 + r\sin\theta_0)$ $\rightarrow (x_p, y_0 + r\sin\theta_0) \rightarrow (x_p, y_p)$
M5	P6$(x_0, y_0, (-\frac{\pi}{2}, -\pi))$ P8$(x_0, y_0, -\pi)$	$(x_0, y_0) \rightarrow (x_0 + r\cos\theta_0, y_0 + r\sin\theta_0) \rightarrow$ $(x_p - 2r, y_0 + r\sin\theta_0) \rightarrow (x_p - 2r, y_0 + r\sin\theta_0 + 2r)$ $\rightarrow (x_p, y_0 + r\sin\theta_0 + 2r) \rightarrow (x_p, y_p)$
M6	P7 (x_0, y_0, π)	$(x_0, y_0) \rightarrow (x_p, y_0) \rightarrow (x_p, y_p)$

3.3   Multi-vehicle Cooperative Parking

Based on the proposed model and the waypoint selection strategy, a multi-vehicle cooperative parking framework is proposed to solve a "class" of problems. Consider several teams of vehicles entering the parking zone by passing a road segment (Figure 5). They aim at different parking bays. Figure 6 and Figure 7 give the function hierarchy and inter car coordination level for the leaders of each team. The leader of the first team L_1 is the master of the communication group and can trigger other leaders. By inter car coordination level, the vehicles will park team by team and the vehicles in the same team will park into by formation reconfiguration. The trajectory for the leader is generated by the waypoint selection method provided before. The transition phase model introduced before will be used here for easy formation reconfiguration. "Shifted" domains can help (plans can be transmitted between leaders).

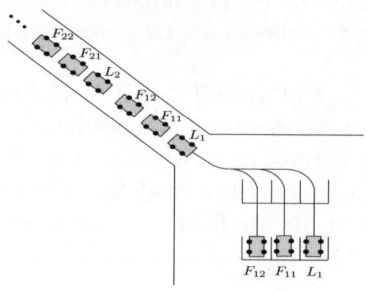

Fig. 5. Cooperative parking

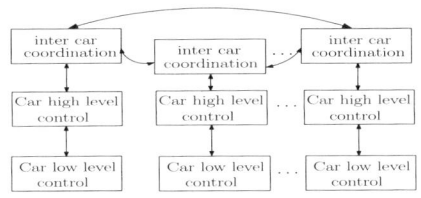

Fig. 6. Functional hierarchy

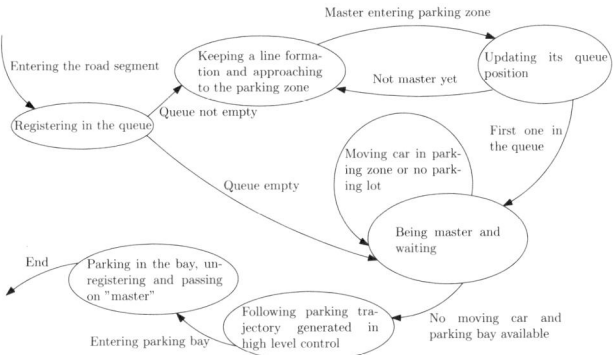

Fig. 7. Inter car coordination level

4   Simulation Results

To illustrate the validity of the above design, several typical parking trajectories are provided. In the following figures, z axis denotes the angle (rad) between vehicle heading direction and parking lot direction. Figure 8 shows the parking trajectory in configuration space and in the $x - y$ plane for initial position in S1 and heading angle $\in$ P2. Vehicle configuration converging to the origin means it enters the parking bay.

Figure 9 shows two different cases in which the initial position is in S1 and S6 respectively. In the first case, the vehicle entered the parking bay by method M1. While in the second one, the vehicle traveled to left half plane first. By symmetry, similar maneuver methods are defined in this area just as M1 to M6 in right half plane. Once the vehicle entered the left half plane, the waypoints will be selected by searching the maneuver in corresponding indexed cubic cell. Figure 10 shows the parking process in configuration space. The vehicle made use of the open area to adjust its heading direction along its trajectory and entered the parking bay at the required angle.

Two other cases are shown in Figure 11. One of them is a special case in which straight reverse is unavoidable. At the intersection point, though the two trajectories have the same position, they are in different z level (different heading

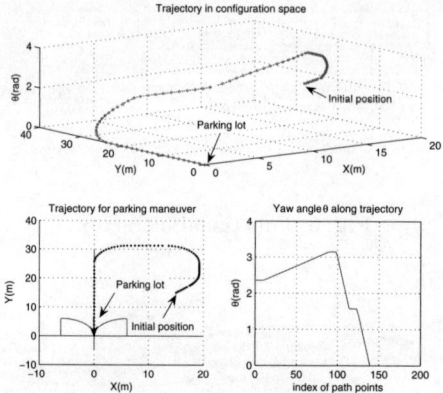

Fig. 8. Parking maneuver

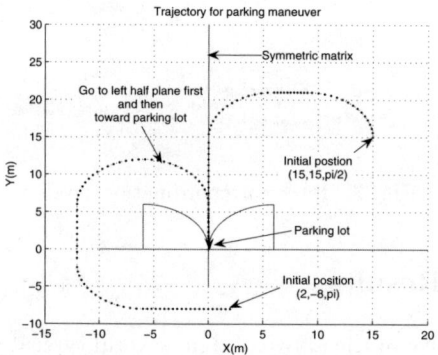

Fig. 9. Waypoint selection in x-y plane

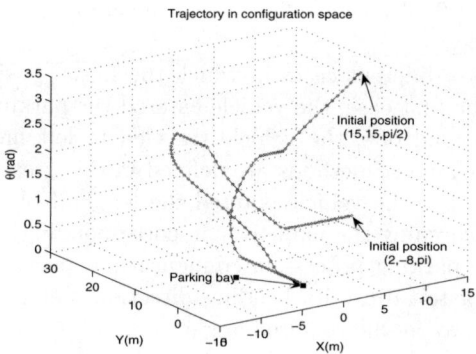

Fig. 10. Parking maneuver configuration space

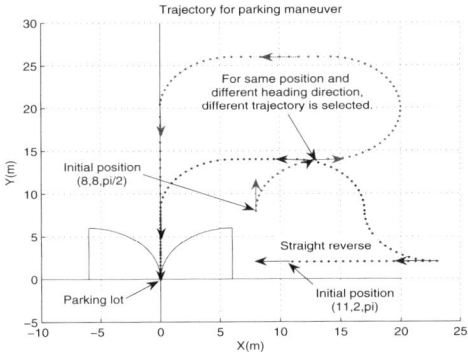

Fig. 11. Waypoint selection in x-y plane

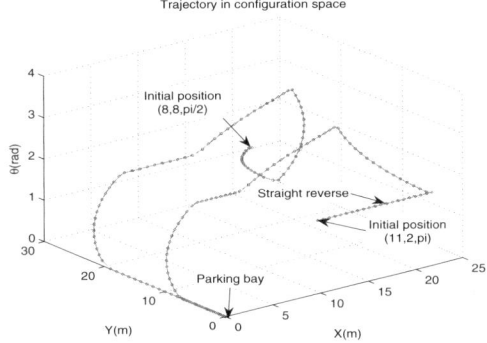

Fig. 12. Parking maneuver configuration space

direction) and will have different waypoint selection method which can be seen from configuration space clearly (Figure 12).

5 Conclusions

This paper presents a new framework for multiple vehicle system modeling and control, emphasizing the team behavior in a multi-level, multi-resolution way. After the search space is modeled by a coarse grid, with fixed cells, a complete algorithm is developed for waypoint selection taking into account the dimensions of the free space and practical aspects of motion generation. The trajectory generated from these waypoints can be fed as reference inputs to leader vehicles of each team. Multiple vehicle cooperative parking is among the most meaningful application of the proposed algorithm. Following a typical hybrid system design procedure as illustrated, a "class" of problems can be solved by formation reconfiguration based on the proposed transition phase model and waypoint selection methods.

References

1. R. W. Beard E. P. Anderson and T. W. McLain. Real-time dynamic trajectory smoothing for unmanned air vehicles. *IEEE Transactions on Control Systems Technology*, 13(3):471–477, 2005.
2. F. Gómez-Bravo F. Cuesta and A. Ollero. Parking maneuvers of industrial-like electrical vehicles with and without trailer. In *IEEE Transactions on Industrial Electronics*, volume 51, pages 257–269, April 2004.
3. J. P. Laumond F. Lamiraux. Smooth motion planning for car-like vehicles. *IEEE Transactions on Robotics and Automation*, 17(4):498–502, 2001.
4. D. N. Godbole J. Lygeros and S. Sastry. Verified hybrid controllers for automated vehicles. *IEEE Transactions on Automatic Control*, 43(4):522–539, 1998.
5. K. Jiang and L. D. Seneviratne. A sensor guided autonomous parking system for nonholonomic mobile robots. *Proceedings of the 1999 IEEE International Conference on Robotics and Automation*, pages 311–316, May 1999.
6. I. E. Paromtchik and C. Laugier. Motion generation and control for parking an autonomous vehicle. In *Proceedings of the 1996 IEEE International Conference on Robotics and Automation*, pages 3117–3122, Minneapolis, Minnesota, April 1996.
7. U. Ozguner Q. Chen. A hybrid system model and overlapping decomposition for vehicle flight formation control. *Proceedings of the 42nd IEEE Conference on Decision and Control*, pages 516–521, December 2003.
8. W. Ren. Cooperative control design strategies with local interactions. In *IEEE International Conference on Networking, Sensing and Control*, pages 451–456, April 2006.
9. U. Ozguner S. D. Waun. Controlling formations of multiple mobile robots. In *Proceedings of IEEE International Conference on Robotics and Automation*, pages 2864–2869, 1998.
10. S. Sastry S. Zelinski, T. J. Koo. Hybrid system design for formations of autonomous vehicles. *Proceedings of the 42nd IEEE Conference on decision and control*, pages 1–6, December 2003.
11. U. Ozguner Y. Zheng. Modelling of grouped vehicles within a cellular spatial structure. In *Proceedings of the 2006 American Control Conference*, pages 4951–4956, Minneapolis, Minnesota, June 2006.

Cooperative Formation Flying in Autonomous Unmanned Air Systems with Application to Training

Hongliang Yuan[1], Vivian Gottesman[2], Mark Falash[2], Zhihua Qu[1], Etyan Pollak[2], and Jiangmin Chunyu[1]

[1] School of EECS, University of Central Florida, Orlando FL 32826
[2] L-3 Communications, Link Simulation and Training, Strategic Technologies, Orlando, FL 32826

Abstract. The study of unmanned aerial systems (UAS) has been an active research topic in recent years due to the rapid growth of UAS real-world applications driven by the Global War on Terrorism (GWOT). UAS are defined as a complete unmanned system including control station, data links, and vehicle. Unmanned aerial vehicle (UAV) refers to the vehicle element of the UAS. Currently UAS operate standalone, independent of neighboring UAS and used primarily for reconnaissance. However UAS roles are expanding to the point where UAV swarms will operate as cooperative autonomous units. The reason is that cooperatively controlled multiple UAS have the potential to complete mission critical complicated tasks with the higher efficiency and failure tolerance, such as coordinated navigation to a target, coordinated terrain exploration and search and rescue operations.

This chapter presents study results associated with real-time trajectory planning and cooperative formation flying algorithms for use in performing multi-UAV cooperative operations. Closed form analytical and simulation results were used along with a UAS simulation test bed for evaluating and verifying these algorithms in multi-UAV cooperative scenarios. The full kinematics constraints of the UAV model is explicitly used, ensuring the planned trajectories and formations are feasible. Two operational modes are implemented for every UAV, one corresponding to the search phase, the other corresponding to the cooperative flying phase. Each phase is executed upon receiving commands. Finally this chapter discusses the use of this simulation environment for multi-UAV cooperative operator training.

1   Introduction

In order to provide a comprehensive solution for the trajectory planning problem, it should be recognized that the motion-planning of robots is analogous to the real-time trajectory planning of a group of UAV. Therefore, leveraging past research efforts devoted to the motion-planning problem of robots is directly applicable. Some popular approaches among them are potential fields, splines, and numerical methods such as the D* and A* search algorithm.

M.J. Hirsch et al. (Eds.): Adv. in Cooper. Ctrl. & Optimization, LNCIS 369, pp. 203–219, 2007.
springerlink.com   © Springer-Verlag Berlin Heidelberg 2007

For the potential fields approach in [1] and [2], the trajectories are expelled away from obstacles by pre-built repulsive potential fields around the obstacles, and the goal is surrounded by an attractive potential field. This approach generally has multiple local minima and requires massive computational resources when applied to 3D applications.

To illustrate this, consider the repulsive potential field:

$$U(r) = \frac{1}{r^2}.$$

The attractive potential field is defined as

$$U(r') = r'^2,$$

where r, r' are the corresponding distances. A robot is to reach its goal along the gradient direction of its overall potential, that is,

$$U(r, r') = U(r) + U(r') = \frac{1}{r^2} + r'^2.$$

This scalar field has local minima close to the goal point. If a robot approches a local minima, it will become stuck. When multiple obstacles are injected into the scenario, the potential becomes more complicated.

For the splines approach in [3], a sequence of splines is used to generate a path through a given set of waypoints. However, prior knowledge of the waypoints may not be available due to the unknown environment and the kinematic constraints of the robots are not considered in splines. Thus, the trajectory may not be applicable to a specific robot.

In a common cubic spline method, each section of the path could be described by the parametric equations:

$$x(u) = a_x u^3 + b_x u^2 + c_x u + d_x$$
$$y(u) = a_y u^3 + b_y u^2 + c_y u + d_y,$$

where $u \in [0 \ \ 1]$. This type of parameterization concentrates on the smooth property at the connection of various segments, rather than the kinematic constraints of the robot. The trajectory obtained by this method may not be feasible for specific types of robots.

In search based methods, A* (proposed in [4]) utilizes a heuristic function to guide the search direction to the goal, thus making it more efficient than the Dijkstra algorithm and ensures an optimal solution from initial point to end point can be found, if one exists. However it requires all of the map information. To deal with dynamic environments, it needs to do a complete recalculation each time the map information is updated, making it inefficient. A typical heuristic index used in A* is:

$$f(n) = h(n) + g(n),$$

where $f(n)$ is the overall cost for a node, $h(n)$ is the cost already spent from the start node to the current node, and $g(n)$ is the estimated cost from the current node to the end node. Generally $g(n)$ can be taken as the Euclidean distance between the current and end nodes.

One improvement of the A* method is found in D* (proposed in [5] and [6]). The D* search algorithm does not require all of the map information. It starts with a priori map and at each time the map data is updated, it invokes a localized A* search to make incremental changes to the path. Its performance is compromised relative to the performance of the A* search. Both search algorithms require heavy computational resources and do not take a kinematic model into account.

By acknowledging the limitations of these techniques, we can improve on these methods by leveraging this information and create an approach that determines a real-time collision-free path for a UAV. In this chapter a parametric solution is proposed to address the limitations of the above techniques. The kinematic constraints are considered, resulting in a class of smooth trajectories. A solution is proposed to design a local decentralized cooperative control for a group of UAV to fly along an arbitrary set of waypoints.

2   Trajectory Planning

The objective of trajectory planning is to find a feasible and smooth trajectory that leads the UAV along its starting waypoint to its final waypoint. In this chapter, trajectory planning is based on the following kinematic model of UAV:

$$
\begin{aligned}
\dot{r}_x &= v_{r1} \cos(r_\theta) \\
\dot{r}_y &= v_{r1} \sin(r_\theta) \\
\dot{r}_\theta &= v_{r2},
\end{aligned}
\tag{1}
$$

where (r_x, r_y) are the world coordinates of the UAV, r_θ is the heading angle, v_{r1} is the longitudinal velocity, and v_{r2} is the angular velocity.

2.1   Parameterized Feasible Trajectories

By analyzing the kinematics model described by (1), it can be established that the trajectory is defined by some smooth function $r_y = f(r_x)$. Given initial and final conditions $q_0 = (r_x^0, r_y^0, r_\theta^0)$ and $q_f = (r_x^f, r_y^f, r_\theta^f)$, the model imposes four constraints on the trajectory. That is, the position and first derivative of each end has to match the boundary value. Thus, when the trajectory is parameterized by a polynomial, it should have at least four coefficients. To achieve a class of trajectories, the coefficients could be more than four. In this application, the trajectory is parameterized by a *4th* order polynomial. That is,

$$
r_y = a_0 + a_1 r_x + a_2 r_x^2 + a_3 r_x^3 + a_4 r_x^4,
\tag{2}
$$

where a_0, a_1, a_2, a_3 are coefficients to be solved and a_4 is free. Given the boundary conditions q_0 and q_f the solution to the coefficients are:

$$\begin{bmatrix} a_0 \\ a_1 \\ a_2 \\ a_3 \end{bmatrix} = (B)^{-1}(Y - Aa_4),$$

where

$$B = \begin{bmatrix} 1 & r_x^0 & (r_x^0)^2 & (r_x^0)^3 \\ 0 & 1 & 2r_x^0 & 3(r_x^0)^2 \\ 1 & r_x^f & (r_x^f)^2 & (r_x^f)^3 \\ 0 & 1 & 2r_x^f & 3(r_x^f)^2 \end{bmatrix}, \quad Y = \begin{bmatrix} r_y^0 \\ \tan(r_\theta^0) \\ r_y^f \\ \tan(r_\theta^f) \end{bmatrix}, \quad A = \begin{bmatrix} (r_x^0)^4 \\ 4(r_x^0)^3 \\ (r_x^f)^4 \\ 4(r_x^f)^3 \end{bmatrix}.$$

2.2 Trajectory Planning for Avoiding Dynamic Obstacles

To deal with the changing environment, as the new obstacle information becomes available, the parameterized trajectory given by (2) may require updates. The updating could be satisfied by a piecewise polynomial parametrization. Let T be the time for a UAV to complete its maneuver from the initial configuration q_0 to its final configuration q_f, and T_s be the sampling period, such that $\bar{k} = T/T_s$ is an integer. When $k = 0$, the initial condition is q_0. For $\bar{k} > k > 0$, the initial condition is given by $q_k = (r_x^k, \ r_y^k, \ r_\theta^k)$, the terminal condition is always q_f. By using the new initial condition, the path planning method described in the previous subsection can be used for real-time replanning as k increases. In the latter part of this chapter, all the notations with superscript k or subscript k indicate they are in the kth sampling period.

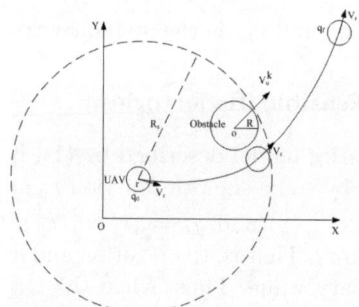

Fig. 1. A UAV in the presence of moving obstacle

Figure 1 illustrates a UAV moving from q_0 to q_f. The radius of the UAV envelope, r, and the sensing range, R_s are known. At the beginning of the kth sampling period, there is a moving obstacle in the sensing range of the UAV, the radius of the obstacle envelop is R with center at (x_k, y_k). In this sampling period, its velocity is v_o^k and assumed to be linear, and the values for the data is obtained by onboard sensors. The trajectory (2) is rewritten as:

$$r_y = a_0^k + a_1^k r_x + a_2^k r_x^2 + a_3^k r_x^3 + a_4^k r_x^4. \tag{3}$$

The obstacle avoidance criterion is:

$$(r_y - y_k - v_{o,y}^k \tau)^2 + (r_x - x_k - v_{o,x}^k \tau)^2 \geq (r + R)^2, \tag{4}$$

where $\tau = t - (t_0 + kT_s)$ for $t \in [t_0 + kT_s, \; t_0 + T]$.

According to the results in Sect. 2.1,

$$[a_0^k \; a_1^k \; a_2^k \; a_3^k]^T = (B^k)^{-1}(Y^k - A^k a_4^k), \tag{5}$$

where

$$B^k = \begin{bmatrix} 1 & r_x^k & (r_x^k)^2 & (r_x^k)^3 \\ 0 & 1 & 2r_x^k & 3(r_x^k)^2 \\ 1 & r_x^f & (r_x^f)^2 & (r_x^f)^3 \\ 0 & 1 & 2r_x^f & 3(r_x^f)^2 \end{bmatrix}, \quad Y^k = \begin{bmatrix} r_y^k \\ \tan(r_\theta^k) \\ r_y^f \\ \tan(r_\theta^f) \end{bmatrix}, \quad A^k = \begin{bmatrix} (r_x^k)^4 \\ 4(r_x^k)^3 \\ (r_x^f)^4 \\ 4(r_x^f)^3 \end{bmatrix}.$$

It is not necessary to consider the collision avoidance criterion for all $t \in [t_0 + kT_s, \; t_0 + T]$. Since the collision may only happen when the UAV's x (or y) coordinate is within a certain range. Specifically, the potential collision range obtained from the x coordinate is when $r_x \in [x_k + v_{o,x}^k \tau - r - R, \; x_k + v_{o,x}^k \tau + r + R]$. From this condition, a potential collision time interval could be solved as $[\underline{t}^*, \; \bar{t}^*]$. It is only in this time interval that the collision avoidance condition is checked.

Substituting (3) and (5) into (4), one obtains the following inequality:

$$g_2(r_x, k)(a_4^k)^2 + g_1(r_x, k, \tau)a_4^k + g_0(r_x, k, \tau)|_{\tau = t - t_0 - kT_s} \geq 0, \tag{6}$$

for all $t \in [\underline{t}^*, \; \bar{t}^*]$, where

$$g_2(r_x, k) = [r_x^4 - h(r_x)(B^k)^{-1}A^k]^2$$
$$g_1(r_x, k, \tau) = 2[r_x^4 - h(r_x)(B^k)^{-1}A^k][h(r_x)(B^k)^{-1}Y^k - y_k - v_{o,y}^k \tau]$$
$$g_0(r_x, k, \tau) = [h(r_x)(B^k)^{-1}Y^k - y_k - v_{o,y}^k \tau]^2 + (r_x - x_k - v_{o,x}^k \tau)^2 - (r + r)^2$$
$$h(r_x) = [1 \; r_x \; r_x^2 \; r_x^3].$$

Inequality (6) describes the adjustable coefficient a_4^k, and as long as the chosen a_4^k satisfies this inequality, the obstacle is avoided. For multiple moving obstacles, each obstacle would impose a constraint similar to (6) on a_4^k. When a_4^k satisfies all the constraints simultaneously, all obstacles are avoided.

Figure 2 shows the actual path that the UAVs travelled during the search phase, where the small dots represent static obstacles.

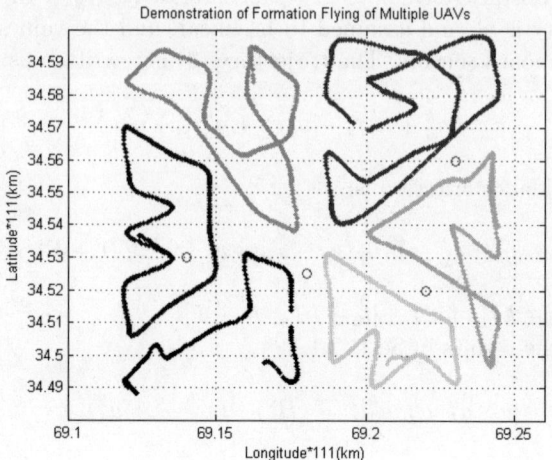

Fig. 2. Trajectory generated in search phase

3   Cooperative Control

In recent years there has been rapid progress in the study of cooperative and formation control for a group of mobile autonomous robots. The reason for this is that cooperatively controlled multiple robots have the potential to complete complicated tasks with a higher efficiency and failure tolerance, such as coordinated navigation to a target, coordinated terrain exploration and search and rescue operations.

Motivated by the flocking behavior of birds in flight, Reynolds introduced a computer animation model for cohesion, separation, and alignment in [10]. Subsequently, a simple discrete-time model (Vicsek model) was given in [11] for the heading alignment of autonomous particles moving in the plane. Simulation results verified the correctness of the Vicsek model. More recently, a theoretical explanation of Vicsek's model was presented in [12] using results from graph theory. The conditions on the connectivity of undirected sensor graphs are given for overall system convergence. This result was extended to networks with directed sensor graphs in [13], [14].

One recent development on designing decentralized local cooperative control is based on matrix theory. Up until now, less restrictive, but successful results have been established in [7]. Given a group of robots that can be feedback linearized into a certain form and their sensing communication matrix satisfies a sequentially complete condition, their production results in a matrix with identical rows, where all the state errors of the group of robots converge, and thus cooperative control is achieved.

3.1   Objectives of Cooperative Control

In general, the control objective for cooperative control is to make the states (or error states) of a group of dynamical systems converge to the same steady state. When applied to formation flying, a group of UAVs converge to a formation when the error states from a group of desired trajectories converge to zero. This is because the states of the group of UAVs in the kinematics model are exactly their position and heading.

3.2   Cooperative Control Algorithm

In order to simplify the design procedure, define the following diffeomorphic state and control transformations

$$\phi_1 = r_x + L\cos(r_\theta), \quad \phi_2 = r_y + L\sin(r_\theta),$$

and

$$\begin{bmatrix} v_1 \\ v_2 \end{bmatrix} = \begin{bmatrix} \cos(r_\theta) & -L\sin(r_\theta) \\ \sin(r_\theta) & L\cos(r_\theta) \end{bmatrix} \begin{bmatrix} v_{r1} \\ v_{r2} \end{bmatrix}.$$

The UAV model can be transformed into the single integrator model as follows with the stable internal dynamics

$$\dot{\phi} = v, \tag{7}$$

where $\phi = [\phi_1, \phi_2]^T$ and $v = [v_1, v_2]^T$.

For an arbitrary path H, a formation can be defined by using its Frenet frame $F_H(t)$, which moves with the path. Let $e_1(t) \in \Re^2$ and $e_2(t) \in \Re^2$ be the orthonormal base of $F_H(t)$, and $\psi^d(t) = [\psi_1^d(t), \psi_2^d(t)] \in \Re^2$ be the origin of $F_H(t)$. A formation consists of q UAVs in $F_H(t)$, denoted by $\{P_1, \cdots, P_q\}$, where

$$P_i = d_{i1}(t)e_1(t) + d_{i2}(t)e_2(t), \quad i = 1, \cdots, q$$

with $d_i(t) = [d_{i1}(t), d_{i2}(t)] \in \Re^2$ being the desired coordinates for the ith robot in $F_H(t)$. It is clear that the rigid formation can be modeled when $d_i(t)$ is constant. The desired position for the ith robot is then

$$\psi_i^d(t) = \psi^d(t) + d_{i1}(t)e_1(t) + d_{i2}(t)e_2(t). \tag{8}$$

To map (7) into the canonical form proposed in [7], define the following decentralized state transformation

$$x_i(t) = \phi_i - \psi_i^d, \quad v = \dot{\psi}_i^d - \phi_i + \psi_i^d + u_i.$$

It follows that

$$\dot{x}_i = A_i x_i + B_i u_i, \quad y_i = C_i x_i,$$

where u_i is the cooperative control for ith robot, and

$$A_i = \begin{bmatrix} -1 & 0 \\ 0 & -1 \end{bmatrix}, \quad B_i = \begin{bmatrix} 1 & 0 \\ 0 & 1 \end{bmatrix}, \quad C_i = \begin{bmatrix} 1 & 0 \\ 0 & 1 \end{bmatrix}.$$

To capture the nature of information flow, define the following sensing/communication matrix:

$$S(t) = \begin{bmatrix} S_1(t) \\ S_2(t) \\ \vdots \\ S_q(t) \end{bmatrix} = \begin{bmatrix} s_{11} & s_{12}(t) & \cdots & s_{1q}(t) \\ s_{21}(t) & s_{22} & \cdots & s_{2q}(t) \\ \vdots & \vdots & \vdots & \vdots \\ s_{q1}(t) & s_{q2}(t) & \cdots & s_{qq} \end{bmatrix},$$

where $s_{ii} \equiv 1$; $s_{ij}(t) = 1$ if the states of the jth UAV is known by the ith UAV at time t; otherwise $s_{ij}(t) = 0$. The general class of cooperative controls are given in the following expression: for $i = 1, \cdots, q$,

$$u_i = \sum_{j=1}^{q} G_{ij}(t)[s_{ij}(t)y_j], \qquad (9)$$

where $s_{ij}(t)$ is the entry in the sensing/communication matrix, G_{ij} is a 2×2 block in gain matrix G that reflects the influence of jth UAV's output to the control of ith UAV, it could be designed in the following form:

$$G_{ij}(t) = \frac{s_{ij}(t)}{\sum_{\eta=1}^{q} s_{i\eta}(t)} K_c, \qquad j = 1, \cdots, q, \qquad (10)$$

where the design parameter $K_c \in \Re^{2 \times 2}$ is a constant, nonnegative, and row stochastic matrix.

3.3   Trajectory Parameterization for Arbitrary Waypoints

In Sect. 3.2, a formation in the Frenet frame $F_H(t)$ is proposed. In most applications, the path of the frame H is not given. Instead, it is desired that the group of UAVs fly through a set of specified waypoints. Suppose a set of waypoints $(w_i, z_i), i = 0, 1, 2, 3$ is given. The following parameterization approach can be used to find the path H.

$$z = z_0 \frac{(w - w_1)(w - w_2)(w - w_3)}{(w_0 - w_1)(w_0 - w_2)(w_0 - w_3)} + z_1 \frac{(w - w_0)(w - w_2)(w - w_3)}{(w_1 - w_0)(w_1 - w_2)(w_1 - w_3)}$$
$$+ z_2 \frac{(w - w_0)(w - w_1)(w - w_3)}{(w_2 - w_0)(w_2 - w_1)(w_2 - w_3)} + z_3 \frac{(w - w_0)(w - w_1)(w - w_2)}{(w_3 - w_0)(w_3 - w_1)(w_3 - w_2)}.$$

Assuming the origin of the frame has a constant overall velocity V (which means the corresponding UAV in the formation has a constant cruise speed), and first waypoint is (w_0, z_0), with a start time at t_0, then the whole timing profile of the Frenet frame $\psi^d(t)$ can be obtained as the following:

$$\psi_1^d(t) = w_0 + \int_{t_0}^{t} \frac{V}{\sqrt{1 + (dz/dw)^2}} dt$$

$$\psi_2^d(t) = z_0 + \int_{t_0}^{t} \frac{V}{\sqrt{1 + (dw/dz)^2}} dt.$$

Combining with (8), the desired trajectory of all the UAVs can be determined.

4   Simulation

The simulation program was developed in Microsoft Visual Studio .Net 2003 and used with an evaluation copy of the Qt Class library provided by Trolltech. Figure 3 is a flow chart of the simulation program that describes how the code and modules are organized.

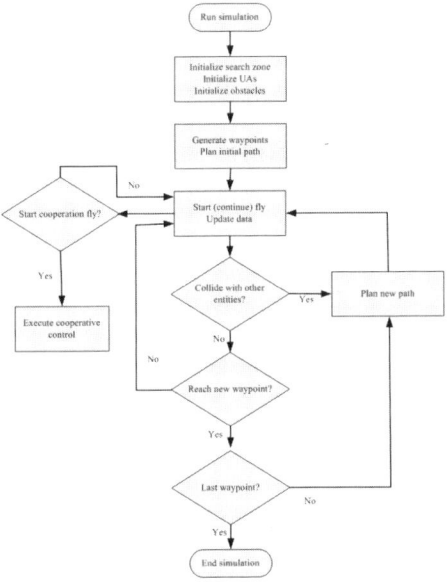

Fig. 3. Simulation platform

4.1   Simulation Prototype

The simulation scenario consists of six UAVs searching within a rectangular area. In the first phase (search phase) the six UAVs perform a complete coverage search over the entire area. In the second phase (cooperative control phase) waypoints are sent to the UAVs via XMPP protocol from the Human Machine Interface (HMI). The six UAVs will converge to a triangular formation and fly along these waypoints.

Table 1. Geographical coordinates of Kabul city

	Longitude(DEG)	Latitude(DEG)
Low-left	69.122650	34.491754
Up-left	69.121734	34.583547
Up-right	69.246323	34.583034
Low-right	69.245865	34.492267

Table 2. Initial configuration of UAVs

	Longitude(DEG)	Latitude(DEG)	Heading(RAD)
UAV 1	69.121734	34.537651	$7\pi/12$
UAV 2	69.163263	34.583547	$-11\pi/12$
UAV 3	69.204793	34.583547	$-11\pi/12$
UAV 4	69.246323	34.537651	$-5\pi/12$
UAV 5	69.204793	34.491754	$\pi/12$
UAV 6	69.163263	34.491754	$\pi/12$

Table 3. Position of static obstacle

	Longitude(DEG)	Latitude(DEG)	Radius(meter)
OBS 1	69.23	34.56	600
OBS 2	69.22	34.52	600
OBS 3	69.14	34.53	600
OBS 4	69.18	34.525	600

Table 4. Waypoints in cooperative fly

	Trajectory 1		Trajectory 2	
WP 1	69.122192	34.5376505	69.1173846	34.5314035
WP 2	69.147018	34.5461272	69.1438258	34.5559345
WP 3	69.196671	34.5648806	69.1938001	34.5779071
WP 4	69.246323	34.5830341	69.2183311	34.5579971

The geographical coordinates of the search area are listed in Table 1. To model static obstacles, moving obstacles with a velocity of zero were used. The initial configuration of the UAV and obstacles are listed in Tables 2 and 3.

In the cooperative flying phase, the two sets of waypoints that the formation should pass are listed in Table 4.

The sensing/communication pattern in the simulation are randomly changing among the following three matrices at each sampling period:

$$S_1 = \begin{bmatrix} 1 & 0 & 0 & 0 & 0 & 0 \\ 1 & 1 & 0 & 0 & 0 & 0 \\ 0 & 1 & 1 & 0 & 0 & 0 \\ 0 & 0 & 1 & 1 & 0 & 0 \\ 0 & 0 & 0 & 1 & 1 & 0 \\ 0 & 0 & 0 & 0 & 1 & 1 \end{bmatrix} \quad S_2 = \begin{bmatrix} 1 & 0 & 0 & 0 & 0 & 0 \\ 1 & 1 & 0 & 0 & 0 & 0 \\ 1 & 0 & 1 & 0 & 0 & 0 \\ 1 & 0 & 0 & 1 & 0 & 0 \\ 1 & 0 & 0 & 0 & 1 & 0 \\ 1 & 0 & 0 & 0 & 0 & 1 \end{bmatrix} \quad S_3 = \begin{bmatrix} 1 & 0 & 0 & 0 & 0 & 0 \\ 1 & 1 & 0 & 0 & 0 & 0 \\ 0 & 1 & 1 & 0 & 0 & 0 \\ 1 & 0 & 0 & 1 & 0 & 0 \\ 0 & 1 & 0 & 0 & 1 & 0 \\ 0 & 0 & 0 & 1 & 0 & 1 \end{bmatrix}$$

The design parameter K_c in (10) is:

$$\begin{bmatrix} 0 & 1 \\ 1 & 0 \end{bmatrix}$$

4.2   Simulation Results

To determine the route in the search phase, a minimum number of circles that fit the sensing range of the UAVs are placed in area [9] with the union of these circles covering the entire area. The centers of these circles are the waypoints that will be traveled along to form the route. Next, each UAV determines the waypoints to use. This is done by a Voronoi algorithm, which means each waypoint belongs to the nearest UAV. Finally, each UAV will choose the nearest waypoint as its first waypoint, and then by applying a computational geometry algorithm, the UAV will travel to the nearest unvisited waypoint relative to its current position (going clockwise). This method forces the UAVs to travel in a counterclockwise path. This phase uses the path planing and obstacle avoidance algorithms discussed in Sect. 2.

One case for the search phase and two cases for the cooperative flying phase are simulated. The results for the search phase are shown in Fig. 2. Figures 4 and 5 show the results for the cooperative control phase. By sending different sets of waypoints, the six UAVs fly on different paths in the given triangular formation. After receiving the waypoints, the program first uses the approach discussed in Sect. 3.3 to parameterize the desired trajectories through the set of waypoints, then applies the algorithm presented in Sect. 3. The two figures of the cooperative control phase illustrate that after a transient process, the UAVs gradually converges to their desired trajectory.

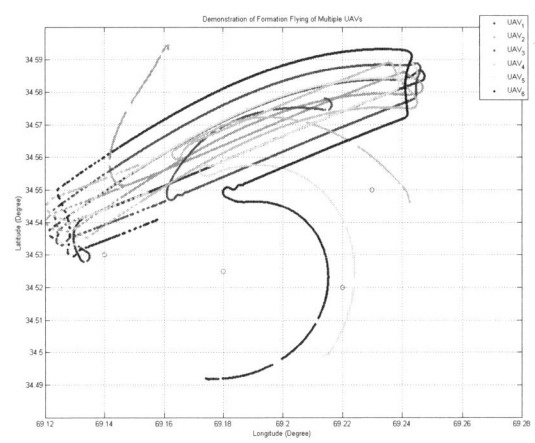

Fig. 4. Trajectory generated in formation fly phase

5   UAS Test Bed

The Unmanned Aerial System Test Bed (UAS Test Bed) is a web-based infrastructure for UAS operational "what if" assessments and development of training strategies developed by L-3 Communications, Link Simulation and Training. It

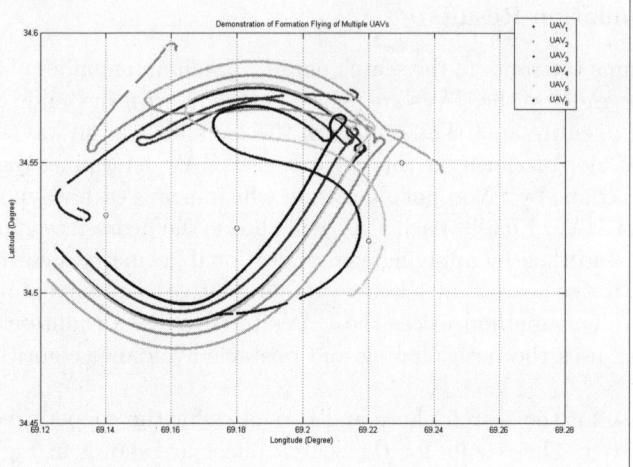

Fig. 5. Trajectory generated in formation fly phase

is built on commercial web, gaming and open source technologies. The Physics-Based Urban Environment provides and manages a common view or state of the virtual environment that the UAS application interacts with. The Multi-Trainer Servers host the UAV, Ground Control Station (GCS) and communication models, which interface to the Physics-Based Urban Environment through a set of APIs. The GCS interfaces to the Human Computer Interface (HCI) through web-based services and protocols. The GCS interfaces to the UAV models through Data Links in the communications layer.

The HCI component provides a tailored trainee interface depending on GCS configuration and desired training position (e.g. vehicle or sensor operator). The HCI's primary displays are used for situation assessment with secondary displays used to assess vehicle or subsystem health. An example generic web based GCS HCI is used to demonstrate the test bed vehicle controls, sensor controls and situation assessment. The HCI is connected to the server side GCS logic, which in turn communicates to UAV models through protocols such as STANAG 4586. Figure 6 illustrates the basic infrastructure in the UAS test bed framework, and Fig. 7 shows the components in the HCI.

The user interacts with this framework and connects to the HCI through a web portal. As a trainee, a user can download and run the HCI while logged into the UAVS web portal, which is their gateway to the simulation. Prior to launching the HCI, the trainee can configure their training session by providing initial parameters and attributes for the vehicle and GCS, and selecting a specific mission. Once the configuration is complete, the trainee can join the simulation as any position they have permissions for. The HCI contains multiple components that are controlled by various roles or positions. A trainee's role is dependent on initial registration parameters and determines group permissions while logged into the web portal. User permissions or authority is based on the training role, and used to determine the available features accessable through the web portal

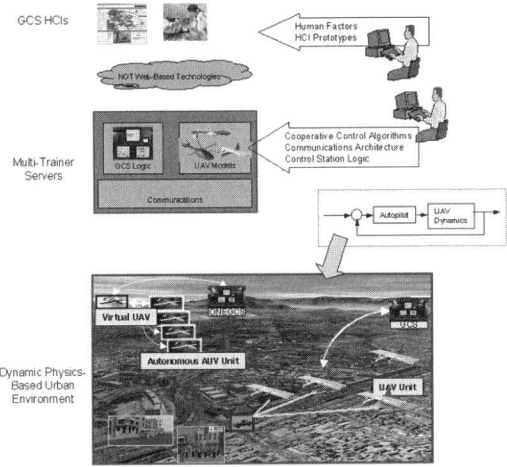

Fig. 6. UAS test bed framework infrastructure

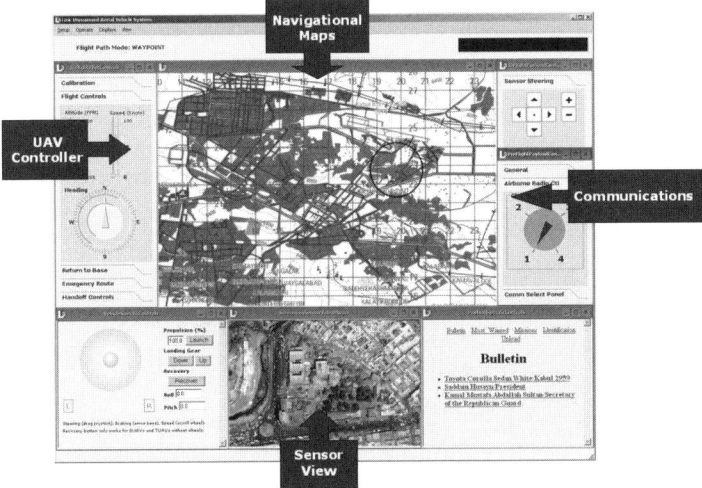

Fig. 7. UAS human control interface

and HCI. Instructors and mission commanders can manipulate all areas of the HCI. Vehicle operators drive the flight controls and execute routes and loiter zones for a single UAV. Sensor operators control the payload steering and zoom for a single UAV. Operators can handoff control to drive other UAVs in the simulation.

Observers can view the UAVs on a map, as well as their sensor output, while planners can only view the map. The map displays all UAVs in the simulation, their inertial states, and all routes and loiter zones that can be executed by the vehicle operators. All users can communicate via voice-over Internet protocol (VOIP) on various channels.

6   UAS Cooperative Control Test Bed Framework

For the UAS Cooperative Control Test Bed, an interface was created, that incorporated the aforementioned cooperative control algorithms, and integrated them into the UAS Test Bed for evaluating their effectiveness using multi-UAV based scenarios. The main areas of integration within the HCI were the moving-map application and the Core UAV Control System (CUCS) communication interface. For the moving-map application, a table of inertial states was added and the ability to interpret additional configuration files. These configuration files provide the locations of static obstacles and targets that are to be displayed on the map. A data translator is used to map the inertial states provided by the cooperative control interface to the UAS interface. Search areas, static obstacles, targets, and pre-defined routes are defined in XML files to be used during the exercise. Figure 8 illustrates the components after integrating the algorithm into the testbed.

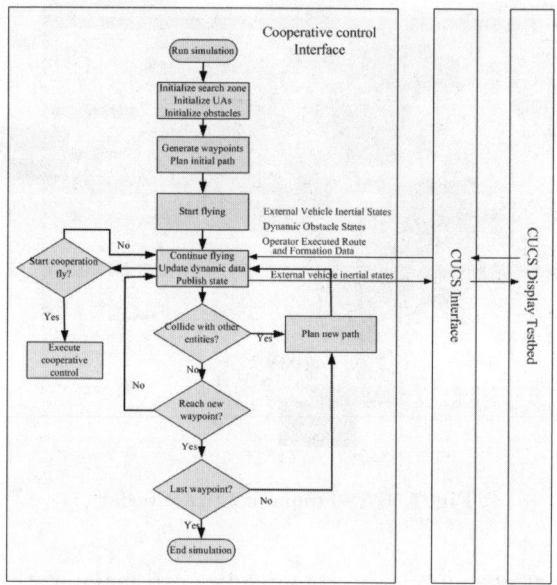

Fig. 8. Integrated simulation platform

Once the program is executed, the UAVs begin searching their respective areas and communicate using the STANAG 4586 [15] standard messages transmitted using the Jabber protocol [16]. The vehicle operator controls the UAVs through a virtual UAV, which can be any one of the UAVs in the formation or the GCS.

This virtual UAV communicates with the formation and they will collectively cooperate to achieve their goal. During the real-time simulation, the following capabilities can be demonstrated:

– Planning the Coverage Search Path

Once the exercise begins, a search area and the location of static obstacles and targets are uploaded to each UAV. The UAV then generates an initial route that will cover their entire search area, while avoiding the static obstacles. If there are any known dynamic obstacles in the path, the UAV's initial route will reflect this and avoid these entities. Figure 9 shows a snapshot of the working scenario of HCI during the searching phase. The trajectories of the UAVs correspond to Fig. 2.

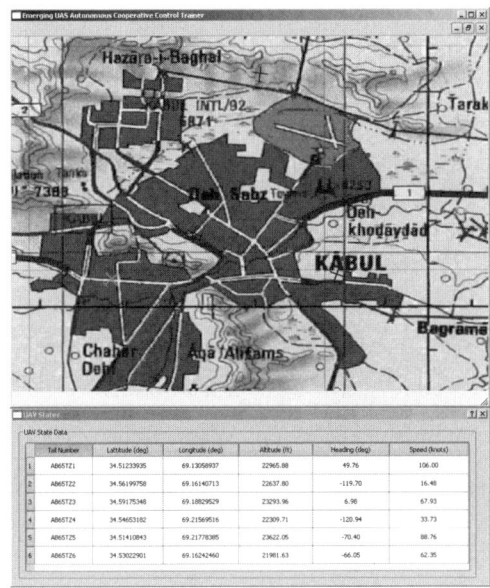

Fig. 9. UAVs flying pre-defined search area

– Real-Time Trajectory Generation

As the UAVs search they continually communicate with each other their current inertial state and obtain inertial states from other vehicles. Using this information, the UAVs re-plan their route to avoid flying into one another while avoiding obstacles.

– Dynamic and Static Obstacle Avoidance

During the initial planning and re-planning of the search route, the UAV trajectory will avoid known static obstacles in its path. Other vehicles' inertial states, not limited to UAVs, are also communicated to the HCI's CUCS interface. The UAV uses these locations and velocities to continuously re-plan their route to avoid these dynamic obstacles during search and formation flying.

– Convergence to Specified Targets

The descriptions of specified targets are uploaded to each UAV upon start of the simulation. The UAV sensor will scan the search area for these targets. Once the target is identified, the UAV will communicate this to the other UAVs. After either all targets have been located or all UAVs have searched the area, the UAVs will converge to the specified targets and begin a loiter pattern.

– Formation Fly Along a Specified Route

Routes are described in XML and can be uploaded to the virtual UAV and executed. Once the route is executed, the UAVs will generate a new path to fly this route in a formation specified by the XML. The route planning is a cooperative process between all the UAVs in the formation. They communicate their routes to the virtual UAV and re-plan the routes collectively in order to successfully fly the specified route in formation while avoiding obstacles. Figure 10 shows how the group of UAVs flow along a specified set of waypoints, their trajectories correspond to Fig. 4.

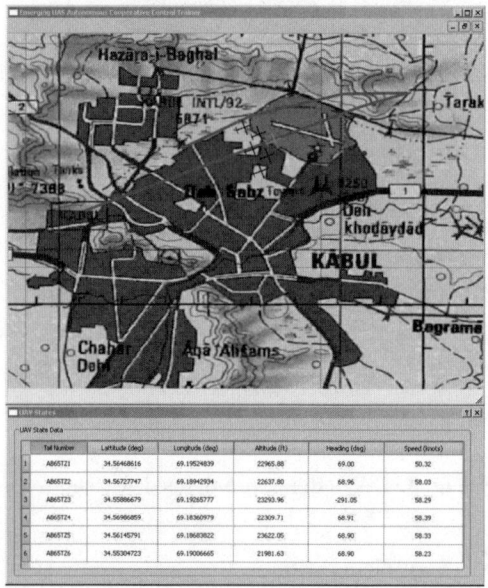

Fig. 10. UAVs flying in formation along a pre-defined route

Given autonomous and cooperative control of multiple UAVs is an emerging capability, there are no real world examples illustrating the operational concepts or training procedures. The integration of cooperative control algorithms with the UAS test bed gives us a platform and ability to perform "what if" scenarios necessary to understand the implications of these capabilities to assess mission performance, operating policy and potential training gaps.

References

1. Jen-Hui Chuang.: Potential-based modeling of three-dimensional workspace for obstacle avoidance. IEEE Trans. on Robotics and Automation, **14** (1998) 778-785
2. K.J.Kyriakopoulos, P.Kakambouras, and N.J.Krikelis.: Potential fields for non-holomic vehicles. Proceedings of the IEEE International Symposium, (1995) 461-465
3. Judd K.B. and Mclain T.W.: Spline based path planning for unmanned air vehicles. AIAA Guidance, Navigation, and Control Conference and Exhibit, volume AIAA-2001-4238, Montreal, Canada, Aug 2001
4. Nilsson, N.J.: Principles of Artificial Intelligence. Tioga Publishing Company, 1980
5. A. Stentz.: Optimal and efficient path planning for partially-known environments. IEEE International Conference on Robotics and Automation, May 1994
6. A. Stentz.: The Focussed D* Algorithm for Real-Time Replanning. Proceedings of the International Joint Conference on Artificial Intelligence, August 1995
7. Z. Qu, J. Wang, and R. A. Hull.: Cooperative control of dynamical systems with application to autonomous vehicles. Submitted to IEEE Transactions on Automatic Control
8. Z. Qu, J. Wang, and C.E.Plaisted.: A New Analytical Solution to Mobile Robot Trajectory Generation in the Presence of Moving Obstacles. IEEE Transactions on Robotics, **20** (2004) 978-993
9. Yi Guo, Z Qu.: Coverage control for a mobile robot patrolling a dynamic and uncertain environment. 5th World Congress on Intelligent Control and Automation, Hangzhou, China, Jan. 2004
10. Reynolds,C.W.: Flocks, herds, and schools: a distributed behavioral model. Computer Graphics (ACM SIGGRAPH Conference Proceedings), **21** (1987) 25-34
11. Vicsek, T., Czirok, A., Jacob, E.B., Cohen, I. and Shochet, O.: Novel type of phase transition in a system of self-driven particles. Physical Review Letters, **75** (1995) 1226-1229
12. Jadbabaie, A., Lin, J., and Morse, A.S.: Coordingation of groups of mobile autonomous agents using nearest neighbor rules. IEEE Trans. on Automatic Control, **48** (2003) 988-1001
13. Lin, Z., Brouchke, M., and Francis, B.: Local control strategies for groups of mobile autonomous agents. IEEE Trans. on Automatic Control, **49** (2004) 622-629
14. Moreau, L.: Leaderless coordination via bidirectional and unidirectional time-dependent communication. Proceedings of the 42nd IEEE Conference on Decision and Control, Maui, Hawaii 2003
15. North Atlantic Treaty Organization: Standard Interfaces of the UAV Control System (UCS) for NATO UAV Interoperability. April, 2004
16. Dodson Catherine: Jabber Technical White Paper. Jabber.com, Inc, April, 2000

Bib references.

Virtual Leader Based Formation Control of Multiple Unmanned Ground Vehicles (UGVs): Control Design, Simulation and Real-Time Experiment

Wenchuan Cai, Liguo Weng, R. Zhang, M. Zhang, and Y.D. Song

Electrical and Computer Engineering
North Carolina A & T State University
Greensboro, NC 27411, USA
{wcai, lweng, rzhang, mzhang, songyd}@ncat.edu

Abstract. Closed formation control for multiple unmanned ground vehicles (UGVs) is studied in this chapter. The leading-following strategy with a virtual leader is applied to coordinate the whole formation group so that only local information is sufficient for every UGV to maintain close formation. Error-shaping memory-based control is designed for formation path tracking. The salient feature of this approach lies in its simplicity in design and implementation, and no need for detailed information on external disturbances and uncertainties. The performance of the proposed method is verified via real-time experiment on various formation operations.

Keywords: formation control, leader-follower strategy, memory-based control, real-time experiment.

1 Introduction

The problem of coordinating the motion of multiple agents has attracted significant attention in recent years. Nature is abundant in marvelous examples of cooperative behavior, such as animal aggregation, fish schooling, bird flocking, bee dancing, ant colony, traffic flow evolution, cell migration, etc. Bio-inspired by such behaviors, formation control of multiple autonomous vehicles has been an important theme and considerable efforts have been devoted to this research for the possible applications in broad areas like formation flying of unmanned aerial vehicles (UAVs) [2]-[5], coordinated movement of a cluster of Unmanned Ground Vehicles (UGVs) [6]-[9], satellite formation and Multiple spacecraft formation flying (MSFF) [10].

The "Leader-Follower" Strategy is one of three conventional methods of formation control, in which an agent acted as a group leader and other agents would just follow the cohesion/ separation/ alignment rules, resulting in leading-following [1]. The advantage of this strategy is that it ensures all agents eventually align with each other and group into a tight formation with common heading direction, no collision and no need for global knowledge and computation. However, the leader-follower strategy always involves a singularity that results in uncontrollability as the followers are driven to move in parallel with the leader [2],[9]. This work addresses the idea of

M.J. Hirsch et al. (Eds.): Adv. in Cooper. Ctrl. & Optimization, LNCIS 369, pp. 221–230, 2007.
springerlink.com © Springer-Verlag Berlin Heidelberg 2007

virtual leader based formation control [11],[12], in which a virtual leader is applied to replace the real leader and all mobile vehicles are considered as "follower" so that all vehicles can group into any formation shape.

In fact, the problem of formation control is a two-leveled hybrid architecture, which includes: 1) motion coordination level, like leader-follower coordination strategy, where the trajectory planning is achieved; and 2) control level, where the developed control algorithm is applied. In the control level, this work addresses error shaping memory-based control, which improves the memory-based control algorithm mentioned in [13]-[15]. The fundamental idea behind this method is to use certain gathered information (i.e., current tracking error, most recent tracking errors, and previous control experiences) to generate new control commands. The salient feature of the proposed approach lies in its simplicity in design and implementation, and no need for detailed information on external disturbances and uncertainties.

In this work, we have successfully implemented the leader-follower coordination strategy and memory based control algorithm for formation control of multiple UGVs. A series of simulation and real-time experiments are developed to verify the effectiveness of virtual leader-follower coordination strategy and error shaping memory-based control in what follows.

2   Real-Time Testbed of UGV and Its Modeling

Unanticipated challenges and issues often are exposed in test fields before a theoretical control algorithm is applied in a real-world scenario. Real-time verification is essential for any control algorithm before its practical implementation. Therefore, we have built the real-time testbed "P3-DS1401" based on Pioneer 3 robot, shown in Figure 1, developed by the Center of Cooperative Systems at North Carolina Agricultural and Technical State University, in cooperation with dSPACE Inc. This robot is an agile, versatile intelligent mobile robotic platform, which is equipped with three kinds of powerful sensing tools, a high-performance driving module and an intelligent real-time rapid control prototyping tool-MicroAutoBox(DS1401) [16].

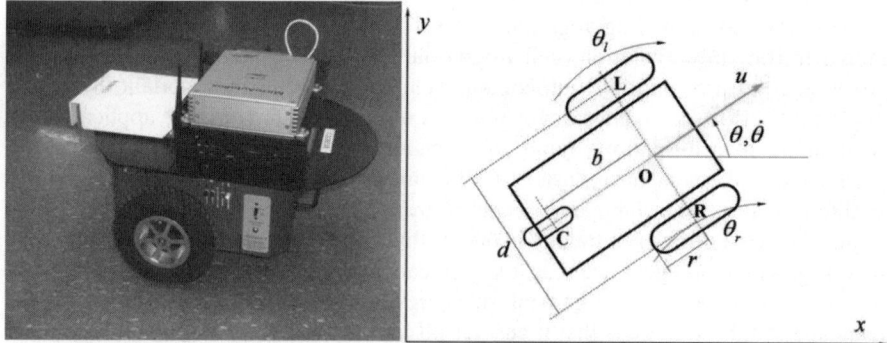

Fig. 1. A three wheeled robots and its schematic.

The dynamic equations for the motion of the three wheeled robot are given by [17][18]

$$m\dot{u} = F_r + F_l + F_{cx} \tag{1}$$

$$J_o\ddot{\theta} = (F_r - F_l)d/2 - F_{cy}b \tag{2}$$

where u is the heading speed of the robot, θ is its orientation angle; m is the mass of the robot, J_o is its moment of inertia with respect to the point O, F_r and F_l are the forces generated by right and left driving wheels respectively, F_{cx} and F_{cy} stand for the friction forces of the caster in x_b and y_b directions, d denotes the distance of two driving wheels, and b is the distance from the caster to axle of two driving wheels, shown in figure 1.

The dynamics of wheel-DC motor assembly can be described as

$$J_w\ddot{\theta}_r = K_t I_{mr} - F_r r - \eta\dot{\theta}_r \tag{3}$$

$$J_w\ddot{\theta}_l = K_t I_{ml} - F_l r - \eta\dot{\theta}_l \tag{4}$$

where θ_r and θ_l are rotating angles of two driving wheels, J_w is the moment of inertia of the driving wheel, r denotes the radius of the driving wheel, η is the friction coefficient, K_t is the motor torque constant, I_{mr} and I_{ml} are motor currents of right and left driving motors respectively. For the P3-DS1401 testbed, the motor current of the driving motor is proportional to the control current; i.e. $I_{mr} = K_{am}I_{cr}$ and $I_{ml}=K_{am}I_{cl}$, where K_{am} is the current amplification factor, I_{cr} and I_{cl} are control currents of two driving motors.

Geometrically, $u, \dot{\theta}$ and θ_r, θ_l have the following relation

$$u = (\dot{\theta}_r + \dot{\theta}_r - \dot{\theta}_{sr} - \dot{\theta}_{sl})r/2 \tag{5}$$

$$\dot{\theta} = (\dot{\theta}_r - \dot{\theta}_r - \dot{\theta}_{sr} + \dot{\theta}_{sl})r/d \tag{6}$$

where θ_{sr} and θ_{sl} denote the slip angle of the two driving wheels.

From (1)-(6), consequently, we have

$$\begin{bmatrix} \dot{u} \\ \ddot{\theta} \end{bmatrix} = K_t K_{am} \begin{bmatrix} m_1 & 0 \\ 0 & m_2 \end{bmatrix}^{-1} \begin{bmatrix} 1 & 1 \\ 1 & -1 \end{bmatrix} \begin{bmatrix} I_{cr} \\ I_{cl} \end{bmatrix} + Q \tag{7}$$

with

$$m_1 = (2J_w/r + mr), m_2 = J_w d/r + 2J_o r/d \tag{8}$$

$$Q = \begin{bmatrix} m_1 & 0 \\ 0 & m_2 \end{bmatrix}^{-1} \begin{bmatrix} F_{cx} - \eta(\dot{\theta}_r + \dot{\theta}_l) - J_w(\ddot{\theta}_{sr} + \ddot{\theta}_{sl}) \\ -2rbF_{cy}/d - \eta(\dot{\theta}_r - \dot{\theta}_l) - J_w(\ddot{\theta}_{sr} - \ddot{\theta}_{sl}) \end{bmatrix} \tag{9}$$

3 Virtual Leader Based Formation

The "leader-Follower Strategy" is adopted since it is easy to control multiple vehicles in a desired formation and it is suitable for describing the vehicle formation. The formation geometry of UGVs can be determined by the relative position between the virtual leader and the followers, which is illustrated as Figure 2, where (x,y) are the ground coordinates, (x_{VL}, y_{VL}) and (x_F, y_F) are the global positions of the virtual leader and the follower, respectively, u_{VL} and u_F are leader's and follower's linear velocities, θ_{VL} and θ_F are their orientation angles respectively, ρ and φ are the follower's relative distance and orientation angle with respect to the virtual leader. The objective of the formation control is to drive the relative position and angle of the followers to the desired value, i.e., $\rho \to \rho_d$ and $\varphi \to \varphi_d$ (subscript 'd' means desired value).

Based on this coordination strategy, the separation errors of follower j in the x and y directions referring to the global frame are easily determined via

$$e_x^j = x_{VL} - x_F^j - \rho_d^j \cos(\theta_F^j + \varphi_d^j) \tag{10}$$

$$e_y^j = y_{VL} - y_F^j - \rho_d^j \sin(\theta_F^j + \varphi_d^j) \tag{11}$$

The kinematic equations of the virtual leader and follower j are given as

$$\dot{x}_{VL} = u_{VL} \cos \theta_{VL}, \dot{y}_{VL} = u_{VL} \sin \theta_{VL} \tag{12}$$

$$\dot{x}_F^j = u_F^j \cos \theta_F^j, \dot{y}_F^j = u_F^j \sin \theta_F^j \tag{13}$$

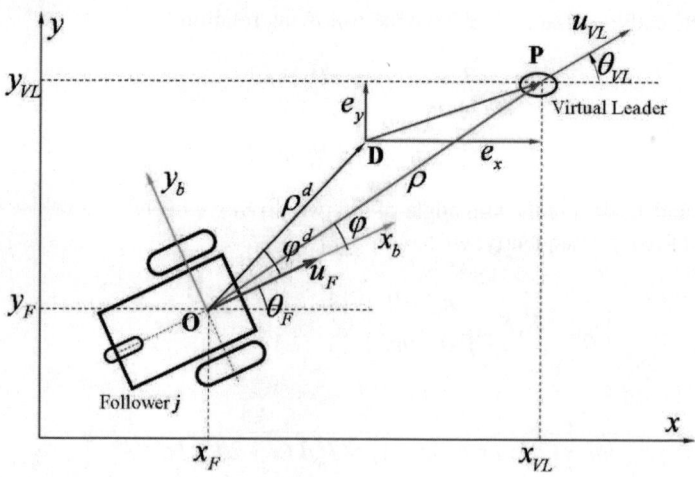

Fig. 2. Virtual leader based Leading-following configuration

Then, the error rates can be expressed as

$$\begin{bmatrix} \dot{e}_x^j \\ \dot{e}_y^j \end{bmatrix} = \begin{bmatrix} u_{VL} \cos \theta_{VL} \\ u_{VL} \sin \theta_{VL} \end{bmatrix} + C^j \begin{bmatrix} u_F^j \\ \dot{\theta}_F^j \end{bmatrix} \tag{14}$$

with

$$C^j = \begin{bmatrix} -\cos \theta_F^j & \rho_d^j \sin(\theta_F^j + \varphi_d^j) \\ -\sin \theta_F^j & -\rho_d^j \cos(\theta_F^j + \varphi_d^j) \end{bmatrix} \tag{14}$$

Note that det(C)=$\rho_d \cos\varphi_d$, and the matrix C is invertible only if $\varphi_d \neq \pm\pi/2$. This requirement is easily satisfied because ρ_d and φ_d can be chosen arbitrarily for the case of virtual leader based formation.

Consequently, the formation error dynamic equation of the follower j is derived as:

$$\begin{bmatrix} \ddot{e}_x^j \\ \ddot{e}_y^j \end{bmatrix} = C^j \begin{bmatrix} \dot{u}_F^j \\ \ddot{\theta}_F^j \end{bmatrix} + \dot{C}^j \begin{bmatrix} u_F^j \\ \dot{\theta}_F^j \end{bmatrix} + \frac{d}{dt} \begin{bmatrix} u_{VL} \cos \theta_{VL} \\ u_{VL} \sin \theta_{VL} \end{bmatrix} \tag{16}$$

Combining (16) with (7) leads to

$$\ddot{E}^j = C^j G I^j + H^j(.) \tag{17}$$

where $E^j = [e_x^j, e_y^j]^T$, $I^j = [I_{cr}, I_{cl}]^T$ and

$$G = K_t K_{am} \begin{bmatrix} m_1 & 0 \\ 0 & m_2 \end{bmatrix}^{-1} \begin{bmatrix} 1 & 1 \\ 1 & -1 \end{bmatrix} \tag{18}$$

$$H^j(.) = C^j Q + \dot{C}^j \begin{bmatrix} u_F^j \\ \dot{\theta}_F^j \end{bmatrix} + \frac{d}{dt} \begin{bmatrix} u_{VL} \cos \theta_{VL} \\ u_{VL} \sin \theta_{VL} \end{bmatrix} \tag{19}$$

Therefore, the objective of formation control for P3-DS1401 robots can be converted to make efforts to design the control current I to drive the formation error E to zero based on the error dynamic equation (17).

4   Error Shaping Memory-Based Control

Apparently, (17) denotes a highly nonlinear system, where H, including the side slip of two driving wheels, motion states, external disturbances and system uncertainties, is too complex to determine its precise information; the matrix G also involves uncertainties, arising from mass change, imprecise measurement of moment of inertia J_o and J_w, and estimation of motor torque constant K_t and current amplification factor K_{am}. Therefore, the control design should be very challenging to deal with such disturbances and uncertainties. In our work, memory-based control [15] is applied and successfully implemented to real-time formation control of multi-UGVs. The main idea behind this method is to use certain gathered information (i.e., current tracking error, most recent tracking errors, and previous control experiences) to generate new control commands without the need for detailed information on external disturbances and uncertainties.

However, offline simulation often showed extremely large control signals maybe appear for the memory-based control at very beginning of control if the initial control error is large, which limits this control algorithm to be applied to real-time control cases. To eliminate such limitation, an error shaping method is developed, which can be expressed as: design memory-based control algorithm based on the dynamic equation of the shaping error, which is defined as

$$S^j(t) = (\dot{E}^j + \beta E^j) - [\dot{E}^j(h) + \beta E^j(h)]\exp(-\alpha t) \tag{20}$$

where α and β are any positive numbers, h denotes the sampling period, $\dot{E}(h)$ and $E(h)$ are the first sampling values of $\dot{E}$ and E, respectively. Note that, the shaping error has three properties:

1) $\lim_{t \to 0} S^j(t) = 0$, that means the $S(t)$ has very small initial value;

2) If $S^j(t) \to 0$, then, $E^j(t) \to E^j(h)\exp(-\alpha t)$ and $\dot{E}^j(t) \to \dot{E}^j(h)\exp(-\alpha t)$, which implies the formation error E and error rate $\dot{E}$ will approach ideal error response $E^j(h)\exp(-\alpha t)$ and $\dot{E}^j(h)\exp(-\alpha t)$, that is the source of the name-"Error shaping" method;

3) If $S^j(t) \to 0$, then $\lim_{t \to 0} E^j(t) = 0, \lim_{t \to 0} \dot{E}^j(t) = 0$ so that $S^j(t) \to 0$ can ensure both formation error and error rate approach to zero as time goes to infinity.

That is to say, if the memory-based controller is designed based on the shaping error (20) to achieve $\|S^j(t)\|$ goes to zero, then the objective of formation control can be ensured without effect of large initial error. Therefore, the control objective can be converted to design the controller based on (20).
From (17) and (20), we have

$$\dot{S}^j = C^j G I^j + H^j(.) + \zeta^j(.) \tag{21}$$

with

$$\zeta^j(.) = \beta \dot{E}^j + \alpha[\dot{E}^j(h) + \beta E^j(h)]\exp(-\alpha t) \tag{22}$$

According to the control algorithm in [15], the first order memory-based controller can be determined as

$$I_k^j = (C_k^j G)^{-1}[C_{k-1}^j G I_{k-1}^j - (2/h)S_k^j + (1/h)S_{k-1}^j - (\zeta_k^j - \xi_{k-1}^j)] \tag{23}$$

and the error performances of such controller are

$$\lim_{t \to \infty} \|E_{k+1}^j\| \le \frac{h^2}{\beta} \max_t \left\| \frac{d}{dt} H^j(.) \right\| \tag{24}$$

$$\lim_{t\to\infty}\left\|\dot{E}_{k+1}^{j}\right\|\leq 2h^2 \max_t\left\|\frac{d}{dt}H^j(.)\right\| \qquad (25)$$

Note that, the boundaries of (24) and (25), generally, are very small. For example, if the sampling period $h=0.001$sec, then 100m/sec speed of $H(.)$ only involves 0.1mm/β formation error and 0.2mm/sec error rate. That means high precise control can be achieved by the first order memory-based controller (23). Meanwhile, extremely large signals are eliminated according to the property 1) of the shaping error.

5 Simulation and Real-Time Experiment

To verify the effectiveness of the proposed virtual leader based coordination strategy and the error shaping memory-based control method, three sets of experiments are developed on formation control of three agents (one virtual leader, two mobile robots). Some parameters of the P3-DX robot and simulation parameters are listed in Table 1. experiment 1 presents two robots driven by a virtual leader to follow a circular path, experiment 2 proposes two robots go out to Lab and return again, and experiment 3 tests formation change from parallel formation to leading-following formation and back to parallel formation. All results of experiment 1, including error performance and control signals of offline simulation, error performance and control signals of real-time experiment, and real-time control performance, are shown as Figure 3-Figure 5. For experiment 2 and 3, only real-time control performances are posted in this work because all three experiments have similar error performance and control signals.

The figure 3 and figure 4 show that the error shaping memory-based control works well. Both formation errors of two robots in offline simulation and real-time experiments converge to zero smoothly and quickly, and no extremely large signal appears in offline simulation, with control signals of both robots bounded and smooth. The figures 5-7 are real-time control performances of formation motion, including fixed formation and dynamic formation (formation change). These three experiments show that the virtual leader based formation control is successfully applied to real-time formation control cases.

Table 1. Physical parameters of the P3-DX robot and simulation parameters

Robot Parameters	Value	Simulation Parameters	Value
d	32 [cm]	α	0.8
r	9.5 [cm]	β	1.5
b	30 [cm]	h	0.001[sec]
m	10 [kg]		
Jo	0.520 [kg. m^2]		
Jw	0.013 [kg. m^2]		
Kt	10 [N/A]		
Kam	20		

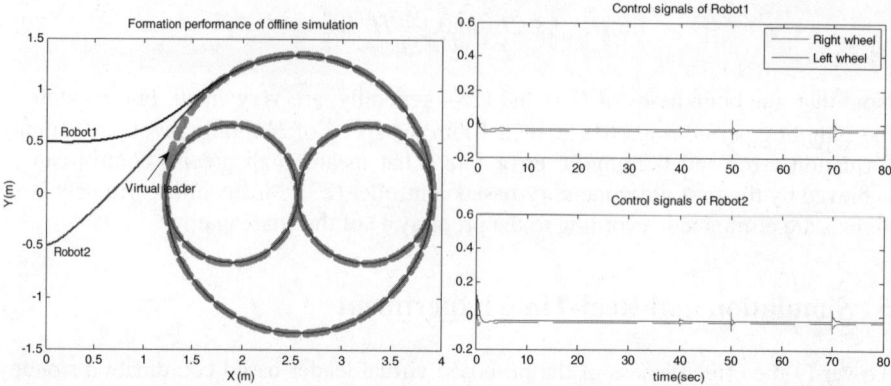

Fig. 3. Control signals and formation performance for offline simulation of experiment 1

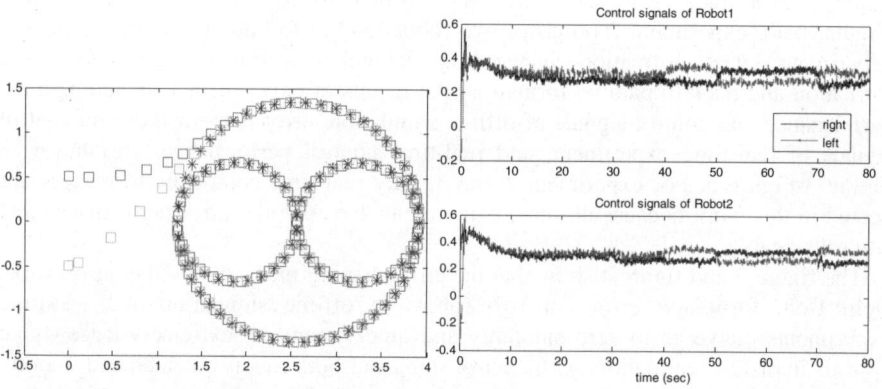

Fig. 4. Formation performance and control signals for real-time control case of experiment 1

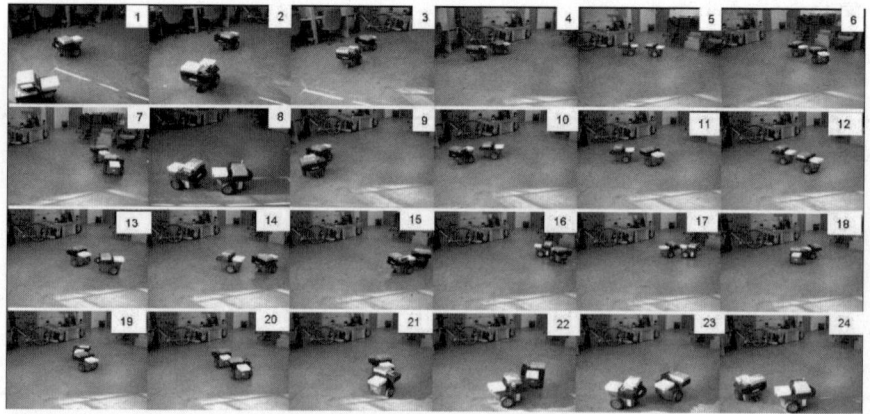

Fig. 5. Experiment 1: real-time control performance of two UGVs following circular path

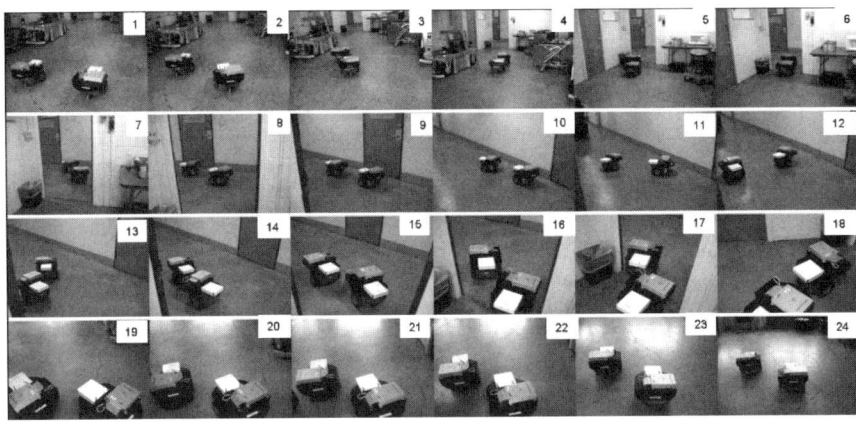

Fig. 6. Real-time control performance of experiment 2: two UGVs going out of door and return

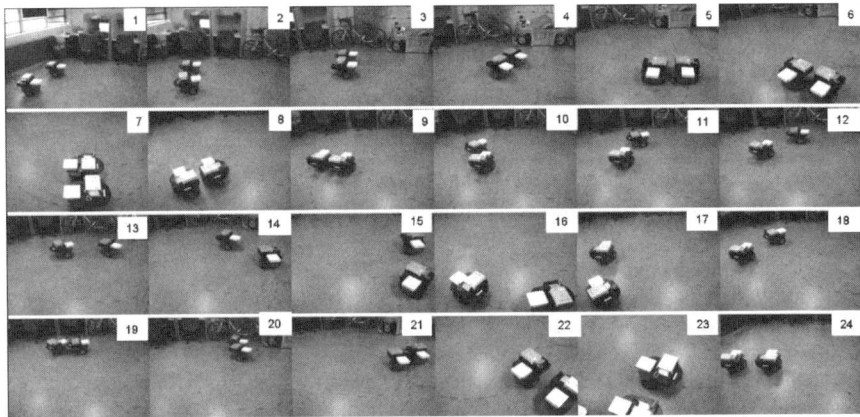

Fig.7. Real-time control performance of experiment 3: formation change (parallel, leading-following, position switching)

6   Conclusions

This work addressed the implementation of virtual leader based formation control of multi-UGVs in real-time control cases. A virtual leader was applied to coordinate a group of mobile vehicles so that all vehicles can group into any desired formation shape. Error shaping memory-based control was developed to deal with the problem of virtual leader based formation control. More importantly, both virtual leader based coordination strategy and error shaping memory-based control were successfully implemented in real-time experiments in this work, and the experiment results have shown virtual leader coordination strategy is more flexible to manage a formation group and error shaping method is an effective approach to convert the theoretical memory-based control algorithm to real-time control field.

References

1. H. Tanner, A. Jadbabaie and G. J. Pappas, "Stable Flocking of Mobile Agentsrt I: Fixed Topology," in Proc. Conf. Decision and Control, Maui, HI, Dec. 2003, pp. 2010-2015.
2. Song, Y.D.; Yao Li; Liao, X.H.; "Orthogonal transformation based robust adaptive close formation control of multi-UAVs", American Control Conference, 2005. Proceedings of the 2005 8-10 June 2005 Page(s):2983 - 2988 vol. 5.
3. Galzi, D.; Shtessel, Y.; "UAV formations control using high order sliding modes", American Control Conference, 2006 14-16 June 2006.
4. Gu, Y.; Seanor, B.; Campa, G.; Napolitano, M.R.; Rowe, L.; Gururajan, S.; Wan, S.; "Design and Flight Testing Evaluation of Formation Control Laws", Control Systems Technology, IEEE Transactions on Volume 14, Issue 6, Nov. 2006 Page(s):1105–1112.
5. Regmi, A.; Sandoval, R.; Byrne, R.; Tanner, H.; Abdallah, C.T.; "Experimental implementation of flocking algorithms in wheeled mobile robots" American Control Conference, 2005. Proceedings of the 2005 8-10 June 2005 Page(s):4917-4922 vol. 7
6. Takahashi, H.; Ohnishi, K.; "Autonomous decentralized control for formation of multiple mobile-robots considering ability of robot", Industrial Electronics Society, 2003. IECON '03. The 29th Annual Conference of the IEEE Volume 3, 2-6 Nov. 2003.
7. Zhi-Dong Wang; Takano, Y.; Hirata, Y.; Kosuge, K.; "A pushing leader based decentralized control method for cooperative object transportation", Intelligent Robots and Systems, 2004. (IROS 2004). Proceedings. 2004 IEEE/RSJ International Conference on Volume 1, 28 Sept.-2 Oct. 2004 Page(s):1035 - 1040 vol.1.
8. Regmi, A.; Sandoval, R.; Byrne, R.; Tanner, H.; Abdallah, C.T.; "Experimental implementation of flocking algorithms in wheeled mobile robots", American Control Conference, 2005. Proceedings of the 2005 8-10 June 2005 Page(s):4917-4922 vol.
9. Zhao Sun, Bin Li, XH Liao, Wenchuan Cai, Liguo Weng, Long Ni, and Y.D. Song. "Robust Adaptive Cooperative Control of Multiple UGVs". 6th International Conference on Cooperative Control and Optimization, 1630–1700, 2006.
10. Liguo Weng; Wenchan Cai; Ran Zhang; Song, Y.D.; "Bio-Inspired Control Approach to Multiple Spacecraft Formation Flying", e-Science and Grid Computing, 2006. e-Science '06. Second IEEE International Conference on Dec. 2006 Page(s):120-120.
11. Leonard, N.E.; Fiorelli, E.; Virtual leaders, artificial potentials and coordinated control of groups, Decision and Control, 2001. Proceedings of the 40th IEEE Conference, Dec. 2001
12. Hong Shi; Long Wang; Tianguang Chu; "Virtual Leader Approach to Coordinated Control of Multiple Mobile Agents with Asymmetric Interactions", Decision and Control, 2005 and 2005 European Control Conference. CDC-ECC '05. 44th IEEE Conference Dec. 2005.
13. Y. D. Song, "Memory-Based Control Methodology for Nonlinear Systems", Technical Report EE-9, North Carolina A&T State University, 1995.
14. Deng, X.Y.; Song, Y.D.; Anderson, J.N.; "Performance of a Memory-Based Approach to the Control of Robotic Manipulators", System Theory, 1997 Proceedings of the Twenty-Ninth Southeastern Symposium on 9-11 March 1997.
15. Song, Y.D.; "Memory-based Control of Nonlinear Dynamic Systems Part I - Design and Analysis"; Industrial Electronics and Applications, 2006 1ST IEEE Conference,May 2006.
16. W.C. Cai, L.G. Weng, R. Zhang, B. Li , F. Stewart, A. Dhaliwal and Y. D. Song, "Development of Real-time Control Test-bed for Unmanned Mobile Vehicles," in Proc. 32 *Annu*. Conf. IEEE Industrial Electronics Paris, FRANCE, November 7 - 10, 2006.
17. Yulin Zhang, Daehie Hong, Jae H. Chung, and Steven A. Velinsky, "Dynamic Model Based Robust Tracking Control of a Differentially Steered Wheeled Mobile Robot", in Proceedings of the American Control Conference in Philadephia, Pennsylvania, June 1998.
18. Z. Sun, W. C. Cai, X. H. Liao, T. Dong and Y. D. Song, " Adaptive Path Control of Unmanned Ground Vehicle," System Theory, 2006 Proceeding of the Thrity-Eighth Southeastern Symposium on March 5, 2006 Page(s):335-339.

Cooperative Control of Multiple Agents and Search Strategy

Vitaliy A. Yatsenko[1,*], Michael J. Hirsch[2], and Panos M. Pardalos[3]

[1] Space Research Institute of NASU and NSAU, Kiev, 03022, Ukraine
yatsenko@ufl.edu
[2] Raytheon, Inc., Net-Centric Systems, P.O. Box 12248, St. Petersburg, FL, 33733
U.S.A.
michael_j_ hirsch@raytheon.com
[3] Center for Applied Optimization, Department of Industrial and Systems
Engineering, University of Florida, Gainesville, FL 32611 U.S.A.
pardalos@ufl.edu

Abstract. This chapter discusses problems dealing with cooperative control of multiple agents moving in a region. An appropriate search strategy for the whole system can be embodied: hierarchical, coordinated, or cooperative. Geometrical and computational aspects of many-target search problems are considered. Nonlinear and bilinear processes of search for moving objects are proposed. Search problems of ecological danger objects and detection of biological and chemical agents using multi-spectral information are also considered.

Multiagent coordination problems are studied in detail. This problem is addressed for a class of targets for which control Lyapunov functions can be derived. We describe such a multiagent system by a hierarchical structure, which can be simplified using a fiber bundle. Then, using geometrical techniques, we study controllability, observability, and optimality problems. In addition, we also consider a cooperative problem when the agents motions must satisfy a separation constraint throughout the encounter to be conflict-free. A classification of maneuvers based on different commutative diagrams is introduced using their fiber bundle representation. In the case of two agents, these optimality conditions allow us to construct optimal maneuvers geometrically.

1 Introduction

Modern game theory basically deals with dynamical systems on smooth manifolds. However, many practical systems like multiple agents do not have such structures. Axiomatic control theories should adequately be reflected, in terms of their internal language of notions and control problems [1]. In terms of these theories, the control structures can make up various hierarchies. According to [2], the most general structure is represented by a controllability-reachability structure

---

* Corresponding author.

M.J. Hirsch et al. (Eds.): Adv. in Cooper. Ctrl. & Optimization, LNCIS 369, pp. 231–263, 2007.
springerlink.com   © Springer-Verlag Berlin Heidelberg 2007

over which the optimal control structure is built. This approach regarding the structure of optimal control and Yang–Mills Fields was discussed in [3] and [4].

In this chapter, the multiagent coordination problem is studied. This problem is addressed for a class of targets in which control Lyapunov functions can be found. The main result is a suite of prepositions about formation maintenance, task completion time, and formation velocity. It is also shown how to moderate the requirement that, for each individual target, there exists a control Lyapunov function.

We discuss mathematical aspects of Unified Game Theory (UGT) and the Theory of Control Structures (TCS). We consider a game as a hierarchical structure. It is assumed that each agent can be described by a fiber bundle. A joint maneuver has to be chosen to guide each agent from its starting position to its target position while avoiding conflicts. Among all the conflict-free joint maneuvers, we aim to determine the one with the least overall cost. The cost of an agents maneuver is its energy, and the overall cost is a weighted sum of the maneuver energies of all individual agents, where the weights represent priorities of the agents.

As an example, we consider the hierarchical structure of such multi-agent system on Figure 1. Each agent of the system can be described by a stochastic or deterministic differential equation with a control. In this chapter, we first reduce the model to a hierarchical geometric representation using fiber bundles. Then we consider an integrated geometrical model where the separated model of agents are integrated into a single model. For example, the interaction between six robots, as seen in Figure 2, can be described by a hierarchical structure. This integrated model allows for solving of controllability, observability, and cooperative control problems.

In Section 2, we demonstrate the power of the satisficing solution methodology for cooperative control problems regarding many-target search. An appropriate search strategy for the whole system can be embodied: hierarchical, coordinated, or cooperative. Geometrical and computational aspects of many-target search problems are considered. In Section 3, we analyze in detail the relationship between gauge fields, identification problems, and control systems. We consider a Lie group related to Yang–Mills gauge groups. We show that the estimation algebra of the identification problem is a subalgebra of the current algebra. Section 4 focuses on nonlinear control systems and Yang–Mills fields. This section is devoted to geometric models of multiagent systems as controlled dynamical-information objects. It is shown that these systems can be described by commutative diagrams which allow analysis of symmetries. Conclusions are drawn in Section 5.

2   Coordination for Different-Type Objects and Search Strategies

This section is dedicated to the development of methods for solving the problems of interception of multiple mobile targets by a group of unmanned vehicles.

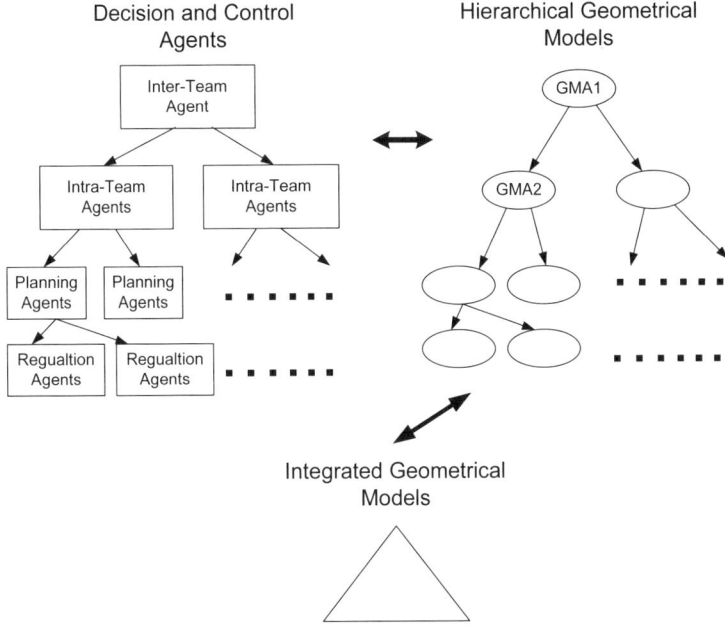

Fig. 1. Hierarchical structure of multiple agents

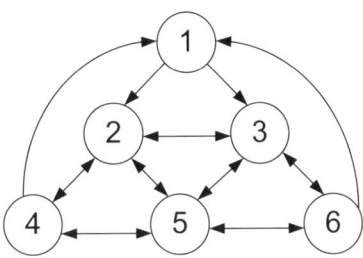

Fig. 2. Hierarchical structure of multiple robot

Emphasis will be paid on the following aspects: Interaction of controlled object groups; Active coordination for different-type objects; Implementation of new pursuit strategies; Investigation of group pursuit problems; and Search strategies.

2.1 Interaction of Controlled Object Groups

Methods and strategies will be devised for interception of multiple mobile targets (evaders), on the basis of the by-interval decomposition principle. This principle assumes that at the initial instant of time the interceptors (pursuers) and the targets are divided into subgroups, each consisting of either multiple pursuers and single target or single pursuer and multiple targets. Such targets' distribution can be performed either on the basis of certain experience or by using discrete optimization methods. As a result, the complicated process of a groups interaction is decomposed into a number of independent subproblems of group or successive pursuit, whereby the term 'group pursuit' is meant the pursuit of a single evader by multiple pursuers, and by the term 'successive pursuit' we mean the pursuit of multiple evaders by a single pursuer.

Fig. 3. Pursuit along the 'pursuit curve'

Let us fix the first instant of time when one of the mobile targets is intercepted and therefore can be excluded from further analysis. As a result, the newly freed pursuers can be included into other subgroups. At the instant t_k, let us perform a new decomposition of the pursuers and the remaining targets into subgroups, analogous to the first step. Analyzing the obtained problems of group and successive pursuit, we find the next instant t_{k+1} of interception of next target(s). At the instant t_{k+1}, a new target distribution is performed, and the process repeats.

It this manner, the process of optimization of controlled object interaction is reduced to the iterative procedure, which assumes solving the following typical problems:

1. Target distribution problem.
2. Group pursuit problem.
3. Successive pursuit problem.

The suggested procedure is particularly advantageous for sufficiently large numbers of mobile targets and pursuers, because in this case it reduces a complex original problem into a number of considerably less complicated processes, evolving in parallel, and makes it feasible for computer simulation on parallel computers. The problems of group and successive pursuit can be solved by using the Method of Resolving Functions (MRF) [5,6].

This method proved to be efficient in solving the group pursuit problem. It makes it feasible to study all known (classical) methods of pursuit from the unified standpoint. In particular, the MRF fully substantiates Parallel Pursuit Guidance Law, well known to designers of rocket and aerospace techniques [6].

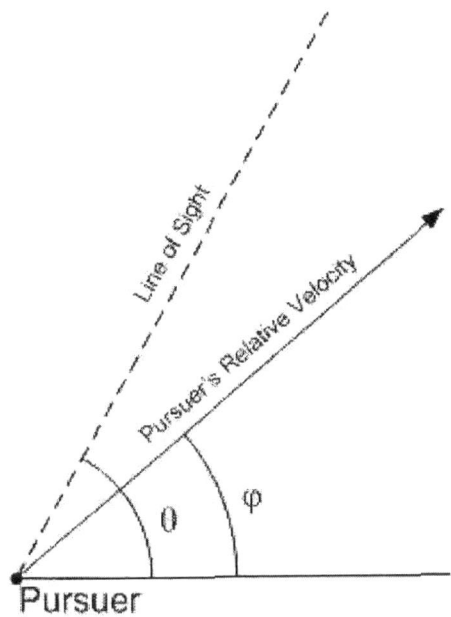

Fig. 4. Proportional navigation guidance law

Line of Sight (LOS) Guidance Law. This strategy has been long known from Euler's time [7]. It implies that at each instant of time the pursuer's velocity is directed along the Line of Sight (LOS) (Figure 3). The problem of finding the form of a trajectory of the pursuer, moving in the plane under the LOS Guidance Law was first formulated by Leonardo da Vinci [7]. It was solved by Pierre Bouguer in 1732 [8]. Despite simplicity in realization, this strategy fails to account for possible mistakes of the evader and frequently yields capture times significantly longer then optimal. The LOS strategy appears as a specific case of the Extremal Targeting Rule (ETR) in [9].

It is possible to formulate ETR in terms of support functions that essentially facilitate constructing the pursuer control and make it feasible to present the latter in explicit form. A modified ETR version for solving the problems of group and successive pursuit is discussed in [5].

Proportional Navigation (PN) Guidance Law. This method is well known to engineers involved in design of aerospace techniques [6]. The basic idea is that the angular velocity of the bearing, φ', varies proportionally with the angular velocity of the LOS, θ', $\varphi' = k\theta'$, where k is a navigation constant. The geometry behind this method is shown on Figure 4.

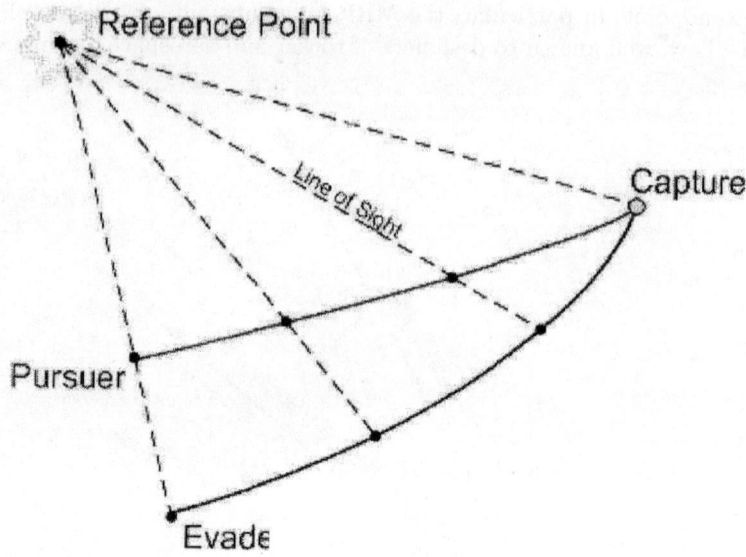

Fig. 5. Parallel pursuit

Note that when $k = 1$, the PN and LOS guidance laws coincide. Furthermore, as k approaches infinity the PN guidance law turns into the strategy of Parallel Pursuit (PP) [10]. The PP strategy implies that the lines of sight are parallel in the course of pursuit.

It is known that, in the case of simple motions, the strategy of parallel pursuit yields the optimal capture time (Figure 5) [5,6]. The MRF approach allows one to extend the ideology of parallel pursuit to wide classes of pursuit problems, where parallel pursuit is meant in a generalized sense. On the MRF basis, the authors have obtained important results concerning both the group and the successive pursuit [6]. In addition, necessary and sufficient conditions for solvability of the group pursuit problem were derived, together with explicit formulas for controlling functions [6].

2.2 Actions Coordination for Different-Type Objects

In the case of unmanned aerial vehicles (UAV) and unmanned ground vehicles (UGV), this problem becomes more complicated in view of state constraints, as one group of objects is moving in the air, while the other in the plane. This difficulty was successfully overcome in solving the problem of soft landing (e.g., airplanes landing on an aircraft carrier) [11].

2.3 Implementation of New Pursuit Strategies

In practice, when pursuing a moving target it is sometimes important for the pursuer not only to rapidly intercept the target, but also to conceal its approach.

This requirement is ensured by the pursuit strategy called *Motion Camouflage* (Figure 6). *Motion camouflage* is observed in insects, especially dragonflies [12].

In this strategy the pursuer camouflages itself against a fixed background object (reference point) so that the evader observes no relative motion between the pursuer and the fixed object (e.g., the sun). The pursuer simply remains on the line between the evader and the reference point, so it seems to be stationary from the evaders perspective [13,14]. If the reference point is at infinity, we obtain the parallel pursuit strategy described above. The motion camouflage strategy can be immediately applied to autonomous system control. For example, low observability behaviors have obvious applications in UAVs and guided missiles.

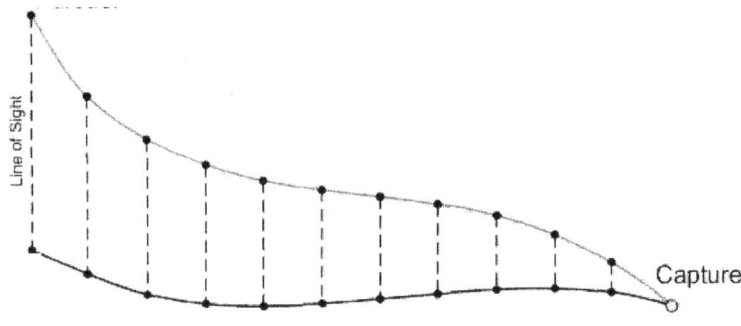

Fig. 6. Motion camouflage

2.4   Investigation of Group Pursuit Problems

The problems of search and observation for mobile targets constitute an important branch of the theory of conflict-controlled processes. Fundamental studies are provided in and [15,16,17,18]. The main feature of such problems is that only information on probability density of the current target state is available to the pursuing object(s), rather than the exact position. Using the probability density evolving according to the Fokker-Planck-Kolmogorov equation [17,18], we developed an approach (cell model of search), which is based on discretization of the search process both temporally and spatially. This process is bilinear and may appear as a Markovian or semi-Markovian chain. The detection probability or the average detection time are utilized as the performance criteria. The Pontryagin's discrete maximum principle and the Bellman dynamic programming method, respectively, were used to optimize the performance criteria. Game problems for the processes, described by the Ito equation, are studied in [18].

The problem of determining sea clutter dynamics and the application of reconstruction methodology in detection and classification of small targets has been considered in [19]. We explore the use of dynamical system techniques, optimization methods and statistical methods to estimate the dynamical characteristics of sea clutters. We assume that radar information is in a form of nonlinear time series. Hence we employ a dynamical approach for characterizing a radar signal, based on nonlinear estimation of dynamical characteristics, by forming a vector of these characteristics, and modeling the evolution of dynamical processes over time.

For the Navy domain, [6] created a decision support system tailored for the search for submarines in various tactical episodes. It is based on the above mentioned scheme of searching for mobile targets. Cases of discrete, continuous, and cyclic search, in their number conducted by a tactical group, were treated, as well as search performed in a hidden way, with the use of contemporary tools (e.g., UAV, UGV). For searches, performed by a tactical group, cases of information exchange within a group and individual search were analyzed. The problem of search for multiple targets was also studied. In this framework, we are planning to apply the achieved theoretical results and gained experience for the creation of methods and algorithms of search for multiple mobile targets by multiple-agent unmanned vehicles (both aerial and ground).

2.5   Cellular Search Model

Let us consider a search region which can be divided into a finite number of cells (states) $i = 1, \ldots, n$. A pursuer in state i at time t is able to move with probability $p_i(t)$, thus

$$p_i(t) \geq 0 \ \forall \ i = 1, \ldots, n \tag{1}$$

$$\sum_{i=1}^{n} p_i(t) = 1, \ \ t = 0, 1, \ldots . \tag{2}$$

Denote $p(t) = \left(p_1(t), \ldots, p_n(t) \right)$. The dynamics of the pursuer can be described by the discrete differential equation

$$p(t+1) = U^*(t)p(t), \ \ t = 0, 1, \ldots , \tag{3}$$

where $U(t)$ is a stochastic square matrix of order n, and $U^*(t)$ is the conjugate matrix, which play the role of control parameters and satisfy the constraints

$$u_{i,i_1}(t) \geq 0 \ \forall \ i, i_1 = 1, \ldots, n \tag{4}$$

$$\sum_{i_1=1}^{n} u_{i,i_1}(t) = 1 \ \forall \ i = 1, \ldots, n. \tag{5}$$

Suppose that an evasion object can be found in any state $j = 1, \ldots, m$ at time t with probability $q_j(t)$, i.e.,

$$q_j(t) \geq 0 \ \forall \ j = 1, \ldots, m \tag{6}$$

$$\sum_{j=1}^{m} q_j(t) = 1, \ \ t = 0, 1, \ldots . \tag{7}$$

Denote $q(t) = \left(q_1(t), \ldots, q_m(t) \right)$. The dynamics of the evasion object can be described by the discrete differential equation

$$q(t+1) = V^*(t)q(t), \ t = 0, 1, \ldots, \tag{8}$$

where $V(t)$ is a stochastic square matrix of order m, and $V^*(t)$ is the conjugate matrix elements, which play the role of control parameters and satisfy the constraints

$$v_{jj_1}(t) \geq 0, \ \forall \ j, j_1 = 1, \ldots, m \tag{9}$$

$$\sum_{j_1=1}^{m} v_{j,j_1}(t) = 1, \quad j = 1, \ldots, m. \tag{10}$$

The problem of optimal probability detection can be reduced to a conflict control problem of finite state

$$W_0(T) = (c, x(T)), \ c = (0, \ldots, 0, 1) \tag{11}$$

of the bilinear discrete process

$$x(t+1) = A(U(t), V(t))x(t), \quad t = 0, 1, \ldots, \tag{12}$$

where $W_0(T)$ is the probability of detection for time T.

Let r_{ij} be the probability of detection of the evasion object for the ith pursuer state and jth evasion state. Then the joint probability of evasion transition from j to j_1 at the moment t under undetected condition of the evasion object until time t is determined by the equation

$$f(i, i_1, j, j_1) = u_{ii_1}(t)v_{jj_1}(t)(1 - r_{ij}). \tag{13}$$

Denote by $F(u(t), v(t))$ the matrix function of dimension $m \cdot n$ with elements $f(i, i_1, j, j_1, t)$, where $u(t)$ is vector function with n^2 components $\{u_{ii_1}(t)\}$, $v(t)$ is vector function with m^2 components $\{v_{jj_1}(t)\}$.

This problem can be described by the optimization model

$$\omega_+ = \min_V \max_U W_0(T)$$

$$= \min_{V(0)} \max_{U(0)} \ldots \min_{V(T-1)} \max_{U(T-1)} W_0(T) \tag{14}$$

$$\omega_- = \max_U \min_V W_0(T)$$

$$= \max_{U(0)} \min_{V(0)} \ldots \max_{U(T-1)} \min_{V(T-1)} W_0(T) \tag{15}$$

where $U = U(0), \ldots, U(T-1)$, $V = V(0), \ldots, V(T-1)$, and $W_0(T)$ is the detection probability in time T.

The mean value of the target detection time is determined by the equation

$$\tau(u, v) = (W(0), N\xi), \tag{16}$$

where $N = \sum_{t=0}^{\infty} F^t(u, v)$, and ξ is the column vector with all components equal to one. It is evident that matrix $N = (1 - F(u, v))^{-1}$ exists and can be considered the problem of optimization of the mean target detection time.

The deviation of target detection for fixed control is given by the equation

$$D(u, v) = (W(0), (2N - E)N\xi - (N\xi)_{sq}), \tag{17}$$

where E is the single matrix and $(N\xi)_{sq}$ is the vector with components which equal the square of components of vector $N\xi$.

3   Geometrical Aspects of Multiagent Coordination

Investigations of controlled multiagent objects have been under active development for last few years. Despite the achievements that have been made in this area, effective mathematical methods for investigating such systems have not yet been developed. One possible approach is the differential geometry methods of system theory [20,21]. This section is devoted to one of the problems of this area of research, that of developing a method for analyzing a class of mathematical models of symmetric controlled processes. Assuming that the process is described by a commutative diagram [20,21] which is based on the lamination concept, we propose a geometric method for "identifying" its hidden structure.

Investigation of geometrical aspects of multiagent coordination is one of the most essential stages in the creation of new strategies. The goal of the experimental and theoretical research is the implementation of optimal strategy using complex structure non-equilibrium processes in such systems. To investigate these processes it is required to develop the corresponding mathematical methods. In this context we propose an approach, which is based on the assumption that one can use models from mathematical system theory to adequately describe informational processes. The essence of this approach is in the following.

Some dynamic system, S, which implements a transformation, F, or an input informational action, U, into an output one, X, is considered. It is assumed that one can affect the information-transforming process by a reconfiguring action that changes the dynamic behavior, structure, symmetry, etc. of the process. We refer to the objects described in the preceding S as dynamic information-transforming agents (DITA).

The connection between the input and output actions is necessary for obtaining answers to questions about the method of programming the entire system, optimizing the flow of informational signals, and the interconnections among the global system properties (stability, controllability, etc.) and the corresponding local properties of the various subsystems. One has to answer those questions also when solving pattern–recognition problems, constructing an associative memory. A generalized description of an DITA that contains a large number of subsystems (e.g., a neural network) is postulated in this section: the controlled process in the DITA is described adequately by a commutative diagram which generalizes the concept of a nonlinear controlled dynamic system on a manifold. Taking into account the symmetry concept which is characteristic of classical mechanics

[22], one has to transfer it to the DITA, "identify" the hidden structure of the informational process, and demonstrate that the proposed model admits local and/or global decompositions into smaller dimensionality feedback subsystems.

We note that the decomposition idea was first applied to discretely symmetric automatic control systems [23]. Continuous symmetry group dynamic systems were considered by [20]. While substantive results on the decomposability of systems with symmetries have been obtained [24], this question remains open for DITAs.

In this section, we investigate the problem of how to coordinate a collection of targets in such a way that they maintain a given formation relative to each other. The main assumption about the dynamics of the individual robots that we initially make in this paper is that they have control Lyapunov functions (CLFs). Based on this assumption, an abstract and theoretically sound coordination strategy can be developed.

3.1   Necessary Concepts and Definitions

Some definitions and concepts that are necessary for describing the DITA structure and the conditions for its decomposability are presented in this section. The necessary notions about manifolds, connectedness, and distributions are given in [25]. We introduce the definition of a nonlinear DITA.

Definition 1. *We refer a triple, $F(B, M, \psi)$, where B is a smooth fiber over M with the projection $\pi : B \rightarrow M$; π_M is the natural projection of TM on M; and ψ is a smooth mapping such that the diagram presented in Figure 7 is commutative, by a "geometrical model of the agent".*

We interpret the M manifold as the state space and the $\pi^{-1}(x) \in B$ layer as the space of input action values which depends in the general case on the current system state. If one chooses the coordinates (x, u), which correspond to the B_x layer, then this definition of the agent, F, corresponds locally to the nonlinear transformation $\psi : (x, u) \rightarrow (x, \psi(x, u))$ and the dynamic system

$$\dot{x}(t) = \psi(x(t), u(t)), \quad u(t) \in U, \tag{18}$$

where x is the state vector, $u = (u^1, u^2)$ are the control actions, $u^1(\cdot, \cdot)$ is the vector of the coded input informational action which depends in general on time and on the current state, and $u^2(\cdot, \cdot)$ is the action used to reconfigure the dynamic properties of the agents and to train it.

The control algorithm, u^2, inputs to the system the capability of transforming the set of input actions into a set of output signals that allows one to identify the input images uniquely. In essence, it realizes the decoding process, which identifies the input images. In the simplest case, it can be realized on the basis of the successive input action segmentation method. Such a method facilitates a unique separation of the input images by the use of the simplest binary decoding rule.

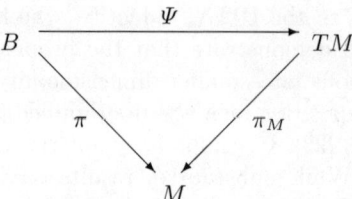

Fig. 7. Diagram of a nonlinear controlled DITA

Our primary object of study is a collection of targets, whose dynamics can be described by the following set of controlled differential equations:

$$\dot{x}_i(t) = \psi(x(t), u(t)) = f_i(x_i) + g_i(x_i)u_i, \quad i = 1, \ldots, n, \tag{19}$$

where $f_i, g_i \in C^\infty$, $x_i \in \mathbb{R}^n$, and $u_i = \mathbb{R}^{p_i}$. Now, a desired formation in $\mathbb{R}^{nm}$ is simply a set $(x_{10}, \ldots, x_{m0}) \in \mathbb{R}^{nm}$, and we define this set implicitly through the null set of a so-called formation function.

3.2   Coordinated Control

By using the Lyapunov formation functions derived from the individual target, we can now shift our attention to actually controlling the evolution of the formation. The one parameter that we can control is the time evolution of the desired positions. We do this by specifying the trajectory that we want the so-called virtual leader, $x_0(s(t))$, to follow.

This nonphysical leader is a reference point in the state space with respect to which we can define the rest of the formation. We denote the trajectory executed by the virtual leader as $x_0(s(t)) = p(s(t))$. Intuitively, one might want to set $s(t) = t$. But, due to robustness considerations, we incorporate error feedback into the time evolution of s and let $\dot{s}$ be given by

$$\dot{s} = \min\left[\frac{v_0}{\delta + \|\frac{\partial p(s)}{\partial s}\|}, \frac{-(\frac{\partial F}{\partial x})^T}{\delta + |\frac{\partial F}{\partial s}|}\left[\frac{\sigma(F_U)}{\sigma(F(s, x))}\right]\right]. \tag{20}$$

Here, $\delta > 0$ is a small positive constant that prevents $\dot{s}$ from becoming singular, and F_U is the bound or something smaller chosen by the user. It can be shown to be an upper bound on the Lyapunov formation function $F(s, x)$. The idea is that the formation is being respected as long as $F(s, x) \leq F_U$. Furthermore, v_0 is the nominal velocity that we want the formation to move with, and it holds that $\|\dot{x}_0(s(t))\| \approx m_0$ when small.

3.3   Symmetry of Multiagent Coordination

Definition 2. *Let M be a smooth manifold. We say that the smooth mapping $Q : G \times M \to M$ such that: i) $Q(e, x) = x$ for all $x \in M$, and ii) $Q(g, Q(h, x)) = Q(gh, x)$ for any $g, h \in G$, and all $x \in M$, is the left action (or G-action) of the G Lie group on M.*

We fix one of the variables for various time instants and examine the Q action as a function of the remaining variables. Let $Q_g : M \to M$ denote the function $x \mapsto Q(g, x)$ and $Q_x : G \to M$ denote the function $g \mapsto Q(g, x)$. We note that since $(Q_g)^{-1} = Q_g^{-1}$, Q_g is a diffeomorphism. We introduce the definition of group action on a manifold.

Definition 3. *Let Q be the action of G on M. We say that the set $G \cdot x = \{Q_g(x) | g \in G\}$ is the orbit (Q-orbit) of the point $x \in M$. The action is free at x if $g \mapsto Q_g(x)$ is one-to-one. It is free on M if and only if it is free at all $x \in M$.*

We now introduce the concept of global symmetry of a controlled DITA.

Definition 4. *Let $\hat{F}(B, M, \psi)$ be a nonlinear controlled DITA, and θ and Q be actions of G on B and M, respectively. Then, F has symmetry (G, θ, Q) if the diagram presented in Figure 8 is commutative for all $g \in G$.*

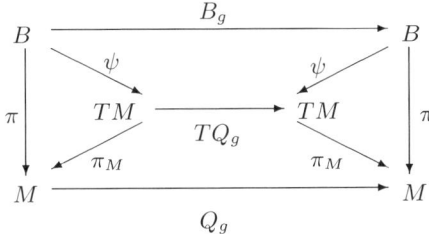

Fig. 8. A commutative diagram of an DITA with symmetries

We consider, within the framework of the presented definition, the special case in which the symmetry lies "entirely within the state space".

Definition 5. *Let $B = M \times U$, where U is some manifold. Then, (G, Q) is a symmetry of the state space of system $\hat{F}(B, M, \psi)$ if (G, θ, Q) is a symmetry of $\hat{F}$ for $\theta_g = (Q_g, Id_U) : (x, u) \to (Q_g(x), u)$.*

Global state space symmetry can be defined only for a DITA B_x, which is a trivial lamination, since otherwise the input spaces would depend on the state and the problem is made substantially more complicated. We introduce the definition of local symmetry.

Definition 6. *We assume that $Q : G \times M \to M$ is an action and that $\varepsilon \in T_e G$. Then, $Q^\xi(R \times M \to M) : (t, x) \mapsto Q(\exp t\xi, x)$, where $\exp : T_e G \to G$ is the usual exponential mapping, is the $\mathbb{R}$-action on M, and Q^ξ is the complete flow on M. We say that the corresponding vector field on M, which is defined by the expression*

$$\xi_m(x) = \frac{d}{dt} Q(\exp t\xi, x)\Big|_{t=0}, \tag{21}$$

is the infinitesimal action generator, which corresponds to ξ.

Let X_t denote the flow of the vector field X, i.e., $X_t = F_t(X_0)$. It is obvious from the definition of the infinitesimal generator that if (G, θ, Q) is a symmetry of the $\hat{F}(B, M, \psi)$ system, then the diagram presented in Figure 9 is commutative for all $t \in \mathbb{R}$ and $\xi \in T_eG$.

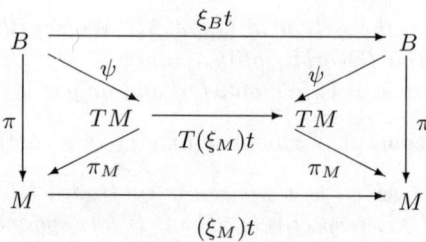

Fig. 9. Diagram of a symmetric DITA

On the basis of the local commutativity property we present the following definition of infinitesimal DITA symmetry.

Definition 7. *Let $\hat{F}(B, M, \psi)$ be a nonlinear DITA. Then, (G, θ, Q) is an infinitesimal symmetry of F if, for each $x_0 \in M$, there exist an open neighborhood $\hat{O}$ of the point x_O and $\xi > 0$ such that*

$$(\xi_M)_t * \psi(\xi) = \psi((\xi_b)_t(b)), \tag{22}$$

for all $b \in \pi^{-1}(\hat{O})$, $|t| < \xi$, and $\| \xi \| < 1$, $\xi \in T_eG$, where $\| \cdot \|$ is an arbitrary fixed norm on T_eG.

One can define an infinitely small state space symmetry for nontrivial laminations of the input actions manifold when one can introduce integrable connectivity. For this we introduce Definition 8.

Definition 8. *Let $H(\cdot)$ be an integrable connectivity on B and (G, θ, Q) be a symmetry of F. Then, (G, θ, Q) is an infinitesimal state space symmetry if $\xi_B(b) \in H(b)$ for all $\xi \in T_eG$, that is, the infinitesimal generators θ are horizontal.*

We introduce a definition of feedback equivalence of two DITAs in analogy with [20].

Definition 9. *A system, $F(B, M, \psi)$, is feedback equivalent to a system, $F'(B, M, \widetilde{\psi})$, if there exists an isomorphism, $\gamma : B \to B$, such that the diagram presented in Figure 10 is commutative.*

Isomorphism means that, for $x \in M$, γ_x is a mapping from the layer over x' into the layer over x', and it is a diffeomorphism. Consequently, this corresponds to a "control feedback".

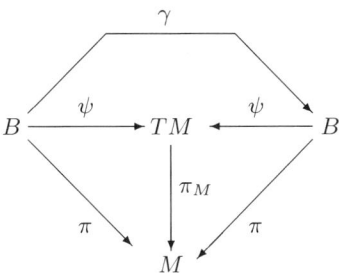

Fig. 10. Diagram of feedback-equivalent DITAs

3.4 The Local Structure of DITAs with Symmetries

Since we are interested in the local structure of an DITA, we have to assume that the system has an infinitesimal symmetry, which satisfies some nonsingularity condition. For this, we set the dimensionality of M to n and that of G to k, where $k < n$. We note that the action $Q : G \times M \to M$ is free at the point $m \in M$ if $Q_m : G \to M$ is one-to-one. This is equivalent to saying that the tangent mapping Q is of full rank, that is, rank $Q = \dim G$. Hence, Q is free on M if and only if it is free in some neighborhood of m. We say that an action which satisfies this condition is nonsingular at the point m.

The basic result of this section is that the existence of an infinitesimal symmetry in a neighborhood of a singular point in an DITA makes it possible to decompose the system into a cascade union of simpler subsystems. The structure of these subsystems depends, in general, on the symmetry group G. If, for example, G has a nontrivial center, then one of the subsystems is in fact a quadrature subsystem.

In addition, let $C = \{h \in G \mid hg = gh \ \forall \ g \in G\}$ be the center of the G group to which the kernel, C_+, of the Lie semialgebra $T_e G$, which has the same dimensionality as C, corresponds. Hence, if G has an l-dimensional center, there exist linearly independent vectors $\xi^1, \ldots, \xi^k \in T_e G$ such that $[\xi^i, \xi^j] = 0$ for all $1 \leq i \leq l$ and $1 \leq j \leq k$.

Using the results in [20,26] that deal with the properties of systems with symmetries as applied to DITAs, one can formulate the following theorems.

Theorem 1. *Let us assume that $\hat{F}(B, M, \xi)$ is a controlled DITA with an infinitesimal state space symmetry, (G, θ, Q), that G has an l-dimensional center, and that Q is nonsingular at the point $m \in M$. Then, the B coordinates $(x_1, \ldots, x_n, u)$ in a neighborhood of m exist such that $\hat{F}$ is given in these coordinates by the expression.*

Using the obtained results for systems with infinitesimal state space symmetries, one can propose the structure of the decomposed system. It suffices to demonstrate that the decomposed system with infinitesimal symmetry is locally feedback-equivalent to the original system with infinitesimal state space symmetry.

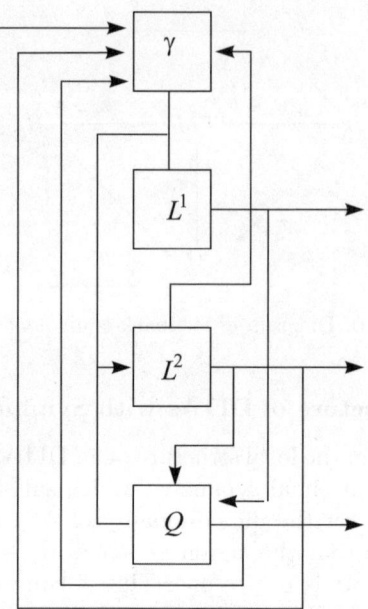

Fig. 11. Local structure of DITA with infinitesimal symmetries

Definition 10. *Let $\hat{F}(B, M, \psi)$ be a controlled DITA and $\hat{O}$ be an open subset of M. Then, we say that a system of the form $\hat{F}(\pi^{-1}(\hat{O}), \hat{O}, \psi)|\pi^{-1}(O)$ is $\hat{F}|\hat{O}$ ($\hat{F}$ bounded on $\hat{O}$).*

Theorem 2. *Let $\hat{F}(B, M, \psi)$ have an infinitesimal symmetry (G, θ, Q) and Q be nonsingular at the point m. There exists a neighborhood of m and a system F with infinitesimal symmetry (G, θ, Q) such that $\hat{F}|\hat{O}$ is feedback equivalent to the system $\hat{F}$.*

Let $\hat{F}(B, M, \psi)$ be a controlled DITA with symmetry (G, θ, Q) and Q be nonsingular at the point m. Then, in a neighborhood of m, $\hat{F}$ is feedback-equivalent to $\hat{F}$ with infinitesimal symmetry and has the structure shown in Figure 11, where γ is the feedback function, the L^i are nonlinear subsystems of dimensions $n - k$ and $k - l$, respectively, and Q is an l-dimensional "quadrature" system

$$\dot{x}_i = f_i(x_1, \ldots, x_{n-k}, u), \quad i = 1, \ldots, n - k \tag{23}$$

$$\dot{x}_j = f_j(x_1, \ldots, x_{n-1}, u), \quad i = n - k + 1, \ldots, k. \tag{24}$$

3.5   The Global Structure of DITA

The decomposability of a DITA with global symmetries is the result of factoring the DITA state space, which follows from the properties of a symmetry. We introduce the definition of proper action.

Definition 11. *Let Q be a G-action on M. We say that Q acts properly if $(g, m) \to m$ is a proper mapping, that is, if the pre-images of compact sets are compact.*

This definition is equivalent to the following assertion: whenever x_n converges on M and $Q_{g_n}(x_n)$ converges on M, g_n includes a subsequence, which converges in G. Hence, if G is compact, this condition is satisfied automatically. Membership in the same Q-orbit is an equivalence relation on M. Let M/G be the set of equivalence classes and $p : M \to M/G$ be specified by the relation $p(m) = Gm$. We introduce on M/G a relations topology, that is, $V \subset M/G$ is open if and only if $p^{-1}(V)$ is open on M. In general, M/G can be a rather poor space.

If G acts freely and properly on M, then M/G is a smooth manifold and $p : M \to M/G$ is the principal lamination with Lie group G. We introduce the following constraints on the principal lamination:

1. p is a smooth full-rank function;
2. $p : M \to M/G$ has a cross section (that is, a smooth mapping $\sigma : M/G \to M$ such that $p \cdot \sigma$ is the identity mapping on M/G if and only if M is equivalent to $M/G \times G$;
3. the topological conditions which guarantee the existence of a section, that is, if M/G or G is a contraction mapping, a cross section must exist, are specified.

We formulate a theorem, which is necessary for obtaining a global factorization of the DITA state space. Let $Q_m : G \to G \cdot m$ be specified by $g \to Q(G, m)$. The following result about the global structure of a DITA with symmetries holds.

Theorem 3. *We assume that $\hat{F}(M \times U, M, \psi)$ is a controlled DITA with a state space symmetry (C, Q). Then, if Q is free and proper, and $p : M \to M/G$ has a cross section σ, then $\hat{F}$ is isomorphic to the system*

$$\dot{y} = \Psi(y, u) \tag{25}$$

$$\dot{g} = (T_e L_g)(T_e Q_{\sigma(y)})^{-1} \left[\Psi(\sigma(y), u) - (T_y \sigma) \Psi(y, u) \right], \tag{26}$$

defined on $M/G \times G$.

Assertion 1. *Let the DITA $F(M \times U, M, \psi)$ have a symmetry (G, θ, Q) such that Q is free and proper. Then, there exists a system F with symmetry (G, Q) to which F is feedback equivalent under the condition that $p : M \to M/G$ has a cross section σ.*

Combining Theorem 3 and Assertion 1, we obtain the following corollary:

Corollary 1. *Let DITA $\hat{F}(M \times U, M, \psi)$ have a symmetry (G, θ, Q), Q be free and proper, and $p : M \to M/G$ have a cross section. Then, there exists a model of DITA F with state space symmetry (G, Q) to which $\hat{F}$ is feedback-equivalent. Consequently, F has a global structure.*

3.6   The Feasibility of Applying the Results to the Investigation of Agents

It is of interest to investigate the decomposability of DITAs composed of neural-like agents that are described by the system

$$\dot{x}(t) = \psi(x(t), u(t)). \tag{27}$$

One can define for Equation (27) a decomposed system L as a nontrivial cascade of subsystem L^1 and L^2. If the Lie algebra $\hat{L}(L)$ is the semidirect sum of finite-dimensional subalgebra L^1 and the ideal of L^2, it has a nontrivial cascade decomposition into subsystems L^1 and L^2 such that $\hat{L}(L^1) = L^1$, and $\hat{L}(L^2) = L^2$. Using this fact and Levy's theorem one can demonstrate that if $\hat{L}(L)$ is finite-dimensional, the DITA admits a nontrivial decomposition into a parallel cascade of L^i systems with simple Lie algebras followed by a cascade of one-dimensional systems, L^j. As a result, the basic informational transformation is done in subsystems with simple Lie algebras. The state space, M, of the original system, L, is adopted here as the state space of these systems. Therefore, despite the fact that the system has been partitioned into simpler parts, the overall dimensionality of these parts is, in general, larger than that of the original system. (One can reduce at the local level this dimensionality by replacing the L^i system by matrix equivalents defined on the exponential functions of the Lie algebras that correspond to them.) These results can be compared with the conditions for decomposability obtained by analyzing the DITA symmetries described in this section for which the subsystem dimensionality equals that of the original system. No assumptions about the finite dimensionality of the Lie algebra are required here. We consider a class of neural nets described by the linear-analytic equations

$$\dot{x}(t) = f(x) + \sum_{i=1}^{k} u_i g_i(x). \tag{28}$$

One can formulate the necessary and sufficient conditions for parallel-cascade decomposability by Lie algebras. In doing so, one can pose the condition that each component of the input action be applied to only one of the subsystems, that is, the decomposition procedure partition the inputs into disjoint subsets. However, such an approach cannot be applied to the decomposition of an DITA with scalar input.

If DITA $\hat{F}(B, M, \psi)$ has an infinitesimal symmetry (G, θ, Q), local commutativity of the diagram means that $\psi * \varepsilon_B = \varepsilon_m$ and $\pi * \varepsilon_B = \varepsilon_n$. Let $\Delta_B = \text{span}\{\varepsilon \mid \varepsilon_B \in T_e G\}$ and the same hold for Δ_m. Then, $\psi * \Delta B \subset \Delta_m$, $\pi * \Delta_B = \Delta$, and Δ_m is a controlled invariant distribution. Models of neural networks, including affine ones, have invariant distributions that induce decompositions of the system into simpler subsystems. However, since the symmetry conditions are constraints, the decompositions are obtained as more detailed and structured.

A class of dynamic information-transforming systems that are described by a commutative diagram is examined in this section. Constraints on systems with symmetry under which one can expose, explicitly the hidden structure of the controlled process are formulated. We show that the effect of the DITA on the information-transforming process depends substantially on the type of system symmetry. The informational process is subject here to the action of cascade groups, transformations, or the action of a dynamic-transformation operator with feedback. The obtained results can be expanded to adaptive learning systems by introducing the corresponding optimization models. When doing so, one can expect that a DITA, of which the quality functional is invariant in symmetry-conserving transformations, will be described adequately by a nonlinear system with optimal feedback and will have a differential-geometric structure, which is of interest from the point of view of applications.

4   Fiber Bundles and Observability

In the last decade, important work has been done on a differential geometric approach to nonlinear input state-output systems, which in local coordinates have the form

$$\dot{x} = g(x, u), \quad y = h(x), \tag{29}$$

where x is the state of the system, u is the input, and $\dot{y}$ is the output. Most of the attention has been directed to the formulation in this context of fundamental system theoretic concepts like controllability, observability, minimality, and realization theory.

In spite of some very natural formulations and elegant results that have been achieved, there are certain disadvantages in the whole approach, from which we summarize the following points:

1. Normally, the equations

$$\dot{x} = g(x, u) \tag{30}$$

   are interpreted as a family of vector fields on a manifold parameterized by u; i.e., for every fixed $\bar{u}$, $g(\cdot, \bar{u})$ is a globally defined vector field. We propose another framework by looking at (30) as a coordinization of the following diagram.

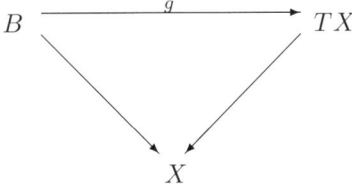

   where B is a *fiber bundle* above the state space manifold X and the fibers of B are the *state dependent* input spaces, while TX is as usual the tangent bundle of X (the possible velocities at every point of X).

2. The "usual" definition of *observability* has some drawbacks. In fact, observability is defined as *distinguishable*; i.e., for every $x_1, x_2 \in X$, there exists a *certain* input function (in principle, dependent on x_1 and x_2) such that the

output function of the system starting from x_1 under the influence of this input function is different from the output function of the system starting from x_2 under the influence of the same input function. Of course, from a practical point of view this notion of observability is not very useful, and also is not in accord with the usual definition of observability or reconstructibility for general systems.

3. In the class of nonlinear systems (29), *memoryless* systems

$$y = h(u) \tag{31}$$

are not included. Of course, one could extend the system (29) to the form

$$\dot{x} = g(x, u), \quad y = h(x, u), \tag{32}$$

but this gives, if one wants to regard observability as distinguishability, the following rather complicated notion of observability. As can be seen, distinguishability of (32) with $y \in \mathbb{R}^p$, $u \in \mathbb{R}^m$ and $x \in \mathbb{R}^n$ is equivalent to distinguishability of

$$\dot{x} = g(x, u), \quad \overline{y} = \overline{h}(x), \tag{33}$$

where $\overline{h} : \mathbb{R}^n \to (\mathbb{R}^p)^{\mathbb{R}^m}$ is defined by $\overline{h}(x)(u) = h(x, u)$. Also, checking the Lie algebra conditions for distinguishability for the system (33) is not very easy.

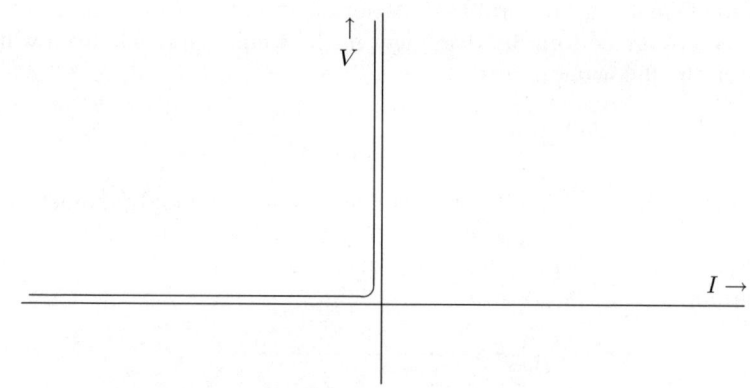

Fig. 12. Ideal diode for the $I - V$ characteristic

4. It is often not clear how to distinguish a priori between inputs and outputs. Especially in the case of a nonlinear system, it could be possible that a separation of what we call *external variables* in input variables and output variables should be interpreted only *locally*. An example is the (nearly) ideal diode given by the $I - V$ characteristic in Figure 12. For $I < 0$ it is natural to regard I as the input and V as the output, while for $V > 0$ it is natural to see V as the input and I as the output. An input-output description should

be given in the scattering variables $(I - V, I + V)$. Moreover, in the case of nonlinear systems it can happen that a global separation of the external variables in inputs and outputs is simply not possible. This results in a definition of a system, which is a generalization of the usual input-output framework. It appears that various notions like the definitions of autonomous (i.e., without inputs), memoryless, time-reversible, Hamiltonian and gradient systems are very natural in this framework.

4.1 Nonlinear Model of Agents

The (say C^∞) agents can be represented in the commutative diagram

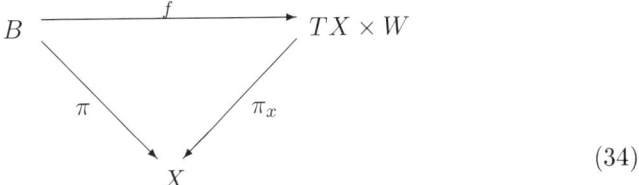

(34)

where (all spaces are smooth manifolds) B is a fiber bundle above X with projection π, TX is the tangent bundle of X, π_x the natural projection of TX on X and f is a smooth map. W is the space of external variables (think of the inputs *and* the outputs). X is the state space and the fiber $\pi^{-1}(x)$ in B above X represents the space of inputs (to be seen initially as *dummy* variables), which is state dependent e.g., forces acting at different points of a curved surface.

This definition formalizes the idea that at every point $x \in X$ we have a set of possible velocities, elements of TX, and possible values of the external variables, elements of W, namely the space

$$f(\pi^{-1}(x)) \subset T_x X \times W. \tag{35}$$

We denote the system (34) by $\Sigma(X, W, B, f)$. It is easily seen that in local coordinates x for X, v for the fibers of B, w for W, and with f factored as $f = (g, h)$, the system is given by

$$\dot{x} = g(x, v), \quad w = h(x, v). \tag{36}$$

Of course one should ask how this kind of system formulation is connected with the usual input-output setting. In fact, by adding more and more assumptions successively to the very general formulation (34) we shall distinguish among three important situations, of which the last is equivalent to the "usual" interpretation of system (29).

1. Suppose the map h restricted to the fibers of B is an *immersive* map into W (this is equivalent to assuming that the matrix $\partial h / \partial v$ is injective). Then:

   Lemma 1. *Let h, restricted to the fibers of B, be an immersion into W. Let $(\overline{x}, \overline{v})$ and $\overline{w}$ be points in B and W respectively such that $h(\overline{x}, \overline{v}) = \overline{w}$. Then* *locally around* $(\overline{x}, \overline{v})$ *and* $\overline{w}$ *there are coordinates* $(x, v) \in B$, *coordinates* $(w_1, w_2) \in W$ *and a map* $\overline{h}$ *such that h has the form*

$$(x, v) \gg h > (w_1, w_2) = (\overline{h}(x, v), v). \tag{37}$$

Proof. The lemma follows from the implicit function theorem. Hence, *locally* we can interpret a part of the external variables, i.e., w_1, as the outputs, and a complementary part, i.e., w_2, as the inputs. If we denote w_1 by y and w_2 by u, then system (36) has the form, only locally,

$$\dot{x} = y(x, u), \quad y = \overline{h}(x, u). \tag{38}$$

2. Now we not only assume that $\partial h / \partial v$ is injective, which results in a *local* input-output parametrization (38), but we also assume that the output set denoted by Y is *globally* defined. Moreover, we assume that W is a fiber bundle above Y, so that $p : W \to Y$, and that h is a bundle morphism (i.e., maps fibers of B into fibers of W). Then:

Lemma 2. *Let* $h : B \to W$ *be a bundle morphism, which is a diffeomorphism restricted to the fibers. Let* $\overline{x} \in X$ *and* $\overline{y} \in Y$ *be such that* $h(\pi^{-1}(\overline{x})) = p^{-1}(\overline{y})$. *Take coordinates* $x \in X$ *around* $\overline{x}$ *and coordinates* $y \in Y$ *around* $\overline{y}$. *Let* $(\overline{x}, \overline{v})$ *be a point in the fiber above* $\overline{x}$ *and let* $(\overline{y}, \overline{u})$ *be a point in the fiber above* $\overline{y}$ *such that* $h(\overline{x}, \overline{v}) = (\overline{y}, \overline{u})$. *Then there are local coordinates* v *around* $\overline{v}$ *for the fibers of* B, *coordinates* u *around* $\overline{u}$ *for the fibers of* W *and a map* $\overline{h} : X \to Y$ *such that h has the form*

$$(x, v) \gg h > (y, u) = (\overline{h}(x), v). \tag{39}$$

Proof. Choose a locally trivializing chart $(0, \phi)$ of W around $\overline{y}$. Then $\phi : p^{-1}(0) \to 0 \times U$, with U the standard fiber of W. Take local coordinates u around $\overline{u} \in U$. Then (y, u) forms a coordinate system for W around $(\overline{y}, \overline{u})$. Because h is a bundle morphism, it has the form

$$(x, \overline{v}) \gg h > (y, u) = (\overline{h}(x), h'(x, \overline{v})), \tag{40}$$

where $(x, \overline{v})$ is a coordinate system for B around $(\overline{x}, \overline{v})$. Now adapt this last coordinate system by defining

$$v = (h')^{-1}(x, u) \quad \text{with } x \text{ fixed.} \tag{41}$$

Because h restricted to the fibers is a diffeomorphism, v is well defined and (x, v) forms a coordinate system for B in which h has the form

$$(x, v) \gg h > (y, u) = (\overline{h}(x), u). \tag{42}$$

Hence under the conditions of Lemma 2 our system is locally (around $\overline{x} \in X$ and $\overline{y} \in Y$) described by

$$\dot{x} = g(x, u), \quad y = \overline{h}(x). \tag{43}$$

This input-output formulation is essentially the same as the one proposed by Brockett and Takens, who take the input spaces as the fibers of a bundle above a globally defined output space Y. In fact, this situation should be regarded as the normal setting for nonlinear control systems.

3. Take the same assumptions as in 2 and assume moreover that W is a *trivial* bundle, i.e., $W = Y \times U$, and that B is a trivial bundle, i.e., $B = X \times V$. Because h is a diffeomorphism on the fibers, we can identify U and V. In this case the output set Y and the input set U are *globally* defined, and the system is described by

$$\dot{x} = g(x, u), \quad y = \overline{h}(x), \tag{44}$$

where for each fixed $\overline{u}, g(\cdot, \overline{u})$ is a globally defined vector field on X. This is the "usual" interpretation of (29).

Some remarks are in order:

Remark 1. *When h restricted to the fibers of B is* not *an immersion we have a situation where we could speak of "hidden inputs". In fact, in this case there are variables in the fibers of B which can affect the internal state behavior via the equation $\dot{x} = g(x, v)$ but which cannot be directly identified with some of the external variables.*

Remark 2. *The splitting of the external variables into inputs and outputs as described in Lemma 1 is of course by no means unique. This fact has interesting implications, even in the linear case, which is beyond the scope of this chapter.*

Remark 3. *From Lemma 2 it is clear that the coordinization of the fibers of the bundle W uniquely determines, via h, the coordinization of the fibers of B. It should be remarked that a coordinization of the fibers of W is locally equivalent to the existence of an (integrable) connection on the bundle W, and that one coordinization is linked to another by what is essentially an output feedback transformation, i.e., a bundle isomorphism from W into itself.*

Remark 4. *A beautiful example of this kind of system is the Lagrangian system. Here the output space is equal to the configuration space Q of a mechanical system. The state space X is the configuration space with the velocity space, so $X = TQ$. The space W is equal to T^*Q (the cotangent bundle of Q), with the fibers of T^*Q representing the external forces. When we denote the natural projection of TQ on Q by ρ, then B is just $\rho^* T^*Q$ (the pullback bundle via ρ). Now given a function $L : TQ \to \mathbb{R}$ (called the Lagrangian) we can construct a symplectic form $d(\partial L/\partial \dot{q}) \wedge dq$ (with $(q, \dot{q})$ coordinates for TQ) on TQ, which uniquely determines a map $g : B \to TTQ$. Finally, in coordinates the system is given by*

$$\ddot{q} = F(q, \dot{q}) + \sum_j u_j Z_j(q, \dot{q}), \quad y = q, \tag{45}$$

with the vector fields $F(q, \dot{q})$ and $Z_j(q, \dot{q})$ satisfying certain conditions. Moreover the vector fields Z_j commute, i.e., $[Z_i, Z_j] = 0$ for all i, j, a fact which has a very interesting interpretation.

Remark 5. *Most cases where B can be taken as trivial are generated by a space X such that TX is a trivial bundle. For instance, when X is a Lie group TX is automatically trivial.*

4.2   Minimality

We want to give a definition of minimality for a general nonlinear agent.

Definition 12. *Let $\Sigma(X, W, B, f)$ and $\Sigma'(X', W, B', f')$ be two smooth systems. Then we say $\Sigma' \leq \Sigma$ if there exist surjective submersions $\phi : X \to X', \Phi : B \to B'$ such that the following diagram commutes.*

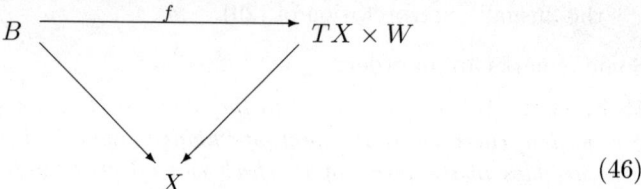

$$(46)$$

Σ is called *equivalent* to Σ' (denoted $\Sigma \sim \Sigma'$) if ϕ and Φ are diffeomorphisms. We call Σ *minimal* if $\Sigma' \leq \Sigma \Rightarrow \Sigma' \sim \Sigma$.

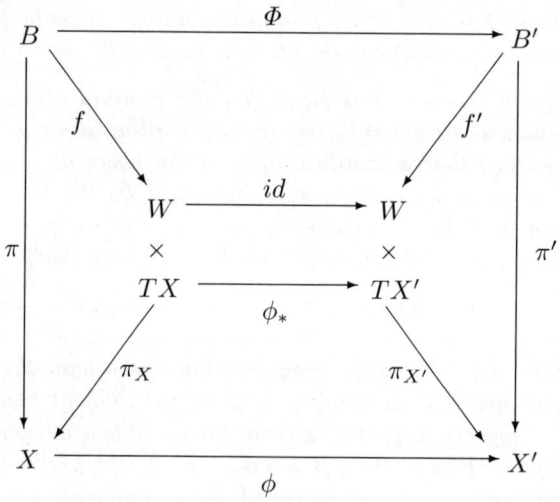

Remark 6. *This definition formalizes the idea that we call Σ' less complicated than Σ $(\Sigma' \leq \Sigma)$ if Σ' consists of a set of trajectories in the state space, smaller than the set of trajectories of Σ, but which generates the same external behavior. (The external behavior Σ_e of $\Sigma(X, W, B, f)$ consists of the possible functions $w : \mathbb{R} \to W$ generated by $\Sigma(X, W, B, f)$. Hence, when we define $\Sigma := \{(x, w) : \mathbb{R} \to X \times W|_x$ that are absolutely continuous and $(\dot{x}(t), w(t)) \inf(\pi^{-1}(x(t)))$ a.e.$\}$, then Σ_e is just the projection of Σ on $W^{\mathbb{R}}$).*

Remark 7. *Notice that we only formalize the* regular *case by asking that Φ and ϕ be surjective as well as submersive. In fact we could, for instance, allow that at isolated points ϕ or Φ are not submersive. However, we do not discuss this problem here, and treat only the regular case as described in Definition 12.*

Remark 8. *Notice that $\Sigma_1 \leqq \Sigma_2$ and $\Sigma_2 \leqq \Sigma_1$ need not imply $\Sigma_1 \sim \Sigma_2$. This fact leads to very interesting problems, which again are out of scope for this chapter.*

Of course, Definition 12 is an elegant but rather abstract definition of minimality. From a differential geometric point of view it is very natural to see what these conditions of commutativity mean *locally*. In fact, we will see in Theorem 5 that locally these conditions of commutativity do have a very direct interpretation. But first we have to state some preparatory lemmas and theorems.

Let us look at Diagram (46). Because Φ is a submersion it induces an involutive distribution D on B given by

$$D := \{Z \in TB | \Phi_* \dot{Z} = 0\} \tag{47}$$

(the foliation generated by D is of the form $\Phi^{-1}(c)$ with c constant). In the same way ϕ induces an involutive distribution E on X. Now the information in the diagram (46) is contained in three subdiagrams (we assume $f = (g, h)$ and $f' = (g', h')$):

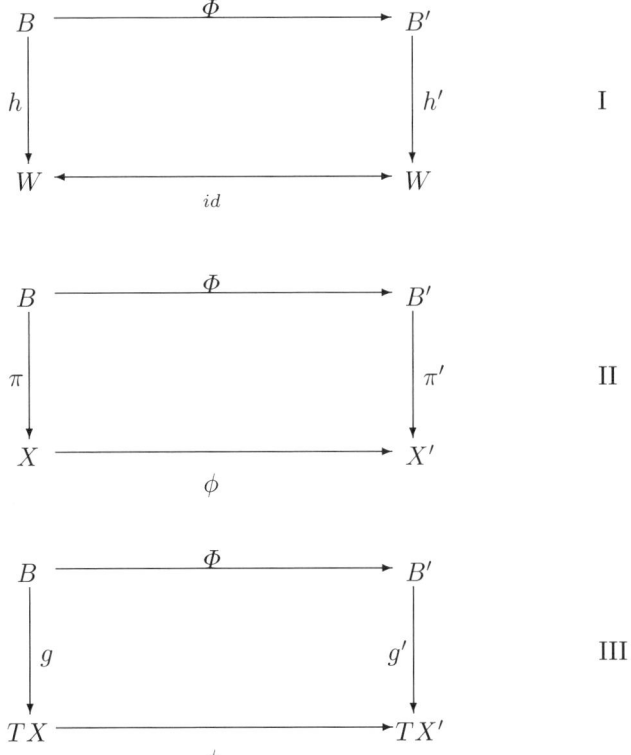

Lemma 3. *Locally the diagrams I, II, III are equivalent, respectively, to*

$$I' : \quad D \subset \ker dh \tag{48}$$

$$II' : \quad \pi_* D = E \tag{49}$$

$$III' : \quad g_* D \subset TE = T\pi_*(D). \tag{50}$$

Proof. I' and II' are trivial. For III' observe that, when ϕ induces a distribution E on X, then ϕ_* induces the distribution TE on TX.

Now we want to relate conditions I', II', III' with the theory of nonlinear disturbance decoupling. Consider in local coordinates the system

$$\dot{x} = f(x) + \sum_{i=1}^{m} u_i g_i(x) \quad \text{on a manifold } X. \tag{51}$$

We can interpret this as an affine distribution on manifold.

Theorem 4. *Let $D \in A(\Delta_0)$. Then the condition*

$$[\Delta, D] \subseteq D + \Delta_0 \tag{52}$$

(we call such a $D \in A(\Delta_0)\Delta(\mathrm{mod}\,\Delta_0)$ invariant) is equivalent to the two conditions: a) there exists a vector field $F \in \Delta$ such that $[F, D] \subseteq D$ and b) there exist vector fields $B_i \in \Delta_0$ such that span $\{B_i\} = \Delta_0$ and $[B_i, D] \subset D$.

With the aid of this theorem the disturbance decoupling problem is readily solved. The key to connecting our situation with this theory is given by the concept of the *extended system*, which is of interest in itself.

Definition 13. (Extended system). *Let*

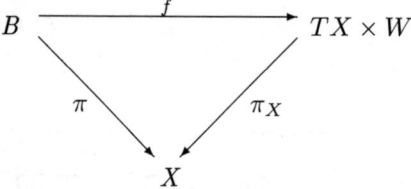

Then we define the extended system of $\Sigma(X, W, B, f)$ as follows: We define Δ_0 as the vertical tangent space of B, i.e.,

$$\Delta_0 := \{Z \in TB | \pi_* Z = 0\}. \tag{53}$$

Note that Δ_0 is *automatically involutive*. Now take a point $(\overline{x}, \overline{v}) \in B$. Then $g(\overline{x}, \overline{v})$ is an element of $T_{\overline{x}}X$. Now define

$$\Delta(\overline{x}, \overline{v}) := \{Z \in T_{(\overline{x}, \overline{v})} | \pi_* Z = g(\overline{x}, \overline{v})\}. \tag{54}$$

So $\Delta(\overline{x}, \overline{v})$ consists of the possible lifts of $g(\overline{x}, \overline{v})$ in $(\overline{x}, \overline{v})$. Then it is easy to see that Δ is an affine distribution on B, and that $\Delta - \Delta = \Delta_0$. We call the affine system (Δ, Δ_0) on B constructed in this way, together with the output function $h : B \to W$, the extended system $\Sigma^e(X, W, B, f)$. We have the following:

Lemma 4. *i) Let D be an involutive distribution on B such that $D \cap \Delta_0$ has constant dimension. Then $\pi_* D$ is a well-defined and involutive distribution on X if and only if $D + \Delta_0$ is an involutive distribution. ii) Let D be an involutive distribution on B and let $D \cap \Delta_0$ have constant dimension. Then the following two conditions are equivalent: $a)\pi_* D$ is a well-defined and involutive distribution on X, and $g_* D \subset T\pi_* D$ and $b)$ $[\Delta, D] \subset D + \Delta_0$.*

Proof. *i)* Let $D + \Delta_0$ be involutive. Because D and Δ_0 are involutive this is equivalent to $[D, \Delta_0] \subset D + \Delta_0$. Applying Theorem 4 to this case gives a basis $\{Z_1, \ldots, Z_k\}$ of D such that $[Z_i, \Delta_0] \subseteq \Delta_0$. In coordinates (x, u) for B, the last expression is equivalent to $Z_i(x, u) = (Z_{ix}, Z_{iu}(x, u))$, where Z_{ix} and Z_{iu} are the components of Z_i in the x- and u-directions, respectively. Hence $\pi_* D =$ span $\{Z_{1x}, \ldots, Z_{kx}\}$ and is easily seen to be involutive. The converse statement is trivial.

ii) Assume i); then there exist coordinates (x, u) for B such that $D = \{\partial/\partial x_1, \ldots, \partial/\partial x_x\}$ (the integral manifolds of D are contained in the sections $u = \text{const}$). Then $g_* D \subset T\pi_* D$ is equivalent to

$$\left(\frac{\partial g}{\partial x_i} \right)_{j^e \text{comp}} = 0 \tag{55}$$

with $i = 1, \ldots, k$ and $j = k + l, \ldots, n$ (n is the dimension of X). From these expressions $[\Delta, D] \subset D + \Delta_0$ readily follows. The converse statement is based on the same argument.

Now we are prepared to state the main theorem of this section. First we have to give another definition.

Definition 14. (Local minimality). *Let $\Sigma(X, W, B, f)$ be a smooth system. Let $\overline{x} \in X$. Then $\Sigma(X, W, B, f)$ is called locally minimal (around $\overline{x}$) if when D and E are distributions (around $\overline{x}$) which satisfy conditions I', II', III' of Lemma 3, then D and E must be the zero distributions.*

It is readily seen from Definition 12 that minimality of $\Sigma(X, W, B, f)$ *locally* implies local minimality (locally every involutive distribution can be factored out). Combining Lemma 3, Definition 13 and Lemma 4 we can state:

Theorem 5. *$\Sigma(X, W, B, f = (g, h))$ is locally minimal if and only if the extended system $\Sigma^e(X, W, B, f = (g, h))$ satisfies the condition that there exist no nonzero involutive distribution D on B such that*

$$i) \quad [\Delta, D] \subset D + \Delta_0, \tag{56}$$

$$ii) \quad D \subset \ker dh. \tag{57}$$

Remark 9. *It is very surprising that the condition of minimality locally comes down to a condition on the extended system, which is in some sense an infinitesimal version of the original system.*

Remark 10. *Actually there is a conceptual algorithm to check local minimality.*
Define

$$\Delta^{-1}(\Delta_0 + D) := \{vector\ fields\ Z\ on\ B\ ||[\Delta, Z] \subseteq \Delta_0 + D\}. \tag{58}$$

Then we can define the sequence $\{D^\mu\}, \mu = 0, 1, 2, \ldots$ as follows:

$$D^0 = \ker\ dh, \tag{59}$$

$$D^\mu = D^{\mu-1} \cap \Delta^{-1}(\Delta_0 + D^{\mu-1}), \quad \mu = 1, 2, \ldots. \tag{60}$$

Then $\{D^\mu\}$, $\mu = 0, 1, 2, \ldots$, is a decreasing sequence of involutive distributions, and for some $k \geq \dim(\ker\ dh)D^k = D^\mu$ for all $\mu \geq k$. Then D^k is the *maximal* involutive distribution which satisfies

$$i) \qquad [\Delta, D^k] \subset D^k + \Delta_0, \tag{61}$$

$$ii) \quad D^k \subset \ker\ dh. \tag{62}$$

From Theorem 5 it follows that $\Sigma(X, W, B, f)$ is locally minimal if and only if $D^k = O$.

4.3   Observability

It is natural to suppose that our definition of minimality has something to do with controllability and observability. However, because the definition of a non-linear system (34) also includes autonomous systems, (i.e., no inputs), minimality cannot be expected to imply, in general, some kind of controllability. In fact an autonomous linear system

$$\dot{x} = Ax, \quad y = Cx \tag{63}$$

is easily seen to be minimal if and only if (A, C) is observable. Moreover, it seems natural to define a notion of *observability* only in the case that the system (34) has at least a local input-output representation; i.e., we make the standing assumption that $(\partial h / \partial v)$ is injective (see Lemma 1). Therefore, *locally* we have as our system

$$\dot{x} = g(x, u), \quad y = \overline{h}(x, u) \tag{64}$$

for every possible input-output coordinization (y, u) of W. For such an input-output system local minimality implies the following notion of observability, which we call *local distinguishability*.

Proposition 1. *Choose a local input-output parametrization as in (64). Then local minimality implies that the only involutive distribution E on X which satisfies i) $[g(\cdot, u), E] \subset E$ for all u (E is invariant under $g(\cdot, u)$) and ii) $E \subset \ker\ d_x h(\cdot, u)$ for all u ($d_x \overline{h}$ means differentiation with respect to x) is the zero distribution.*

Proof. Let E be a distribution on X which satisfies *i)* and *ii)*. Then we can lift E in a trivial way to a distribution D on B by requiring that the integral manifolds of D be contained in the sections $u = \text{const}$. Then one can see that D satisfies $[\Delta, D] \subset D + \Delta_0$ and $D \subset \ker dh$. Hence $D = 0$ and $E = 0$.

Corollary 2. *Suppose there exists an input-output coordinization*

$$\dot{x} = g(x, u), \quad y = \overline{h}(x). \tag{65}$$

Then local minimality implies local weak observability.

Proof. As can be seen from Proposition 1, local minimality in this more restricted case implies that the only involutive distribution E on X which satisfies *i)* $[g(\cdot, u), E] \subset E$ for all u and *ii)* $E \subset \ker d\overline{h}$, is the zero distribution. It can be seen that the biggest distribution which satisfies *i)* and *ii)* is given by the null space of the codistribution P generated by elements of the form

$$L_{g(\cdot, u^1)} L_{g(\cdot, u^2)} \cdots L_{g(\cdot, u^k)} d\overline{h}, \quad \text{with} \quad u^j \text{ arbitrary.} \tag{66}$$

Because this distribution has to be zero, the codistribution P equals $T_x^* X$, in every $\in X$. This is, apart from singularities (which we don't want to consider), equivalent to local weak observability.

Moreover, let (65) be locally weakly observable. Then all feedback transformations $u \mapsto v = \alpha(x, u)$ which leave the form (65) invariant (i.e., y is only the function x) are exactly the output feedback transformations $u \mapsto v = \alpha(y, u)$. It can be easily seen in local coordinates that after such output feedback is applied, the modified system is still locally weakly observable.

In Proposition 1 and its corollary we have shown that local minimality implies a notion of observability, which generalizes the usual notion of local weak observability. Now we will define a much stronger notion. Let us denote the (defined only locally) vector field $\dot{x} = g(x, \overline{u})$ for fixed $\overline{u}$ by $g^{\overline{u}}$ and the function $\overline{h}(x, \overline{u})$ by $h^{\overline{u}}$ (with g and $\overline{h}$ as in (64)).

Definition 15. *Let $\Sigma(X, W, B, f) = (g, h)$ be a smooth nonlinear system. It is called strongly observable if for every possible input-output coordinization (64) the autonomous system*

$$\dot{x} = g^{\overline{u}}(x), \quad y = h^{\overline{u}}(x) \tag{67}$$

with $\overline{u}$ constant is locally weakly observable, for all $\overline{u}$.

Remark 11. *Let $\Sigma(X, W, B, f = (g, h))$ be strongly observable. Take one input-output coordinization (y, u). The system has the form (in these coordinates)*

$$\dot{x} = g(x, u), \quad y = \overline{h}(x, u). \tag{68}$$

Because the system is strongly observable, every constant *input-function (constant in this coordinization) distinguishes between two nearby states. However, in*

*every other input-output coordinization every constant (i.e., in this coordiniza-
tion) input function also distinguishes. This implies that in the first coordinization every C^∞ input function distinguishes. Because the C^∞ input functions
are dense in a reasonable set of input functions, every input function in this
coordinization distinguishes.*

Proposition 2. *Consider the Pfaffian system constructed as follows:*

$$P = dh^{\overline{u}} + L_{g^{\overline{u}}}dh^{\overline{u}} + L_{g^{\overline{u}}}(L_{g^{\overline{u}}}dh^{\overline{u}}) + \cdots + L_{g^{\overline{u}}}^{n-1}dh^{\overline{u}}, \qquad (69)$$

*with n the dimension of X and $L_{g^{\overline{u}}}$ the Lie derivative with respect to $g^{\overline{u}}$. As is
well known, the condition that the Pfaffian system P as defined above satisfies
the condition $P_x = T_x^* X$ for all $x \in X$ (the so called observability rank condition)
implies that the system*

$$\dot{x} = g^{\overline{u}}(x), \quad y = h^{\overline{u}}(x) \qquad (70)$$

*is locally weakly observable. Hence, when the observability rank condition is sat-
isfied for all u, the system is strongly observable.*

We will call the Pfaffian system P the *observability codistribution.*

Remark 12. *As is known, local weak observability of the system*

$$\dot{x} = g^{\overline{u}}(x), \quad y = h^{\overline{u}}(x) \qquad (71)$$

*implies that the observability rank condition (i.e., $\dim P_x = T_x^* X$) is satisfied
almost everywhere (in fact, in the analytic case everywhere). Because we don't
want to go into singularity problems, for us local weak observability and the
observability rank condition are the same.*

Remark 13. *It is easily seen that when for one input-output coordinization the
observability rank condition for all u is satisfied, then for every input-output
coordinization the observability rank condition for all u is satisfied. This follows
from the fact that the observability rank condition is an open condition.*

4.4   Controllability

The aim of this section is to define a kind of controllability which is "dual" to the
definition of local distinguishability (Proposition 1). The notion of controllability
we shall use is the so-called "strong accessibility".

Definition 16. *Let $\dot{x} = g(x, u)$ be a nonlinear system in local coordinates. Define $R(T, x_0)$ as the set of points reachable from x_0 in exactly time T; in other
words,*

$$R(T, x_0) := \{x_1 \in X \mid \exists \text{ state trajectory } x(t) \text{ generated by } g$$
$$\ni x(0) = x_0 \wedge x(T) = x_1\}. \qquad (72)$$

We call the system *strongly accessible* if for all $x_0 \in X$, and for all $T > 0$ the set $R(T, x_0)$ has a nonempty interior.

For systems of the form (in local coordinates)

$$\dot{x} = f(x) + \sum_{i=1}^{m} u_i g_i(x) \tag{73}$$

(i.e., affine systems) we can define A as the smallest Lie algebra which contains $\{g_1, \ldots, g_m\}$ and which is invariant under f (i.e., $[f, A] \subset A$). It is known that $A_x = T_x X$ for every $x \in X$ implies that the system (73) is strongly accessible. In fact, when the system is analytic, strong accessibility and the rank condition $A_x = T_x X$ for every $x \in X$, are equivalent. We call A the *controllability distribution* and the rank condition the controllability rank condition. Now it is clear that for affine systems (73) this kind of controllability is an elegant "dual" of local weak observability.

It is well known that the extended system (see Definition 13) is an affine system. Hence for this system we can apply the rank condition described above. This makes sense because the strong accessibility of $\Sigma(X, W, B, f)$ is very much related to the strong accessibility of $\Sigma^e(X, W, B, f)$, which can be seen from the following two propositions.

Proposition 3. *If $\Sigma^e(X, W, B, f = (g, h))$ is strongly accessible, then $\Sigma(X, W, B, f = (g, h))$ is strongly accessible as well.*

Proof. In local coordinates the dynamics of Σ^e and Σ are given by

$$I \quad \dot{x} = g(x, u) \quad (\Sigma), \tag{74}$$
$$II \quad \dot{x} = g(x, v) \quad (\Sigma^e), \tag{75}$$
$$\dot{v} = u. \tag{76}$$

It is easy to show that if for Σ^e one can steer to a point x_1 then the same is possible for Σ (even with an input that is smoother).

The converse is more difficult to prove:

Proposition 4. *Let $\Sigma(X, W, B, f = (g, h))$ be strongly accessible. In addition, let the fibers of B be connected. Then $\Sigma^e(X, W, B, f = (g, h))$ is strongly accessible.*

Proof. Consider the same representation of Σ and Σ^e as in the proof of Proposition 3. Let $x_0 \in X$ and x_1 be in the (nonempty) interior of $R_\Sigma(x_0, T)$ (the reachable set of system Σ). Then it is possible to reach x_1 from x_0 by an input function $v(t)$ which cannot be generated by the differential equation $\dot{v} = u$. However, we know that the set of the v generated in this way is dense in L^2. (For this we certainly need that the fibers of B are connected.) Because we only have to prove that the interior of a set is nonempty, this makes no difference. Now it is obvious from the equations

$$\dot{x} = g(x, v), \quad \dot{v} = u \tag{77}$$

that if we can reach an open set in the x-part of the (extended) state, then it is surely possible in the whole (x, v)-state.

5 Conclusions

In this chapter, cooperative control of multiple agents was studied. Methods and algorithms were explored for solving the problem of vehicle group interaction, when one group of vehicles is moving in a plane (UGV) and another in a halfspace (UAV-s). We have already analyzed an analogous situation, when one object (a pursuer) is moving in a halfspace while the other (an evader) - in a plane, in solving the problem of "soft meeting". Nonlinear and bilinear Markovian models are proposed for solution of the game theoretic problem of searching for a moving object in discrete time over a finite set of states.

The multiagent coordination problem has been studied. This problem is addressed for a class of targets for which control Lyapunov functions can be found. The main result is a suite of propositions about formation maintenance, task completion time, and formation velocity. It is also shown how to moderate the requirement that, for each individual target, there exists a control Lyapunov function.

The connection between cooperative control and Yang–Mills fields has been established. A geometric model of a controlled agent as dynamic information-transforming system was examined. A description of the information-transforming system within the framework of the geometric formalism was also proposed. After a classification of the fiber bundle types of conflict and conflict-free maneuvers, a weighted energy can be proposed as the cost function to select the optimal one. Various local and global controllability and observability conditions are derived. For the general multi-agent case, a convex optimization algorithm is proposed to find the optimal multi-legged maneuvers. To completely characterize the optimal conflict-free maneuvers, many issues remain to be addressed.

Possible directions of future research include the analysis of the proposed mathematical models in terms of its performance and its robustness with respect to uncertainty of the agents positions and velocities, and a more realistic study for the agent dynamics. Summing up, we can say that the combined problems of "search and tracking" and "pursuit and evasion" for multiple different-type pursuing objects and multiple evaders will be solved in the next step.

References

1. Cressman, R.: Evolutionary Dynamics and Extensive Form Games. MIT Press (2003)
2. Kalman, R., Falb, S., Arbib, M.: Topics in Mathematical System Theory. McGraw Hall (1969)
3. Yatsenko, V.: Control systems and fiber bundles. Avtomatika **5** (1985) 25–28
4. Butkovskiy, A., Samoilenko, Y.: Control of Quantum-Mechanical Processes and Systems. Kluwer Academic Publishers (1990)

5. Chikrii, A.: Differential games with multiple pursuers. In: of the Banach International Mathematical Centre. Volume 14. (1985) 81–107
6. Chikrii, A.: Conflict-Controlled Processes. Kluwer Academic Publishers (1997)
7. Gall, S.: Search Games. Academic Press (1980)
8. Larrie, F.: Ships and Science: The Birth of Naval Architecture in the Scientific Revolution, 1600-1800. MIT Press (2007)
9. Krasovskii, A., Subbotin, A.: Game-theoretical control problems. Springer-Verlag (1988)
10. Imado, F.: The features of optimal avoidance in two dimensional pursuit-evasion dynamic games. Technical report, Information, Technology, and Management (ITEM) (2002)
11. Albus, J., Meystel, A., Chikrii, A., Belousov, A.: Analytic method for solution of the game problem of soft landing for moving objects. Cybernetics and Systems Analysis **37**(1) (2001) 75–91
12. Mizutani, A., Chahl, J., Srinivasan, M.: Insect behavior: Motion camouflage in dragonflies. Nature **423**(6940) (2003) 604
13. Matichin, A., Chikriy, A.: Motion camouflage in differential games of pursuit. Journal of Automation and Information Science **37**(3) (2003) 1–5
14. Matichin, I., Chikriy, A.: Motion camouflage in differential games of pursuit (in Russian). Journal of Automation and Information Science **37**(3) (2005) 1–5
15. Koopman, B.: The theory of search. Operations Research **4** (1956) 324–346
16. Stone, L.: Theory of optimal search. Academic Press (1975)
17. Hellman, O.: Optimization of search for an object drifting in outer space. Journal of Spacecraft and Rockets **7**(7) (1970) 886–889
18. Hellman, O.: On the optimal search for a randomly moving target. SIAM Journal on Applied Mathematics **22**(4) (1972) 545–552
19. Pardalos, P., Yatsenko, V., Grundel, D.: Nonlinear dynamics of sea clutter and detection of small targets. In Pardalos, P., Murphey, R., Butenko, S., eds.: Recent Developments in Cooperative Control and Optimization. Kluwer Academic Publishers (2004) 407–426
20. Van der Shaft, A.: Controllability and observability for affine nonlinear hamiltonian systems. IEEE Transactions on Automatic Control **27** (1982) 490–494
21. Van der Shaft, A.: Equations of motion for Hamiltonian systems with constraints. Journal of Physics A **11** (1987) 3271–3277
22. Arnold, V.: Mathematical Methods of Classical Methods. Springer–Verlag (1983)
23. Samoilenko, Y.: Reduction to the elementary cell of the linear system with discrete symmetry. Cybernetics and Computer Techniques **3** (1970) 48–53
24. Krener, A.: On the equivalence of control systems and the linearization of nonlinear systems. SIAM Journal of Control **11** (1973) 670–676
25. Griffiths, P., Coates, J., Helgason, S.: Exterior Differential Systems and the Calculus of Variations. Birkhauser Verlag (1983)
26. Marcus, L.: General theory of global dynamics. In Mayne, D., Brockett, R.W., eds.: Geometric Methods in System Theory. (1973) 150–158

Real-Time Optimal Time-Critical Target Assignment for UAVs*

Yoonsoo Kim, Da-Wei Gu, and Ian Postlethwaite

Department of Engineering, University of Leicester, Leicester, LE1 7RH, U.K.
{yk17, dag, ixp}@le.ac.uk

Abstract. In the literature, e.g. [10], one can find the so-called *basic* UAV mission target assignment in which m UAVs each with a capacity limit q visit n targets in a cooperative manner (and return to their departure points) such that the cost incurred by each UAV's travel is minimized. In [10], we proposed a mixed integer linear program (MILP) formulation which exactly solves the problem, as well as four alternative MILP formulations which are computationally less intensive (and therefore suited for real-time purposes) yet yield a theoretically guaranteed sub-optimal solution. In this chapter, we further consider *timing constraints* imposed on some p of the targets, so-called *prime targets*. This consideration is often required for scenarios in which prime targets must be visited in a pre-defined time interval, and mathematically results in the addition of several integer linear constraints to the previous MILP formulation making the problem computationally intractable. We propose a novel procedure of adding these cumbersome timing constraints to the previous MILP formulation, in order to avoid increasing too much computational cost under practically valid assumptions. We first show that the proposed procedure still guarantees the previously claimed theoretical solution quality associated with the basic mission. We then show through extensive numerical simulations that under certain conditions, our algorithms return solutions which are still computationally manageable.

Keywords: Unmanned aerial vehicles (UAV); target assignment; mixed integer linear program; timing constraints.

1 Introduction and Problem Statement

In [10], for a given number m of UAVs U_i ($i = 1, 2, \ldots, m$, $m \geq 2$) at corresponding positions T_0^i, and a number n of targets T_j ($j = 1, 2, \ldots, n$, $n \geq m$) within a terrain $\mathbf{X}$, we consider a mission in which the UAVs visit all the targets in a cooperative manner (and return to where they departed from) such that the cost (reflecting UAV operating time and risk) incurred by each UAV's travel is

* This research has been supported by the UK Engineering and Physical Sciences Research Council and BAE Systems. The authors truly appreciate the valuable comments and suggestions by the editors and anonymous reviewers.

M.J. Hirsch et al. (Eds.): Adv. in Cooper. Ctrl. & Optimization, LNCIS 369, pp. 265–280, 2007.
springerlink.com © Springer-Verlag Berlin Heidelberg 2007

minimized whilst keeping the number of targets visited by a single UAV below a certain limit q. More precisely, we would like to calculate $\mathbf{T}^*$

$$\mathbf{T}^* \equiv \min_{j} \max_{\mathcal{A}(j)_i} \mathbf{T}(j)_i^*, \tag{1}$$

where $\mathbf{T}^*$ is the least maximum cost among all UAVs in visiting their assigned targets (and returning to their departure points), $\mathcal{A}$ is the set of feasible target assignments to UAVs, $\mathcal{A}(j)$ ($\in \mathcal{A}$) is one of the feasible assignments, $\mathcal{A}(j)_i$ is the sub-assignment given to the ith UAV within $\mathcal{A}(j)$ and finally $\mathbf{T}(j)_i^*$ is the optimal cost of completing the sub-assignment $\mathcal{A}(j)_i$ by the ith UAV. We note that the objective functional to be minimized is appropriate for balancing workloads across UAVs. In this chapter, we further consider the same problem with practical *timing constraints*. This is basically due to the frequent presence of so-called *prime targets* that must be visited in a fixed time interval in many UAV applications. As a result, we require that the solution assignment to (1) be chosen such that a UAV visits a prime target T_k within a given time window $[t_k^\alpha, t_k^\beta]$.[1] In addition, the total number of prime targets is limited by p, and the maximum number of prime targets which a UAV is capable of handling is limited by q' ($\leq q$), in order to increase the probability of mission success.

There is a large number of papers dealing with various target assignment problems. These include Weapon-target assignment [1,14], timetabling [20,22], the celebrated Travelling Salesman Problem [19] and more generally capacity-limited vehicle routing problems [9,17]. We note that these problems are slightly different from the problem in the present context, in that (i) we may not require UAVs to return to their starting positions; (ii) we minimize the individual *tour* cost for balanced workload, not the total cost incurred by the whole mission; (iii) UAVs do not necessarily depart from the same depot. There is also much literature available on coordinated target assignment of UAVs, for example [2,3,4,5], and some of which add the timing and precedence constraints to the original problem [3,11,15,16]. As the underlying problem is known to be NP-hard, it is often fruitless to approach the problem in a direct or exact manner. Nevertheless, as many papers have shown, direct MILP formulations offer a promising way forward in terms of providing an optimal solution to the problem in spite of the computational demands [3,22]. As an example of the MILP approaches, we note the petal algorithm introduced in [2,5]. It considers all feasible task assignments (so-called petals), i.e. identifies all possible sequences of waypoints, for every UAV subject to its capabilities, and subsequently constructs the shortest paths connecting the waypoints as well as avoiding threats. Then, a MILP formulation is employed to find the best assignment in terms of the underlying cost. As implied by the numerical tests shown in [5], the petal algorithm becomes computationally problematic for a large number of targets, e.g. $n > 12$, due to the exhaustive consideration of all feasible petals. As an alternative method,

---

[1] In the literature, one may find *precedence constraints*, i.e. some target must be visited before other targets. However, we here assume that this can be viewed as a special case of the aforementioned timing constraints.

tabu search based methods are also useful for this purpose [17,18]. When time is critical, heuristic or non-exact methods have been considered, even if global optimality may not be achieved [4,9]. Among many heuristics, we note the *Iterated Optimal Tour Partitioning* (IOTP) algorithm proposed in [9,13], mainly for multi-vehicle-single-depot routing with capacity constraints. With the IOTP algorithm, it is claimed that one can obtain a tour whose cost is at most $2 - 1/q$ times of the optimal tour cost, where q is the capacity of vehicles.

In conclusion, what would be desirable is a direct MILP formulation combined with a non-exact method in such a way that the advantages of each are enjoyed. As inspired by the solution strategy introduced in [10] for the target assignment problem without timing constraints, we reduce the possibly large MILP associated with the original time-critical target assignment problem down to smaller MILPs while minimizing the loss of optimality, followed by solving the smaller MILPs exactly. Since this approach involves only small MILPs, it is therefore computationally tractable. We note that a similar approach has been recently proposed in [21] in which the problem in question is interpreted as a mixed version of the minimum cost network flow problem and the travelling salesman problem. The two problems are then approached by a promising heuristic employing linear programming, MILP and tabu search in order to reduce the computational complexity. However, no formal analysis on the algorithm performance, for example something like (16), is given, and therefore the algorithm may not be suitable for particular problem parameters. Our main challenges (contributions) are to ensure the previously claimed performance bounds in [10] and to allow only a slight increase of computational cost even after adding timing constraints. To this end, we first briefly describe two algorithms, which were introduced in [10], and their performance bounds in Section 2.1. We then show in Section 2.2 and 2.3 how the timing constraints in question can be incorporated into the existing MILPs with the performance bounds unchanged. In Section 3, we examine the computational aspect of our modified algorithms through extensive simulations. Concluding remarks are presented in Section 4.

2 Algorithms

2.1 Two Algorithms

The two algorithms (denoted by $\mathbf{H}_2$ and $\mathbf{H}_3$) introduced in [10] need a feasible network of UAVs' flying routes as their input. One can create such a feasible network of UAVs' flying routes by defining significant waypoints and links connecting the waypoints, and assigning cost (again, reflecting UAV operating time and risk) to the links. For illustration, Fig. 1 shows two UAV starting positions (T_0^1 and T_0^2), two targets (T_1 and T_2) and two obstacles (dashed objects). In order to identify significant waypoints, the two obstacles are approximated by two rectangles. The corner points of the rectangles then become part of the set of waypoints along with T_0^1, T_0^2, T_1 and T_2. The feasible network of the two UAVs' flying routes is the set of links connecting the waypoints. Each link carries the cost of travel based on its length and possible risks on it.

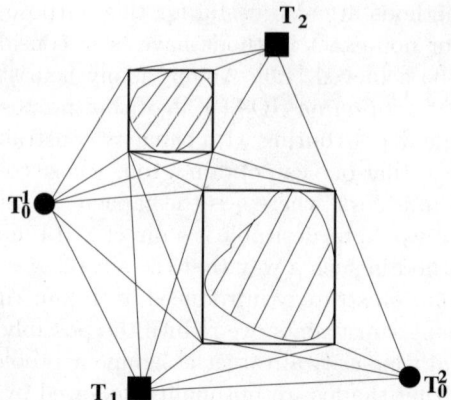

Fig. 1. A rectangular cover of two objects (the dashed area) and the corresponding construction of links

Given the network of UAVs' flying routes, the algorithm $\mathbf{H}_2$ (respectively, $\mathbf{H}_3$) is devised for the problem without (respectively, with) the UAV's return constraint. The basic principle behind the algorithms is that the original problem of possibly large size is handled by two solution steps each of which involves a problem of relatively small size. The first step in the present context solves a MILP for finding m groups of targets such that each group is disjoint and contains exactly one UAV and less than or equal to q targets. The second step finds an optimal order of visiting the assigned targets for each UAV by solving another MILP. These two MILPs are computationally manageable as long as q is small (≤ 4). The first grouping step is crucial in terms of both solution quality and computational complexity. The algorithms $\mathbf{H}_2$ and $\mathbf{H}_3$ employ the objective functionals which minimize the T_j-to-T_k cost and the T_0^i-to-T_j-to-T_k cost for each j, k, respectively, where T_j or T_k is the starting position of the ith UAV or a target position to be covered by the ith UAV. The following are the formal descriptions of $\mathbf{H}_2$ and $\mathbf{H}_3$. For detailed explanation on the constraints in the MILPs below, see [10].

Step 1. *Sub-optimal partitioning:* Consider an optimization problem $\mathcal{F}_2$ (respectively, $\mathcal{F}_3$), as shown below, that solves for x_{ij} ($x_{ij} = 1$ if the jth target is assigned to the ith UAV) to partition the underlying set of targets into m subsets $\mathcal{T}_i$ ($i = 1, 2, \ldots, m$) such that (i) each $\mathcal{T}_i$ contains at most q elements; (ii) the travelling cost T_j-to-T_k (respectively, T_0^i-to-T_j-to-T_k) for each j, k is minimized, where T_j or T_k is the starting position of the ith UAV or belongs to $\mathcal{T}_i$; (iii) each target is covered by exactly one UAV.

Step 2. *Optimal path-planning:* For each $\mathcal{T}_i$, consider $\mathcal{F}_E$ (respectively, $\mathcal{F}_{Eret}$), as shown below, that solves for a_{ij}^k ($a_{ij}^k = 1$ if the ith UAV visits the kth target after $j - 1$ targets, so that the ith UAV visits $T_{x_{i1}}, T_{x_{i2}}, \ldots$ in turn, where

$x_{ij} = \sum_{k=1}^{n} a_{ij}^k)$ to obtain the optimal path of visiting all the targets contained in $\mathcal{T}_i$ by the ith UAV.

$\mathcal{F}_E$: minimize r

subject to

$$\sum_{k=1}^{n} a_{ij}^k \leq 1 \quad \forall i, j \tag{2}$$

$$\sum_{i=1}^{m} \sum_{j=1}^{q} a_{ij}^k = 1 \quad \forall k \tag{3}$$

$$a_{ij}^k \in \{0, 1\} \quad \forall i, j, k \tag{4}$$

$$\sum_{k=1}^{n} a_{i(j+1)}^k \leq \sum_{k=1}^{n} a_{ij}^k \quad \forall i, j \tag{5}$$

$$a_{ij}^v + a_{i(j+1)}^w + a_{ij}^w + a_{i(j+1)}^v = 2\, y_{i\eta(v,w)}^j + \tilde{y}_{i\eta(v,w)}^j \quad \forall i, j, v, w \tag{6}$$

$$y_{i\eta(v,w)} = \sum_j y_{i\eta(v,w)}^j \quad \forall i, v, w \tag{7}$$

$$y_{i\eta(v,w)}^j \in \{0, 1\}, \quad \tilde{y}_{i\eta(v,w)}^j \in [0, 1] \quad \forall i, j, v, w \tag{8}$$

$$\sum_{k=1}^{n} C_0(i, k)\, a_{i1}^k + \sum_{v=1}^{n-1} \sum_{w=v+1}^{n} c(\eta(v, w))\, y_{i\eta(v,w)} \leq r \quad \forall i \tag{9}$$

where $i \in \{1, 2, \ldots, m\}$, $j \in \{1, 2, \ldots, q\}$ for (2)-(4) or $j \in \{1, 2, \ldots, q-1\}$ for (5)-(8), $k \in \{1, 2, \ldots, n\}$, $v, w \in \{1, 2, \ldots, n\}$ $(v < w)$, $\eta(v, w) = (v-1)n - v(v-1)/2 + w - v$. In (9), $C_0(i, k)$ (respectively, $c(\eta(v, w))$) is the travelling cost from T_0^i to T_k (respectively, from T_v to T_w).

$\mathcal{F}_{Eret}$: minimize r

subject to (2)–(8) and

$$a_{i1}^k = b_{i1}^k \quad \forall i, k$$

$$b_{i(j-1)}^k - \sum_{k=1}^{n} a_{ij}^k \leq b_{ij}^k \leq b_{i(j-1)}^k + \sum_{k=1}^{n} a_{ij}^k \quad \forall i, j, k$$

$$a_{ij}^k \leq b_{ij}^k \leq a_{ij}^k + (1 - \sum_{k=1}^{n} a_{ij}^k) \quad \forall i, j, k$$

$$\sum_{k=1}^{n} C_0(i, k)\, (a_{i1}^k + b_{iq}^k) + \sum_{v=1}^{n-1} \sum_{w=v+1}^{n} c(\eta(v, w))\, y_{i\eta(v,w)} \leq r \quad \forall i$$

where $i \in \{1, 2, \ldots, m\}$, $j \in \{2, 3, \ldots, q\}$, $k \in \{1, 2, \ldots, n\}$.

$\mathcal{F}_2$: minimize r

subject to

$$\sum_{j=1}^{n} x_{ij} \leq q \quad \forall i \tag{10}$$

$$\sum_{i=1}^{m} x_{ij} = 1 \quad \forall j \tag{11}$$

$$y_{i\eta(j,k)} \leq \frac{x_{ij} + x_{ik}}{2} \leq y_{i\eta(j,k)} + \frac{1}{2} \quad \forall i, j, k \ (j < k) \tag{12}$$

$$C_0(i,j)\, x_{ij} \leq r \quad \forall i \tag{13}$$

$$c(\eta(j,k))\, y_{i\eta(j,k)} \leq r \quad \forall i, j, k \ (j < k) \tag{14}$$

$$x_{ij}, y_{i\eta(j,k)} \in \{0, 1\} \quad \forall i, j, k \ (j < k) \tag{15}$$

where $i \in \{1, 2, \ldots, m\}$, $j, k \in \{1, 2, \ldots, n\}$ and $\eta(j,k) = (j-1)n - j(j-1)/2 + k - j$.

$\mathcal{F}_3$: minimize r

subject to

$$\sum_{j=1}^{n} x_{ij} \leq q \quad \forall i$$

$$\sum_{i=1}^{m} x_{ij} = 1 \quad \forall j$$

$$y_{i\eta} \leq \frac{x_{ij} + x_{ik}}{2} \leq y_{i\eta} + \frac{1}{2} \quad \forall i, j, k \ (j < k)$$

$$2\, C_0(i,j)\, x_{ij} \leq r \quad \forall i, j$$

$$C_0(i,j)\, x_{ij} + c(\eta(j,k))\, y_{i\eta(j,k)} \leq r \quad \forall i, j, k \ (j < k)$$

$$x_{ij}, y_{i\eta(j,k)} \in \{0, 1\} \quad \forall i, j, k \ (j < k)$$

where $i \in \{1, 2, \ldots, m\}$, $j, k, l \in \{1, 2, \ldots, n\}$ and $\eta(j,k) = (j-1)n - j(j-1)/2 + k - j$.

The performance bound for $\mathbf{H}_2$ is obtained by the two facts that (i) $\mathbf{T}^*$ is less than the T_j-to-T_k cost for any i, j; (ii) one can create a UAV's feasible path which sequentially visits all the (at most q) targets within the assigned group from the first grouping step. Similarly, the performance bound for $\mathbf{H}_3$ is due to the facts that (i) $\mathbf{T}^*$ is less than the T_0^i-to-T_j-to-T_k cost for any i, j, k when the UAV's return constraint is imposed; (ii) one can create a feasible path such that a UAV goes back to its departure point every time after it sequentially visits two targets within the associated group. As a result, $\mathbf{H}_2$ and $\mathbf{H}_3$ guarantee

$$1 \leq \frac{\mathbf{T}}{\mathbf{T}^*} \leq q \tag{16}$$

and

$$1 \leq \frac{\mathbf{T}}{\mathbf{T}^*} \leq 2 \left\lceil \frac{q}{3} \right\rceil - \kappa, \tag{17}$$

respectively, where $\mathbf{T}$ is the maximum of actual costs incurred by each UAVs' travel using $\mathbf{H}_2$ or $\mathbf{H}_3$, and $\kappa = 1$ if $q = 3k + 1$ ($k = 0, 1, \ldots$); otherwise $\kappa = 0$. Note that the performance bound (17) is guaranteed only if every UAV can go back to its departure point every time after visiting two targets, because of the second fact used for deriving the bound. This assumption may fail when timing constraints are imposed in the next section. See the proof of Proposition 1.

2.2 Incorporation of Timing Constraints into the Existing Framework

For incorporation of timing constraints into the existing framework, we need two kinds of T_0^i-to-T_j and T_v-to-T_w costs over the same network of UAVs' flying routes. One kind, denoted by $C_0(i,j)$ and $C(v,w)$, is used for being minimized and reflects both flight time and risk information due to threats. The other kind, denoted by $\underline{C_0}(i,j)$ and $\underline{C}(v,w)$, is used in concert with timing constraints and solely contains flight time information. The latter is computationally not as cumbersome as the former because algorithms to be introduced do not require T_0^i-to-T_j and T_v-to-T_w costs for *all* j, v, w. In fact, for the ith UAV the first grouping step of the new algorithms needs $C_0(i,j)$ and $C(v,w)$ for all j, v and w, but $\underline{C_0}(i,j')$ and $\underline{C}(v',w')$ only for j', v' and w' such that $T_{j'}$, $T_{v'}$ and $T_{w'}$ are prime targets.

Clearly, the first grouping step of $\mathbf{H}_2$ or $\mathbf{H}_3$ must be modified such that prime targets are assigned to a UAV such that the UAV can actually reach the assigned prime targets within the required time intervals. For this purpose, as the targets are grouped by the same technique used for $\mathbf{H}_2$ or $\mathbf{H}_3$, we impose the additional constraint such that each UAV visits its assigned prime targets prior to non-prime targets. As a result, the new grouping step provides each UAV with a feasible assignment accounting for timing constraints, along with an explicit order of visiting prime targets. Note that the additional constraint, which forces each UAV to visit prime targets prior to non-prime targets in the first step, is however neglected in the second step in which the optimal path for each UAV to visit all the assigned targets is computed. The second step is basically the same as before, in the sense that one exactly solves the time-critical target assignment problem but now of small size. For an illustration of the new algorithm, as shown in Fig. 2, suppose the new grouping step returns the assignment such that a UAV at T_0 covers two non-prime targets, T_1 and T_2, and two prime targets, T_3 and T_4, with their associated time windows, $[0, 10]$ and $[0, 50]$, respectively. Then, although the new grouping step directs the UAV to visit T_3 and T_4 (dotted line) prior to T_1 and T_2, the second step disregards the

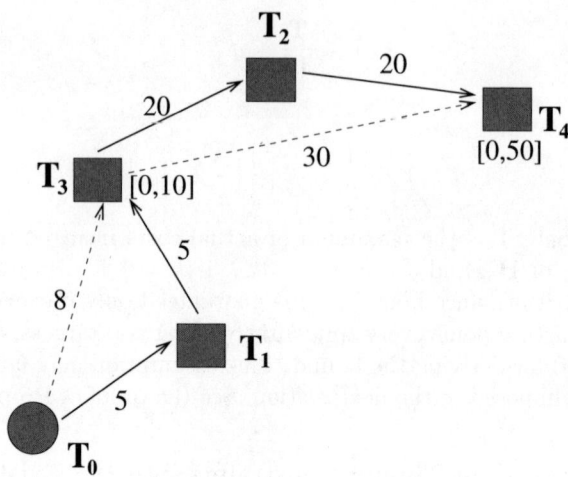

Fig. 2. An illustration of the new algorithm

direction and yields the path (solid line), $T_1 \to T_3 \to T_2 \to T_4$. This is because T_0 is closer to T_1 than T_3, and also T_3 is closer to T_2 than T_4. Note that this change is allowed because the timing constraints are still not violated. Based on the brief description of the new algorithms $\widetilde{\mathbf{H}}_2$ and $\widetilde{\mathbf{H}}_3$, the following result immediately follows. The formal description of $\widetilde{\mathbf{H}}_2$ will be given later.

Proposition 1. *If $t_k^\alpha = 0$ for every k, the algorithm $\widetilde{\mathbf{H}}_2$ guarantees the bound (16) for every positive integer q', where q' is the maximum number of prime targets visited by a UAV. However, $\widetilde{\mathbf{H}}_3$ guarantees the bound (17) only for $q' < 3$.*

Proof. First note that under the condition that $t_k^\alpha = 0$, no feasible assignments are *lost* by the new grouping step. In fact, for any feasible target T_k assignment to a UAV, one can always construct a feasible path such that the UAV visits prime targets prior to non-prime targets as long as $t_k^\alpha = 0$ for all k. This immediately implies that the new algorithm $\widetilde{\mathbf{H}}_2$ guarantees the same bound as (16). However, when $q' \geq 3$, this violates the underlying assumption used to derive (17) that one can create a feasible path such that a UAV goes back to its departure point every time after the UAV sequentially visits two targets within the associated group. This proves the claim.

As noted before, it is often desirable to have small q' in order to increase the probability of mission success, in which case the new algorithms are still applicable in practice.

2.3   MILPs Including Timing Constraints

One can find the necessary (integer linear) constraints for the grouping with no timing constraints in [10]. We thus focus here on developing methods of converting timing constraints into integer linear constraints.

To begin, let us first define integer grouping variables x_{ij} ($i \in \{1, 2, \ldots, m\}$ and $j \in \{1, 2, \ldots, n\}$) which represent the relationship between the ith UAV and the jth target, i.e. $x_{ij} = 1$ if the jth target is assigned to the ith UAV; $x_{ij} = 0$ otherwise. For the consideration of timing constraints, we then consider q' *rooms* (numbered from 1 to q') for each UAV. Each room may hold at most one identifier (ID) of prime target and has to be filled in ascending order, so that the resultant path of the ith UAV becomes $T_0^i \rightarrow T_{z_1} \rightarrow T_{z_2} \ldots \rightarrow T_{z_{\tilde{q}'}}$, where z_j ($j \in \{1, 2, \ldots, p\}$) is the ID of the jth prime target and $\tilde{q}' \leq q'$. For this purpose, we define other integer variables a_{ij}^k (similar to the one previously defined in $\mathcal{F}_E$), where i, j and k run from 1 to m, 1 to q' and 1 to p, respectively, in order to capture the relationship between the ith UAV, its jth room and the kth prime target. Similarly, we also define integer variables z_{ij}^k, where i, j and k run from 1 to m, 1 to q' and 1 to n, respectively. The difference between a and z is that a pertains to the ID (numbered from 1 to p) assigned to a prime target amongst only p prime targets, whereas z pertains to the ID (numbered from 1 to n) assigned to a prime target amongst all n targets. This seemingly unnecessary definition of z becomes useful when the connection between a and x need to be made later. The following are the integer linear constraints for implementing the aforementioned verbal expressions:

$$\sum_{k=1}^{p} a_{ij}^k \leq 1 \quad \forall i, j \quad \text{and} \quad \sum_{i=1}^{m} \sum_{j=1}^{q'} a_{ij}^k = 1 \quad \forall k \tag{18}$$

for $i \in \{1, 2, \ldots, m\}$, $j \in \{1, 2, \ldots, q'\}$ and $k = \{1, 2, \ldots, p\}$, and

$$\sum_{k=1}^{p} a_{i(j+1)}^k \leq \sum_{k=1}^{p} a_{ij}^k \quad \forall i, j \tag{19}$$

for $i = \{1, 2, \ldots, m\}$ and $j = \{1, 2, \ldots, q' - 1\}$.

Next, we consider the flight time from T_0^i to the first room for the ith UAV, and from the jth room to the $(j+1)$th room. The former is simply $\sum_{k=1}^{p} \underline{C_0}(i, k) a_{i1}^k$, and thus we need

$$\sum_{k=1}^{p} t_k^\alpha a_{i1}^k \leq \sum_{k=1}^{p} \underline{C_0}(i, k) a_{i1}^k \leq \sum_{k=1}^{p} t_k^\beta a_{i1}^k$$

where $i \in \{1, 2, \ldots, m\}$, for satisfying the associated timing constraints. However, the latter is not trivial.[2] The MILP expression of the flight time between targets requires the introduction of the additional auxiliary variables $d_{i\eta(v,w)}^j$ and $\tilde{d}_{i\eta(v,w)}^j$ which are defined through the following equality:

$$a_{ij}^v + a_{i(j+1)}^w + a_{ij}^w + a_{i(j+1)}^v = 2 d_{i\eta(v,w)}^j + \tilde{d}_{i\eta(v,w)}^j \tag{20}$$

$$d_{i\eta(v,w)}^j \in \{0, 1\}, \quad \tilde{d}_{i\eta(v,w)}^j \in [0, 1] \tag{21}$$

---

[2] The present approach to the latter is similar to the one in [10], but the derivation of the linear inequality constraints corresponding to timing constraints is novel in the chapter.

for all i, j, v, w, where $i \in \{1, \ldots, m\}$, $j \in \{1, \ldots, q'-1\}$, $v, w \in \{1, 2, \ldots, p\}$ $(v < w)$ and $\eta(v, w) = (v-1)p - v(v-1)/2 + w - v$. The flight time from T_0^i to T_{z_j} (via $T_{z_1}, \ldots, T_{z_{(j-1)}}$) then becomes

$$\sum_{k=1}^{p} \underline{C_0}(i, k) a_{i1}^k + \sum_{u=1}^{j-1} \sum_{v=1}^{p-1} \sum_{w=v+1}^{p} \underline{c}(\eta(v, w)) d_{i\eta(v,w)}^u$$

where $\underline{c}$ is a one-dimensional form of two-dimensional $\underline{C}$. Equality (20) enables $d_{i\eta(v,w)}^j$ to be 1 only if $a_{ij}^v = a_{i(j+1)}^w = 1$ or $a_{ij}^w = a_{i(j+1)}^v = 1$. In other words, $d_{i\eta(v,w)}^j$ is set to 1 only if the vth and wth targets are assigned to the consecutive jth and $(j+1)$th (or $(j+1)$th and jth) rooms of the ith UAV.

As a result, the timing constraints imposed on the targets occupying the second to last rooms for the ith UAV may be now represented as the following:

$$\sum_{k=1}^{p} t_k^{\alpha} a_{ij}^k \leq \sum_{k=1}^{p} \underline{C_0}(i, k) a_{i1}^k + \sum_{u=1}^{j-1} \sum_{v=1}^{p-1} \sum_{w=v+1}^{p} \underline{c}(\eta(v, w)) d_{i\eta(v,w)}^u \leq \sum_{k=1}^{p} t_k^{\beta} a_{ij}^k \quad (22)$$

for each $j \in \{1, 2, \ldots, q'\}$. The left inequality is fine, but the right inequality causes a problem when some of the rooms for a UAV are empty, i.e. less than q' prime targets are assigned to a UAV, thereby forcing the a variables corresponding to unoccupied rooms to be zero. For this reason, we use the following method:

$$\sum_{k=1}^{p} \underline{C_0}(i, k) a_{i1}^k + \sum_{u=1}^{j-1} \sum_{v=1}^{p-1} \sum_{w=v+1}^{p} \underline{c}(\eta(v, w)) d_{i\eta(v,w)}^u \leq \sum_{k=1}^{p} t_k^{\beta} a_{ij}^k + M \left(1 - \sum_{k=1}^{p} a_{ij}^k \right) \quad (23)$$

where $M > 0$ is a large constant number. This makes the right inequality vacuous whenever a room is unoccupied, i.e. $\sum_{k=1}^{p} a_{ij}^k = 0$ for some i, j.

The final task for grouping targets is to make the relationship between variables a and x. The main difficulty in doing this is that x is defined for all n targets, but a for only p prime targets. In order to resolve this problem, we recall the integer variable z, as defined at the beginning of this section, and consider the following linear constraints:

$$\sum_{k=1}^{p} (ID)_k a_{ij}^k = \sum_{k=1}^{n} k z_{ij}^k \quad \forall i, j \quad (24)$$

$$\sum_{k=1}^{n} z_{ij}^k \leq 1 \quad \forall i, j \quad (25)$$

$$\sum_{j=1}^{q'} z_{ij}^u \leq x_{iu} \quad \forall i, u \quad (26)$$

where i, j and u run from 1 to m, 1 to q' and 1 to n, respectively, and the constant one-dimensional array $(ID)_k$ $(k \in \{1, 2, \ldots, p\})$ contains a unique identifier (number from 1 to n) for each prime target. The equality (24) and inequality (25) basically perform the function: if the ith UAV's jth room is occupied by the target with ID $(ID)_k$, then z_{ij}^k is enabled and subsequently x_{ik} is set to 1 by (26).

By putting all the aforementioned constraints together, $\widetilde{\mathbf{H}}_2$ can be described as follows:

Step 1. *Sub-optimal partitioning*: Consider an optimization problem $\widetilde{\mathcal{F}}_2$, as shown below, that solves for x_{ij} ($x_{ij} = 1$ if the jth target is assigned to the ith UAV) to partition the underlying set of targets into m subsets $\mathcal{T}_i$ ($i = 1, 2, \ldots, m$) such that (i) each $\mathcal{T}_i$ contains at most q elements; (ii) the travelling cost T_j-to-T_k for each j, k is minimized, where T_j or T_k is the starting position of the ith UAV or belongs to $\mathcal{T}_i$; (iii) each target is covered by exactly one UAV.

Step 2. *Optimal path-planning*: For each $\mathcal{T}_i$, consider $\widetilde{\mathcal{F}}_E$ (see the remark below) in order to obtain the optimal path of visiting all the targets contained in $\mathcal{T}_i$ by the ith UAV.

$\widetilde{\mathcal{F}}_2$: minimize r

subject to (10)–(15), (18)–(21), the left inequality of (22) and (23)–(26).

We do not further elaborate on the MILPs associated with the first step of $\widetilde{\mathbf{H}}_3$, in which the UAV's return constraints are considered, and the second steps of both $\widetilde{\mathbf{H}}_2$ and $\widetilde{\mathbf{H}}_3$ and their associated programs $\widetilde{\mathcal{F}}_E$ and $\widetilde{\mathcal{F}}_{Eret}$, because all these MILPs can be easily constructed using the aforementioned techniques. Instead, we remark that the numbers of binary variables newly introduced to accommodate timing constraints for the first step and the second step of both algorithms are $mq'(p(p+1)/2+n) - mp(p-1)/2$ and 0, respectively, and the numbers of newly added constraints are approximately $mq'(p(p-1)+6) + mn + p$ and $2mp$, respectively. In the next section, we investigate via extensive numerical simulations how much these added variables and constraints affect the performance of the algorithms.

3   Numerical Simulations

We first present an introductory example showing how timing constraints change the solution to a basic target assignment problem. Figure 3 shows an example scenario and solution paths chosen when no timing constraints are considered. The scenario consists of 3 UAVs, 4 non-prime targets, 5 prime targets, and both the maximum number of targets and the maximum number of prime targets visited by a single UAV are 3, i.e. $m = 3$, $n = 9$, $p = 5$, $q = 3$ and $q' = 3$. The scenario also contains five threats which are each marked as "x". The threats create a joint probabilistic risk distribution (contour lines in the figure) and the risk is determined using the deterministic formulae found in [10]. In the figure, ten contour lines representing the risks ranged from 0.1 to 1 are plotted for each threat. The closer a line is to a threat source the higher the risk. The targets

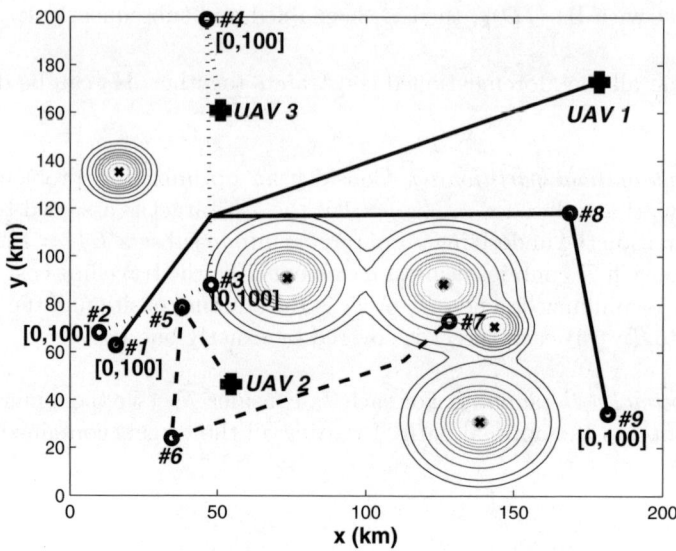

Fig. 3. Example solution paths (solid line for UAV 1, dashed line for UAV 2 and dotted line for UAV 3) when no timing constraints are considered

labeled with #1, #2, #3, #4 and #9 are supposed to be prime targets which are required to be visited by a UAV before 100 (min) after the mission starts, but their associated timing constraints are neglected at this time.

Following the aforementioned procedure of creating a network of UAVs' flying routes, we identify *nodes* including the UAV starting positions, the target position and the corner points of the smallest rectangles which cover the risky area due to the threats. We then associate each segment (joining two distinct nodes) with a cost weighting of 90% to the total risk (scaled to 1) along the segment and 10% to the length (scaled to 1) of the segment, and subsequently compute necessary travelling costs C_0, C, $\underline{C_0}$ and $\underline{C}$. Under the assumption that UAVs fly at a constant speed of 2 (km/min) and an altitude of 2 (km) in the operational range of $[0, 200] \times [0, 200]$ (km), algorithm $\mathbf{H}_2$ returns the solution paths shown in Fig. 3. The first UAV first visits #1, then #8 and finally #9, the second UAV visits #5, #6 and #7, and the third UAV #4, #3 and #2 in turn. As easily noticed, two prime targets (#2 and #9) are visited after, not before, 100 (min). However, the modified algorithm $\widetilde{\mathbf{H}}_2$ accounting for the timing requirements returns completely different solution paths, as shown in Fig. 4. This time, the first UAV first visits #8, then #9 and finally #5, the second UAV visits #3, #1 and #7, and the third UAV visits #4, #2 and #6 in turn. Note that every prime target is visited within the required time window.

Next we proceed with investigating the effect of timing constraints on the total computational time needed for executing $\widetilde{\mathbf{H}}_2$ and $\widetilde{\mathbf{H}}_3$. We first recall from the results in [10] that $\mathbf{H}_2$ and $\mathbf{H}_3$ show the performance of yielding solutions within

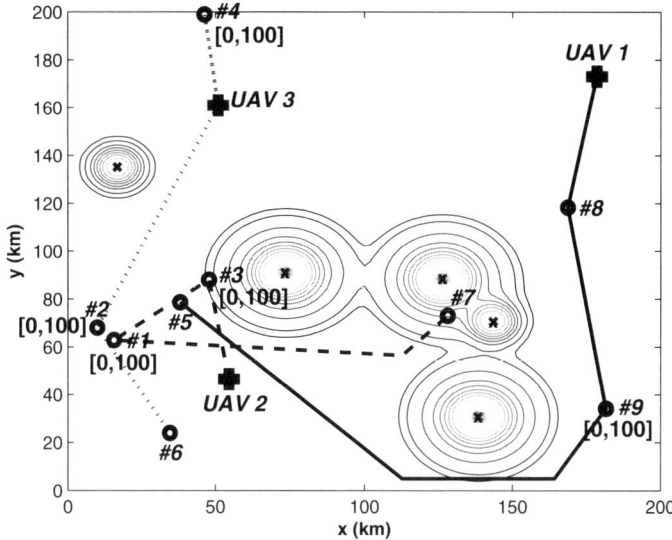

Fig. 4. Example solution paths (solid line for UAV 1, dashed line for UAV 2 and dotted line for UAV 3) when timing constraints are considered: prime targets (labeled with #1, #2, #3, #4 and #9) are required to be visited by a UAV before 100 (min) after the mission starts

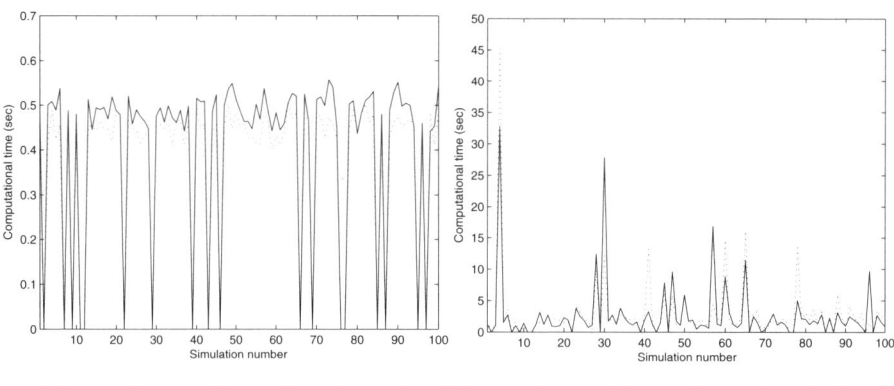

(a) $m = 3$, $n = 10$, $p = 0$ and $q = 4$ (b) $m = 3$, $n = 10$, $p = 6$, $q = 4$ and $q' = 2$

Fig. 5. The total computation times needed for executing $\widetilde{\mathbf{H}}_2$ (solid line) and $\widetilde{\mathbf{H}}_3$ (dotted line) when $m = 3$ and $n = 10$

6 seconds on average as well as guaranteeing $\mathbf{T}/\mathbf{T}^* < 1.5$ for up to $m = 5$, $q = 4$ and $n = 20$. As timing constraints may greatly decelerate the speed of finding $\mathbf{T}^*$, i.e. solving the problem exactly, we here especially focus on investigating the computational performance of $\widetilde{\mathbf{H}}_2$ and $\widetilde{\mathbf{H}}_3$ versus various m, n and p. To this end, for fixed m, n, p, q and q' we create one hundred random scenarios with various UAV starting positions ($\in \mathbf{X}_2$ (km)), where $\mathbf{X}_h = [0, 200] \times [0, 200] \times h$ (km),

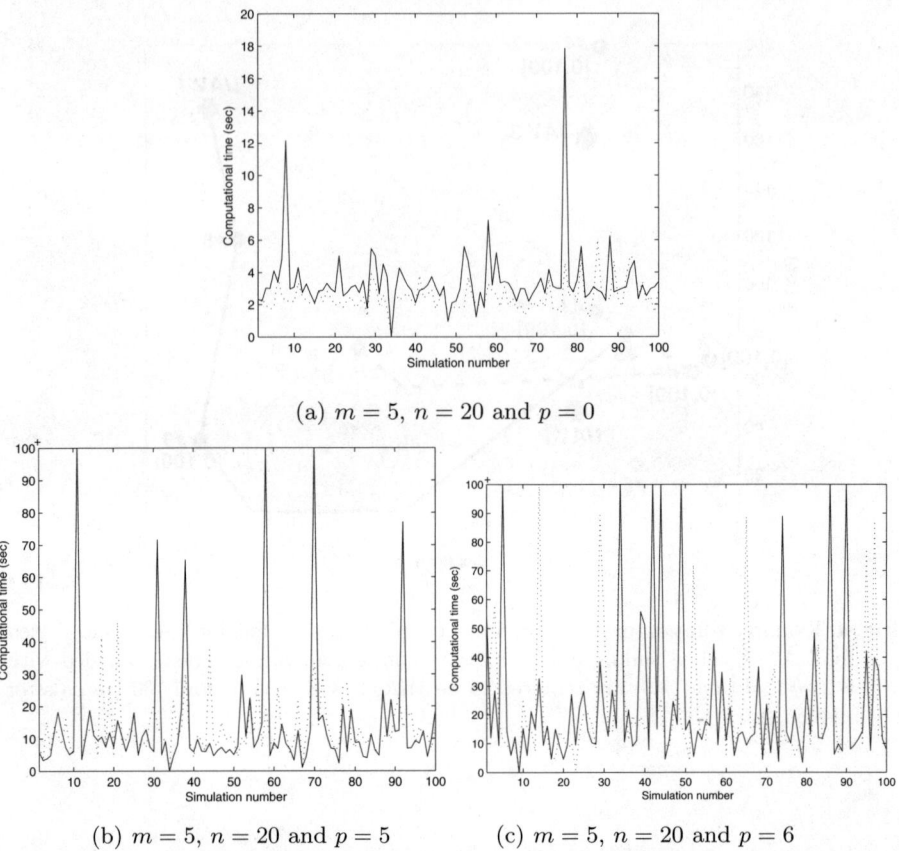

(a) $m = 5$, $n = 20$ and $p = 0$

(b) $m = 5$, $n = 20$ and $p = 5$ (c) $m = 5$, $n = 20$ and $p = 6$

Fig. 6. The total computation times needed for executing $\widetilde{\mathbf{H}}_2$ (solid line) and $\widetilde{\mathbf{H}}_3$ (dotted line) when $m = 5$ and $n = 20$; 100^+ denotes the number greater or equal to 100

the target positions ($\in \mathbf{X}_2$), the number of threats ($\in \{5, 6, \ldots, 10\}$), the threat locations ($\in \mathbf{X}_0$) and the threat ranges (7 or 25 (km)). We again assume that UAVs fly at a constant speed of 2 (km/min), and every prime target must be visited within 100 (min) from the mission starting. All numerical tests are done with a personal computer equipped with an Intel(R) Pentium 4 CPU 3.40GHz.

Figure 5(a) shows the results when the scenarios consist of 3 UAVs, 10 non-prime targets, no prime target and the maximum number of targets visited by a single UAV set to 4. The dotted (respectively, solid) line shows the total computational times when the UAV's return constraint is (respectively, not) considered. Note that all the computations are done in 0.6 (sec). However, when the number of prime targets gets increased to 6 and the maximum number of prime targets visited by a single UAV is set to 2, one can see several peaks, as shown in Fig. 5(b). Although many cases (93%) are handled within less than 10 (sec), the figure shows that timing constraints can greatly complicate the *basic* assignment problem. Similar observations follow from Fig. 6. As expected, for the cases in which 5 UAVs cover 20 targets with no timing constraints and each UAV is restricted to visit at most 4 targets, the proposed algorithms manage to deal with all the cases within 6 (*sec*) on average, as depicted in Fig. 6(a). However, when 5 of the 20 targets become prime targets and each UAV is further restricted to visit at most two prime targets, Fig. 6(b) suggests that the computational burden dramatically increases for several cases. The situation becomes worse when the number of prime targets gets increased to 6, as seen in Fig. 6(c). However, in spite of the presence of such *unpleasant* scenarios, we note that for more than 90% of the tested cases, $\widetilde{\mathbf{H}}_2$ and $\widetilde{\mathbf{H}}_3$ return solutions within 20 (*sec*) when the sum of the numbers of UAVs and targets (including prime targets) is less than or equal to 25, the maximum number of targets visited by a single UAV is less than or equal to 4 and the number of prime targets is less than or equal to 5 or 6 depending on the numbers of UAVs and targets.

4 Concluding Remarks

We have considered the UAV-to-target assignment problem especially focused on the presence of time-critical (prime) targets. Our main challenges (contributions) in doing this are to keep the previously guaranteed theoretical bounds (16) and (17) and to allow only a slight increase of computational cost after including time-critical targets. We analytically show that by means of adding several integer linear constraints to the previous MILP formulation, the bounds still hold under a mild condition ($q' < 3$). In the numerical experiments, for more than 90% of the tested cases the newly proposed algorithms returned solutions within 20 (*sec*) when the sum of the numbers of UAVs and targets (including prime targets) is less than or equal to 25, the maximum number of targets visited by a single UAV is less than or equal to 4 and the number of prime targets is less than or equal to 5 or 6 depending on the numbers of UAVs and targets.

References

1. Ahuja, R., Kumar, A., Jha, K., Orlin, J.: Exact and heuristic methods for the weapon target assignment problem. Oper. Res. (to appear)
2. Alighanbari, M., How, J.: Decentralized task assignment for unmanned aerial vehicles. Proc. 44th IEEE Conf. Decision Control (2005) 5668–5673
3. Alighanbari, M., Kuwata, Y., How, J.: Coordination and control of multiple UAVs with time constraints and loitering. Proc. IEEE Amererican Control Conf. (2003) 5311–5316
4. Beard, R., McLain, T., Goodrich, M., Anderson, E.: Coordinated target assignment and intercept for unmanned aer vehicles. IEEE Trans. Robot. Automat. **18** (2002) 911–922
5. Bellingham, J., Tillerson, M., Richards, A., How, J.: Multi-Task Allocation and Trajectory Design for Cooperating UAVs. Proc. 2nd Annual Conf. Cooperative Control Optimization. (2001) 1–19

6. Christofides, N., Mignozzi, A., Toth, P.: The vehicle routing problem. Combinatorial Optimization. Wiley and Sons (1979) 315–338
7. Gu, D.-W., Kamal, W. A., Postlethwaite, I.: A UAV Waypoint Generator. Proc. AIAA 1st Intelligent Systems Technical Conf. **AIAA-2004-6227** (2004)
8. Gu, D.-W., Postlethwaite, I., Kim, Y.: A Comprehensive Study on Flight Path Selection Algorithms. Proc. IEE Seminar Target Tracking (2006) 77–89
9. Haimovich, M., Kan, R., Stougie, L.: "Analysis of heuristics for vehicle routing problems" in Vehicle Routing: Methods and Studies. Elsevier (1988) 87-127
10. Kim, Y., Gu, D.-W., Postlethwaite, I.: Real-time optimal mission scheduling and flight path selection. IEEE Trans. Automat. Contr. (to appear)
11. Kingston, D. B., Schumacher, C. J.: Time-dependent cooperative assignment. Proc. IEEE American Control Conf. (2005) 4084–4089
12. Korte, B., Vygen, J.: Combinatorial Optimization: Theory and Algorithms. Springer (2002)
13. Li, C., Simchi-levi, D.: Worst-case analysis of heuristics for multidepot capacitated vehicle routing problem. ORSA J. Comput. **2** (1990) 64–73
14. Manne, A.: A target-assignment problem. Oper. Res. **6** (1958) 346–351
15. MacLain, T. W., Beard, R. W.: Coopertative path planning for timing-critical missions. Proc. IEEE American Control Conf. (2003) 296–301
16. Mitrovic-minic, S., Krishnamurti, R.: The multiple travelling salesman problem with time windows: bounds for the minimum number of vehicles. Simon Fraser University **TR-2002-11** (2002)
17. Marinakis, Y., Migdalas, A.: Heuristic solutions of vehicle routing problems in supply chain management. Working Paper (2001)
18. Ryan, J., Bailey, G., Moore, J., Carlton, W.: Reactive tabu search in unmanned aerial reconnaissance simulations. Proc. Winter Simulation Conf. (1998) 873–879
19. Savelsbergh, M.: Local search in routing problems with time windows. Ann. Oper. Res. **4** (1985) 285–305
20. Schaerf, A.: A survey of automated timetabling. Artif. Intell. Rev. **13** (1999) 87–127
21. Shetty, V., Sudit, M., Nagi, R.: Priority-based assignment and routing of a fleet of unmanned combat aerial vehicles. Comp. & Oper. Res. **doi: 10.1016/j.cor.2006.09.013** (2006)
22. White, C., Lee, Y., Kim, Y., Thomas, R., Perkins, P.: Creating weekly timetables to maximize employee preferences. The UMAP J. **25** (2004) 5–25

Sequential Inspection Using Loitering[*]

Meir Pachter[1], Swaroop Darbha[2], and P. Chandler[3]

[1] Professor, Dept. of Electrical and Computer Engineering
[2] Visiting Professor, AFRL/VACA, & Associate Professor, Department of
Mechanical Engineering, Texas A&M University at College Station
[3] Air Force Research Laboratory, AFRL/VACA, Department of Electrical and
Computer Engineering, Air Force Institute of Technology, Wright-Patterson Air Force
Base, OH 45433

Abstract. A set of objects of interest is to be sequentially inspected by
a Micro Aerial Vehicle (MAV) equipped with a camera. Upon arriving
at an object of interest, an image of the object is sent to a human operator, who, upon inspecting the image, sends his feedback to the MAV.
The feedback from the operator may consist of the pose angle of the
object and whether he has seen any distinguishing features of the object.
Upon receiving the feedback, the MAV uses this information to decide
whether it should perform a secondary inspection of the object of interest
or continue to the next object. A secondary inspection has a reward (or
value or information gain) that is dependent on the operator's feedback.
There is an associated cost of reinspection and it depends on the delay
of the operator's feedback. It seems reasonable to let the MAV loiter for
a while near the most recently inspected object of interest so that it ex-
pends a small amount of endurance from the reserve after receiving the
feedback from the operator. The objective is to increase the information
and hence, the total expected reward about the set of objects of interest.
Since the endurance of the MAVs is limited, the loiter time near each
object of interest must be carefully determined. This paper addresses
the determination of the optimal loiter time through the use of Stochastic dynamic programming. Numerical results are presented that show the
optimal loiter time is a function of the maximum expected operator delay.

1 Introduction

The following inspection scenario is considered. A set of n objects of interest O_i,
$i = 1, ..., n$, is sequentially visited by an MAV equipped with a camera. Upon arriving at an object of interest, an image of the object is sent to a human operator
for classification. The operator, upon inspecting the image, sends his feedback,
e.g., the object's pose angle and whether he has seen a distinguishing feature in
the object's image, to the MAV. When the operator's feedback is received, the
MAV must make a decision whether it should revisit the object for a secondary

---

[*] The views expressed in this article are those of the authors and do not reflect the
official policy or position of the United States Air Force, Department of Defence, or
U.S. Government.

M.J. Hirsch et al. (Eds.): Adv. in Cooper. Ctrl. & Optimization, LNCIS 369, pp. 281–291, 2007.
springerlink.com   © Springer-Verlag Berlin Heidelberg 2007

inspection. The information gain (reward) associated with a secondary inspection is dependent on the feedback from the operator. Each MAV has a finite endurance reserve and a revisit of an object requires expenditure from the reserve. This expenditure is a function of the operator's delay and the action that the MAV decides to take. The operator functions as a sensor/classifier in the inspection loop and the MAV decides on the course of action. The objective of the decision making is to maximize the total expected reward given the constraints on the endurance.

The MAV makes decisions *sequentially* based on the information available to it - operator's delay and feedback about the object, the number of objects left to be visited by the MAV, and the current reserve. While the operator's delay, τ_i, associated with the ith object is a random variable (whose probability density function (p. d. f) $f(\tau_i)$ is known), we emphasize that the realized value of this random variable may only be known at the time of decision making. We do not allow for the possibility of the MAV revisiting an object more than once or revisiting an object after it has decided to go to the next object in the sequence. At the time of making a decision, the actions that we allow the MAVs to take are the following: loiter around the object; move onto the next object; or revisit the object. Associated with the ith object, there is a continuous decision variable, u_i, which indicates the maximum allowable loitering time and a binary decision variable, v_i, which indicates whether the object should be revisited.

The motivation for the introduction of loitering is as follows: If the MAV were to move away from the object after the first visit, then the time (and hence, expenditure of the reserve) for a revisit is at least twice the operator's delay; by allowing the MAV to loiter near the object, the time to get back is shortened. If the operator's feedback is received by the MAV before the maximum allowable loiter time, only then does the MAV take a decision about revisiting the object; otherwise, it will go to the next object in the sequence. Since the objective is to increase the information about the set of objects of interest and since the MAV's endurance reserve c_1 is limited, the maximum allowable loiter time associated with the objects must be carefully determined.

The paper is organized as follows. In Section 2, a stochastic optimal control problem which models the class of decision scenarios at hand is formulated. In Section 3, the method of Dynamic Programming is brought to bear on the sequential decision problem and numerical results corroborating the methodology presented for the decision problems considered in this paper are provided. In Section 4, a generalization of the present formulation is explored. Conclusions are drawn in Section 5.

We use the following notation throughout the paper:

i Index of stage in a Stochastic Dynamic Program (SDP)

p *a priori* probability

r_i Running reward for the ith stage

u_i Decision variable at the ith stage

τ_i Delay in communicating the first observation

$f(\tau_i)$ Probability density function (p.d.f) of the operator delay τ_i

τ_0 Fixed communication delay

2   Stochastic Optimal Control

The Dynamic Program (DP) has n stages and one state variable, c_i - the endurance reserve on arrival to O_i. The operator's delay at O_i, $0 \leq \tau_i \leq \tau_{max}$ is a random variable whose realization may *not* be known at the time of making a decision. The p.d.f. of τ_i is $f(\tau_i)$ and is assumed known. The decision/control variable is u_i, the *maximal* loiter or waiting time at O_i.

The nonlinear dynamics are driven by the control variable u and by the random variable τ:

$$c_{i+1} = c_i - \min (u_i, \ \tau_i) \quad i = 1, ..., n \tag{1}$$

The initial reserve, c_1, is known. The control variable is constrained according to

$$0 \leq u_i \leq \min(c_i, \tau_{\max}) \tag{2}$$

and the random variable τ_i is characterized by its p.d.f. $f(\tau)$.

The running payoff is

$$r_i(u_i, \tau_i) = \begin{cases} 1 & \text{if } \tau_i < u_i \\ 0 & \text{otherwise} \end{cases} \tag{3}$$

The payoff function

$$J(u_1, ..., u_n; c_1) = \mathcal{E}_{\tau_1,...,\tau_n} \left(\sum_{i=1}^{n} r_i(u_i, \tau_i) \right) \tag{4}$$

The optimal strategy is a state feedback control law $u_i^*(c_i)$, $i = 1, ..., n$.

3   Dynamic Programming Recursion

The stochastic optimal control problem (1)-(4) is solved using the method of Dynamic Programming (DP). We emphasize that the realization of the random variable may not be known at the time of making a decision.

We shall require the following definition:

$$p(u) \equiv \int_{u}^{\tau_{max}} f(\tau)d\tau$$

The term $p(u_i)$ is the probability that the MAV comes out empty handed, that is, the MAV leaves the vicinity of O_i after waiting for a time u_i without receiving the operator's feedback.

The value function $V_i(c_i)$ is the maximal expected reward at the time of making a decision concerning O_i, given the endurance reserve of the MAV at O_i is c_i.

The DP recursion is established as follows: For each $u_i \in [0, \min\{c_i, \tau_{\max}\}]$,

$$V_i(c_i|u_i, \tau_i) = r_i(u_i, \tau_i) + V_{i+1}(c_i - \min(u_i, \tau_i))$$
$$\Rightarrow V_i(c_i|u_i) = \mathcal{E}_{\tau_i}[r_i(u_i, \tau_i) + V_{i+1}(c_i - \min(u_i, \tau_i))]$$
$$= [0 + V_{i+1}(c_i - u_i)]p(u_i) + \int_0^{u_i} [1 + V_{i+1}(c_i - \tau)]f(\tau)d\tau$$

$$V_i(c_i) = \max_{0 \le u_i \le \min(c_i, \tau_{\max})} \mathcal{E}_{\tau_i}\left(r_i(u_i, \tau_i) + V_{i+1}(c_{i+1}(c_i, u_i, \tau_i)) \right)$$

$$= \max_{0 \le u_i \le \min(c_i, \tau_{\max})} \{[0 + V_{i+1}(c_i - u_i)]p(u_i) + \int_0^{u_i} [1 + V_{i+1}(c_i - \tau_i)]f(\tau_i)d\tau_i\}$$

$$= \max_{0 \le u_i \le \min(c_i, \tau_{\max})} [p(u_i)V_{i+1}(c_i - u_i) + 1 - p(u_i) + \int_0^{u_i} V_{i+1}(c_i - \tau_i)f(\tau_i)d\tau_i]$$

Hence, the DP recursion is

$$V_i(c_i) = 1 + \max_{0 \le u_i \le \min(c_i, \tau_{\max})} [p(u_i)V_{i+1}(c_i - u_i) - p(u_i) + \int_0^{u_i} V_{i+1}(c_i - \tau_i)f(\tau)d\tau] \quad (5)$$

Assuming that the value function, V_{i+1}, is known, the above recursion allows one to compute the value function V_i. The optimal control,

$$u_i^*(c_i) = \arg \max_{0 \le \min c_i, \tau_{\max}} [p(u_i)V_{i+1}(c_i - u_i) - p(u_i) + \int_0^{u_i} V_{i+1}(c_i - \tau_i)f(\tau)d\tau]. \quad (6)$$

For one to begin the recursion, the value function V_n must be specified. This is presented in the next subsection.

3.1   The Boundary Condition

Obviously, at time n

$$u_n^*(c_n) = c_n$$

Hence, if $c_n > \tau_{\max}$, $r_n = 1$ and therefore $V_n(c_n) = 1$. If, however, $c_n \le \tau_{\max}$, then

$$V_n(c_n) = \mathcal{E}_{\tau_n}\left(r_n(c_n, \tau_n) \right)$$
$$= 0 \cdot p(c_n) + 1 \cdot (1 - p(c_n))$$
$$= 1 - p(c_n)$$

Thus, the boundary condition is

$$V_n(c_n) = \begin{cases} 1 & if \ c_n > \tau_{max} \\ 1 - p(c_n) & if \ 0 \le c_n \le \tau_{max} \end{cases} \quad (7)$$

3.2   Computing Suboptimal Value Functions

In the examination of sensitivity of the value function to a perturbation in the probability distribution of the operator's feedback delay, one is interested in the computation of suboptimal value functions. Let $U_i(c)$ be the sub-optimal value function for a given strategy. Let $f_p(\tau)$ denote the perturbed distribution and $\tilde{\tau}_{max}$ denote the maximum value of the corresponding delay. Let $\tilde{p}(c)$ denote the integral

$$\int_c^{\tilde{\tau}_{\max}} f_p(\tau)d\tau.$$

The function $U_i(c)$ may be computed recursively as follows:

$$U_n(c) = \begin{cases} 1 - \tilde{p}(c) & c \in [0, \tilde{\tau}_{max}], \\ 1 & c \geq \tilde{\tau}_{max} \end{cases}$$

and

$$U_i(c|u_i^*(c)) = 1 + [\tilde{p}(u_i^*)U_{i+1}(c_i - u_i^*) - \tilde{p}(u_i^*) + \int_0^{u_i^*} U_{i+1}(c_i - \tau_i)f_p(\tau)d\tau].$$

3.3   Numerical Implementation

For the purposes of implementation, we discretize the cumulative density function, $f_c(\tau)(:= \int_0^\tau f(\eta)d\eta)$ and deal with the corresponding discrete probability density function. We specifically assume the following: the maximum reserve $c_{\max}$ and the maximum number of objects of interest $(N_{\max})$ are known a priori. Further, we assume the reserve to be an integer multiple of a fixed increment of reserve, Δ, i.e., $c = k\Delta$ for some $k \geq 0$, and that the fixed delay, τ_0 and the delay, τ are also integral multiples of a fixed increment of reserve, i.e., $\tau = l\Delta$ for some integral $l, l_0 \geq 0$. Since the delay can only take discrete values (which are integral multiples of Δ), one may approximate the continuous p.d.f by a discrete p.d.f. for $f(\tau)$ as: $P(\tau = l\Delta) = p_l$ and hence,

$$f(\tau) = \sum_{j=1}^{\infty} p_j \delta(\tau - j\Delta),$$

where $\delta(k) = 1$ if $k = 0$ and is 0 otherwise.

The value function $V_n(k\Delta)$ may be readily computed from the discretization of the cumulative distribution function as follows:

$$V_n(k\Delta) = \sum_{l \leq k} p_l.$$

Clearly, if $l\Delta > \tau_{\max} = D_{\max}\Delta$, then $V_n(l\Delta) = 1$. The optimal decision, u_n^* is specified by the following equation:

$$u_n^*(k\Delta) = k\Delta.$$

The recursive equations for the value functions, $V_i(k\Delta)$, $i < n$, can be expressed as follows:

$$V_i(k\Delta) = 1 + \min_{0 \le j \le \min\{k, D_{max}\}} \left(\sum_{l>j} V_{i+1}((k-j)\Delta) - \sum_{l>j} p_l + \sum_{l \le j} V_{i+1}(k-l)p_l \right).$$

The corresponding optimal decisions are:

$$u_i^*(k\Delta) = \Delta \arg \min_{0 \le j \le \min\{k, D_{max}\}} \left(\sum_{l>j} V_{i+1}((k-j)\Delta) - \sum_{l>j} p_l + \sum_{l \le j} V_{i+1}(k-l)p_l \right).$$

Real-time Implementation: Once the $V_i(c)$ and $u_i^*(c)$ is computed for each $c \in [0, C_{max}]$ and $i = 1, \ldots, N_{max}$, it is stored as two matrices in the MAV's on-board processor. From the knowledge of c and the number of objects to visit, one can compute the relevant optimal decision (waiting time) from the table. If the operator does not provide any feedback before the optimal waiting time, then the MAV moves onto the next object; if he does provide feedback, it will revisit the object.

3.4   Numerical Results

We have considered the following case: The maximum number, N_{max} of objects to visit is 20, the maximum reserve, $c_{max} = 1000$ *units*, the maximum delay is 200 *units* and $\Delta = 1$ *unit*. The operator delay is initially assumed to be uniform. The corresponding value functions and optimal decisions are shown in Figures 1 and 2.

From Figure 1, we can see that the expected number of revisits increases monotonically with the reserve and the number of objects to visit. From Figure 2, for any fixed reserve (smaller than the maximum operator's feedback delay), the optimal wait time decreases with the number of the objects to visit. This is consistent with our intuition. In particular if the number of objects is arbitrarily large, then the optimal waiting time is the minimum delay of 1 unit and the maximum number of revisits that are possible is c if the initial reserve is c units.

In order to examine the sensitivity of the optimal strategy to the distribution of the operator's feedback delay, we perturbed the distribution. When we refer to a sub-optimal value function, we decide on a revisit based on the optimal waiting time obtained when the operator's delay is uniformly distributed between 1 and 200 seconds. An optimal value function will correspond to the optimal waiting time associated with the perturbed distribution. We considered four perturbed distributions and associated with each perturbed distribution there is a sub-optimal revisit function and a sub-optimal value function. The case when the operator's feedback delay is randomly (as opposed to uniformly) distributed between 1 and 200 units of reserve shows little difference with the optimal (uniform) revisit and value functions.

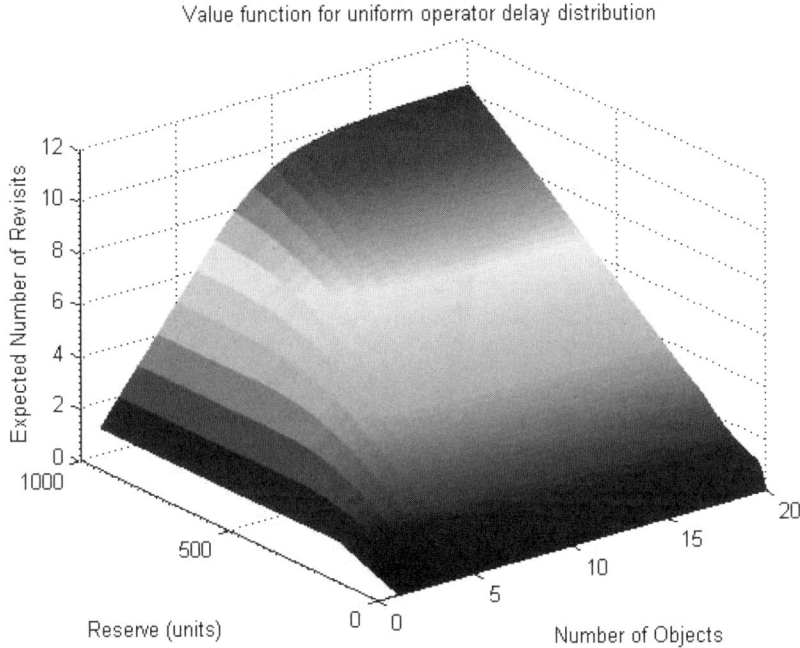

Fig. 1. Value function for uniform operator delay

Three other cases with varying degrees of randomness in operator delay were tested and similar results were found. The trends remained the same with minor variations in the expected number of revisits and the optimal waiting time. While the overall performance (expected number of revisits) is sensitive to the maximum operator delay assumed in the derivation of the optimal decisions, it is not sensitive to the exact distribution of the delay. In particular, if the maximum delay was underestimated, the performance deteriorated (i.e., the expected number of revisits was significantly smaller than the case corresponding to the knowledge of the exact value of the maximum delay) and if the maximum delay was overestimated, the degradation in the performance was insignificant. Essentially, it is better to overestimate the maximum value of the operator's feedback delay as opposed to underestimating it.

4   Generalization: Including an Endurance Cost When Revisiting an Object

Suppose there are L different ways to revisit an object. Suppose there is an overhead cost of τ_{ij} units of reserve when the ith object is revisited in the jth way. Let v_{ij} be a binary variable that takes a value 1 if the ith object is revisited in the jth way and is 0 otherwise. Then, the governing constraints may be expressed as:

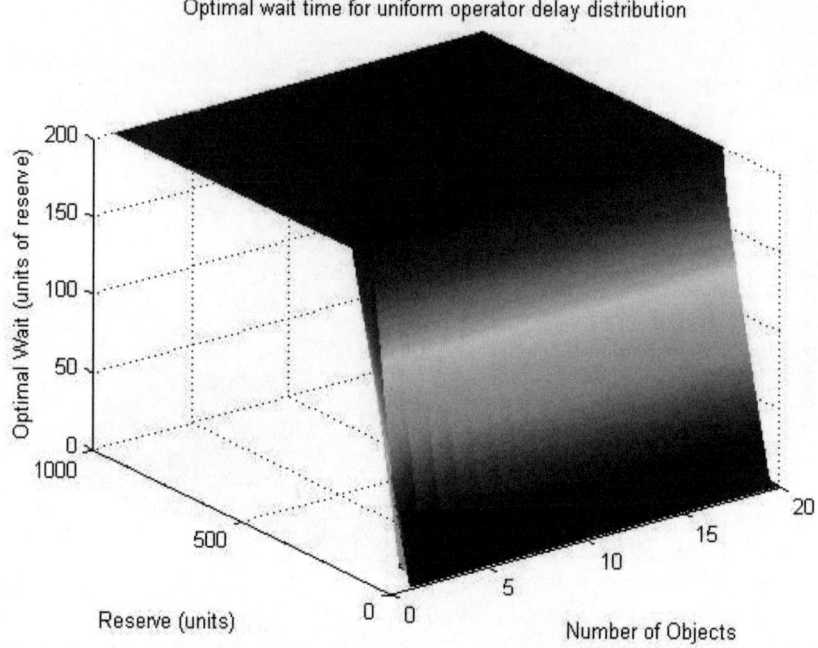

Fig. 2. Optimal wait for uniform operator delay

$$c_{i+1} = c_i - \min(u_i, \tau_i) - \sum_{j=1}^{L} \tau_{ij} v_{ij} \qquad (8)$$

$$0 \le c_i - \min(u_i, \tau_i) - \sum_{j=1}^{L} \tau_{ij} v_{ij}, \qquad (9)$$

$$\sum_{j=1}^{L} v_{ij} = \begin{cases} 1 & \text{if } u_i \ge \tau_i \text{ and } \tau_i \le c_i - \min_j \tau_{ij} \\ 0 & \text{otherwise.} \end{cases} \qquad (10)$$

The last constraint indicates that a MAV will always revisit an object in exactly one way if the operator's feedback delay is smaller than the waiting time associated with the object provided it has sufficient reserve and will not revisit otherwise.

In plain words, the problem may be stated as follows: The MAV can revisit objects after it receives a feedback from the operator in one of L ways. The MAV does not revisit if the operator does not provide his feedback within the optimal waiting time associated with each object in the specified sequence. A revisit of the ith object in the jth way fetches a reward or payoff of β_{ij}. At the time of receiving the feedback, if it happens to be smaller than the optimal waiting time, the MAV must decide which of the L following ways it should revisit so as to maximize the total expected payoff.

One can think of a different waiting time for each way of revisiting the object; however, for simplicity, we do not consider such a scheme in this paper.

Mathematically, the payoff for visiting the ith object in the kth way is feasible only if the operator's feedback delay is smaller than the waiting time set for that object and there is sufficient reserve to revisit the object of interest. Hence,

$$r_i(u_i, \tau_i | v_{ik} = 1) = \begin{cases} \beta_{ik} & \text{if } \tau_i \leq \min\{c_i - \tau_{ik} v_{ik}, u_i, \tau_{max}\} \\ 0 & \text{otherwise.} \end{cases}$$

We shall assume, henceforth, the following: $\beta_{i1} > \beta_{i2} > \ldots > \beta_{iL}$ and correspondingly, $\tau_{i1} > \tau_{i2} > \ldots > \tau_{iL}$. This assumption implies that to get a higher payoff, one must pay a higher overhead cost (reserve).

Let

$$J = \mathcal{E}_{\tau_1,\ldots,\tau_n}[\sum_{i=1}^{n}\sum_{j=1}^{L}\underbrace{r_i(u_i,\tau_i|v_{ij}=1)v_{ij}}_{r_i(u_i,\tau_i,v_{ij})}].$$

The term J indicates the total expected payoff for any given set of decisions, $u_i, v_{ij}, j = 1, \ldots, L, i = 1, \ldots, n$. The objective of the optimization is to maximize the expected payoff, J, over the possible set of decisions, $u_i, i = 1, \ldots, n$ and $v_{ij}, j = 1, \ldots, L, i = 1, \ldots, n$.

Let

$$V_i(c_i) := \max_{u_k, v_{kj},\, i \leq k \leq n,\, 1 \leq j \leq L} \mathcal{E}_{\tau_i,\ldots,\tau_n}[\sum_{k=i}^{n}\sum_{j=1}^{L} r_k(u_k, \tau_k, v_{ik})]. \tag{11}$$

One can then use DP to get the following recursion:
$V_i(c_i | u_i, \tau_i) =$

$$\begin{cases} V_{i+1}(c_i - u_i) & \text{if } \tau_i > \min\{u_i, c_i - \tau_{iL}\}, \\ \max_{1 \leq j \leq L}\{\beta_{ij} + V_{i+1}(c_i - \tau_{ij} - \tau_i) : c_i - \tau_i - \tau_{ij} \geq 0\} & \text{otherwise.} \end{cases}$$

Let $\mathcal{D}_{ij} := \{\tau : j = \arg\max_{1 \leq k \leq L}\{\beta_{ij} + V_{i+1}(c_i - \tau_{ik} - \tau_i) : c_i - \tau_i - \tau_{ik} \geq 0\}\}$. Therefore,

$$V_i(c_i | u_i) = \mathcal{E}_\tau(V_i(c_i | u_i, \tau_i))$$

$$= \int_{\min(u_i, c_i - \tau_{iL})}^{\tau_{max}} V_{i+1}(c_i - u_i) f(\tau) d\tau +$$

$$\sum_{j=1}^{L} \int_{\tau \in \mathcal{D}_{ij}} (\beta_{ij} + V_{i+1}(c_i - \tau_{ij} - \tau_i)) f(\tau) d\tau.$$

Hence,

$$V_i(c_i) = \max_{u_i \in (0, \min(c_i, \tau_{max}))} V_i(c_i | u_i).$$

The boundary condition that completes the recursion and enables the determination of all the value functions. Clearly, the waiting time if there is only one object to visit is equal to the reserve—$u_n^*(c_n) = c_n$ and correspondingly:

$$V_n(c_n|\tau_n) = \begin{cases} \beta_{nj} & \tau_n \in (c_n - \tau_{n,j-1}, c_n - \tau_{nj}] \\ 0 & \tau_n > c_n - \tau_{nL}. \end{cases}$$

Let $\tau_{n,0} = c_n$ and

$$V_n(c_n) = \sum_{j=1}^{L} \int_{c_n - \tau_{n,j-1}}^{c_n - \tau_{n,j}} \beta_{nj} f(\tau) d\tau.$$

Once the value functions are computed, the computation of optimal waiting time is straight forward:

$$u_i^*(c_i) = argmax_{u_i \in (0, \min(c_i, \tau_{max}))} V_i(c_i|u_i), \ i = 1, \ldots, n-1,$$
$$u_n^*(c_n) = c_n.$$

The optimal decisions to revisit are as follows:

$$v_{ij}^*(c_i, \tau_i) = \begin{cases} 1 & \text{if } \tau_i \in \mathcal{D}_{ij}(c_i, u_i^*(c_i)), \\ 0 & \text{otherwise.} \end{cases}$$

5   Conclusion

We have observed in numerical simulations that the performance (i.e., the expected number of revisits) of the sequential inspection decision system is sensitive to the assumed value of the maximum operator delay, but not sensitive to the actual distribution of the delay. The structure of the strategy is reasonably simple for its actual real-time implementation on the MAVs: We store the optimal wait time as a function of the reserve and the number of objects to visit and based on the operator's delay, decide on the future course of action. If the operator's delay is smaller than the optimal wait time associated with the MAV's reserve and the number of objects to revisit, an appropriate action for revisiting the object is taken; otherwise, it is optimal for the MAV to continue to the next object in the sequence. The optimal loiter time comes from solving the stochastic dynamic programming problem.

References

1. Gross D, Rasmussen S, Chandler PR, Feitshans G. "Cooperative Operations in UrbaN TERrain (COUNTER)," *2006 SPIE Defense & Security Symposium*, Orlando, FL, 17-21 April 2006.
2. Derman, C., Lieberman G. J., and Ross, S. M.: "A Sequential Stochastic Assignment Problem", Management Science, Vol. 18, No. 7, Theory Series, pp. 349-355, March 1972.

3. Ross SM. *Introduction to Stochastic Dynamic Programming*, Academic Press, 1982.
4. Pachter M, Chandler PR, Darbha S. "Optimal Control of an ATR Module Equipped MAV–Human Operator Team," to appear in Vol. 588 of *Lecture Notes in Economics and Mathematical Systems*, Springer 2007.
5. Girard AR, Pachter M, Chandler PR. "Decision Making Under Uncertainty and Human Operator Model for Small UAV Operations," *Proceedings of the AIAA Guidance, Navigation and Control Conference*, Keystone, CO, August 2006.
6. Pachter, M., P. R. Chandler and Swaroop Darbha, "Optimal Sequential Inspections," *Proceedings of the IEEE Conference on Decision and Control*, December 2006.
7. Pachter, M., P. R. Chandler and Swaroop Darbha, "Robust Control of Unmanned Aerial Vehicles in Uncertain Environments," to appear in *International Journal of Robust and Nonlinear Control*, 2007.
8. R. J. Gallager: "Information Theory and Reliable Communication", Wiley 1968, pp. 18.
9. Parzen, E. *Modern Probability Theory and its Applications*, Wiley 1960.

Distributed Cooperative Systems with Human Operator-in-the-Loop*

Pavlo A. Krokhmal[1] and David E. Jeffcoat[2]

[1] Department of Mechanical and Industrial Engineering, The University of Iowa,
Iowa City, IA, 52242
krokhmal@engineering.uiowa.edu
[2] Air Force Research Lab/Munitions Directorate, Eglin AFB, FL 32542
david.jeffcoat@eglin.af.mil

Abstract. We study the evolution of distributed multi-agent search systems where the autonomous agents may cooperate among each other, and/or with a human operator, in order to achieve the system's objective. The cooperation is facilitated by means of information sharing among the autonomous agents and/or human operator, which has the purpose of improving the effectiveness of the autonomous agents. The evolution of cooperative systems is modeled using discrete-state, continuous-time Markov chains, and a technique for measuring and quantification of cooperation within such systems is proposed.

Keywords: Cooperation; Markov chains; target detection.

1 Introduction

In this chapter we study the evolution of distributed systems whose constituents, or autonomous agents, may cooperate among each other, and/or with human operator, in order to achieve the system's objective. The cooperation is facilitated by means of information sharing among the autonomous agents and/or human operator, which has the purpose of improving the effectiveness of the autonomous agents. Although the presented approach and the analysis are quite general, we focus on a particular type of cooperative distributed system, namely, a search system.

The present endeavor is a continuation of the ongoing research efforts of the authors [1,2], where the Markov chain framework was applied to modeling of two types of cooperative search systems, with the main emphasis being placed on their asymptotical properties. The objective of the present work is twofold: firstly, we extend and apply the techniques introduced in [1,2] to modeling and analysis of cooperative search systems where the autonomous agents are assisted/controlled by human operators. Secondly, we develop an approach to analysis and quantification of the degree of cooperation within distributed systems, i.e., to answering the question "just how cooperative is the given system?"

* This work was partially supported by the U.S. Air Force grants FA8651-06-1-0008, FA8651-07-1-0001, and LRIR 99MN01COR.

The chapter is organized as follows. In the next section we introduce several types of cooperative systems, where the autonomous searchers may cooperate with each other and/or human operator by sharing information. The systems are modeled as discrete-state continuous-time Markov chains, and the corresponding sets of Chapman-Kolmogorov differential equations are derived. Section 3 presents numerical results on the models considered in Section 2. Finally, Section 4 discusses an approach to measuring the degree of cooperation within a distributed system, and illustrates it on the examples of the search systems discussed in this chapter.

2 Modeling of Cooperative Systems: A Search Mission Example

The particular type of distributed cooperative system that we focus on in this work is the search system, where N autonomous agents, or searchers, are given the objective of discovering (detecting) a certain kind of objects of interest, or targets.

In the simplest case, each of N searchers may assume only two states: *search* or *detect* (see Figure 1). Once a searcher reaches the *detect* state (i.e., detects a target), it never returns to state *search* (i.e., never resumes the search again). Such a setup is common to search-and-rescue missions, where, upon detecting a target searchers would try to perform a rescue operation instead of continuing the search. The effectiveness of individual searchers is characterized by the detection rate θ, i.e. the probability of detecting a target within time interval Δt:

$$\mathsf{P}\{\text{searcher } i \text{ detects a target during time } \Delta t\} = \theta\Delta t + o(\Delta t), \quad i = 1, \ldots, N. \tag{1}$$

Then, the conditional probability of target detection at time t, given that the searcher is in the *search* state, has an exponential distribution with mean $1/\theta$:

$$f(t;\theta) = 1 - e^{-\theta t}, \quad t \geq 0. \tag{2}$$

Cooperation within the search system is facilitated via cueing, or information sharing among the searchers, and has the purpose of increasing their detection capabilities. The informational content of the cueing signals is not important in the presented framework; instead, we are interested in the degree by which cueing impacts the search capabilities of individual agents in a cooperative system.

Below we discuss several cooperative search systems with various forms of cueing, which serves as a mechanism to achieve a better system performance.

2.1 Measures of Effectiveness

One of the important measures of effectiveness (MOE) of a cooperative search system is the *probability of detection $P_D(t)$*

$$P_D(t) = \mathsf{P}\{\text{search mission is completed by time } t\}$$
$$= \mathsf{P}\{\text{all } N \text{ searchers have detected targets by time } t\} \tag{3}$$

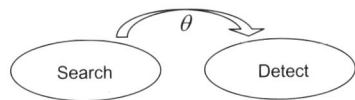

Fig. 1. States of an autonomous searcher

Oftentimes it is preferable to operate with a "single-number" characteristic that describes the performance of a search system. Such a MOE, employed in this work, is the *time to engage* T_α, or the time needed for all N searchers to detect and engage targets with probability $\alpha \in (0,1)$:

$$T_\alpha = P_D^{-1}(\alpha).$$

In our studies, the parameter α is taken to be $\alpha = 0.95$, which defines $T_{0.95}$ as the time by which the search mission is completed with 95% probability.

2.2   A Basic Cooperative Search Model

We start with the simplest search system, where the autonomous agents perform search without assistance and/or control of a human operator. It is assumed that upon detecting a target and transitioning to state *detect*, the searcher instantaneously cues all other searchers that are still in the *search* state, thereby potentially increasing their detection capabilities. Namely, it is assumed that the detection rates may change values in time as

$$\theta \equiv \theta_0 \to \theta_1 \to \cdots \to \theta_{N-1}, \tag{4}$$

where θ_k is the detection rate common to $N - k$ searchers that are in state *search*, and θ_0 is the initial "uncued" detection rate. It is natural to presume that cooperation in the form of cueing generally leads to improvement of search capabilities: $\theta_k \geq \theta$, $1 \leq k \leq N - 1$.

By introducing the traditional assumptions of independence of target detections on non-overlapping time intervals etc., one can model the outlined cooperative system as a continuous-time discrete-state Markov chain, or, more precisely, as a death process (see, among others, [3,5]). The states of the system can be designated by the number k of searchers that are in the state *detect* at time t, so that $\mathcal{S}_{ik}$ is a state in which there are k searchers that made detection, and $i = N - k$ agents that are still searching. Noting that there are $\binom{N}{k}$ different states $\mathcal{S}_{ik}$, and defining probabilities $P_{ik}(t)$ as

$$P_{ik}(t) = \mathsf{P}\{\text{system is in one of the states } \mathcal{S}_{ik} \text{ at time } t\},$$

one can write the system of backward Chapman-Kolmogorov ODEs governing these probabilities:

$$\frac{\mathrm{d}}{\mathrm{d}t} P_{ik}(t) = -i\theta_k P_{ik}(t) + k\theta_{k-1} P_{i+1,k-1}(t), \quad i + k = N. \tag{5}$$

Noting that index i can be dropped due to the one-to-one mapping $i = N - k$, system (5) can be rewritten as

$$\frac{d}{dt} P_k(t) = -(N - k)\, \theta_k P_k(t) + k\theta_{k-1} P_{k-1}(t), \quad k = 0, \ldots, N. \tag{6}$$

Equations (6) have a simple interpretation: the probability of the system being in a state $\mathcal{S}_k$ decreases at the rate $(N - k)\theta_k P_k(t)$ as each of $N - k$ active searchers may detect a target and turn the system into a state $\mathcal{S}_{k+1}$. On the other hand, probability $P_k(t)$ increases at the rate $k\theta_{k-1} P_{k-1}(t)$ as there are exactly $\binom{k}{1} = k$ states $\mathcal{S}_{k-1}$ that may lead to a (given) state $\mathcal{S}_k$.[1] Initial conditions for equations (6) reflect the fact that at $t = 0$ the system is in the state $\mathcal{S}_0$ with probability 1:

$$P_k(0) = \delta_{k0}, \quad k = 0, \ldots, N, \tag{7}$$

where δ_{ij} is the Kroneker delta. Using (6) and (7) it is straightforward to verify that the probabilities $P_k(t)$ satisfy the identity

$$\sum_{k=0}^{N} \binom{N}{k} P_k(t) = 1, \quad t \geq 0. \tag{8}$$

Equations of type (6) are well known in the area of queueing systems, where they are usually solved via introduction of the generating function

$$G(z, t) = \sum_{n} P_n(t) z^n,$$

whereby the probabilities $P_n(t)$ can be reconstructed by differentiating $G(z,t)$ with respect to z. In the context of cooperative systems, we are primarily interested in the probability $P_N(t)$ of mission completion by time t, and therefore resort to direct application of the Laplace transform [6] to equations (6). Assuming without loss of generality that the eigenvalues of the matrix of system (6) are all different:

$$(N - i)\theta_i \neq (N - j)\theta_j, \quad 0 \leq i < j \leq N - 1, \tag{9}$$

a closed-form solution of (6) is obtained in the form

$$P_k(t) = \sum_{i=0}^{k} \frac{k!\, \theta_0 \cdots \theta_{k-1}}{\prod\limits_{\substack{j=0 \\ j \neq i}}^{k} \left[(N - j)\theta_j - (N - i)\theta_i \right]} e^{-(N-i)\theta_i t}, \quad k = 0, \ldots, N. \tag{10}$$

---

[1] Indeed, let $A_k = \{i_1, \ldots, i_k\}$ be any set containing k inactive searchers. Trivially, A_k can be represented in k different ways as $A_k = A_{k-1}^j \cup \{i_j\}$, where A_{k-1}^j is a subset of $k - 1$ inactive searchers: $A_{k-1}^j = A_k \backslash \{i_j\} \subset A_k$, $j = 1, \ldots, k$. Thus, a state $\mathcal{S}_k$ with k inactive searchers can only be obtained from exactly k states $\mathcal{S}_{k-1}$ with $k - 1$ inactive searchers.

2.3   A Coordinated Search Model

In this subsection we consider a search system where N autonomous search agents are assisted by a human operator. It is assumed that a human operator conducts a search for targets independently of the autonomous searchers, and, upon discovery of appropriate objects of interest, immediately cues one of the searchers. In effect, the operator has two states: *search* and *detect & cue* (see Figure 2), where the transition *search* → *detect & cue* occurs at rate λ, which is the operator's search rate. After sending a cue to one of the autonomous searchers, the operator immediately resumes search, which is denoted by the infinite transition rate from the state *detect & cue* to the state *search*.

The states and transition rates for the autonomous searchers are shown in the right portion of Figure 2. Each searcher has three possible states, *search uncued*, *search cued*, and *detect*, where the first two states refer to searching before and after receiving a cue from the human operator, respectively. The transition rate from *search uncued* to *search cued* depends on the detection rate λ of the operator. It is assumed that the operator will cue only one searcher at a time, and that the cues are equally distributed to the uncued searches. This amounts to the transition rate from *search uncued* to *search cued* being equal to λ/i, where i is the number of "uncued" autonomous searchers, i.e., the searchers in the state *search uncued*. After a searcher receives a cue, its detection rate changes from the initial "uncued" rate θ to the "cued" rate θ_c. The autonomous searchers also have the ability to search independently, thus they can make a direct transition from the *search uncued* to the *detect* state at rate θ. It is natural to assume that $\theta_c \geq \theta$.

Similarly to the exposition of Section 2.2, let $\mathcal{S}_{ijk}$ be a state in which there are i searchers in the state *search uncued*, j searchers in the state *search cued*, and k searchers in the state *detect*, $i+j+k = N$. It is easy to see that for a given triple (i, j, k) there are $\left(\begin{smallmatrix} & N & \\ i & j & k \end{smallmatrix} \right) = \frac{N!}{i!\,j!\,k!}$ different states $\mathcal{S}_{ijk}$. Further, it is important to note that there are $\binom{N+3-1}{N} = \frac{(N+2)(N+1)}{2}$ different triplets (i, j, k) such that $i+j+k = N$. By defining the probability of the search system occupying a state $\mathcal{S}_{ijk}$ at time t as $P_{ijk}(t)$, one can describe the corresponding Markov model with a finite number of states via the following system of Chapman-Kolmogorov equations:

$$
\frac{\mathrm{d}}{\mathrm{dt}} P_{ijk}(t) = - \bar{\delta}_{kN} \Big[i\theta + j\theta_c + \bar{\delta}_{i0}\lambda \Big] P_{ijk}(t)
$$
$$
+ \bar{\delta}_{iN}\bar{\delta}_{j0} \left[\frac{j\lambda}{i+1} \right] P_{i+1,\,j-1,\,k}(t) + \bar{\delta}_{iN}\bar{\delta}_{k0} \big[k\theta \big] P_{i+1,\,j,\,k-1}(t) \qquad (11)
$$
$$
+ \bar{\delta}_{jN}\bar{\delta}_{k0} \big[k\theta_c \big] P_{i,\,j+1,\,k-1}(t), \quad i+j+k = N,
$$

where factors $\bar{\delta}_{ij}$ represent the negation of the Kroneker delta, $\bar{\delta}_{ij} = 1 - \delta_{ij}$, and have the obvious function of handling the extreme cases of i, j, or k being equal to 0 or N. Let us present the interpretation of equations (11). In the most general case, a state $\mathcal{S}_{ijk}$ with i uncued searchers, j cued searchers, and k inactive searchers can be obtained

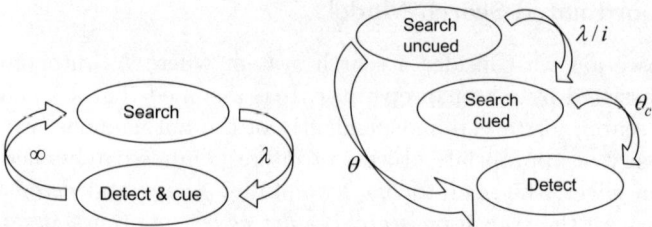

Fig. 2. State diagrams for the human operator (left) and autonomous searcher (right) in the coordinated search system

- from a state $S_{i+1,j-1,k}$ due to a transition *search uncued → search cued*, i.e. when one of the $i+1$ uncued searchers receives a cueing signal from the operator. Since each of the $i+1$ uncued searchers is being cued at rate $\frac{\lambda}{i+1}$, and there are $j = \binom{j}{1}$ states $S_{i+1,j-1,k}$ that can result in a given state S_{ijk}, transitions *search uncued → search cued* increase the probability $P_{ijk}(t)$ at the rate $\frac{j\lambda}{i+1}P_{i+1,j-1,k}(t)$. This amounts to the second term in equation (11).
- from a state $S_{i+1,j,k-1}$ due to a transition *search uncued → detect*, i.e., when one of the $i+1$ uncued searchers detects a target without cue from the operator. The search rate of an uncued agent is θ, and there are $k = \binom{k}{1}$ different states $S_{i+1,j,k-1}$ that can lead to a given state S_{ijk}. Thus, due to transitions *search uncued → Detect* the probability $P_{ijk}(t)$ increases at the rate $k\theta P_{i+1,j,k-1}(t)$, which amounts to the third term in (11).
- from a state $S_{i,j+1,k-1}$ due to a transition *search cued → detect*, when one of the $j+1$ cued searchers detects a target. The search rate of a cued agent is θ_c, and there are $k = \binom{k}{1}$ different actual states $S_{i,j+1,k-1}$ that can lead to a given state S_{ijk}. Thus, due to transitions *search cued → Detect* the probability $P_{ijk}(t)$ increases at the rate $k\theta_c P_{i,j+1,k-1}(t)$, which amounts to the fourth term in (11).
- finally, the first term in the right-hand side of (11) accounts for the possibility of transition from the given state S_{ijk} to states $S_{i-1,j,k+1}$, $S_{i,j-1,k+1}$, and $S_{i-1,j+1,k}$ correspondingly.

At $t = 0$ the system is in the state S_{N00}, thus the initial conditions for equations (11) are

$$P_{ijk} = \delta_{iN}, \quad i + j + K = N, \tag{12}$$

and, similarly to (8), solutions of (11) satisfy the identity

$$\sum_{\substack{j+j+k=N \\ i,\,j,\,k\,\geq\,0}} \binom{N}{i \quad j \quad k} P_{ijk}(t) = 1, \quad t \geq 0. \tag{13}$$

Application of the Laplace transform method to equations (11) leads to unwieldy expressions for the state probabilities $P_{ijk}(t)$. Thus, we employ an alternative procedure to derive the solution of equations (11) by representing them in matrix form, with a lower-triangular matrix. A lower-triangular form of equations (11) is obtained by introducing the notation $P_{ijk}(t) = \tilde{P}_\ell(t)$, where the index ℓ runs from 0 to $L = \frac{(N+2)(N+1)}{2} - 1 = \frac{N(N+3)}{2}$ and is determined by the indices i, j, and k as

$$\ell = \sum_{r=0}^{j+k-1} (r+1) + k = \frac{(j+k)(j+k+1)}{2} + k \quad \text{for all } 0 \leq j+k \leq N, \quad (14)$$

with the inverse relations given by

$$\begin{pmatrix} i \\ j \\ k \end{pmatrix} = \begin{pmatrix} N - \varpi \\ \frac{1}{2}\varpi(\varpi+3) - \ell \\ \ell - \frac{1}{2}\varpi(\varpi+1) \end{pmatrix}, \quad \text{where} \quad \varpi = \left\lfloor \frac{-1 + \sqrt{1+8\ell}}{2} \right\rfloor. \quad (15)$$

The above relation between $P_{ijk}(t)$ and $\tilde{P}_\ell(t)$ can be explicitly enumerated as

$$\begin{aligned}
\tilde{P}_0(t) &= P_{N00}(t), & \tilde{P}_5(t) &= P_{N-2,0,2}(t), \\
\tilde{P}_1(t) &= P_{N-1,1,0}(t), & &\cdots \\
\tilde{P}_2(t) &= P_{N-1,0,1}(t), & &\cdots \\
\tilde{P}_3(t) &= P_{N-2,2,0}(t), & \tilde{P}_{L-1}(t) &= P_{0,1,N-1}(t), \\
\tilde{P}_4(t) &= P_{N-2,1,1}(t), & \tilde{P}_L(t) &= P_{0,0,N}(t).
\end{aligned}$$

It is easy to see that such a correspondence between $P_{ijk}(t)$ and $\tilde{P}_\ell(t)$ allows one to represent equations (11) in the matrix form

$$\frac{\mathrm{d}}{\mathrm{d}t}\tilde{\mathbf{P}} = \mathbf{M}\tilde{\mathbf{P}}, \quad (16)$$

where the matrix $\mathbf{M} \in \mathbb{R}^{(L+1)\times(L+1)}$ is lower-triangular. Similarly to (7), the initial conditions for the above system are formulated as

$$\tilde{P}_\ell(0) = \delta_{\ell 0}, \quad 0 \leq \ell \leq L. \quad (17)$$

Then, the solution to the Cauchy problem (16)–(17) has the form

$$\tilde{P}_\ell(t) = \sum_{i=0}^{\ell} a_{i\ell} e^{m_{ii}t}, \quad \ell = 0, \ldots, L, \quad (18)$$

where the coefficients $a_{i\ell}$ are determined in a recursive manner

$$a_{i\ell} = \sum_{j=i}^{\ell-1} \frac{m_{\ell j} a_{ij}}{m_{ii} - m_{\ell\ell}}, \quad i < \ell, \quad (19a)$$

$$a_{ii} = -\sum_{j=0}^{i-1} a_{ji}, \quad a_{00} = \tilde{P}_0(0) = 1. \quad (19b)$$

Throughout this section it is assumed that the eigenvalues of the matrix $\mathbf{M}$ are all different:

$$i\theta_u + j\theta_c + \bar{\delta}_{i0}\lambda \neq i'\theta_u + j'\theta_c + \bar{\delta}_{i'0}\lambda \quad \text{for all } 0 \leq i+j \leq N \text{ and } 0 \leq i'+j' \leq N.$$

The exact solution (18)–(19) of the governing equations (11) allows for a detailed analysis of the cooperative search system. For example, one can easily deduce the expected time $\overline{T}_N$ needed for all N searchers to engage targets, $\overline{T}_N = \sum_{i=0}^{L-1} a_{iN}/m_{ii}$, and so on.

2.4   Coordinated Cooperative Search Model

Here we consider a generalization of the search systems presented in Sections 2.2 and 2.3, with the autonomous searchers being capable of receiving cues from an operator as well as sending cues to each other upon target detection. Similarly to the coordinated search model of Section 2.3, the autonomous searchers have three states, *search uncued*, *search cued*, and *detect* (see Figure 3). The states *search uncued* and *search cued* refer to the autonomous searcher conducting search before and after receiving a cue from the human operator. Let i, j, and k denote the number of autonomous searchers in states *search uncued*, *search cued*, and *detect*, respectively. Then, the detection rate of each of i searchers in state *search uncued* is equal to θ_k, and detection rate of the j searchers in state *search cued* equals θ_c (Figure 3). In other words, the autonomous searchers in state *search uncued* cooperate via cueing, as described in Section 2.2; in addition, a searcher in state *search uncued* may receive a cue from the operator, which causes the searcher's transition into state *search cued* and changes its detection rate from θ_k to θ_c. It is assumed that the informational content of operator's cue is such that θ_c is significantly greater than θ_k for any $k = 0, \ldots, N-1$. However, this improvement in the search rate is local in that it does not affect the rates of other searchers: the detection rate of searchers in state *search uncued* changes from θ_k to θ_{k+1} only when one of the searchers detects a target and transitions to state *detect*. The detection rate θ_c of searchers in state *search cued*, however, does not change. As in the coordinated search model (Section 2.3), the operator distributes cues at rate λ, thus transition rate from state *search uncued* to state *search cued* is λ/i.

Such a dynamics of detection rates implies that the information provided by the operator to autonomous searchers is indeed valuable, but may be not completely accurate: the searcher is able to detect a target much faster, but the system is able to benefit from this (through an improvement of the detection rates of the remaining searchers) only after an actual detection occurs.

Defining the states $\mathcal{S}_{ijk}$ with the corresponding probabilities P_{ijk} in the same way as above, the Chapman-Kolmogorov equations for the described coordinated cooperative search system can be written in the form

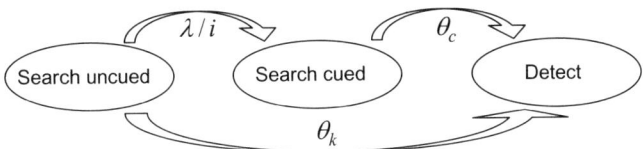

Fig. 3. State diagram for autonomous searchers in the coordinated cooperative search system

$$\frac{\mathrm{d}}{\mathrm{d}t} P_{ijk}(t) = -i\theta_k P_{ijk}(t) + k\theta_{k-1} P_{i+1,j,k-1}(t)$$

$$- \bar{\delta}_{kN} \left[j\theta_c + \bar{\delta}_{i0}\lambda \right] P_{ijk}(t) + \bar{\delta}_{iN}\bar{\delta}_{j0} \left[\frac{j\lambda}{i+1} \right] P_{i+1,\,j-1,\,k}(t) \qquad (20)$$

$$+ \bar{\delta}_{jN}\bar{\delta}_{k0} \left[k\theta_c \right] P_{i,\,j+1,\,k-1}(t), \quad i+j+k = N.$$

Observe that the first two terms in the right-hand side of (20) represent the cooperative part of the system (compare to (5)). The initial conditions are given by (12), and the solutions $P_{ijk}(t)$ do also satisfy the identity (13). An analytical solution of equations (20) can be obtained by (14)–(19).

2.5   Monitored Cooperative Search

Finally, we consider a search system where after detecting a target, the autonomous searchers *engage* (e.g., attack) it, and the decision on engagement of the detected target is made by a human operator. Namely, consider a search system of N autonomous searchers each having three possible states: *search, detect*, and *engage*. Upon detecting a target, an autonomous searcher remains in the state *detect* for as long as it takes for the human operator to decide whether to engage the target or not. If the operator authorizes engagement of the target, the searcher transitions to the state *engage*; otherwise, it resumes search and returns to the state *search*. The corresponding transition rates from state *detect* to states *engage* and *search*, σ_{ijk} and $\bar{\sigma}_{ijk}$, respectively, can be intuitively defined as *decision rates* of the human operator. In general, these decision rates, as well as the search rate θ_{ijk}, can depend on the number i, j, k of searchers currently occupying the states *search, detect*, and *engage*, correspondingly.

As before, we define $\mathcal{S}_{ijk}$ to be a state of the system with i searchers in state *search*, j searchers in state *detect*, and k searchers in state *engage*; the corresponding probability that system is in a state $\mathcal{S}_{ijk}$ at time t is $P_{ijk}(t)$. The differential equations governing the probabilities $P_{ijk}(t)$ read as

$$\frac{\mathrm{d}}{\mathrm{d}t} P_{ijk}(t) = -\bar{\delta}_{kN} \left[i\theta_{ijk} + j(\sigma_{ijk} + \bar{\sigma}_{ijk}) \right] P_{ijk}(t)$$

$$+ \bar{\delta}_{iN}\bar{\delta}_{j0} \left[j\theta_{i+1,j-1,k} \right] P_{i+1,j-1,k}(t)$$

$$+ \bar{\delta}_{i0}\bar{\delta}_{jN} \left[i\bar{\sigma}_{i-1,j+1,k} \right] P_{i-1,j+1,k}(t)$$

$$+ \bar{\delta}_{jN}\bar{\delta}_{k0} \left[k\sigma_{i,j+1,k-1} \right] P_{i,j+1,k-1}(t), \qquad (21)$$

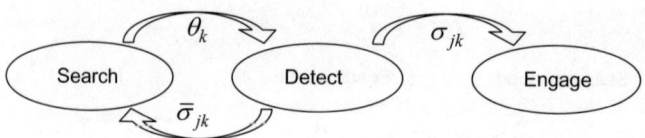

Fig. 4. State diagram for autonomous searchers in the monitored cooperative search system

where $i + j + k = N$. By analogy to the previously considered systems, equations (21) are equipped with initial conditions (12), and their solutions satisfy (13).

3   Numerical Results

In this section we compare the performance of the models developed in Section 2 based on the measures of effectiveness such as probability of detection $P_D(t)$, and time to engage $T_{0.95}$. We assume that the initial, "uncued" detection rate of each searcher is $\theta = 1$, and present numerical results for search systems with $N = 10$ autonomous searchers. The solutions of the Chapman-Kolmogorov equations for the models presented in Sections 2.2–2.5 are *state* probabilities, and will be used in computing the *detection probabilities* $\hat{P}_n(t)$

$$\hat{P}_n(t) = \mathsf{P}\{\text{exactly } n \text{ targets are detected by time } t\}. \tag{22}$$

For example, the introduced MOE *probability of detection* $P_D(t) = \hat{P}_N(t)$.

3.1   Basic Cooperative Search System

In [1] it was shown that in the case when cueing (cooperation) has no effect on detection capabilities of the searchers, $\theta_0 = \ldots = \theta_N = \theta$, the probability of detection $P_D(t)$ is equal to the probability that all N searchers detect targets independently (see also (2))

$$P_D(t) = \hat{P}_N(t) = \left(1 - e^{-\theta t}\right)^N.$$

In general, it is easy to see that the detection probabilities $\hat{P}_n(t)$ of for independent search system are given by

$$\hat{P}_n(t) = \binom{N}{n}\left(1 - e^{-\theta t}\right)^n \left(e^{-\theta t}\right)^{N-n}, \quad n = 0, \ldots, N.$$

The graphs of detection probabilities $\hat{P}_n(t)$, $n = 0, \ldots, N$ for a search system of $N = 10$ independent autonomous searchers are shown in Figure 5(a).[2]

---

[2] The horizontal axis in all graphs of Figure 5 corresponds time t, and the vertical axis measures the probability.

To compare the basic cooperative model (6) with the independent search case, we assumed the detection rates improve due to cueing in the following fashion (see Figure 6(a)):[3]

$$\theta_k = 1 + (\varkappa - 1)\tfrac{2}{\pi} \arctan \tfrac{k}{2}, \quad k = 0, \ldots, N - 1, \quad \text{where} \quad \varkappa = 1.5. \tag{23}$$

Given the solutions $P_k(t)$ of the Chapman-Kolmogorov system (6), the detection probabilities $\hat{P}_n(t)$ can be obtained as

$$\hat{P}_n(t) = \binom{N}{n} P_n(t), \quad n = 0, \ldots, N, \tag{24}$$

(note that $\hat{P}_N(t) = P_N(t)$). The detection probabilities $\hat{P}_n(t)$ of the basic cooperative system with $N = 10$ searchers are shown in Figure 5(b). In particular, the time to engage $T_{0.95} = P_N^{-1}(0.95)$ for the basic cooperative search system equals to

$$T_{0.95} = 3.797,$$

a 28% improvement over the corresponding value for a system of 10 independent searchers (5.275).

3.2   Coordinated Search System

For the coordinated search system described in Section 2.3, we assume that the search rate of the human operator is equal to the uncued search rate of autonomous searchers, and receipt of operator's cue boosts the search rate of the autonomous searcher fivefold:

$$\lambda = 1, \quad \theta_c = 5. \tag{25}$$

The detection probabilities $\hat{P}_k(t)$ are computable via the state probabilities $P_{ijk}(t)$, which solve the Chapman-Kolmogorov equations (11), as

$$\hat{P}_n(t) = \sum_{\substack{k=n \\ i+j+k=N \\ i,j \geq 0}} \binom{N}{i \ j \ k} P_{ijk}(t), \quad n = 0, \ldots, N. \tag{26}$$

The detection probabilities $\hat{P}_n(t)$ for the coordinated search system with parameters (25) are shown in Figure 5(c). The time to engage for the the coordinated search system equals to

$$T_{0.95} = 3.539,$$

which is a 7% improvement over the cooperative model.

---

[3] The vertical axis in Figure 6 corresponds to the search or decision rate that is common to a specific number of agents (horizontal axis).

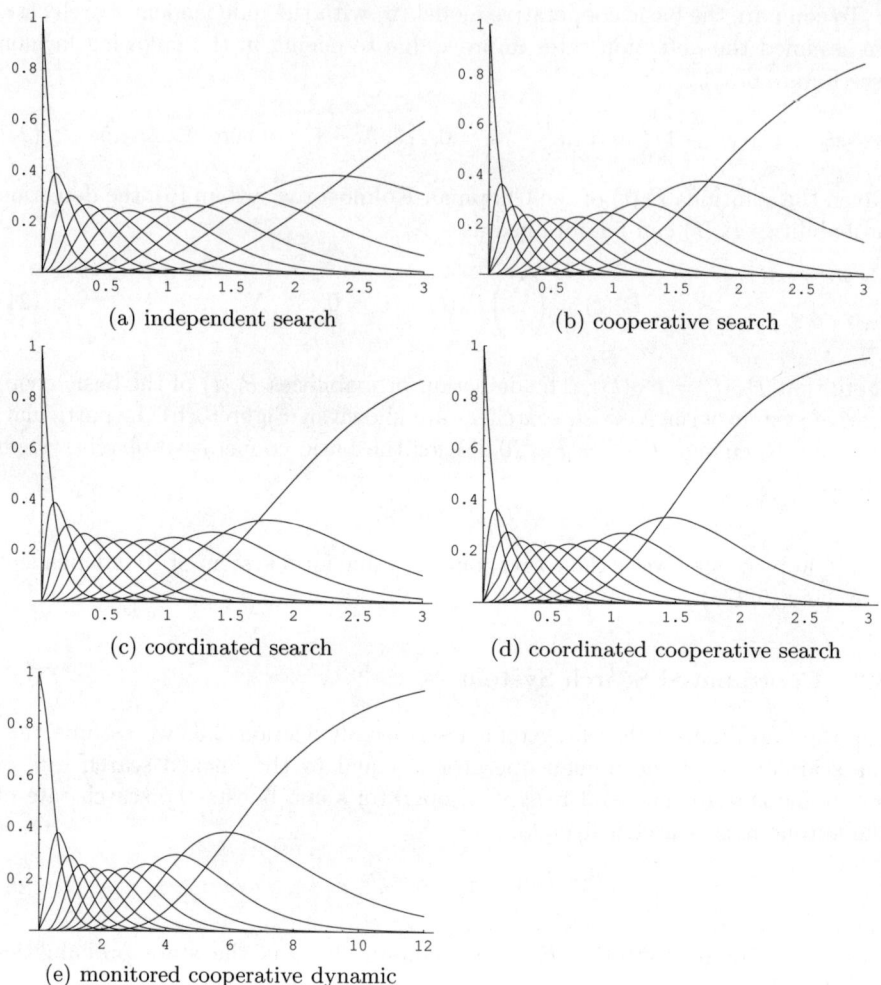

Fig. 5. Detection probabilities in search systems with $N = 10$ autonomous searchers

3.3 Coordinated Cooperative Search

In the coordinated cooperative search model, we assumed that the detection rates improve due to cooperation in accordance to (23); the values for operator's search rate λ and operator-cued rate θ_c are given by (25). The detection probabilities $\hat{P}_n(t)$ are calculated using the solutions $P_{ijk}(t)$ of Chapman-Kolmogorov equations (20) in the same ways as in (26), and are shown in Fig. 5(d). The time to engage in coordinated cooperative system has the value of

$$T_{0.95} = 2.857,$$

a 19% improvement over the coordinated search model.

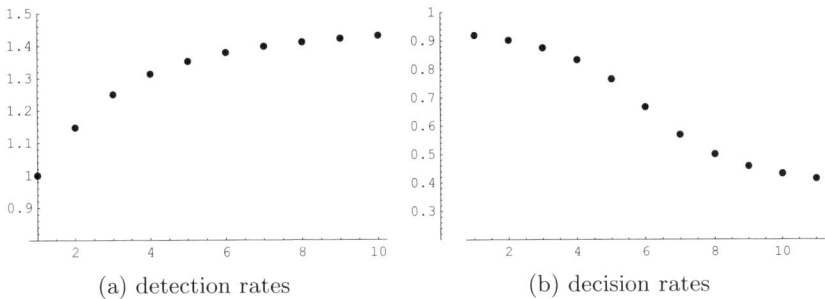

(a) detection rates   (b) decision rates

Fig. 6. Dynamics of detection rates and decision rates in search systems with cooperation

3.4   Monitored Cooperative Search

The equations for the monitored cooperative search model (21) were written in the most general form, with the detection rate θ_{ijk} and the decision rates σ_{ijk}, $\bar{\sigma}_{ijk}$ being dependent on the numbers i, j, and k of autonomous searchers in the states *search*, *detect*, and *engage*, respectively.

To maintain consistency with the previous models, we assume that detection rates are governed by (23). With respect to the decision rates σ and $\bar{\sigma}$, we assume that they depend on j and k as

$$\sigma_{ik} = \bar{\sigma}_{jk} = \varrho(j)\varrho(k), \quad \text{where} \quad \varrho(k) = \frac{2}{3}\left(1 - \frac{1}{\pi}\arctan\frac{k - 2.5N}{N}\right). \quad (27)$$

The graph of function $\varrho(k)$ is presented in Figure 6(b). Dependence (27) of σ_{jk} and $\bar{\sigma}_{jk}$ on the number j of autonomous searchers reflects possible deterioration of decision-making performance of human operators due to stress or anxiety when a large number of searchers made detections and are waiting for permission to engage. Similarly, dependence of the decision rates on the number k of searchers that have engaged targets emulates the effects of fatigue, since k generally increases with the elapsed time.

Similarly to the last two considered systems, the detection probabilities $\hat{P}_n(t)$ of the monitored cooperative search system are computed in the form (26), where $P_{ijk}(t)$ must be taken as the solutions of (21) with parameters $\theta_{ijk} = \theta_k$ given by (23) and $\sigma_{ijk} = \sigma_{jk}$, $\bar{\sigma}_{ijk} = \bar{\sigma}_{jk}$ given by (27). The graphs of detection probabilities $\hat{P}_n(t)$ are shown in Figure 5(e), and the corresponding time to engage is

$$T_{0.95} = 12.586.$$

Clearly, such an increase in $T_{0.95}$ in comparison to the previously considered models can be explained due to additional delays associated with human decision making in state *detect*, as well as due to the possibility of cycling between states *search* and *detect*.

4   Analysis of Cooperation

The introduced measures of effectiveness $P_D(t)$ and $T_{0.95} = P_D^{-1}(0.95)$ allow for quantifying how fast all N searchers accomplish the mission. However, they do not shed light on how the cooperative system evolves, or progresses, *during* the mission. In this section, we attempt to answer the following question: given a cooperative[4] system, how can one estimate its "degree of cooperativeness," or, in other words, how can one distinguish between "more cooperative" and "less cooperative" systems?

4.1   Corresponding Independent Systems

Our approach to measuring the degree of cooperativeness in a given system is based on comparing it with a *corresponding independent* system. A corresponding independent system may be defined as one in which the individual agents accomplish their objectives independently, and their state space and transition rates correspond to those of the agents in the cooperative system. It is important to note that the state space of the agents in the independent system may or may not be identical to that of the cooperative agents.

For example, the agents of the independent search system that corresponds to the basic cooperative system of Section 2.2 are described by the state transition diagram in Figure 1. In this case, the state spaces of cooperative and independent searchers are identical and consist of the states *search* and *detect*; the difference is that the transition rate from *search* to *detect* for cooperative searchers is equal to θ_k, where k is the number of searchers that have completed the mission, and for the independent searchers this transition rate equals to the initial search rate θ.

In the case of the models discussed in Sections 2.3–2.4, the state space of the cooperative searchers includes three states: *search uncued*, *search cued*, and *detect*. Nevertheless, it is easy to see that the state space of the searchers of the corresponding independent system is again described by the diagram in Figure 1. Indeed, the *search cued* state of cooperative searchers embodies cooperation in the system (namely, cooperation between the operator and the searchers), and therefore does not need to be present in the state space of independent searchers.

With respect to the monitored cooperative search model presented in Section 2.5, the choice of the corresponding equivalent system is not as straightforward and depends on the objective of one's analysis. For instance, the state space of independent searchers may consist of the three states *search*, *detect*, and *engage*, with the transition rate between the first two being equal to θ, if the goal is to study the effects of search rate improvement due to cooperation when the searchers have to obtain permission from an operator before engaging a detected target.

In the present work, we are primarily interested in the effect of how an operator-in-the-loop influences the performance of the cooperative system.

---

[4] The term *cooperative* is understood here in a broad sense, not restricted to the context of Section 2.2.

Therefore, we consider independent searchers with the state transition diagram as presented in Figure 1.

4.2 Coefficient of Cooperation and Fitness Function

To quantify the degree of cooperativeness in a multi-agent search system, we introduce the *coefficient of cooperation*, α, as follows. Given a system of N cooperative searchers whose initial detection rate is θ, the coefficient of cooperation α determines the search rate $\alpha\theta$ of independent searchers in a corresponding equivalent system, whose performance would match (as close as possible) the performance of the cooperative system. In other words, $\alpha\theta$ is the rate that independent searchers would need to match the performance of a cooperative (in a broad sense) system. Now, the coefficient of cooperation can be determined by optimizing the fitness function that defines how close one search system is to another.

The problem of quantifying the similarity between two search systems in the present context effectively reduces to measuring the distance between two nonstationary distributions. In general, there is no universally accepted method of quantifying the dissimilarities between probability distributions. Among the most popular ones, we may mention the Kullback-Leibler divergence, or relative entropy [4], Chi-square distance, Kolmogorov-Smirnov distance, etc.

Here we employ a different way of measuring the closeness of two distributions, which is essentially tailored to the proposed approach of comparing the probability distribution $P_k(t)$ of a cooperative system to that of an appropriately chosen independent system. It will be seen below that the probability distribution of the independent search system described above is a mixture of binomial and exponential distributions.

Let $f_\Theta(t)$, where Θ may stand for a vector of parameters, be the probability that a searcher in an independent system detects a target by time t; in our case, $f_\Theta(t) = 1 - e^{-\theta t}$ (see the above discussion). Then, the probability $P_k^*(t)$ that k out of N independent searchers have detected targets by time t is given by the binomial distribution formula

$$P_k^*(t) = \binom{N}{k} \left[f_\Theta(t) \right]^k \left[1 - f_\Theta(t) \right]^{N-k}. \tag{28}$$

Observe that due to the independence assumption, the form of probabilities $P_k^*(t)$ depends solely on $f_\Theta(t)$. Given the form of the detection function $f_\Theta(t)$, the set of time-dependent binomial probabilities $P_k^*(t)$ can be identified by the pairs $\left(t_k^*, P_k^*(t_k^*) \right)$, $k = 1, \ldots, N - 1$, where

$$t_k^* = \arg\max_{t \geq 0} P_k^*(t) = f_\Theta^{-1}\left(\tfrac{k}{N} \right), \quad 1 \leq k \leq N - 1, \tag{29}$$

and

$$P_k^*(t_k^*) = \max_{t \geq 0} P_k^*(t) = \binom{N}{k} \left(\frac{k}{N} \right)^k \left(1 - \frac{k}{N} \right)^{N-k}, \quad 1 \leq k \leq N - 1. \tag{30}$$

It is important to note that $P_k^*(t_k^*)$ are invariant with respect to the form of $f_\Theta(t)$. Similar pairs $(\hat{t}_k, \hat{P}_k(\hat{t}_k))$, $k = 1, \ldots, N-1$, can be computed for the detection probabilities $\hat{P}_k(t)$ (22) of a cooperative system, where

$$\hat{t}_k = \arg\max_{t \geq 0} \hat{P}_k(t). \tag{31}$$

The coefficient of cooperation α is then determined by matching the pairs $(\hat{t}_k, P_k(\hat{t}_k))$ and $(t_k^*, P_k^*(t_k^*))$ for $f_\Theta(t) = 1 - e^{-\alpha\theta t}$ using the least-squares method:

$$\Phi_\alpha = \min_\alpha \sum_{k=1}^{N-1} \left(\hat{P}_k(\hat{t}_k) - \binom{N}{k} \left(1 - e^{-\alpha\theta\hat{t}_k}\right)^k \left(e^{-\alpha\theta\hat{t}_k}\right)^{N-k} \right)^2. \tag{32}$$

Note that the above expression does not take into account the probabilities of no detections, and the detection probability $P_D(t) = \hat{P}_N(t)$.

In the next subsection we report computational results on determining the coefficient of cooperation for the cooperative search models presented in sections 2.2–2.5.

4.3   Numerical Results and Discussion

In the numerical experiments we have used the Markov chain models developed in Section 2 for cooperative search systems with $N = 10$ searchers and detection and decision rates given by (23)–(27). As explained above, the benchmark search system of independent searchers was chosen as the one where each searcher has the probability of detection $f(t) = 1 - e^{-\theta t}$, so that the probabilities $P_k^*(t)$ have the form

$$P_k^*(t) = \binom{N}{k} \left(1 - e^{-\theta t}\right)^k \left(e^{-\theta t}\right)^{N-k}, \tag{33}$$

where $\theta = 1$.

The obtained values of the coefficient of cooperation α and the fitness value Φ_α (see (32)) between the detection probabilities $\hat{P}_n(t)$ of the cooperative systems and the benchmark probabilities (33) are shown in Table 1.

Table 1. Coefficient of cooperation α and the corresponding fitness value Φ_α

Type of system	Coefficient of cooperation α	Fitness value Φ_α
Basic cooperative	1.493	4.4×10^{-5}
Coordinated	0.998	5.9×10^{-6}
Coordinated cooperative	1.622	3.4×10^{-3}
Monitored cooperative	0.316	8.8×10^{-3}

The results for the basic cooperative system indicate that the searchers in the corresponding equivalent independent search system must possess search rates

1.493 times higher than the initial search rate θ of the cooperative searchers in order to match their performance throughout the mission. As expected, the coordinated cooperative system is more efficient, which is reflected by the higher coefficient of cooperation $\alpha = 1.622$.

A more interesting result is that for the coordinated search system: the corresponding value of α is 0.998, which means that the independent searchers do not need an increased detection rate in order to match the performance of the coordinated search system with parameters (25). In other words, the coordinated search system evolves during the mission essentially the same as an independent search system, except for the last stage of the mission (recall that the time to engage $T_{0.95}$ for the coordinated system is 33% less than that of the independent system)!

To explain this result, recall that unlike the other considered systems, the coordinated search system does not allow for a "global" rate improvement among the searchers; the rate improvements $\theta \to \theta_c$ of the searchers that received a cue from the operator are localized in time and bear no effect on the detection rates of other searchers. To put it differently, the searchers in the coordinated system perform their search independently, and occasionally may receive "gift" cues from a human operator. These cues do have an accumulated effect, evidenced by the much shorter time to engage $T_{0.95}$, but during the mission (locally) the coordinated system behaves as an independent one.

Also, it is of interest to note that fitness value Φ_α for the coordinated system is the lowest among all four considered systems, i.e., the coordinated system can be fitted by an independent systems most closely, which reinforces the above conclusions.

The fact that the coefficient of cooperation for the coordinated mission is lower that 1 can be attributed to properties of the employed fitness function that does not take into account probabilities $\hat{P}_0(t)$ and $\hat{P}_N(t)$.

As regards the monitored cooperative system, the value of the coefficient of cooperation $\alpha = 0.316$ can be explained by the additional delays due to decision-making of human operator, and by the possibilities of returning to the *search* state after making a detection. However, due to a high fitness value of 0.0088, it may be concluded the additional delays and cycling caused by the decisions of human operator do not eliminate the cooperative nature of the system - its evolution cannot be fitted closely to that of an independent system (in comparison, for example, with the coordinated system).

5   Conclusions

We presented an approach to modeling and analysis of distributed multi-agent search systems where the autonomous agents may cooperate among each other, and/or with human operator, in order to achieve the system's objective. It was assumed that the cooperation is facilitated by means of information sharing among the autonomous agents and/or human operator, with the purpose of improving the effectiveness of the autonomous agents. The evolution of cooperative systems was modeled using discrete-state, continuous-time Markov chains. We also presented a technique for measuring and quantification of cooperation within such systems.

References

1. Jeffcoat, D., Krokhmal, P., and O. Zhupanska (2006) Effects of cueing in cooperative search, *Naval Research Logistics* **53**(8), 814–821.
2. Jeffcoat, D., Krokhmal, P., and O. Zhupanska (2007) A Markov Chain Approach to Analysis of Cooperation in Multi-Agent Search Missions, *In:* Grundel et al (Eds), *Cooperative Systems: Control and Optimization*, Springer Lecture Notes in Economics and Mathematical Sciences, **588**, 171–184.
3. L. Kleinrock, *Queueing Systems, Volume I: Theory.* John Wiley & Sons (1975).
4. Kullback, S., and Leibler, R. A., (1951) On information and sufficiency, *Annals of Mathematical Statistics* 22, 79-86.
5. A. Papoulis and S. U. Pillai, *Probability, Random Variables, and Stochastic Processes.* 4th Ed. McGraw Hill (2002).
6. H. M. Srivastava, B. R. Kashyap (1982) *Special Functions in Queuing Theory and Related Stochastic Processes.* Academic Press, New York.

Decentralized Extremum-Seeking Control of Nonholonomic Vehicles to Form a Communication Chain

Cory Dixon and Eric W. Frew

Research and Engineering Center for Unmanned Vehicles
University of Colorado, 429 UCB, Boulder, CO 80303, USA

Abstract. Electronic chaining is the formation of a linked communication chain that maximizes the end-to-end throughput using a cooperative team of mobile vehicles in an unknown, dynamic environment. A decentralized extremum seeking (ES) controller, which is an adaptive model free controller, is presented to control the locations of the mobile communication relays based on received signal strength, as opposed to relative position. A communication chain that seeks to maximize a link performance metric such as bandwidth or end-to-end delay does not necessarily form a linear, evenly spaced formation of the nodes when jamming and environmental terrain factors are considered. This chapter presents an ES controller that has been specifically designed for use with teams of unmanned aircraft using a Lyapunov guidance vector field controller and in particular focuses on the performance and stability of the ES algorithm due to the operational limits and constraints of an unmanned aircraft.

1 Introduction

Cooperative electronic chaining is the formation of a linked communication chain using a team of mobile vehicles, acting as communication relays in an ad hoc network, to maximize the end-to-end throughput of the chain while allowing the end nodes of the chain to move independently in an unknown, dynamic environment. Electronic chaining utilizes the fact that with networked mobile vehicles, the quality of a wireless communication link is directly influenced by the motion and location of the vehicles within the radio propagation environment. Thus, controlling the motion of the vehicles using a measure of communication performance can directly influence the performance of the network chain. This chapter presents a decentralized extremum seeking (ES) controller that has been designed for use on nonholonomic vehicles within a cooperating team to maintain an electronic communication chain using the signal-to-noise ratio (SNR) of the communication links.

While there are numerous examples of systems that control the relative position of networked vehicles to maintain communication links (e.g. see [1,2]), position based solutions cannot account for localized noise sources or an unknown radio frequency (RF) environment where there is an unknown noise floor. Thus

M.J. Hirsch et al. (Eds.): Adv. in Cooper. Ctrl. & Optimization, LNCIS 369, pp. 311–322, 2007.
springerlink.com © Springer-Verlag Berlin Heidelberg 2007

an adaptive, model free controller using the SNR of the individual links is required to find a true optimal (in the sense of end-to-end throughput) location for the relay nodes. In previous works by the authors [3,4] extremum-seeking methods using a Lyapunov guidance vector field controller (LGVF) have been adapted for the electronic chaining problem. The contributions of these works were in showing that the natural orbital motion of a vehicle about a virtual center point, such as unmanned aircraft (UA) using a LGVF orbital controller, in a sampled environment provides the required dither and demodulation signals to generate a gradient estimate of performance field in an ES framework.

The specific application of the ES controller presented in this chapter is the maintenance of a solid high-quality connection from a ground station to a single remote unmanned aircraft (UA), which would otherwise be out of communication range, by using the cooperation of other possibly heterogeneous UAs. Of importance are the individual aircraft speed ranges and turning rate performance constraints as these directly influence the responsiveness and convergence of the communication chain to an optimal formation.

The work presented here differs from other work [5,6] in the control of nonholonomic vehicles using ES methods in three primary ways. The first, and foremost, difference is that in the ES algorithm presented, the cyclic (orbital) motion of the vehicle in the sampled environment drives the perturbation signal. Thus unlike other methods, the perturbation signal does not drive the motion of the vehicle. Secondly, the dither signal is taken to be the positional measurement of the vehicle within the environment, and thus requires positional information. This is reasonable as there are numerous GPS and other position measuring solutions for a variety of applications.

The final difference is due to the application of ES with unmanned aircraft (UA) where, because of the dynamic constraints of an aircraft, a bicycle-like kinematic model [7] is used as opposed to a unicycle model. The difference between the two models is that a unicycle model can turn on the spot and in addition is capable of backward and forward translation. A bicycle-like model on the other hand exhibits Dubins' vehicle constraints [8], requiring forward speed to make a turn and has bounded path curvature. In addition, the dynamics and performance limitations of UA introduce additional constraints on the model in terms of bounds on the speed range and turn rate capabilities. These constraints and their impact on the ES algorithm will be discussed in further detail in the following sections.

2   Electronic Chaining Problem Statement

For a linked network chain, independent of the communication protocols used, the achievable chain throughput over time can be directly related to the individual link capacities along the chain. Specifically, the throughput of the chain is limited by the link with smallest capacity along the chain. Figure 1 provides a graphical example of the problem where the link between nodes 3 and 4 is limited to 1 megabit per second (Mbps), either due to distance or environmental noise,

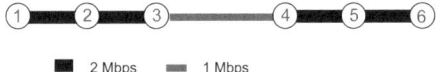

Fig. 1. The overall chain throughput over time is directly limited to the link with the smallest capacity

and the rest of the chain has a 2 Mbps link capacity. It is clear from the figure that even if node 1 tries to transmits at 2 Mbps to node 6, that the link between nodes 3 and 4 will limit the resulting throughput to node 6 to only be 1Mbps. Stated precisely, the bi-directional end-to-end throughput of the communication chain between node 1 and the nth node is given as

$$T = \min_{\substack{i \in R, j \in N \\ |j+i|=1}} \{C_{ij}(\mathrm{p}_i, \mathrm{p}_j), C_{ji}(\mathrm{p}_j, \mathrm{p}_i)\}. \tag{1}$$

where T is the effective chain throughput between nodes 1 and n, $C_{ij}(\mathrm{p}_i, \mathrm{p}_j)$ is the link capacity for node $j \in N$ at position p_j transmitting to node $i \in R$ at position p_i, where $R \subset N$ is the set of relay nodes in the network N, i.e. the set of relay nodes does not contain the two end nodes of the chain which move independent of the chain. The two end nodes represent the users of the communication chain and it is the goal of the mobile relay nodes to position themselves so as to obtain and maintain an optimal communication chain in response to the movements of the two end nodes. Thus nodes 1 and n are allowed to move freely and independently while nodes 2 through n-1 are mobile relays that are controlled by the decentralized ES chaining algorithm.

Maximum bi-directional chain throughput is found by maximizing the minimum individual link capacity by moving the relay nodes, i.e.

$$T^* = \max_{\mathrm{p}_i \in \mathbb{R}^2} \min_{\substack{i \in R, j \in N \\ |j+i|=1}} \{C_{ij}(\mathrm{p}_i, \mathrm{p}_j), C_{ji}(\mathrm{p}_j, \mathrm{p}_i)\}. \tag{2}$$

The Shannon-Hartley Theorem states that the channel capacity C, meaning the theoretical maximum rate of clean (or arbitrarily low bit error rate) data, that can be sent with a given average signal-to-noise ratio (SNR) is [9]

$$C_{ij}(\mathrm{p}_i, \mathrm{p}_j) = B \log_2 (1 + S_{ij}(\mathrm{p}_i, \mathrm{p}_j)), \tag{3}$$

where B is the bandwidth of the channel, and $S_{ij}(\mathrm{p}_i, \mathrm{p}_j)$ is the SNR received at vehicle i from j's transmission. Since link capacity is a monotonically increasing function of SNR, Equation 2 can be restated as

$$S^* = \max_{\mathrm{p}_i \in \mathbb{R}^2} \min_{\substack{i \in R, j \in N \\ |j+i|=1}} \{S_{ij}(\mathrm{p}_i, \mathrm{p}_j), S_{ji}(\mathrm{p}_j, \mathrm{p}_i)\}. \tag{4}$$

Finally, the problem of electronic chaining is to find

$$\mathrm{p}_i^* = \arg \max_{\mathrm{p}_i \in \mathbb{R}^2} \min_{\substack{i \in R, j \in N \\ |j+i|=1}} \{S_{ij}, S_{ji}\} \tag{5}$$

in real time without specific knowledge of the structure or values of the SNR field. Due to the assumption of an unknown structure of the noise field, and that localized noise sources may be present (either due to jamming or faulty nodes), it cannot be assumed that $S_{ij} = S_{ji}$.

While there are numerous examples of systems that control the relative position of networked vehicles to maintain communication links (e.g. see [1,2]), position based solutions cannot account for localized noise sources or an unknown radio frequency (RF) environment where there is an unknown noise environment. Thus an adaptive, model free controller is required to find an optimal (in the sense of end-to-end throughput) location for the relay nodes.

3   Background and Related Work

In this section an overview of the three sub-components of the ES algorithm are presented, including the kinematic bicycle model, extremum seeking, and finally the Lyapunov guidance vector field controller. In addition, constraints that arrive from the use of unmanned aircraft are discussed.

3.1   Vehicle Mobility Model

The algorithms presented here assume the unmanned aircraft are equipped with a low-level flight control system that presents a 2-D kinematic model to the guidance layer of an autopilot system. Let $p_j \in \mathbb{R}^2$, denoted as $p_j = [x_j, y_j]^T$, be the position of vehicle j with inertial speed $[\dot{x}_j, \dot{y}_j]^T \in \mathbb{R}^2$ that evolves according to the standard (cartesian) kinematic bicycle model [7]

$$
\begin{aligned}
\dot{x}_j &= v_j \cos \psi_j \\
\dot{y}_j &= v_j \sin \psi_j \\
\dot{\psi}_j &= \quad v_j c_j
\end{aligned}
\tag{6}
$$

where $[x_j, y_j]^T \in \mathbb{R}^2$ is the two-dimensional inertial position of aircraft j, $\psi_j \in [0, 2\pi)$ is the aircraft yaw angle (compass heading), v_j is the commanded airspeed (held constant in this work), and c_j is the bounded path curvature. The bicycle kinematic model is chosen over a unicycle model because this model covers a wider class of 2D nonholonomic vehicles, moving in only one forward direction and that cannot turn on the spot, such as bicycles, cars, and autonomous underwater and aerial vehicles [7] .

It should be noted that the major difference of the bicycle model is that the heading rate is a function of the vehicle speed and the curvature constraints of the vehicle. For bicycles, the curvature is related directly to the steering angle of the front wheel. For an aircraft in a steady-state coordinated turn, the path curvature is

$$
c(v) = \frac{g \tan \phi}{v^2},
\tag{7}
$$

where ϕ is the UA bank angle and g is the gravitational constant. For simplicity, the velocity of the vehicle is kept at a set (constant) speed and the only input used in the ES chaining algorithm is

$$u_j = v_j c_j(v) = \frac{g \tan \phi}{v_j}. \tag{8}$$

Due to vehicle performance constraints, the control input for a vehicle is bounded by upper and lower limits. For an aircraft at a set speed,

$$\omega_{\max}(v) = \frac{g \tan \phi_{\max}}{v} \tag{9}$$

where $\omega_{\max}$ is the maximum turn rate of the vehicle at speed v_j, for the maximum bank angle $\phi_{\max}$. Thus the input into vehicle j is bounded such that $|u_j| \leq \omega_{\max}$ and gives a minimum orbital radius of

$$r_{\min}(v) = \frac{v}{\omega_{\max}(v)}. \tag{10}$$

It should be noted that aircraft further complicate the design of an ES algorithm due to the limited flight envelope where $0 < v_{\min} \leq v_j \leq v_{\max}$. Thus, $r_{\min}$ cannot be made arbitrarily small.

3.2   Extremum Seeking

Extremum seeking (ES) [10] controllers are adaptive, model free controllers designed to drive the set point of a dynamic system to an optimal, but unpredictable location defined by a performance function that is only known to have an extremum point. That is, given a sufficiently smooth cost function $J : \mathbb{R} \times \mathbb{R}^m \to \mathbb{R}$, ES controllers seek to solve in real time the optimization problem

$$\theta^*(t) = \arg \max_{\theta \in \mathbb{R}^m} J(t, \theta) \tag{11}$$

where J is an unknown, possibly time varying, cost function of the input parameter θ such that $D_\theta J(t, \theta^*) = 0$ and $D_\theta^2 J(t, \theta^*) < 0$ [1].

A typical ES algorithm as presented in [10] has a form as shown in Figure 2 where HPF and LPF stand for the high-pass filter and low-pass filters, respectively. The assumption that $D_\theta^2 J < 0$ is made without loss of generality; the loop feedback gain η in Figure 2 can be replaced by $-\eta$ if $D_\theta^2 J > 0$. The ES algorithm works by generating a measure of the local gradient of the mapping $J(\theta)$ by injecting a perturbation signal, $\alpha \cos(\omega t)$, directly into the plant. The output of the plant will also be sinusoidal, with a DC (or constant) offset that the HPF removes. This signal is then demodulated by $\beta \sin(\omega t - \gamma)$ and low-pass filtered to obtain the gradient estimate. The gradient estimate is then used to update the estimate of the optimal location, $\hat{\theta}$. See [10,11] for formal discussions, including stability proofs and design guidelines, on single and multivariable ES.

---

[1] $D_\theta^i(\cdot)$ denotes the ith directional derivative of J.

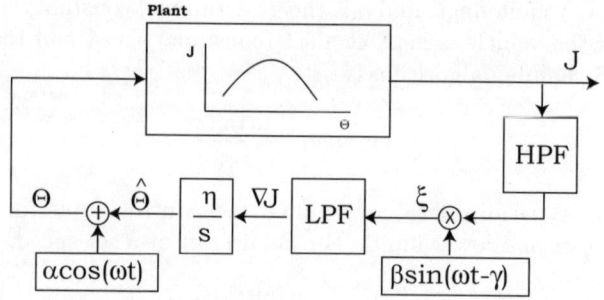

Fig. 2. Typical ES algorithm as presented in [10]

From inspection of Figure 2 it can be seen that the input into the plant is a sinusoidal (cyclic) motion about the current estimate of θ^*. In two dimensions, the input has the appearance of a circular perturbation about a moving (i.e. time varying) orbit center point, $\hat{\theta}$. It is this specific structure that the ES chaining algorithm takes advantage of; some vehicles, like UA, also exhibit a cyclic (circular) motion about a center point when they are station keeping since they must always maintain a forward speed.

3.3   Lyapunov Guidance Vector Field Controller

Because of the constraints that a bicycle-like kinematic vehicle present, it is not practical to drive the vehicle by the ES dither signal directly as done in [6]. Instead a Lyapunov guidance vector field (LGVF) controller [12] is used to provide a circular (cyclic) motion of the vehicle about a virtual center point. The LGVF controller is split into two components, a guidance vector field (GVF) generator and a heading tracker (HT) controller. The heading tracker drives the UA to the desired loiter circle at a radial distance of r_d from the orbit center point $p_{cp} = [x_{cp}, y_{cp}]^T$ as given by the generated vector field

$$
f(p_r) = \begin{bmatrix} \dot{x}_d \\ \dot{y}_d \end{bmatrix} = \beta \begin{bmatrix} -(r^2 - r_d^2) & -2rr_d \\ -2rr_d & -(r^2 - r_d^2) \end{bmatrix} \begin{bmatrix} x - x_{cp} \\ y - y_{cp} \end{bmatrix} + \begin{bmatrix} \dot{x}_{cp} \\ \dot{y}_{cp} \end{bmatrix} \tag{12}
$$

where $r^2 = p_r^T p_r = (x - x_{cp})^2 + (y - y_{cp})^2$ is the squared radial distance of the UA from the loiter center point, p_{cp}, β is a non-negative scalar that guarantees convergence to the desired loiter circle when the center point is moving [12], and $v_{cp} = [\dot{x}_{cp}, \dot{y}_{cp}]^T$ is the center point velocity.

The guidance vector field gives the desired velocity, which is used to generate a turn rate command to the low-level autopilot through the HT. Let $e_\psi = \psi - \psi_d$ where ψ_d is the desired compass heading given as

$$
\psi_d = \arctan\left(\frac{\dot{y}_d}{\dot{x}_d}\right). \tag{13}
$$

The heading angle error is driven to zero by the bounded turn rate command

$$u = \begin{cases} \omega_{\max} & \dot{\psi}_d - \lambda \cdot (\psi - \psi_d) \geq \omega_{\max} \\ \dot{\psi}_d - \lambda \cdot (\psi - \psi_d) & else \\ -\omega_{\max} & \dot{\psi}_d - \lambda \cdot (\psi - \psi_d) \leq -\omega_{\max} \end{cases} \tag{14}$$

where

$$\dot{\psi}_d = \frac{v}{r_d}. \tag{15}$$

Because the turn rate command is bounded by the performance of the vehicle at a given speed, there is a minimum radius for which the vehicle can track and is

$$r_{\min} = \frac{v_j}{\omega_{\max}}. \tag{16}$$

This minimum radius, as will be seen later, is effectively the lower bound on the final error (or distance) of the UA from the optimal communication location, which will be the location of orbit center point for the loiter circle. While the orbit center point can be driven to the location of optimal communication, the UA will always be at best no closer than r_d, with $r_d \geq r_{min}$.

4   Electronic Chaining ES Algorithm

The decentralized ES controller for communication chaining using nonholonomic vehicles is presented in Figure 3. The controller runs locally onboard each vehicle and the only information that needs to be shared is the SNR between the pair-wise links. For simplicity in the figure, the thicker signals represent vector quantities while the thin lines represent scalar values. The feedback loop consists of the 2D kinematic vehicle (with the velocity held at a non-zero constant), the ad hoc network, the ES framework, center point dynamics, and finally the LGVF controller.

From Figure 3 it can be seen by the reader that no external dither signals are introduced into the feedback loop of the system, as is standard in [11,6]. Instead, the amplitude and frequency of the excitation signal is derived from the motion of the vehicle due to the LGVF controller. This type of system has been referred to as a self-exciting ES controller in [13]

The excitation amplitude is related to the desired radial stand off, r_d, of the UA from the center point of the loiter circle and the steady state excitation frequency is given as

$$\omega_j = \frac{v_j}{r_d}. \tag{17}$$

Due to the nature of bicycle like vehicles, excitation amplitude and frequency are not independent choices and are constrained to certain ranges from the individual vehicle performance constraints that are not considered in [10]. For the ES algorithm to be stable the system will need to exhibit three different time scales [13]:

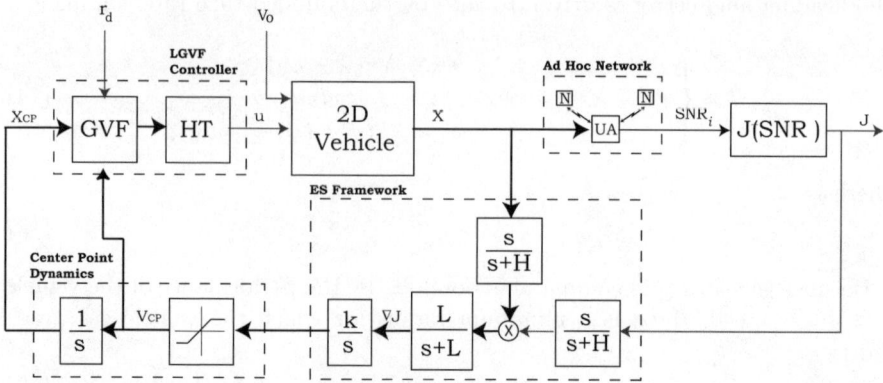

Fig. 3. Decentralized ES chaining algorithm for a 2D kinematic vehicle using a LGVF controller to provide the orbital motion of the UA, including center point dynamics

1. Fast – tracking of the center point
2. Medium – the periodic perturbation
3. Slow – the LPF filter in the ES

Since the dynamics of the periodic perturbation are set by v and r_d, the fast and slow dynamics are also functions of the vehicle performance. For the error dynamics of the LGVF controller to be fast, the motion of the orbit center point must be slow as compared to the UA, i.e. $v_{cp} \ll v_j$. In particular, in the ES chaining algorithm the center point velocity is bounded by V_{cp} so that $v_{cp} \leq V_{cp}$ and $V_{cp} \ll v_j$.

It should be pointed out that center point velocity saturation is required in the loop because even though we can choose k small enough that the speed of the orbit center point remains slow, as compared to the UA for a given environment and performance function, the output of the ES framework depends upon the magnitude and shape of the performance function, which is not necessarily known a priori. Thus, there could be unexpected environments in which if the center point speed was not bounded, it could reach the maximum flight speed of the aircraft. At this point, the motion of the UA about the center point is no longer cyclic and is not generating the periodic signal of the performance function required for the ES framework.

4.1   Performance Function

From Section 2 it was shown that the optimal bi-directional throughput can be found by moving the relay nodes in a manner so as to maximize the minimum SNR of all communication links along the chain. However, it is not desirable to have to share the SNR of every link to every node in the chain, and nor is it required. It is more effecient to share only the SNR of the links that an individual node is responsible for chaining and optimizing. Accordingly, a decentralized performance function for the ES chaining algorithm is given as

$$J_j = \min\{S_{j,j-1}, S_{j-1,j}, S_{j,j+1}, S_{j+1,j}\} \tag{18}$$

and provides a localized measure of the global performance objective of Equation 4 at vehicle j.

Since vehicle j requires a measurement of the SNR as received by its neighbors from its own transmission, e.g. $S_{j\pm1,j}$, there must be information sharing of the SNR among neighbor nodes. Another performance function can be formed that only includes the SNRs that can be measured directly by vehicle j,

$$J_{no-sharing} = \min\{S_{j,j-1}, S_{j,j+1}\}. \tag{19}$$

However, this performance function does not necessarily drive the system to a global extremum when localized noise sources are present.

It should be noted that without any localized noise sources in the environment, $S_{ji} = S_{ij}$, and the performance function (18) reduces to (19). So, for environments where it can be assumed that there are no localized noise sources (such as jamming) the optimal communication chain can be found without any information sharing among the relay nodes.

4.2   Convergence Rate Bounds of the Positional Error

ES controllers are essentially gradient-based ascent (descent) methods [10] and ES algorithms that are not constrained by the dynamics of the plant can exhibit quadratic convergence. However, due to the velocity constraints of the center point, the ES algorithm presented shows linear convergence when the positional error is large, and thus the center point is driven at the saturation value. As the vehicle gets closer to the optimal location then the center point velocity is no longer saturated and the convergence becomes quadratic.

To see how the convergence becomes linear when the center point velocity is saturated, assume

$$\|x_{cp}(0) - x^*\| >> v_{\max}t \tag{20}$$

for some reasonably large amount of time t. Now let $e_k = x_{cp}(k) - x^*$ be the error at time k of the center point to the optimal location. Then at best

$$e_k \le e_{k-1} - V_{cp} \tag{21}$$

and with application of the norm

$$\|e_k\| \le \|e_{k-1} - v_{\max}\| \le \beta \|e_{k-1}\| \tag{22}$$

it can be seen that the error convergence of the system, given bounded velocity constraints on the orbit center point with large initial error, is linear. Results from simulation highlighting these two phases will be presented below.

5   Simulation

In this section, simulations of the ES algorithm for the control of a team of unmanned vehicles are presented. For the simulations, the aircraft are limited to a maximum 30 degree bank angle, flying at 25 m/s, and their ordering is preset and maintained depending upon the starting location of the UA. The maximum center point velocity is set to 5 m/s.

Though not known by the controller, the radios follow the standard exponential decay model

$$P_r = K_r d^{-\alpha} \tag{23}$$

where K_r is the link gain, d is the seperation distance from the transmitter, α is the exponential decay rate, and P_r is the received power. For the simulations the radio values are set to K_r=3822 and α =2. For the simulation with a noise source, the noise source is taken to be a faulty radio transmitting at K_r=382.

Figure 4 shows a simulation run with three UAs and two static end nodes, with Figure 4.a being a top down view of the simulation environment and Figure 4.b the minimum link SNR along the chain at the loiter center point. At the begining

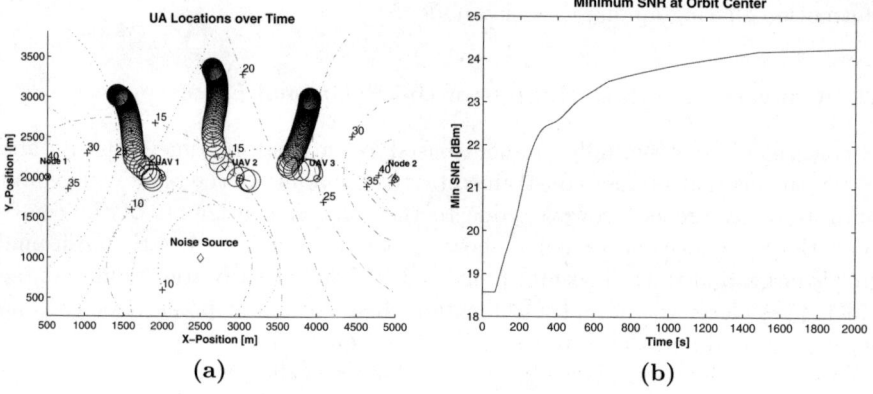

Fig. 4. Simulation of three (3) UA relay nodes reacting to a localized noise source. (a) Motion of UAs within the environment also showing noise source location and the SNR contours of the two end nodes. (b) The minimum SNR value along the chain during the simulation.

of the simulation, the UA relays are aligned along the chain. Then at time $t = 0$s a noise source located at [2500,1000] m is introduced. The figure shows that the UAs react appropriately to the jamming signal due to the noise source and form a bowed communication chain. Figure 4.b shows that the minimum SNR (and total throughput) along the chain is continually improved to a peak value.

Figure 5 shows results from a simulation with a single UA, two end nodes and no localized noise source. In Figure 5.a, the position of the UA and the center point are shown. From this figure one can see that when the UA was far away, it headed directly in the direction of improving minimum SNR (which is the SNR from the far right node) at the maximum speed of the center point. Figure 5.b shows just the X-Y position of the orbit center point to highlight the two different convergence rates of the ES algorithm. For $t \in (50, 500)$s the positional errors (especially on the y-axis) show linear convergence and for $t > 500$s the convergence rate becomes quadratic.

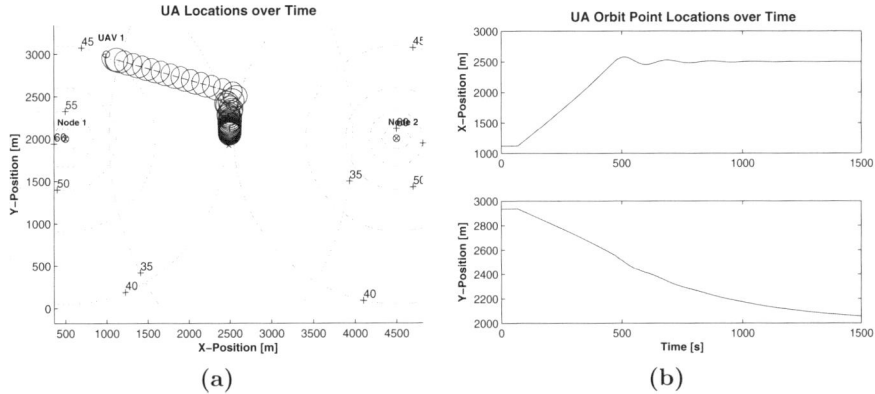

(a)   (b)

Fig. 5. Location of the orbit center point for a single UA and no localized noise showing the linear (from $t = (50, 500)$s) and quadratic convergences (for $t > 500$s) of the UA location to the optimal X-Y location

6   Conclusion

An ES chaining algorithm was presented for optimizing the throughput of a linked communication chain of mobile relay nodes where the key difference in the ES framework is using the physical motion of the vehicle to drive the modulation and demodulation signals. The mobility of the vehicle was modeled as a bicycle-like kinematic model and is chosen over the unicycle model because the model covers a wider class of 2D nonholonomic vehicles, including unmanned aircraft. An orbital motion of the vehicle due to a LGVF controller was applied to extremum seeking in a unique way in that the orbital motion of the vehicle about an orbit center point generated the dither and demodulation signals required by the ES algorithm. A specific application using UAs was presented and simulated to highlight the fact that the performance of the ES algorithm is limited due to the performance constraints and capabilities of the individual vehicles within the chain.

Future work will include varying the aircraft flight speed so as to change the orbiting radius to improve the performance of the ES framework. By slowing the aircraft down when it approaches the optimal location, a smaller orbital radius can be tracked and the aircraft will generate a smaller dither signal and will improve the estimation of the optimal communication location. In addition, since the convergence rate is bounded by the flight speed of the aircraft, it is desirable to have the UA fly close to its maximum flight speed when the center point is far away from the optimal location.

References

1. Basu, P., Redi, J.: Movement control algorithms for realization of fault-tolerant ad hoc robot networks. Network, IEEE **18**(4) (2004) 36–44
2. Goldenberg, D., Lin, J., Morse, A.S., Rosen, B.E., yang, Y.R.: Towards mobility as a network control primitive. In: Mobihoc '04, ACM (2004)
3. Dixon, C., Frew, E.W.: Maintaining a linked network chain utilizing decentralized mobility control. In: AIAA's Guidance, Navigation and Control Conference and Exhibit, AIAA (2006)
4. Dixon, C., Frew, E.W.: Controlling the mobility of network nodes using decentralized extremum seeking. In: 45th IEEE Conference on Decision and Control, IEEE (2006)
5. Ariyur, K.B., Banaszuk, A., Jankovic, M., Rotea, M., Schuster, E., Krstic, M.: Workshop on real-time optimization by extremum seeking control. In: American Control Conference. (2006)
6. Zhang, C., Arnold, D., Ghods, N., Siranosian, A., Krstic, M.: Source seeking with nonholonomic unicycle without position measurement - part I: Tuning of forward velocity. In: 45th IEEE Conference on Decision and Control. (2006)
7. Indiveri, G.: Kinematic time-invariant control of a 2D nonholonomic vehicle. In: 38th Conference on Decision and Control (CDC'99). (1999)
8. Dubins, L.E.: On curves of minimal length with a constraint on average curvature, and with prescribed initial and terminal positions and tangents. American Journal of Mathematics **79**(3) (1957) 497–516
9. Taub, B., Schilling, D.L.: Principles of Communication Systems. McGraw-Hill, New York (1986)
10. Ariyur, K.B., Krstic, M.: Real-Time Optimization by Extremum-Seeking Control. John Wiley and Sons, Inc, Hoboken, NJ (2003)
11. Ariyur, K.B., Krstic, M.: Multivariable extremum seeking feedback: Analysis and design. In: Fifteenth International Symposium on Mathematical Theory of Networks and Systems. (2002)
12. Frew, E.W., Lawrence, D.: Cooperative stand-off tracking of moving targets by a team of autonomous aircraft. In: AIAA Guidance, Navigation, and Control Conference. (2005)
13. Krstic, M., H.Wang, H.: Stability of extremum seeking feedback for general nonlinear dynamic systems. Automatica **36** (2000) 595–601

An Adaptive Sequential Game Theoretic Approach to Coordinated Mission Planning for Aerial Platforms

Dan Shen[1], Genshe Chen[1], Jose B. Cruz, Jr.[2], and Khanh D. Pham[3]

[1] Intelligent Automation, Inc., 15400 Calhoun Dr., Suite 400,
Rockville, MD 20855, USA
{dshen,gchen}@i-a-i.com

[2] Department of Electrical and Computer Engineering, The Ohio State University,
752 Dreese Laboratories, 2015 Neil Avenue Columbus, Ohio 43210-1272 USA
cruz@ece.osu.edu

[3] Air Force Research Laboratory, Space Vehicles Directorate,
Kirtland AFB, NM 87117-5776
Khanh.Pham@kirtland.af.mil

Abstract. Coordinated mission planning is one of the core steps to effectively exploit the capabilities of cooperative control of multiple UAVs. In this chapter, we extend and implement an effective team composition and tasking mechanism and an optimal team dynamics and tactics algorithm for mission planning under a hierarchical adaptive sequential game theoretic framework. Our knowledge/experience based static non-cooperative and non-zero games are used for team composition and tasking to schedule tasks at the mission level and allocate resources associated with these tasks. The dynamic adaptive sequential game model is used for team dynamics and tactics to assign targets and decide the optimal salvo size for each aerial platform to achieve the minimum remaining platforms of red and the maximum remaining platforms of blue at the end of a battle. A simulation software package has been developed to demonstrate the performance of our proposed algorithms.

Keywords: Adaptive Sequential Game; Action-Reaction-Counteraction; Optimal Salvo Size; Nash Equilibrium; Reaction Curve; Coordinated Mission Planning; Hierarchical Framework.

1 Introduction

As in natural systems, cooperation in a team of Unmanned Aerial Vehicles (UAVs) may assume a hierarchical form and the control processes may be distributed or decentralized. Due to the dynamic nature of individuals and interaction between them, the problems associated with cooperative systems usually include many uncertainties. Moreover, in many cases cooperative systems are required to operate in an adversary environment.

M.J. Hirsch et al. (Eds.): Adv. in Cooper. Ctrl. & Optimization, LNCIS 369, pp. 323–338, 2007.
springerlink.com
© Springer-Verlag Berlin Heidelberg 2007

Cooperative mission planning for autonomous vehicle teams is of great interest. A significant amount of current research activities focus on cooperative control of UAVs and some possible research directions in this field are unified in [3,7]. Cooperative real-time search and task allocation algorithms are presented in [8]. A genetic algorithm for task allocation is proposed in [10]. Another mission planning approach is described in [9]. Solutions to general UAV cooperative control problems in adversarial environments can be obtained by solving game problems introduced in [4] and implemented in [1]. Additional game-based works focusing on target assignment of a group of UAVs are discussed in [5,6].

Currently, those game approaches are based on simultaneous game models in which all players move simultaneously, effectively simultaneously. However, sequential games, where players apply their strategies following a certain predefined order and at least some players can observe the moves of other players who make decisions preceding them, are more natural frame-works to address some real problems, such as the "Action-Reaction-Counteraction" paradigm used in military intelligence and advertising campaigns strategies of several competing firms in economics.

The contribution of this chapter is as follows; first, we extended and improved a game theoretic framework [16] of mission planning. Second, we incorporate the idea of expert/knowledge systems. Third, we implement our approach in software with connectivity to the OEP (Open Experimental Platform) [2] from Boeing. The overall architecture is described in section 2. In section 3 the upper level non-cooperative and non-zero game based team composition and tasking (TCT) is reported with details. Then we present our lower level team dynamics and tactics (TDT) with adaptive design for dynamic games in section 4. Simulation results are reported in section 5. Finally, conclusions are in section 6.

2   Framework for Mission Planning

The main goal of coordinated mission planning is to develop and provide an effective team composition and tasking (TCT) mechanism and an optimal team dynamics and tactics (TDT) algorithm to destroy the opposing force combat capabilities. In order to accomplish this, a hierarchical game theoretic framework, as shown in Fig 1, is developed here. Our Coordinated Mission Planning approach is developed and implemented as a software package, which connects to the MICA (Mixed Initiative Control of Automa-Teams) OEP (Open Experimental Platform). The main purpose of our knowledge/experience, based on a static non-cooperative and non-zero sum Nash game, is to develop and provide an algorithm to schedule tasks at the mission level and allocate resources associated with these tasks. First, the Rules of Engagements are included here as constraints. Second, several criteria are considered in order to develop our object (cost) function. Third, the game approach addresses the uncertainty of the presence of an adversarial battle environment. The upper level game provides an algorithm for non-homogenous resource allocation based on the following information: the number of red areas in a scenario, the number of targets in each red area, the air defensibility of each target, the target status, classified as potential, known or unknown, and information ability of each target, such as sensor signatures and communication capability, and the Blue force information as well. Our game framework at this level will estimate the loss of both sides by a probability inner battle field model, which is based the information of both sides. The team composition and tasking information architecture is developed and used at the lower level game. The event based lower level non-cooperative (Nash) game is used to assign targets and decide the salvo size for each aerial platform. Furthermore, the lower game will find an optimal deployment of decoys and avoid collateral damage. Here we assume that the red units are also optimizing and coordinating their targeting strategy against the Blue units and as a result determine the target selection strategy based on a game theoretic approach. At the same time, the lower level game determines an "estimate" of the red team's salvo size strategy. The major accomplishments at this level are as follows: A non-zero-sum non-cooperative game theoretical algorithm has been developed to determine the optimal salvo size to achieve the minimum remaining platforms of red and the maximum remaining platforms of blue at the end of a battle; a de-centralized target assignment algorithm is developed to find optimal aerial platform - red target pairs; a cooperative decoy deployment method has been developed to maximize the total probability of survival of Blue aerial platforms. Fig. 1 shows the relationship between TCT and TDT. They will be described in the following section.

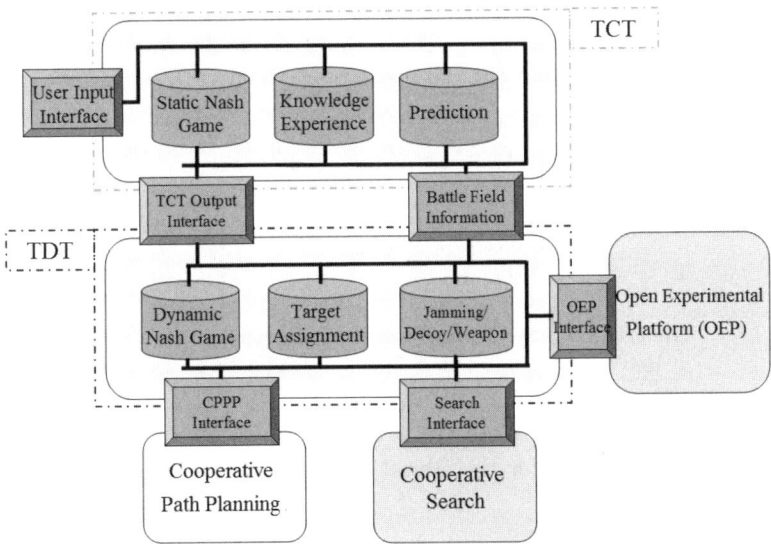

Fig. 1. A game theatric framework for mission planning

3   Team Composition and Tasking

Suppose there are N^{BP} Blue Aerial Platforms, N^{BA} blue force concentration areas and N^{RA} red areas with red forces in a typical scenario. The strategic objectives of blue forces during the fixed time period $[0, T]$ include: to protect the blue concentrations from attack by red surface-to-surface missiles (SSMs), armour and troops; to neutralize the Integrated Air Defense System (IADS) and eliminate red surface-to-air missile (SAM) sites in order to provide safe operations for blue army fixed and rotary aircraft; to eliminate the early warning (EW) Radars, SAMs and command and control (C2) Facilities authorized; to avoid destroying civilians/non-combatants and cities/cultural landmarks (for example, the location of red command and control facilities may be in or near schools, churches and hospitals).

Let N^{BSS}, N^{BSW}, N^{BLS}, N^{BLW}, N^{BCO} denote the number of small sensors, small weapons, large sensors, large weapons and combination blue aerial platforms involved in the battle respectively and P^B be the type number of the blue platforms (in this scenario, $P^B = 5$. There are 5 five types of the blue force). All the blue aerial platforms are equipped with warning sensors which detect SAM tracking radars. However, the weapon aerial platforms only have electronic support measures (ESM) radars, no ground moving target indication (GMTI) and imaging radars. Let N^{RLSAM}, N^{RMSAM}, N^{RT}, N^{ESM}, N^{EW}, N^{RTR}, N^{RSR} denote the total number of the red long SAM sites, red medium SAM sites, re troops, ESMs, EW radars, red tracking radars, red search radars and P^R be the type number of the red platforms.

At this level, the objective of each side is to allocate their forces into $N^{BA} + N^{RA}$ areas to obtain the highest performance score. The score is based on the unit values of the destroyed units. Table 1 and 2 give an example of the target values for the blue force and the red force, respectively.

Table 1. Target Values for Blue Force

Unit Type	Unit Value
ESM sensor, or EW radar	10
Medium SAM	20
Long SAM	30
Red Troop (Mobile HQ, Personnel Carrier)	15
Red Search Radar, Red Tracking Radar	12

The control commands for the Blue Force and the red force are denoted by U^B and U^R respectively.

$$U^B = \left(n_{i,j}^B \right)_{(N^{BA}+N^{RA}) \times P^B} \tag{1}$$

Table 2. Target Values for Red Force

Unit Type	Unit Value
Small Sensor Blue Aerial Platform (SS)	10
Large Sensor Blue Aerial Platform (LS)	20
Small Weapon Blue Aerial Platform (SW)	15
Large Weapon Blue Aerial Platform (LW)	30
Combination Blue Aerial Platform (CO)	20

where $n_{i,j}^B \in \left(I^+ \bigcup \{0\}\right)$, $\sum_{i=1}^{N^{BA}+N^{RA}} n_{i,1}^B = N^{BSS}$, $\sum_{i=1}^{N^{BA}+N^{RA}} n_{i,2}^B = N^{BSW}$,

$\sum_{i=1}^{N^{BA}+N^{RA}} n_{i,3}^B = N^{BLS}$, $\sum_{i=1}^{N^{BA}+N^{RA}} n_{i,4}^B = N^{BLW}$, and $\sum_{i=1}^{N^{BA}+N^{RA}} n_{i,5}^B = N^{BCO}$.

$$U^R = \left(n_{i,j}^R\right)_{(N^{BA}+N^{RA})\times P^R} \tag{2}$$

where $n_{i,j}^R \in \left(I^+ \bigcup \{0\}\right)$, $\sum_{i=1}^{N^{BA}+N^{RA}} n_{i,1}^R = N^{RLSAM}$, $\sum_{i=1}^{N^{BA}+N^{RA}} n_{i,2}^R = N^{RMSAM}$,

$\sum_{i=1}^{N^{BA}+N^{RA}} n_{i,3}^B = N^{RT}$, $\sum_{i=1}^{N^{BA}+N^{RA}} n_{i,4}^B = N^{ESM}$, $\sum_{i=1}^{N^{BA}+N^{RA}} n_{i,5}^B = N^{EW}$,

$\sum_{i=1}^{N^{BA}+N^{RA}} n_{i,6}^B = N^{RTR}$, and $\sum_{i=1}^{N^{BA}+N^{RA}} n_{i,7}^B = N^{RSR}$.

Then the objective functions are defined as

$$J^B(u^B, u^R) = \sum_{i=1}^{N^{BA}+N^{RA}} \sum_{j=1}^{P^R} n_{i,j}^R f_{BR}(u^B, u^R, i, j; T) V_j^R \tag{3}$$

$$J^R(u^B, u^R) = \sum_{i=1}^{N^{BA}+N^{RA}} \sum_{j=1}^{P^B} n_{i,j}^B f_{RB}(u^B, u^R, i, j; T) V_j^B \tag{4}$$

where,

$f_{BR}(u^B, u^R, i, j; T)$ is the experience/knowledge based probability of the event that a type j red force will be killed by the blue force at area i;

$f_{RB}(u^B, u^R, i, j; T)$ is the experience/knowledge based probability of the event that a type j blue force will be killed by the red force at area i;

V_j^R and V_j^B are the platform values of type j of the red force and the blue force respectively.

After each stage of the battle, the real results will be used as feedback information to adjust the two experience/knowledge based probability functions

$f_{BR}(u^B, u^R, i, j; T)$ and $f_{RB}(u^B, u^R, i, j; T)$. At this stage, our upper level knowledge/experience based static non-cooperative and non-zero Nash game is closed loop and self adaptive.

4 Team Dynamics and Tactics

At this level, there are two main issues: how to assign the red targets to each blue platform, and how to optimally terminate the assigned red targets.

4.1 Target Assignment

After getting the updated battle field information from our search approach, all blue platforms in a specific area or team will be assigned or reassigned to red targets. Here we present the de-centralized target assignment algorithm.

Suppose there is a scenario: N^{AR} red targets in an area with two most-likely-type probabilities p^{1st} of P^{1st} and p^{2nd} of P^{2nd} for each red unit. The platform type of the ith blue unit is denoted by P_i^{AB}. For the nth blue platform, define the performance function as

$$\Phi_n = \frac{1}{V_{P_n^{AB}}^B} \left\{ w_n^v \sum_{i=1}^{N^{AR}} \left[\bar{P}_{i,n} + \tilde{P}_{i,n} \right] - w_n^s \delta_n S_n \right\} \tag{5}$$

where,

$$\bar{P}_{i.n} = p_i^{1st} \left[V_{P_i^{1st}}^R P^{BR}(P_i^{1st}, P_n^{AB}) - V_{P_i^{AB}}^B P^{RB}(P_i^{1st}, P_n^{AB}) - w_{i,n} \frac{d_{i,n}}{s_n} \right];$$

$$\tilde{P}_{i,n} = p_i^{2nd} \left[V_{P_i^{2nd}}^R P^{BR}(P_i^{2nd}, P_n^{AB}) - V_{P_i^{AB}}^B P^{RB}(P_i^{2nd}, P_n^{AB}) - w_{i,n} \frac{d_{i,n}}{s_n} \right];$$

w_n^v, $w_{i,n}$ and $w_n^s \in [0, 1]$ are relative weights;

V_Y^R and V_Y^B are the unit values of the Y type red unit and blue unit, respectively;

$P^{BR}(X, Y)$ is the probability of the event that an X type red unit will be destroyed by a Y type blue platform;

$P^{RB}(X, Y)$ is the probability of the event that a Y type blue platform will be destroyed by an X type blue unit;

$d_{i,n}$ is the distance between the ith red unit and the nth blue platform;

s_n is the speed of the nth blue platform;

$\delta_n = 1$, if the nth blue platform has already been assigned a target. Otherwise $\delta_n = 0$;

S_n is the switching cost of the nth blue platform.

In equation (5), $V_{P_i^{1st}}^R PBR(P_i^{1st}, P_n^{AB})$ is the expected score the blue side will obtain if the ith red target is assigned to the nth blue platform. $V_{P_n^{AB}}^B PRB(P_i^{1st}, P_n^{AB})$ is the expected score the blue side will lose if the ith red target is assigned to the nth blue platform. $w_{i,n} \frac{d_{i,n}}{s_n}$ is the weighted time cost. With a consideration of the target-switching cost, we denoted $w_n^v \sum_{i=1}^{N^{AR}} [\bar{P}_{i,n} + \tilde{P}_{i,n}] - w_n^s \delta_n S_n$ as the total "virtual" score the blue side will gain if the nth blue platform is used. Here we use the word "virtual" due to the fact that only one target can be assigned to the nth blue platform. Given that the same scores are gained by two blue units, the less the blue platform value, the better. That is why we put $1/V_{P_n^{AB}}^B$ in front of the final score in the performance function.

Our target assignment approach has two steps. The first one is to give a priority sequence of the N^{AB} blue platforms by the value of the performance function from largest to smallest. This step de-centralizes the whole processing. The second one is to let the Blue platform "greedily" choose their target in the order of the rank created in the first step. In this step, the current k^{th} blue unit will choose the red target (i^{th} red unit) which has the biggest value of the following utility function

$$\Upsilon_k(i) = \frac{1}{V_{P_k^{AB}}^B} \left\{ w_k^v [\bar{P}_{i,k} + \tilde{P}_{i,k}] - w_k^s \delta_k S_k - w_i^r \delta_i^r \right\} \tag{6}$$

where the new variable w_i^r is a relative big number as a weight, and $\delta_i^r = 0$ if the red unit has less than $m(=2)$ blue preys; otherwise, $\delta_i^r = 1$.

In equation (6), $w_k^v [\bar{P}_{i,k} + \tilde{P}_{i,k}] - w_k^s \delta_k S_k$, having the same meaning as in equation (5), is the total expected score the blue side will gain if the ith red target is assigned to the kth blue platform. To prevent the kth blue unit from choosing a target which already has $m(=2)$ blue preys, we subtract a relatively large weight w_i^r if $\delta_i^r = 1$.

4.2   Adaptive Sequential Game Approach for Salvo Size Control

The goal of the salvo size control is to determine optimal salvo-size such that blue force can destroy as many red targets as possible and at the same time save the blue platforms. For the blue weapon platforms, the good salvo size control strategy is important to destroy as many red targets as possible and at the same time to save themselves. We will use sequential games to model the process.

A sequential game is one in which players apply their strategies following a certain predefined order, and in which at least some players can observe the moves of other players who make decisions preceding them. The sequential game is a natural framework to address some real problems, such as the "Action-Reaction-Counteraction" paradigm used in military intelligence and advertising campaigns strategies of several competing firms in economics. Linear quadratic games are researched in [11] from a system dynamic control perspective.

State equation:

$$\begin{cases} X_{k+1}^B = X_k^B - \xi_k^B\left(X_k^B, X_k^R, C_k^B, C_k^R, I_k^B\right) \\ X_{k+1}^R = X_k^R - \xi_k^R\left(X_k^B, X_k^R, C_k^B, C_k^R, I_k^R\right) \end{cases} \tag{7}$$

where X_k^B and X_k^R are the states of blue side and red side, respectively. The platform type, number, and current damage statues are capsulated in the state vectors. I_k^B and I_k^R are the available information sets [12] for blue and red side at time k. ξ_k^B and ξ_k^R are the state transition function of blue and red side, respectively. Both functions are based on the attrition model specified in [4]. C_k^B and C_k^R are the sequential salvo size control of blue and red force.

$$C_k^B = \begin{cases} \gamma_k^B(X_k^B, X_k^R, I_k^B)\ , & k \text{ is even} \\ 0 & , k \text{ is odd} \end{cases} \tag{8}$$

$$C_k^R = \begin{cases} \gamma_k^R(X_k^B, X_k^R, I_k^R)\ , & k \text{ is odd} \\ 0 & , k \text{ is even} \end{cases} \tag{9}$$

γ_k^B and γ_k^R are the action strategies of blue side and red side at time k when it is the player's turn to take the actions.

In general, the system model described by (7)-(9) is nonlinear. By the linearization transformation method introduced in [13], the above nonlinear model can be transformed into a linear system.

$$X_{k+1} = A_x X_k + D_k^B C_k^B + D_k^R C_k^R \tag{10}$$

$$X_k = \begin{bmatrix} X_k^B \\ X_k^R \end{bmatrix} \tag{11}$$

The objective of each side is to minimize its cost function

$$J^B = \underbrace{(X_N^R)^\top Q_N^{BR} X_N^R}_{\text{remaining red units}} - \underbrace{(X_N^B)^\top Q_N^{BB} X_N^B}_{\text{remaining blue units}} - \underbrace{\sum_{k=1}^{N-1} \left(\left[C_k^B\right]^\top W^B C_k^B \right)}_{\text{cost of blue actions}} \tag{12}$$

$$J^R = \underbrace{(X_N^B)^\top Q_N^{RB} X_N^B}_{\text{remaining red units}} - \underbrace{(X_N^R)^\top Q_N^{RR} X_N^R}_{\text{remaining blue units}} - \underbrace{\sum_{k=1}^{N-1} \left(\left[C_k^R\right]^\top W^R C_k^R \right)}_{\text{cost of red actions}} \tag{13}$$

Therefore a linear quadratic sequential game is formed.

With the consideration that the parameters in each side's cost function is not accessible to the other side, we propose to use an adaptation design [11] based on the concept of Fictitious Play (FP) to learn the unknown properties.

As a learning concept, FP was first introduced by G. W. Brown [14] in 1951. Within the learning scheme, each player presumes that the opponents are playing stable (possibly mixed) strategies. Each player starts with some initial belief and

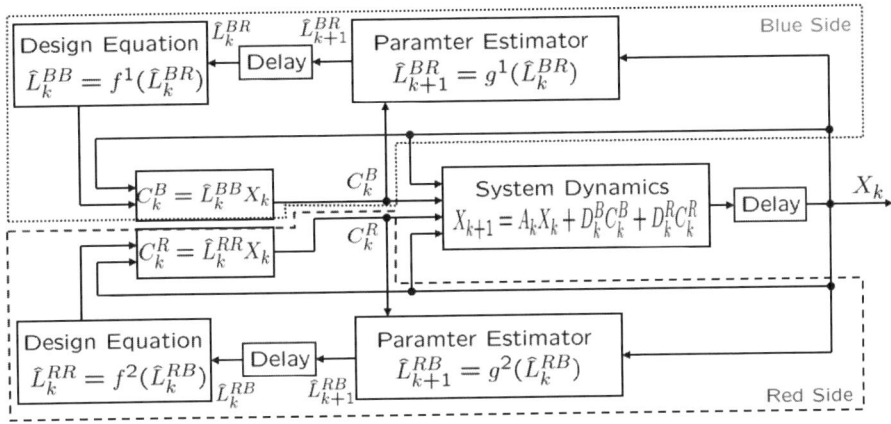

Fig. 2. An adaptation design for dynamic linear quadratic games

chooses a best response to those beliefs as a strategy in this round. Then, after observing their opponents' actions, the players update their beliefs according to some learning rule (e.g. Q-learning or Bayes' rule). The process is then repeated. It is known [15] that if the process converges, then the point of convergence is a Nash equilibrium of the game.

In [11], an adaption scheme, as shown in Fig. 2, is constructed via combining Fictitious Play (FP) [14,15] (in which each player presumes that its opponents are playing Nash strategies, then a best response to those beliefs is chosen as a strategy), and conventional adaptive control techniques such as normalized gradient algorithm and Recursive Least Square (RLS).

5   Simulation and Experiments

To illustrate how our algorithm works, we developed a software package, which can connect to the MICA (Mixed Initiative Control of Automa-Teams) OEP (Open Experimental Platform) [2].

A typical scenario shown in Fig. 3 with several experiments is simulated on our software to evaluate the performance of our proposed Mission Planning algorithm for Multiple Aerial Platforms. The meaning of icons are shown in Fig. 4.

In the scenario, Molian rebels (red force) supported by terrorist organizations and a neighboring country (NNC) are overrunning the country of Molia. Molia has asked the US for support. Friendly Molian forces face numerically superior Molian red force plus volunteers from a neighboring country. Molian Rebels have integrated their air defense into the neighboring country's EW Radars, SAMs and C2 structure. These EW Radars, SAMs and C2 structures are deemed acceptable targets but collateral damage must be avoided. Red forces have overrun

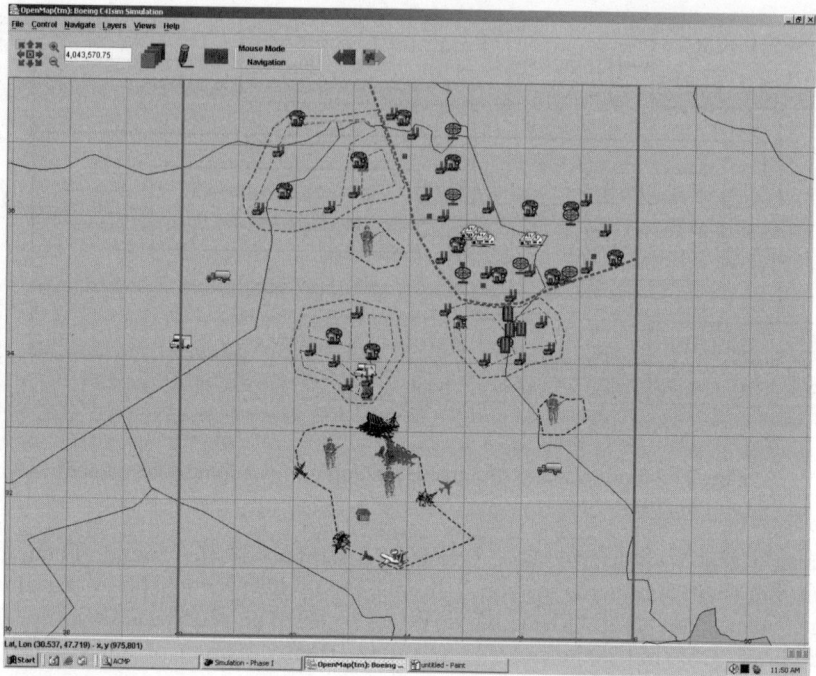

Fig. 3. A typical Scenario

portions of the country. Blue ground forces consists of the Molian Army plus a limited number of US ground forces. The combined blue Force is relying on UAV air power in support of these limited ground forces to contain the situation until additional US ground forces arrive.

For this scenario, after running our upper level knowledge/experience based static non-cooperative and non-zero Nash game, we obtain the following blue force resource allocation results (Note that since the system is stochastic, all the figures below are based on the averages of 30 simulations). In this stage, four missions are assigned: one Close Air Support mission is assigned to team 1 (4 COMB UAVs, 1 Small Sensor UAV, 1 Small weapon UAV) to protect blue Base; one Close Air Support mission is allocated to team 2 (2 COMB UAVs, 1 Small Sensor UAV, 1 Small weapon UAV) to guard blue concentration area #1; one Close Air Support mission is sent to team 3 (2 COMB UAVs, 1 Small Sensor UAV, 1 Small weapon UAV) to guard blue concentration area #2; and one SEAD mission is assigned to team 4 (5 small sensor UAVs, 5 small weapon UAVs, 6 large sensor UAVs, 6 large weapon UAVs) to attack red area #2. No blue force is sent to red area #1, #3 and NNC.

The whole task is divided into three stages. During each stage, we first run the TCT module to optimally and automatically assign resources to the point of interest so that the goal of the stage can be met with the minimum cost. With

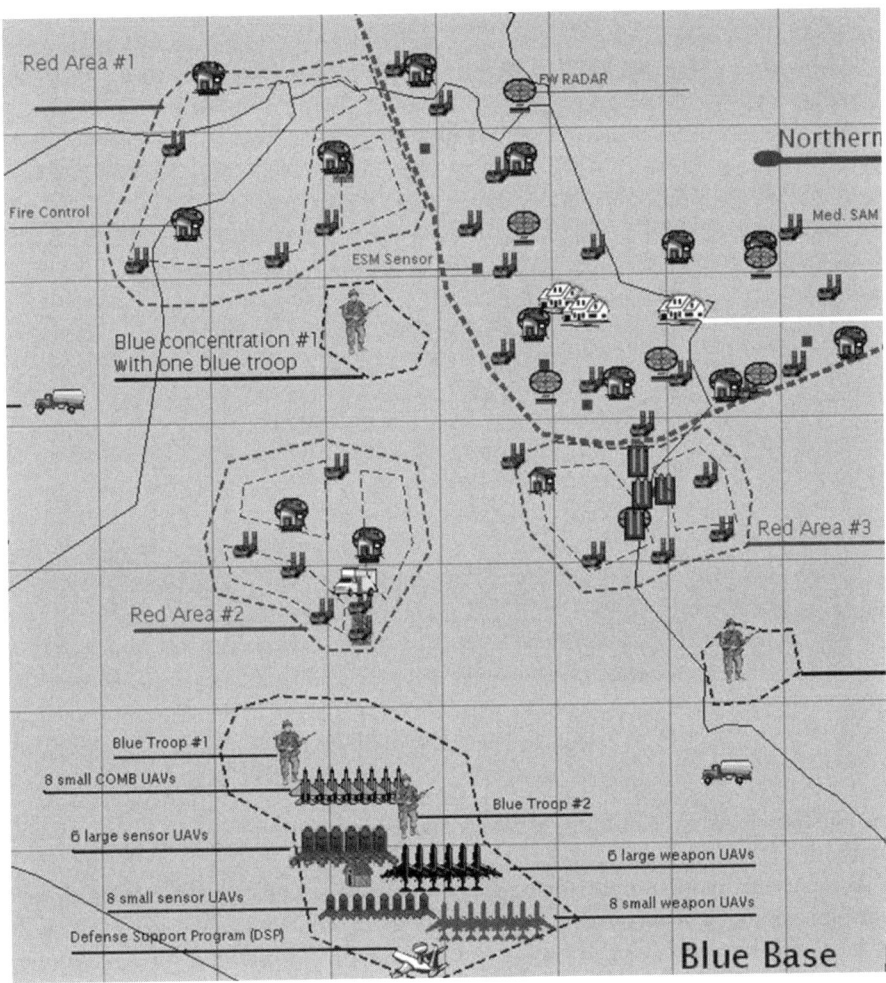

Fig. 4. Icons used in scenarios

the given resource, the lower level TDT module will be called for each unit to dynamically select targets and salvo-size. The salvo-size is optimally determined by our adaptive sequential game algorithm under a Fictitious Play framework so that the linear quadratic cost functions (12)-(13) are minimized for linearized system dynamics (10)-(11).

The result of the stage 1 is shown in Fig. 5. All the red forces (2 long SAM sites, 7 medium SAM sites, 1 red ground troop composed of 4 Tanks, 2 EW radars, 2 Tracking radars and 2 search radar) in red area #2 are terminated with the cost of losing three Large Weapon UAVs and 2 Large Sensor UAVs partially damaged in Team 4 . We also compare the result with one of other

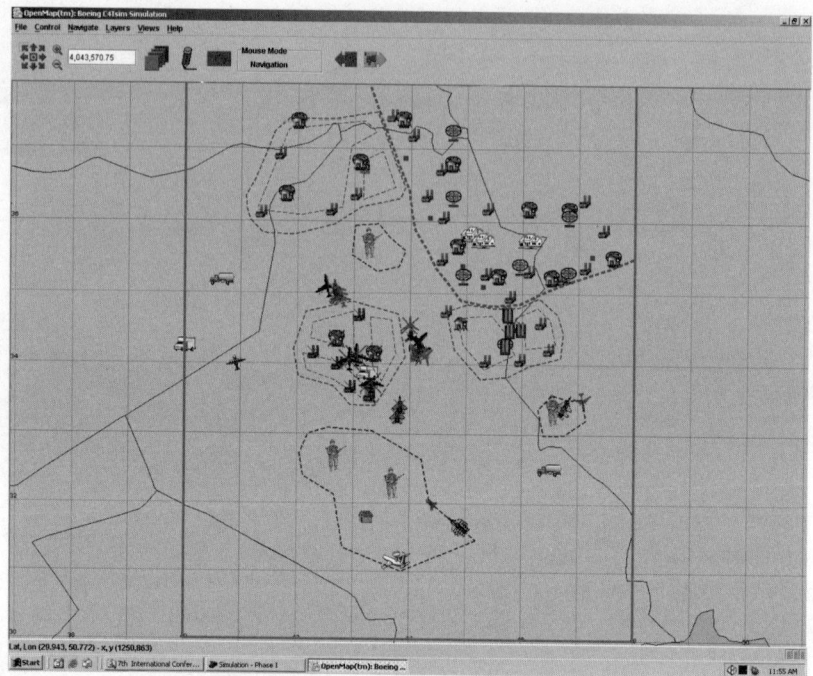

Fig. 5. The result of Stage 1

options, such as no jamming strategy, no decoy approach, and only weapon method.

From the damage comparison results, we can see our proposed jamming-decoy-weapon approach is better than other methods. It also shows that decoy strategy is more efficient in saving the blue forces than the Jamming strategy. We also notice that all blue forces are totally destroyed by the red force in the weapon only approach during the stage 2.

The adaptive scheme for TDT is implemented during each stage. The result of the first stage is shown in Fig. 6. In this plot, $\{\hat{L}_k^{ij}\}_s$ is the sth element of $\hat{L}_k^{ij}$, which is the estimate of state feedback control gain of player j by player i, $i, j = 1, 2$. We can see the convergence of to L^j. which is the actual value of control gain for player j.

After Stage 1, we should call the upper level game again with the updated battle field information. In this step, the experience/knowledge functions will be adjusted too. The blue force allocation chart of stage 2 is shown in Fig. 7. The simulation result of stage 2 is illustrated in Fig. 8.

In Stage 2, there are two SEAD missions. Mission 1 is to attack the red area #1 and mission 2 is to dominate the red area #3. During mission 1, two small weapon UAVs are shot down by the red force.

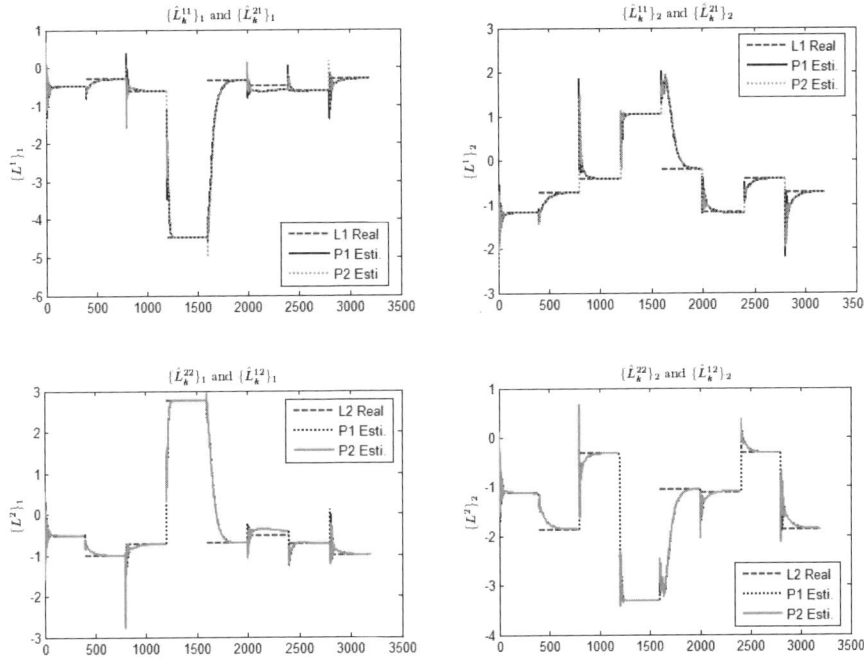

Fig. 6. Results of Adaptive Design for Dynamic Games in TDT

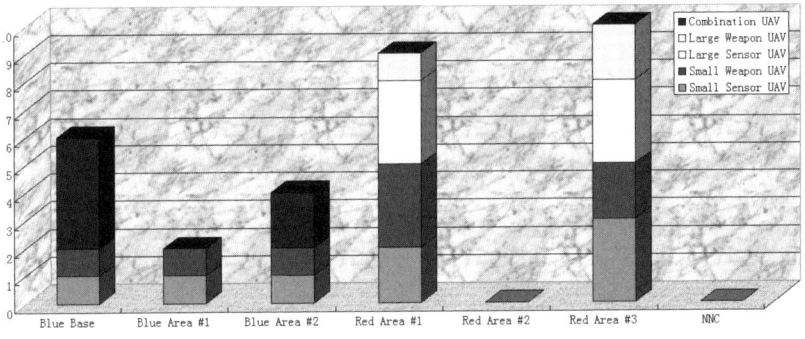

Fig. 7. Blue Force Allocation Chart of Stage 2 (Note: Red Area # 1 and Red Area #3 will be attacked during stage 2)

In the SEAD mission 2, one small sensor UAV and one large sensor UAV are totally destroyed by red Integrated Air Defense System (IADS). In Fig. 8, notice that all the blue force in team 2, which is assigned the Close Air Support (CAS) to protect blue area #1, are destroyed. There is one destroyed small weapon UAV in team 3, which has the task to protect blue Concentration #2. The TCT result of stage 3 is shown in Fig. 9.

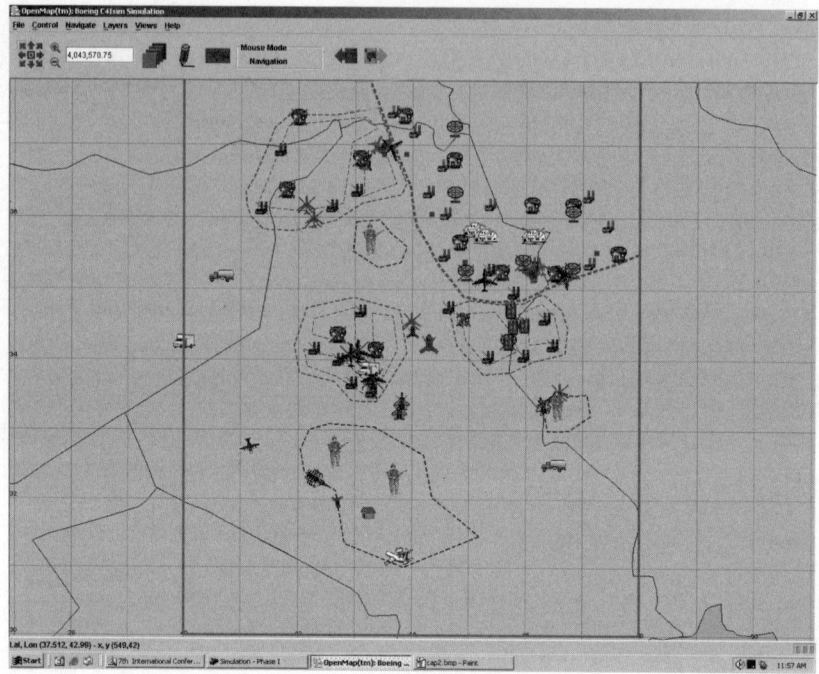

Fig. 8. The result of Stage 2

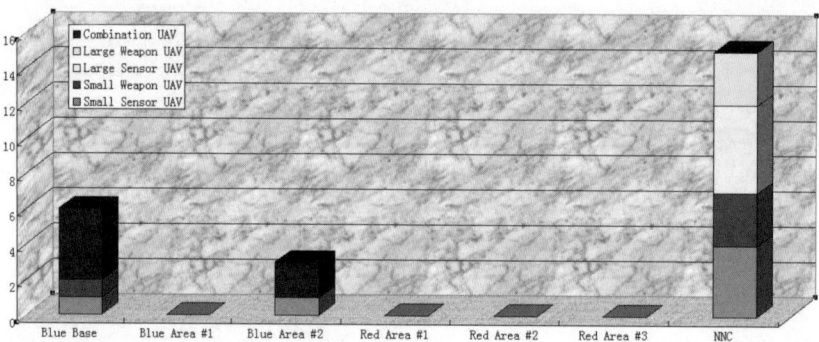

Fig. 9. Blue Force Allocation Chart of Stage 3 (Note: NNC will be attacked during stage 3)

The simulation result of stage 3 is shown in Fig. 10. All the red forces in NNC are destroyed. It is surprisingly good. We think there are two reasons to account for it. One is that we assigned two teams to attack NNC. The other is that most units in NNC are red force support devices such as Radars and ESM sensors.

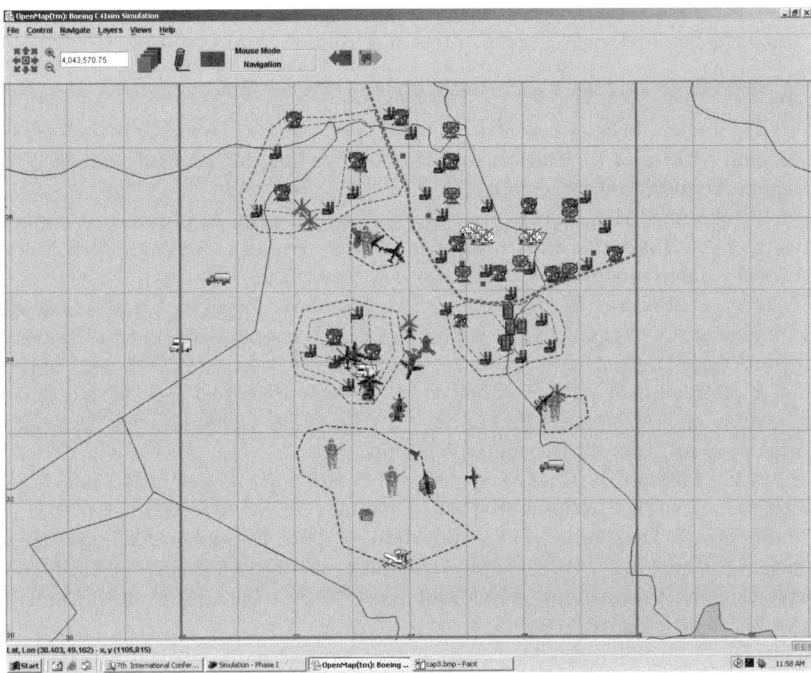

Fig. 10. The result of Stage 3

6   Conclusion

In UAV cooperative control, mission planning is of great importance for the efficient destruction of the opposing force combat capabilities. In this chapter, an effective TCT mechanism and an adaptive TDT algorithm are developed under a hierarchical adaptive and sequential game theoretic framework. The simulations show that the game theoretic algorithm is capable of solving the coordinated mission planning problem.

References

1. Cruz, Jr., J.B., Chen, G., Garagic, D., Tan, X., Li, D., Shen, D., Wei, M., Wang, X.: Team Dynamics and Tactics for Mission Planning. Proceedings of the 42nd IEEE Conf. on Decision and Control, Hawaii, December 2003
2. User Guide for the Open Experimental Platform (OEP), version 1.3, Boeing, Mar, 2003
3. Chandler, P., Pachter, M.: Research issues in autonomous control of tactical UAVs. Proceedings of the American Control Conference, Philadelphia, Pennsylvania, June 24-26, 1998

4. Cruz, Jr., J.B., Simaan, M.A., Gacic, A., Jiang, H., Letellier, B., Li, M., Liu, Y.: Game-Theoretic Modeling and Control of a Military Air Operation. IEEE Transactions on Aerospace and Electronic Systems, Vol. 37, No. 4, October 2001, pp.1393-1405

5. Liu, Y., Simaan, M.A., Cruz, Jr., J.B.: An Application of Dynamic Nash Task Reassignment Strategies to Multi-Teams Military Air Operations. Automatica, Vol. 39, Issue 8, August 2003, pp. 1469-1478

6. Liu, Y., Simaan, M.A., Cruz, Jr., J.B.: Game-Theoretic Approach to Cooperative Teaming and Tasking in the Presence of an Adversary. Proceedings, 2003 American Control Conference, Denver, Colorado, June 4-6, 2003

7. Chandler, P., Pachter, M., Swaroop, D., Fowler, J.M., Howlett, J.K., Rasmussen, S., Schumacher, C., Nygard, K.: Complexity in UAV cooperative control. Proceedings of the American Control Conference, Anchorage, Alaska, May 2002, pp. 1831-1836

8. Jin, Y., Minai, A.A., Polycarpou, M.M.: Cooperative real-time search and task allocation in UAV teams. Proceedings of the IEEE Conference on Decision and Control, Maui, Hawaii, December 2003, pp. 7-12

9. Gil, A.E., Passino, K.M., Ganapathy, S., Sparks, A.: Cooperative scheduling of tasks for networked uninhabited autonomous vehicles. Proceedings of the IEEE Conference on Decision and Control, Maui, Hawaii, December 2003, pp. 522-527

10. Chen, G., Cruz, Jr., J.B.: Genetic Algorithm for Task Allocation in UAV Cooperative Control. Proceedings, AIAA Guidance, Navigation, and Control Conference, Austin, Texas, August 2003

11. Shen, D.: Nash Strategies for Dynamic Noncooperative Linear Quadratic Sequential Games. Ph.D. Dissertation, Advisor: Cruz, Jr., J.B., the Ohio State University, 2006

12. Basar T., Olsder, G.J.: Dynamic Noncooperative Game Theory. SIAM Series in Classics in Applied Mathematics, second ed., January, 1999

13. Sastry, S.S.: Nonlinear Systems: Analysis, Stability and Control, Springer-Verlag, New York, NY, 1999

14. Brown, G.W.: Iterative solutions of games by fictitious play. Activity Analysis of Production and Allocation (T. C. Koopmans, ed.), New York: Wiley, 1951

15. Fudenberg, D., Levine, D.K.: The Theory of Learning in Games. Cambridge, MIT Press, 1998

16. Shen, D., Cruz, Jr., J.B., Chen, G., Kwan, C., Vannevel, A.: A Game Theoretic Approach to Mission Planning for Multiple Aerial Platforms. AIAA Infotech@Aerospace Conference, Arlington, VA, September 26-29, 2005

Characteristics of the Distribution of Hamming Distance Values Between Multidimensional Assignment Problem Solutions

Alla R. Kammerdiner[1], Pavlo A. Krokhmal[2], and Panos M. Pardalos[1]

[1] University of Florida, Gainesville, FL
[2] University of Iowa, Iowa City, IA

Abstract. The Multidimensional Assignment Problem (MAP) is a combinatorial optimization problem that arises in many important practical areas including capital investment, dynamic facility location, elementary particle path reconstruction, multiple target tracking and sensor fusion. Since the solution space of the MAP increases exponentially with the problem parameters, and the problem has exponentially many local minima, only moderate-sized instances can be solved to optimality. We investigate the combinatorial structure of the solution space by extending a concept of Hamming distance. The results of numerical experiments indicate a linear trend for average Hamming distance to optimal solution for the cases where one of the parameters is fixed.

1 Introduction

The Multidimensional Assignment Problem (MAP) is an NP-hard combinatorial optimization problem that can be viewed as a special case of the multi-index transportation problem, as well as a higher dimensional extension of the linear assignment problem (LAP). The MAP has many applications in such important practical areas as data association problems arising in multiple target tracking and sensor fusion, air traffic control, satellite launching, surveillance, dynamic facility location, capital investment, etc [1,2,3].

Most of the MAP solution methods are developed for the three-dimensional MAPs, although a number of important practical problems are modeled by MAP with a higher dimensionality parameter. For example, the problem of reconstructing the path of charged elementary particles produced by the Large Electron-Positron Collider is studied using the five-dimensional MAP as a mathematical model [4]. Due to inherent complexity of the MAP, not only the problem size increases extremely fast with increase in the MAP parameters, but also the mean number of local optima is exponential in the number of dimensions of the MAP [5]. However, all currently known exact methods for solving this immense problem are enumerative in nature, and therefore, such methods are too slow for many practical applications of the MAP. In addition, many exact and suboptimal algorithms developed for the multidimensional assignment problem utilize to a certain degree some kind of a local search procedure. Taking into

M.J. Hirsch et al. (Eds.): Adv. in Cooper. Ctrl. & Optimization, LNCIS 369, pp. 339–352, 2007.
springerlink.com

consideration a large number of possible local optima of the MAP, it becomes clear that the application of local search procedures for solving MAP instances is negatively affected by the structure of the solution space.

In this paper, we investigate some properties of the solution space structure using a natural extension of Hamming distance to evaluate the distance between a feasible solution of the MAP and its optimum. The outcomes of the numerical experiments clearly indicate that the structure of solution space is influenced by the problem parameters. This result is intuitively explained by the combinatorial representation of a feasible solution of the MAP as a collection of permutations. An application of Hamming distance to the solution space allows us to estimate by how much a feasible solution differs from the (unique) optimal solution.

This paper is organized as follows. Section 2 presents a formulation of the MAP as a combinatorial optimization problem, and discusses some important previous results. Section 3 introduces Hamming distance extension to the MAP, and describes the connection between the distance to the optimum and the corresponding LAP. Results of computational experiments calculating Hamming distances are reported in Section 4. Finally, the Section 5 summarizes our investigation into solution space structure of the MAP.

2   MAP Formulations and Related Previous Results

The MAP is often introduced as a higher dimensional extension of the linear assignment problem (LAP). The LAP is usually described as a problem of uniquely assigning each worker in a group a specific task, so that there is only one task for each worker, and each task is completed by only one worker. Notice that the LAP has two groups, a group of workers and a collection of tasks, and that is why its dimensionality parameter is two. The MAP has a similar interpretation, but the dimensionality parameter is increased. For example, we can uniquely assign each worker a task scheduled to be performed at a certain time slot and in a specified location, so every parameter (worker, task, time, location) is assigned to a unique quadruple, such that none of the parameters in the quadruple is in any other assignment. Clearly, the dimensionality of the MAP given by the above example is four.

W.P. Pierskalla initially considered the three-dimensional version of the MAP by extending the LAP in 1966 [6]. The first general formulation of the MAP was also given by W.P. Pierskalla as a zero-one integer programming problem in 1968 [7].

The MAP can be compactly formulated in the following fashion:

$$\min_{x \in \{0,1\}} \sum_{\substack{1 \le i_k \le n, \\ k \in \{1,\dots,d\} \setminus \{j\}}} c_{i_1 \dots i_d} \cdot x_{i_1 \dots i_d}, \tag{1}$$

$$\text{s. t.} \sum_{\substack{1 \le i_k \le n, \\ k \in \{1,\dots,d\} \setminus \{j\}}} x_{i_1 \dots i_d} = 1,\ 1 \le i_j \le n,\ 1 \le j \le d,$$

where $c_{i_1 \dots i_d}$ denote the cost coefficients.

More precisely, formulation (1) represents a d-dimensional "axial" MAP characterized by the same cardinality parameter n for element sets in each dimension.

The MAP has an interesting interpretation in terms of graph theory. It is well known that the LAP can be formulated using a bipartite graph. Analogously, the MAP can be described as a problem of finding a partition of the vertex set of a d-partite graph into n pairwise disjoint cliques of a minimum cost. More formally, the graph-theoretic definition of the MAP can be constructed as follows:

Given d mutually disjoint vertex sets $V_1, \ldots, V_d$ each with cardinality n, and the edge set E, let $G = (V_1, \ldots, V_d; E)$ denote a complete d-partite graph. A subset of the vertex set $V = \cup_{i=1}^{d} V_i$ is called a *clique* if it contains exactly one vertex from each V_i, $1 \leq i \leq d$. Suppose there is real-valued cost function defined on the set of cliques of G. Then the d-dimensional MAP constitutes a problem of determining such a partition of the vertex set V into a collection of n pairwise disjoint cliques, which minimizes the cost function.

The MAP (1) also has an alternative formulation as a combinatorial optimization problem:

$$\text{minimize} \sum_{1 \leq i \leq n} c_{i\pi_1(i)\ldots\pi_{d-1}(i)}, \text{ subject to } \pi_1, \ldots \pi_{d-1} \in \Pi^n, \qquad (2)$$

where $c_{j_1 j_2 \ldots j_d}$, $1 \leq j_k \leq n$, $1 \leq k \leq d$ denote the assignment cost coefficients, and Π^n is the set of all possible permutations of elements in the set $\{1, 2, \ldots, n\}$.

In other words, in order to solve the MAP given by a d-dimensional cubic matrix of cost coefficients of the size n^d, one must find a permutation of the rows and columns of the costs matrix minimizing the sum of the diagonal elements.

The combinatorial formulation (2) allows for a clear and efficient representation of a feasible solution of the MAP as an $n \times d$ matrix with columns given by permutations of $\{1, 2, \ldots, n\}$, i.e.

$$\begin{pmatrix} \pi_1(1) & \pi_2(1) & \ldots & \pi_d(1) \\ \pi_1(2) & \pi_2(2) & \ldots & \pi_d(2) \\ \vdots & \vdots & \ddots & \vdots \\ \pi_1(n) & \pi_2(n) & \ldots & \pi_d(n) \end{pmatrix} = (\pi_1 \ \pi_2 \ \ldots \ \pi_d),$$

where $\pi_i = (\pi_i(1) \ \pi_i(2) \ \ldots \ \pi_i(n))^\top$, for every $1 \leq i \leq d$, are permutations from Π^n.

To ensure that such representation is unique for every feasible solution, we must set the first column to be an identity permutation ι. Actually, it is enough to specify a permutation $\pi_j \in \Pi^n$ of arbitrary single column j of the matrix representation of feasible solutions to obtain a one-to-one correspondence between all feasible solutions of the MAP and their respective representations via permutations.

Also notice that given the matrix representation of a feasible solution above, the associated solution cost is

$$z = c_{\pi_1(1)\pi_2(1)\ldots\pi_d(1)} + c_{\pi_1(2)\pi_2(2)\ldots\pi_d(2)} + \ldots + c_{\pi_1(n)\pi_2(n)\ldots\pi_d(n)}$$

Using the commutativity of the summation, we can switch the terms in the solution cost without altering its value, and therefore, the solution remains the same under any permutation of rows in the solution matrix.

It follows directly from matrix representation of a feasible solution of the MAP as $(\iota\ \pi_2\ \ldots\ \pi_d)$ with the first column fixed as identity permutation ι that the cardinality N of the solution space can be calculated by the formula:

$$N = (n!)^{d-1} \tag{3}$$

In other words, the number of feasible solutions grows at least exponentially with increase in the problem parameters d and n. Nevertheless, all currently known exact methods developed for solving the MAP are enumerative in their nature. In particular, many of them use some variation of branch-and-bound techniques [7,8,9,10]. In addition, the MAP is known to be generally NP-hard, which follows by reduction to three-dimensional matching problem [11]. As a result of the inherent complexity of the problem, most exact methods for solving the MAP are designed specifically for its three-dimensional version.

A number of heuristic approaches have been used to solve the MAP, including simulated annealing [13], greedy randomized adaptive search procedure (GRASP) [12,14,15], and tabu search [16]. Most of these algorithms utilize some type of local neighborhood search. On the other hand, it was shown for the MAP with randomly distributed assignment cost coefficients that the expected number of local minima is exponential with respect to the number of dimensions [5]. Moreover, numerical experiments indicate that large numbers of local minima of the MAP have a statistically significant negative effect on performance of several heuristics that involve local neighborhood search, such as GRASP and simulated annealing [5].

For the remainder of this paper, we examine some characteristics of the structure of the space of all feasible solutions of the MAP with respect to its parameters d and n. For simplicity, we assume that the unique global minimum of the MAP exists. This is true, for example, (at least almost surely) in the case when assignment cost coefficients are randomly generated from a continuous distribution.

3   Hamming Distance

The Hamming distance was first introduced by R.W. Hamming in 1950 as a measure of errors (or substitutions) that transform one string of a binary code into another [17]. The Hamming distance has found applications in various areas, such as coding theory, information theory, cryptography, combinatorial optimization, etc [18]. Given two strings of an equal length with characters from any alphabet (not necessarily binary), the Hamming distance between them is usually defined as the number of positions in which these strings disagree.

The landscape structure of many combinatorial optimization problems can be investigated using the Hamming distance. In particular, the Hamming distance defined on the set of permutations of a given length was applied to study the fitness landscape of the quadratic assignment problem (QAP) [19]. Since any permutation $\pi \in \Pi^n$ can be represented as a string $(\pi(1)\,\pi(2)\,\ldots\,\pi(n))$ of characters $\pi(j) \in \{1, 2, \ldots, n\}$ that does not allow the same character at any two distinct positions, the Hamming distance between two permutations $\pi, \sigma \in \Pi^n$ is defined as

$$d_H(\pi, \sigma) = |\{j : \pi(j) \neq \sigma(j)\}|$$

Although any feasible solution of the QAP can be represented by means of a single permutation, the MAP does not allow such a representation. In fact, as shown in the previous section, any feasible solution of the MAP with parameters d and n can be uniquely represented by a $n \times d$ matrix

$$\begin{pmatrix} 1 & \pi_1(1) & \ldots & \pi_{d-1}(1) \\ 2 & \pi_1(2) & \ldots & \pi_{d-1}(2) \\ \vdots & \vdots & \ddots & \vdots \\ n & \pi_1(n) & \ldots & \pi_{d-1}(n) \end{pmatrix} = (\iota\;\pi_1\;\ldots\;\pi_{d-1}), \tag{4}$$

with the first column given by identity permutation ι, and the other $d-1$ columns given by the permutations $\pi_1, \ldots, \pi_{d-1} \in \Pi^n$. Furthermore, row permutations of the solution matrix do not change the solution. Because of this property, the extension of the Hamming distance to the MAP is not as straightforward as in the case of the QAP. To demonstrate this, let us consider a simple example of the MAP with parameters $d = 3$ and $n = 2$:

$$\pi = \begin{pmatrix} 1\,1\,1 \\ 2\,2\,2 \end{pmatrix}, \quad \sigma = \begin{pmatrix} 1\,2\,2 \\ 2\,1\,1 \end{pmatrix}$$

Clearly, π and σ are two solutions of the above MAP represented in the matrix form (4). If we defined the Hamming distance extension to MAP, simply as the number of positions in which these two matrices disagree, then the distance between π and σ would be 4. However, this approach is incorrect, since it does not take into account that the solution remains the same under the permutation of rows in the solution matrix. Permuting the rows of the matrix σ, we get:

$$\sigma = \begin{pmatrix} 2\,1\,1 \\ 1\,2\,2 \end{pmatrix},$$

which differs from π in only two positions, and hence, the Hamming distance between π and σ is 2, not 4.

Thus, the Hamming distance between any two feasible solutions $\pi = (\pi_1\,\pi_2\,\ldots\,\pi_d)$ and $\sigma = (\sigma_1\,\sigma_2\,\ldots\,\sigma_d)$ of the MAP is defined as the minimum number of positions in which two matrices disagree, i.e.

$$d_H(\pi, \sigma) = \min_{\tau \in \Pi^n} \sum_{k=1}^{d} |\{j : \pi_k(j) \neq \sigma_k(\tau(j))\}|$$

$$= \min_{\tau \in \Pi^n} \sum_{k=1}^{d} \sum_{j=1}^{n} \mathbb{I}_{\{\pi_k(j) \neq \sigma_k(\tau(j))\}} \tag{5}$$

This definition of the Hamming distance for the MAP can be given an alternative formulation in terms of the LAP with the cost coefficients given by the Hamming distances between two rows. Indeed, let us consider the corespondent matrix representations of two feasible solutions π and σ of the MAP:

$$
\pi = \begin{pmatrix} \pi_1(1) & \pi_2(1) & \dots & \pi_d(1) \\ \pi_1(2) & \pi_2(2) & \dots & \pi_d(2) \\ \vdots & \vdots & \ddots & \vdots \\ \pi_1(n) & \pi_2(n) & \dots & \pi_d(n) \end{pmatrix}, \quad \sigma = \begin{pmatrix} \sigma_1(1) & \sigma_2(1) & \dots & \sigma_d(1) \\ \sigma_1(2) & \sigma_2(2) & \dots & \sigma_d(2) \\ \vdots & \vdots & \ddots & \vdots \\ \sigma_1(n) & \sigma_2(n) & \dots & \sigma_d(n) \end{pmatrix},
$$

Since the solutions remains the same under any row permutation $\tau \in \Pi^n$, then σ can also be written as:

$$
\sigma = \begin{pmatrix} \sigma_1(\tau(1)) & \sigma_2(\tau(1)) & \dots & \sigma_d(\tau(1)) \\ \sigma_1(\tau(1)) & \sigma_2(\tau(2)) & \dots & \sigma_d(\tau(2)) \\ \vdots & \vdots & \ddots & \vdots \\ \sigma_1(\tau(n)) & \sigma_2(\tau(n)) & \dots & \sigma_d(\tau(n)) \end{pmatrix}.
$$

For any fixed row j of the solution matrix π, and any given permutation $\tau \in \Pi^n$, let the LAP cost coefficient $C^{\pi\tau}(j, \tau(j))$ be the Hamming distance between row j of the matrix π and row $\tau(j)$ of σ, i.e.

$$
C^{\pi\tau}(j, \tau(j)) = |\{k : \pi_k(j) \neq \sigma_k(\tau(j))\}| = \sum_{k=1}^{d} \mathbb{I}_{\{\pi_k(j) \neq \sigma_k(\tau(j))\}}, \tag{6}
$$

where $1 \leq j \leq n$ and $\tau \in \Pi^n$.

The LAP is a problem of finding the minimum cost assignment for elements from two disjoint sets of equal size. The LAP stated above, which calculates the Hamming distance of the MAP, assigns every row j from the solution matrix π a corresponding row $i = \tau(j)$ of σ matrix so that the total assignment cost given as a sum of cost coefficients (6) is minimized. Obviously, the permutation τ represents a feasible solution of the LAP. More formally, the LAP for computing the Hamming distance between matrices π and σ is formulated as follows:

$$
d_H(\pi, \sigma) = \min_{\tau \in \Pi^n} \sum_{j=1}^{n} C^{\pi\tau}(j, \tau(j)), \tag{7}
$$

where

$$
C^{\pi\tau}(j, \tau(j)) = \sum_{k=1}^{d} \mathbb{I}_{\{\pi_k(j) \neq \sigma_k(\tau(j))\}}
$$

By interchanging the order of summation, we obtain that the Hamming distance between solutions π and σ of the MAP given by (5) is equivalent to the Hamming distance definition via the corresponding LAP (7).

4   Numerical Results

In this section we report the results of computational experiments conducted to examine the solution space structure with respect to the Hamming distance extension for the MAP. We considered the MAP instances with different values for the parameters d and n. First, for every given pair of parameters (d, n), all feasible solutions were constructed using the matrix representation (4). Then, for each solution matrix, we computed the Hamming distance from the global minimum solution by solving the corresponding LAP. The average of the Hamming distances from the global minimum was obtained for each MAP with fixed parameters (d, n). The results of the conducted numerical experiments clearly show that the averages for the Hamming distances to the global minimum exhibit a linear trend when one of the problem parameters, either n or d is fixed, and the other parameter varies.

Let us discuss these experiments in more detail. Since the cardinality of the space of all feasible solutions of the MAP depends explicitly on the problem parameters d and n as shown by formula (3), it is logical to investigate whether the structure of the solution space of the MAP is also influenced by the parameters d and n. Therefore, we consider various values for the dimensionality parameter d and the cardinality of permutation set n. For each pair (d, n) of specified parameters of the MAP, we studied the solution space structure of the problem by constructing every feasible solution in the form of a $n \times d$ matrix representation of the solution given by (4). Since such a representation is comprised by the columns of permutations of the elements in $\{1, 2, \ldots, n\}$, then, in order to build the solution matrices, we first had to produce all possible permutations of n elements.

The following inductive procedure A was implemented to produce the set Π^n of all permutations:

PROCEDURE A.

- Start with a 2×2 matrix $M_0 = \begin{pmatrix} 1 & 2 \\ 2 & 1 \end{pmatrix}$ that stores all possible permutations of 2 elements.
- Set $k = 3$.
- Construct a row r_k of the size $(k - 1)!$ with all elements equal k.
- Produce a $k \times ((k - 1)!)$ matrix M by adding the row r_k at the bottom of the matrix M_0
- For j starting with $j = 1$ until $j = k - 1$ repeat the following:
  - Construct a $k \times ((k - 1)!)$ matrix M_1 by inserting the row r_k between the $(j - 1)$-th and j-th rows of matrix M_0 (i.e. the first $j - 1$ rows of M_1 consists of the first $j - 1$ rows of M_0 respectively, next the row j of M_1 is given by r_k, and finally the rest $k - 1 - j$ rows of M_1 is the corresponding rows j to $k - 1$ of matrix M_0).
  - Change M by appending M by the constructed matrix M_1 so that the modified matrix $M = (M \ M_1)$
  - Increment $j = j + 1$.

- Update the initial matrix $M_0 = M$ (M stores all possible permutations of k elements.
- Increment $k = k + 1$, and repeat the procedure until $k = n + 1$.

This procedure produces the $n \times (n!)$ matrix M_0, which stores all possible permutations $\pi_\alpha \in \Pi^n$, $1 \leq \alpha \leq n!$, as columns of the matrix. Having built all the permutations, we can construct the feasible solutions of the MAP with parameters d and n.

To construct each solution matrix σ, the following simple procedure B was utilized:

PROCEDURE B.

- Set the first column of σ to equal the identity permutation $\imath$.
- For k starting with $k = 2$ until $k = d$, repeat the following:
  - Randomly select a column c_k from $n!$ columns of the matrix M_0 produced by the previous procedure
  - Set the j-th column of the solution matrix σ equal to the chosen permutation c.
  - Increment $k = k + 1$.

The procedure B is based on formula (4), and so it produces a $n \times d$ matrix $\sigma = (\imath \, c_2 \, \ldots \, c_d)$ with $c_k \in \Pi^n$, $2 \leq k \leq d$, which gives a feasible solution of the MAP with parameters d and n.

As mentioned earlier, for the sake of simplicity, we suppose that the global minimum of the MAP is unique. Moreover, without loss of generality, we can assume that the global minimum is attained at a feasible solution $\pi = (\imath \, \imath \, \ldots \, \imath)$ with every column given by the identity permutation $\imath$. In fact, given a matrix representation of the global minimum solution, the above condition can be achieved by properly reordering the elements in each of d mutually disjoint sets. Suppose some column j, $1 \leq j \leq d$, of the global minimum solution π is a non-identity permutation π_j, then we just apply the inverse permutation π_j^{-1} on the elements from the j-th disjoint set.

Notice also that $\pi = (\imath \, \imath \, \ldots \, \imath)$ is one of the feasible solutions generated by the procedure B.

Next, we estimate by how much every feasible solution σ constructed by B differs from the unique global minimum π by applying the Hamming distance. More precisely, for every σ from the solution space, we compute the Hamming distance between σ and the global optimum solution π using definition (7) of the Hamming distance via the corresponding LAP.

Finally, for every considered MAP with specified problem parameters d and n, we calculate the average of the Hamming distances.

All procedures were implemented in C++, and the LAP was solved using shortest augmenting path algorithm for dense LAP proposed by R. Jonker and A. Volgenant [20].

Figures 1 and 2 display the average values of the Hamming distances to the global minimum solution of the MAP for various parameters.

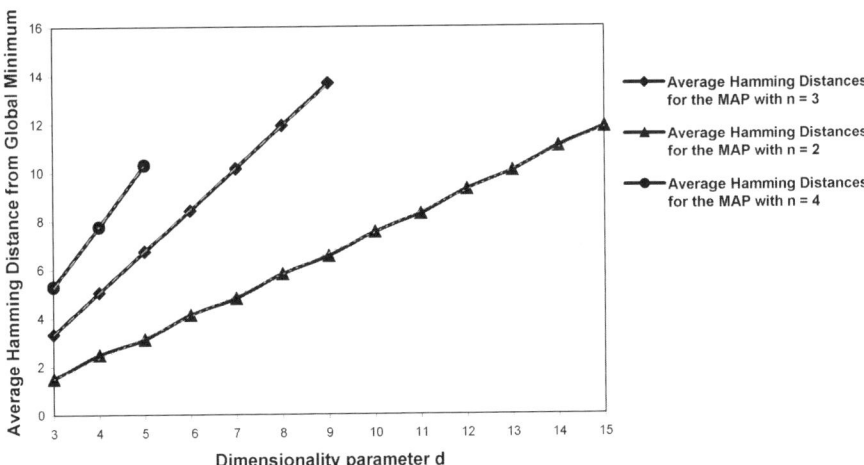

Fig. 1. The average Hamming distances from the global minimum of the MAP as functions of the dimensionality parameter d of the MAP

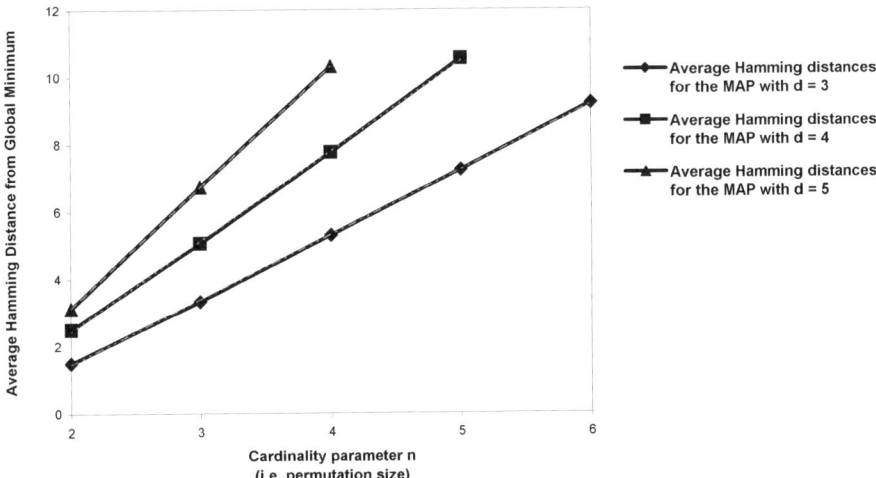

Fig. 2. The average Hamming distances from the global minimum of the MAP as functions of the cardinality parameter n of the MAP

To produce Figure 1, we temporarily fixed the parameter n to some value n_0, and varied the dimensionality of the problem. For each $n = n_0$, $n_0 = 2, 3, 4$, we treated the average Hamming distance to the global minimum as a function of d. By plotting all three functions, we discovered that their graphs clearly display linear trends. Furthermore, fitting a linear trendline to every given average Hamming distance function produced square errors that are very close to 1. In other words, the average Hamming distances from the global minimum solution for the MAP with a specified $n = n_0$ is an approximately linear function with respect to the dimensionality parameter d. The equations for the linear trendlines, which approximate these average Hamming distance functions depending on d, and the respective square errors R^2 are displayed in Table 1.

Table 1. Linear trends for graphs of the average Hamming distances in Figure 1

Parameter n	Equation of trendline	R-squared value
$n = 2$	$y = 0.8623x + 0.6199$	0.9994
$n = 3$	$y = 1.7235x + 1.5846$	0.9999
$n = 4$	$y = 2.5061x + 2.7731$	0.9999

Similar to plotting the average Hamming distances against the dimensionality parameter d for the MAP with a fixed n, we created graphs of the average distance functions with respect to the permutation size parameter n while fixing the dimensionality. We obtained plots of three functions of n, each of the functions corresponding to a specified parameter $d = d_0$, where $d_0 = 3, 4, 5$. Interestingly, for each value d_o, the average distance functions as functions of the disjoint sets size n also exhibit a strong linear trend. In fact, the square error values produced by fitting a linear trendline are extremely close to 1 as indicated in Table 2. The equations of the trendlines are also displayed in Table 2. In order to further investigate the structure of the solution space of the MAP with different problem parameters d and n, we charted the histograms of the Hamming distances between a feasible solution and the global solution. Next, we selected pairs of the MAPs of comparable problem size, and then compared these histograms. The results of such comparisons are presented in Figures 3 and 4.

Table 2. Linear trends for graphs of the average Hamming distances in Figure 2

Parameter n	Equation of trendline	R-squared value
$d = 3$	$y = 1.9303x - 0.4789$	0.9998
$d = 4$	$y = 2.68x - 0.2381$	0.9997
$d = 5$	$y = 3.5894x - 0.4572$	1

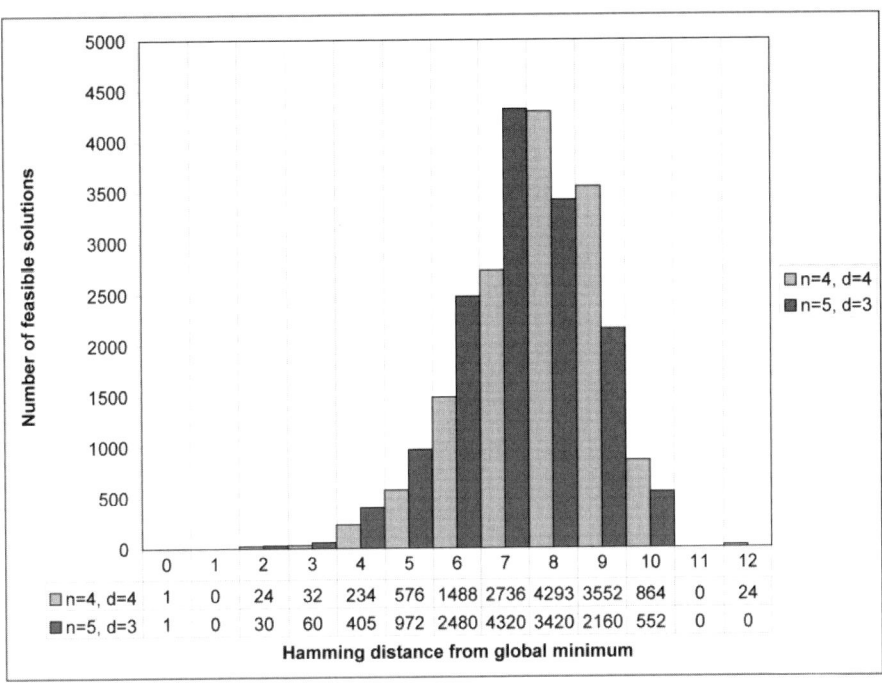

Fig. 3. The histograms of the Hamming distances from the global minimum of the MAP with $n = 4$, $d = 4$, and the MAP with $n = 5$, $d = 3$

For example, in Figure 3 we chose two problems, the MAP with $n = 4$, $d = 4$, and the MAP with $n = 5$, $d = 3$. The respective problem sizes are $N = 13,824$ and $N = 14,400$. Although the range of the first MAP is a little wider, and the two histograms appears slightly shifted, there is some similarity. In fact, both histograms are skewed towards the higher values of the Hamming distance. In other words, among all feasible solutions, there is a larger percentage of those solutions that are further away from the global minimum than those that are closer to the global solution. Hence, if we selected any solution at random, the probability of selecting a solution that is very different from the global minimum solution would be larger.

The patterns displayed in Figure 3 can also be seen in Figure 4, which compares the histograms of another pair of the MAPs. The first MAP has parameters $n = 3$ and $d = 8$, and the problem size $N = 279,936$, and the second MAP has parameters $n = 4$ and $d = 5$, and the problem size $N = 331,776$. In a similar way as before, there are some differences between the histograms of two problems. Specifically, the first MAP has a slightly wider range compared to the second MAP, and the histograms appear shifted with respect to each other. But, more importantly, Figure 4 also shows the previously observed similarity in histograms. Both histograms are characterized by the skewness toward the larger Hamming distance values, i.e. the percentage of feasible solutions, which

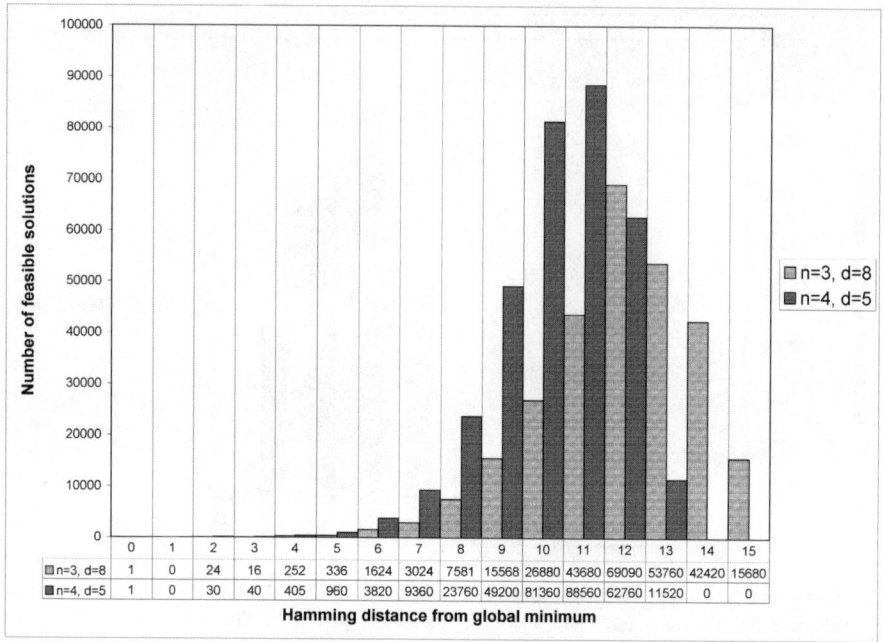

Fig. 4. The histograms of the Hamming distances from the global minimum of the MAP with $n = 4$, $d = 4$, and the MAP with $n = 5$, $d = 3$

differs from the global minimum solution in many positions, is greater than the percentage of the solution that are similar to the global minimum solution.

The examples represented in Figures 3 and 4 are not an exception. For every considered pair of parameters (d, n), we have found that the distribution of the Hamming distances between the feasible solutions and the global solution is rather heavily skewed towards the higher values of Hamming distances. This indicates that the percentage of feasible solutions, which have more differences than similarities with the unique global minimum solution, is definitely greater than the portion of the solution space relatively similar to the global.

5   Conclusions

In this paper, we investigated the solution space structure for the MAPs with various values of problem parameters d and n by means of the Hamming distance extension for the MAP. Since the matrix representation of a feasible solution is invariant to row permutation, we showed that the Hamming distance between two feasible solutions of the MAP can be computed by solving the LAP with the costs given by the Hamming distances between individual rows of two solutions. This result was applied to calculate the Hamming distances between the feasible solutions and the global optimum solution of the MAP, for different problem parameters. Furthermore, we examined the behavior of the average values of the

Hamming distances depending on a different choice of a specified parameter. The histograms of the Hamming distances to the global minimum were analyzed.

The results of the conducted computational experiments clearly indicate a strong linear trend for the Hamming distance as a function of one of the problem parameters, while having the other parameter set to some specified value. In particular, for each given value of the permutation size parameter n, the average Hamming distance to the global minimum can be very closely approximated by a linear function of the dimensionality parameter d. Similarly, for a specified value of the dimensionality parameter d of the MAP, the average of the Hamming distances between the global and a feasible solution is characterized by approximately linear growth with respect to the other problem parameter n. This finding can be used to estimate the Hamming distance for some problem parameters, by extrapolating the known linear trend.

In addition, we discovered that the histograms display certain patterns in the distribution of the Hamming distances that allow us to make some general conclusions about the structure of the solution space of the MAP. More precisely, the skewness of the Hamming distance distribution towards the higher values implies that the solution space consists largely of the solutions that are rather different from the unique global optimum solution, while the solutions that are similar to the global solution comprise a smaller portion of the solution space. A possible direction for further research could include a more detailed statistical analysis of the proportion of the solutions, which are similar to the global minimum, versus those different from the global solution.

References

1. A.B. Poore, *Multidimensional assignment formulation of data association problems arising from multitarget and multisensor tracking, Computation Optimization and Applications*, 3, 27 – 54, 1994
2. W.P. Pierskalla, *The tri-substitution method for the three-dimensional assignment problem*, Journal of Canadian Operation Research Society, 5, 71 – 81, 1967
3. E. Balas, P. Landweer, *Traffic Assignment in Communication Satelites*, Operations Research Letters, 2, 141 – 147, 1983
4. J.F. Pusztaszeri, P.E. Rensing, T.M. Liebling, *Tracking Elementary Particles Near Their Primary Vertex: A Combinatorial Approach*, Journal of Global Optimization, Vol. 9, No. 1, 41 – 64, 1996
5. D. Grundel, P. Krokhmal, C. Oliveira, P. Pardalos, *On the Number of Local Minima for the Multidimensional Assignment Problem*, Journal of Combinatorial Optimization, 2006, to appear
6. W.P. Pierskalla, *The Tri-Substitution Method for Obtaining Near-Optimal Solutions to the Three-Dimensional Assignment Problem*, Tech.Memo.No. 71, Operations Research Group, Case Institute of Technology, Cleveland, Ohio, October 1966
7. W.P. Pierskalla, *The multidimensional assignment problem*, Operations Research, 16, 422–431, 1968
8. E. Balas and M.J. Saltzman, *An algorithm for three-index assignment problem*, Operations Research, 39, 150 - 161, 1991

9. M. Vlach, *Branch and bound method for the three-index assignment problem*, Economicko-Matematicky Obzor, 3, 181 - 191, 1967

10. D. Magos and P. Miliotis, *An algorithm for the planar three-index assignment problem*, European Journal of Operational Research, 77, 141 - 153, 1994

11. M.R. Garey, D.S. Johnson, *Computers and Intractability: A Guide to the Theory of NP-completeness*. W.H. Freedman and Company, 1979

12. R. Murphey, P. Pardalos, L. Pitsoulis, *A greedy randomized adaptive search procedure for the multitarget multisensor tracking problem*, in: DIMACS Series, vol. 40, American Mathematical Society, pp. 277-302, 1998

13. W. Clemons, D. Grundel, and D. Jeffcoat, *Applying simulated annealing on the multidimensional assignment problem*, in Recent Developments in Cooperative Control and Optimization Series: Cooperative Systems , Vol. 3, Butenko, S.; Murphey, R.; Pardalos, P.M. (Eds.), 2004

14. R.M. Aiex, M.G.C. Resende, P.M. Pardalos, G. Toraldo, *GRASP with path relinking for the three-index assignment problem*, INFORMS J. on Computing, vol. 17, no. 2, pp. 224 – 247, 2005

15. N. Lidstrom, P. Pardalos, L. Pistoulis, G. Toraldo, *An approximation algorithm for the three-index assignment problem*, Technical Report, 1997

16. D. Magos, *Tabu search for the planar three-dimensional assignment problem*, Journal of Global Optimization, 8, 35 - 48, 1996

17. R.W. Hamming. *Error Detecting and Error Correcting Codes*, Bell System Technical Journal 26(2):147–160, 1950.

18. R. Matsumoto, K. Kurosawa, T. Itoh, T. Konno, T. Uyematsu, *Primal-dual distance bounds of linear codes with application to cryptography*, IEEE Trans. Inform. Theory, vol. 52, no. 9, 4251 – 4256, Sept. 2006

19. P. Merz, B. Freisleben, *Fitness landscape analysis and memetic algorithms for the quadratic assignment problem*, IEEE Transactions on Evolutionary Computation, 4(4): 337–352, 2000

20. R. Jonker and A. Volgenant, *A Shortest Augmenting Path Algorithm for Dense and Sparse Linear Assignment Problems*, Computing 38, 325-340, 1987

Robust Cooperative Visual Tracking: A Combined NonLinear Dimensionality Reduction/Robust Identification Approach[*]

Vlad I. Morariu[1], Octavia I. Camps[2], Mario Sznaier[2], and Hwasup Lim[3]

[1] Computer Vision Laboratory, University of Maryland, College Park, MD 20742
morariu@umd.edu
[2] Robust Systems Lab, ECE Department, Northeastern University, Boston, MA 02115
{camps,msznaier}@ece.neu.edu
[3] Dept. of Elect. Eng., Penn State University, University Park, PA 16802
hxl211@psu.edu

Abstract. In this chapter we consider the problem of robust visual tracking of multiple targets using several, not necessarily registered, cameras. The key idea is to exploit the high spatial and temporal correlations between frames and across views by (i) associating to each viewpoint a set of intrinsic coordinates on a low dimensional manifold, and (ii) finding an operator that maps the dynamic evolution of points over manifolds corresponding to different viewpoints. Once this operator has been identified, correspondences are found by simply running a sequence of frames observed from one view through the operator to *predict* the corresponding current frame in the other view. As we show in the chapter, this approach substantially increases robustness not only against occlusion and clutter, but also against appearance changes. In addition, it provides a scalable mechanism for sensors to share information under bandwidth constraints. These results are illustrated with several examples.

1 Introduction

In this chapter we consider the problem of robustly tracking multiple targets using several, not necessarily registered, cameras. In principle, tracking targets using multiple cameras should increase robustness against occlusion and clutter since, even if the targets appear largely occluded to some sensors, the system can recover by using the others. Furthermore, examining data from spatially distributed cameras can reveal activity patterns not apparent to single or closely clustered sensors. However, although intuitively appealing, multicamera tracking *does not necessarily improve robustness*. This is illustrated in Figure 1, showing the results of an experiment where a Kalman filter based tracker is implemented using data from two (registered) cameras. Even though the target is always visible in at least one of the cameras, the tracker still loses it, due to occlusion resulting in incorrect data from the other camera.

---

[*] This work was supported in part by AFOSR under grant FA9550-05-1-0437 and NSF under grants IIS-0117387, ECS-0221562 and ITR-0312558.

M.J. Hirsch et al. (Eds.): Adv. in Cooper. Ctrl. & Optimization, LNCIS 369, pp. 353–371, 2007.
springerlink.com   © Springer-Verlag Berlin Heidelberg 2007

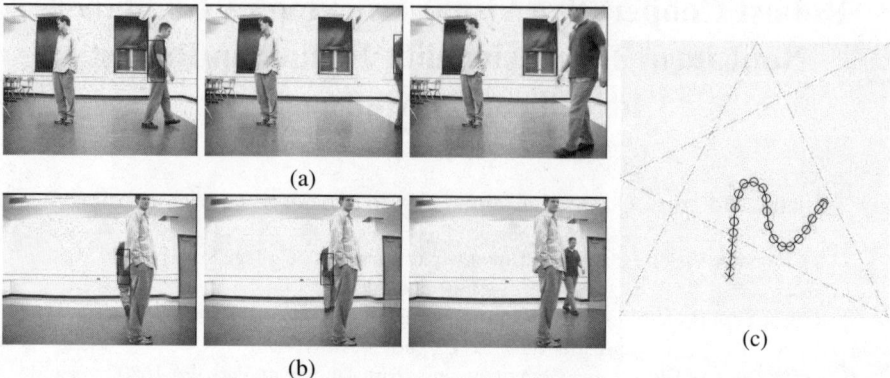

(a)

(b)

(c)

Fig. 1. Multicamera tracking: (a) West view, (b) North view, (c) The trajectory of the target is estimated incorrectly (red crosses) after the target leaves and re-enters the field of view of one of the cameras

Avoiding situations like the one illustrated above requires an efficient coordination mechanism to (i) reject incorrect measurements, and (ii) maintain consistent identity labels of the targets across views. Previous approaches to the "correspondence across views" problem include matching features such as color and apparent height [1; 2; 3; 4], using 3D information from camera calibration [2; 5; 6; 7; 8] or computing homographies between views [9; 10; 11]. More recently, Khan and Shah [12] presented an approach based on finding the limits of the field of view of each camera as visible by the other cameras under the assumption that the world is planar. However, it can be difficult to find matching features across significantly different views, camera calibration information is not always available and planar world hypothesis can be too restrictive.

To avoid these difficulties, in this chapter, we propose a new approach to the problem of cooperative multicamera tracking that does not require feature matching, camera calibration or planar assumptions. The key idea is to exploit the high spatial and temporal correlations between frames and across views by (i) associating to each viewpoint a set of intrinsic coordinates on a low dimensional manifold and (ii) finding an operator that maps the dynamic evolution of points over manifolds corresponding to different viewpoints. Once this operator has been identified, correspondences are found by simply running a sequence of frames observed from one view through the operator to *predict* the corresponding current frame in the other view. It is worth emphasizing that this approach substantially increases robustness not only against occlusion and clutter, but also against appearance changes. In addition, it provides a scalable mechanism for sensors to share information under bandwidth constraints. These results are illustrated with several examples.

2   Notation

$\mathcal{H}_\infty$ denotes the space of functions with bounded analytic continuation inside the unit disk, equipped with the norm: $\|G\|_\infty \doteq ess\sup_{|z|<1} \overline{\sigma}\{G(z)\}$, where $\overline{\sigma}(.)$ is the maximum singular value. ℓ_∞ denotes the space of vector valued sequences $\{\mathbf{x}_i\}$ equipped with the norm: $\|\mathbf{x}\|_\infty \doteq \sup_i \|\mathbf{x}_i\|_\infty$. Similarly, ℓ_2 denotes the space of vector valued sequences equipped with the norm: $\|\mathbf{x}\|_2^2 = \sum_{i=0}^\infty \|\mathbf{x}_i\|^2$, where $\|.\|$ is the usual euclidian norm in R^n. Given a sequence $\{\mathbf{x}_k\}$, $\mathbf{x}(z) \doteq \sum_{i=0}^\infty \mathbf{x}_k z^k$ denotes its z–transform. Finally, given a finite sequence $\{x_k\}_{k=0}^{n-1}$, T_x^n denotes its corresponding (lower triangular) Toeplitz matrix:

$$\mathsf{T}_x^n \doteq \begin{bmatrix} x_o & 0 & \cdots & 0 \\ x_1 & x_o & \ddots & \vdots \\ \vdots & \vdots & \vdots & \ddots & \vdots \\ x_{n-1} & x_{n-2} & \cdots & x_1 & x_o \end{bmatrix}$$

3   Dynamic Identification Based Robust Tracking

In this section we show that robust multicamera tracking can be reduced to a convex optimization problem. For simplicity, in the sequel we first present the main ideas using the simpler single camera case and then extend these ideas to multicamera scenarios. In principle, the location of a target in a video sequence can be predicted using a combination of its (assumed) dynamics, empirically learned noise distributions and past position observations [13; 14; 15; 16]. While successful in many scenarios, these approaches

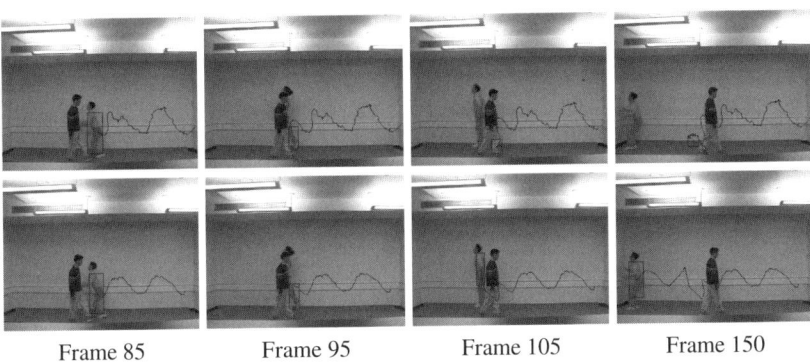

| Frame 85 | Frame 95 | Frame 105 | Frame 150 |

Fig. 2. Tracking in the presence of occlusion. Top: Unscented Particle Filter based tracker loses the target due to occlusion. Bottom: Combination Identified Dynamics/Kalman Filter tracks through the occlusion.

remain vulnerable to model uncertainty and occlusion, as illustrated in the top portion of Figure 2. Following the approach introduced in Camps et al [17] for the single camera case, in this chapter we will address these difficulties by modeling the motion of the target as the output of an operator driven by a stochastic signal. Specifically, consider first the simpler case where the dynamics of the target are approximately linear and start by modelling the evolution of y, the position of a given target feature as:

$$y(z) = \mathcal{H}(z)e(z) + \eta(z) \tag{1}$$

where e_k and η_k represent a suitable input and measurement noise, respectively, $y(z)$, $e(z)$ and $\eta(z)$ denote the corresponding z-transforms, and where the operator $\mathcal{H}$ is not necessarily ℓ_2 stable. For example, in the case of a feature moving with random acceleration, $H(z) = \frac{z^2}{(z-1)^2}$. Further, we will assume that the following *a priori* information is available:

(a) Set membership descriptions $\eta_k \in \mathcal{N}$ and $e_k \in \mathcal{E}$. These can be used to provide deterministic models of the stochastic signals e, η.

(b) $\mathcal{H}$ admits an expansion of the form $\mathcal{H} = \overbrace{\sum_{j=1}^{N_p} p_j \mathcal{H}^j}^{\mathcal{H}_p} + \mathcal{H}_{np}$. Here $\mathcal{H}^j$ are known,

given, not necessarily ℓ_2 stable operators that contain all the information available about possible modes of the target[1].

In this context, the next location of the target feature y_k can be predicted by first identifying the relevant dynamics $\mathcal{H}$ and then using it to propagate the effect of the input e. In turn, identifying the dynamics entails finding an operator $\mathcal{H}(z) \in \mathcal{S} \doteq \{\mathcal{H}(z): \mathcal{H} = \mathcal{H}_p + \mathcal{H}_{np}\}$ such that $y - \eta = \mathcal{H}e$, precisely the class of interpolation problem addressed in [18]. As shown there, such an operator exists if and only if the following set of equations in the variables $\mathbf{p}, \mathbf{h}$ and K is feasible:

$$\mathbf{M}(\mathbf{h}) = \begin{bmatrix} \mathsf{I} & \mathsf{T}_h^T \\ \mathsf{T}_h & K^2\mathsf{I} \end{bmatrix} \geq 0 \tag{2}$$

$$\mathbf{y} - \mathbf{P}\mathbf{p} - \mathbf{h} \in \mathcal{N} \tag{3}$$

where T_h denotes the Toeplitz matrix associated with the sequence $\mathbf{h} = [h_1, \ldots, h_n]$, the first n Markov parameters of $\mathcal{H}_{np}(z)$, and $\mathsf{P} \doteq [f^1\ f^2\ \cdots\ f^{N_p}]$, where f^i is a column vector containing the first n Markov parameters of the *i-th* transfer function $\mathcal{H}^i(z)^2$.

The effectiveness of this approach is illustrated in the bottom portion of Figure 2, showing that a Kalman filter based tracker using the identified dynamics for prediction, instead of a purely assumed simple model such as constant acceleration, is now able to track the target past the occlusion.

Consider now the situation where several (roughly) registered cameras are available. In this case the resulting geometric constraints translate into additional convex constraints that can be added to the identification above. This allows for individual cameras to accurately "guess" the location of a momentarily occluded target by simply translating to the local coordinate system measurements provided by other (non–occluded) cameras and then propagating these measurements through the local model. Figure 3 shows the result of applying the approach outlined above to the same two–camera example used in the introduction. As shown there, the resulting tracker is now capable of continuous

---

[1] If this information is not available the problem reduces to purely non–parametric identification by setting $\mathcal{H}^j \equiv 0$. In this case the proposed approach still works, but obtaining comparable error bounds requires using a larger number of samples.

[2] Here, we have assumed without loss of generality (by absorbing the spectral properties of e into $\mathcal{H}$, if necessary), that $e_k = \delta(0)$, a unit impulse applied at $k = 0$.

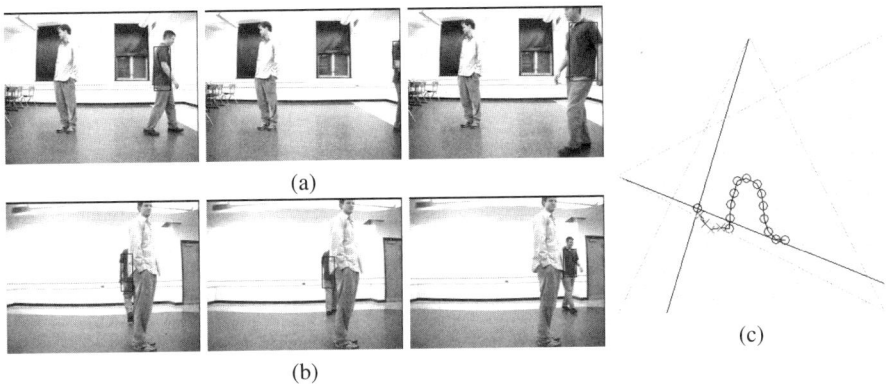

(a)

(b)

(c)

Fig. 3. Dynamics based multicamera tracking: (a) West view, (b) North view, (c) The trajectory of the target is correctly estimated (red crosses) even after the target leaves and re-enters the field of view of one of the cameras

tracking, even when the target is momentarily occluded to one of the cameras. In this example, the experimental information used for identifying the dynamics consisted of centroid position measurements from the first 12 frames, where the target is not occluded. The *a priori* information, estimated from the non–occluded portion of the trajectory is:

1. measurement noise level 5 pixels
2. $\mathcal{H}_p \in \mathrm{span}[1, \frac{1}{z-1}, \frac{z}{z-1}, \frac{z}{(z-1)^2}, \frac{z^2}{(z-1)^2}]$

Using this information, the minimum value of K yielding feasibility of the LMI (2) was found to be $K = 5 \cdot 10^{-4}$, indicating that indeed the relevant dynamics are captured by the parametric portion $\mathcal{H}_p$. During operation of the tracker, the target in each camera was segmented by the backprojection method using the hue histogram and occlusion was detected by changes in its size. In this event, the camera used information from the second sensor, when available, together with the local dynamics, to predict the position of the target.

4   Handling Nonlinear Dynamics and Computational Complexity

As illustrated with the simple example above, the approach outlined in the previous section has the potential to exploit multicamera information to accomplish robust tracking in the presence of severe occlusion. However, extending this approach to realistic, more complex scenarios requires addressing the issues of (i) nonlinear target dynamics and (ii) the computational complexity entailed in combining data from multiple sensors, due to the poor scaling properties of LMI based identification algorithms[3]. As we show in this section, both issues can be addressed by using nonlinear dimensionality reduction methods to project features to points on low dimensional manifolds where the dynamics

---

[3] Recall that the computational complexity of conventional LMI solvers scales as the number of decision variables raised to the 10th power [19].

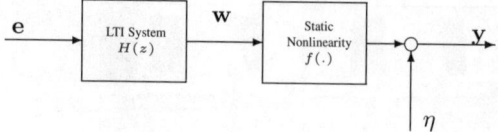

Fig. 4. Wiener System Structure

are linear. Computationally efficient camera coordination can be achieved by having the cameras share projections onto these manifolds (and associated dynamical models), rather than high dimensional raw video streams. Since the projection onto the lower dimensional manifold can be modelled as a static nonlinearity, this approach leads naturally to a Wiener system structure of the form illustrated in Figure 4, consisting of the interconnection of a LTI system $H(z)$ and a memoryless nonlinearity $f(.)$. Identification of the linear dynamics on the manifold can be accomplished using essentially the same methods described in Section 3; identification of the nonlinearity $f(.)$ is addressed next.

4.1 Nonlinear Manifold Learning

Correlation of image sets has been extensively used in image compression, object recognition and tracking [20; 21; 22; 23; 24]. In these applications, images are viewed as high dimensional vectors that can be represented as points in lower dimensional subspaces without much loss of information. Principal component analysis (PCA) is the tool most often used to extract the linear subspaces in which the data has the highest variance. More recently, low-dimensional linear subspace models have been proposed to predict an image sequence from a related image sequence [25; 26] and to model dynamic texture [27].

However, image data does not usually lie in a linear subspace, but instead on a low dimensional nonlinear manifold within the higher dimensional space [28; 29; 30; 31; 32; 33; 34; 35; 36; 37; 38; 39]. As a result, images that are far apart can have similar representations when they are projected onto a linear subspace using a PCA decomposition.

Thus, in this chapter we propose to use a nonlinear dimensionality reduction technique to obtain low dimensional mappings that preserve the spatial and temporal neighborhoods of the data. There are various techniques that can be used for this purpose. Methods such as [36; 38; 39; 40; 41; 42] seek to find an embedding of the data which preserves some relationship between the datasets, without providing an explicit mapping function.

Ideally, we would like to use a nonlinear manifold learning technique such as [28; 30; 37; 43] that provides both the mapping and the embedding of our training set. However, such luxury comes at extra computational cost and algorithm complexity. Thus, in order to obtain algorithms compatible with real time operation, in this chapter we use the locally linear embedding (LLE) algorithm to find the embedding of the data [36]. Though LLE does not directly provide a mapping from the high dimensional image space to the embedding space, methods similar to those described in [36] can approximate the mapping.

Given a set of images $X = [x_1 \ldots x_n] \in \mathbb{R}^{D \times n}$, where x_i is the view of an object at time i, we want to find an embedding $Y = [y_1 \ldots y_n] \in \mathbb{R}^{d \times n}$ such that $d \ll D$. The

LLE algorithm finds an embedding where data point relationships in the high dimensional space are preserved in the embedding.

To learn a locally linear embedding of X, we seek to represent each sample x_i as a linear combination of k neighbors. We define $i \sim j$ to be true if i is a neighbor of j. Thus, we want to find the weights W_{ij} so that for each sample x_i

$$W = \underset{W}{\mathrm{argmin}} \sum_i |x_i - \sum_j W_{ij}x_j|^2 \qquad (4)$$

so that $\sum_j W_{ij} = 1$ and $W_{ij} = 0$ if x_i and x_j are not neighbors. Using these weights we then find the embedding Y so that

$$Y = \underset{Y}{\mathrm{argmin}} \sum_i |y_i - \sum_j W_{ij}y_j|^2 \ . \qquad (5)$$

Letting
$$L = (I - W)^T(I - W), \qquad (6)$$

the solution is found by calculating the eigenvalues and eigenvectors of L. Because it can be shown that the smallest eigenvalue is zero, the embedding coordinates are given by $Y = [v_2 \ldots v_{d+1}]^T$, where v_i is the eigenvector corresponding to the ith smallest eigenvalue of L.

Fig. 5. Representative frames from a walking sequence

To map a new vector x_{new} into the embedding, we use the method described in [36]. We find the k nearest neighbors of x_{new} in the training set X, and compute the weights corresponding to the neighbors which best approximate x_{new}. Using these weights we combine the values in Y corresponding to the neighbors to get an approximation of the new coordinates in the embedding, y_{new}. A similar approach can be used to map from the embedding coordinates to the initial high dimensional space. The values needed for k and d depend on the intrinsic dimensionality of the input dataset, so there is no preset value. The problem of finding acceptable values for k and d is explored in more depth by Saul and Roweis [36]. The constraints we place on the weights also have an effect on the embeddings. For example, we can allow the weights to be negative values to give us an affine reconstruction, or we can force the weights to be positive to give a convex reconstruction. Affine weights can be found in closed form and they do not cause the embedding corners to be rounded. Convex weights provide more robustness to noise, but are found by solving a convex quadratic programming problem [36]. In our experiments, we found that convex weights result in a lower normalized error. Affine reconstruction weights resulted in very high normalized error in cases where the weights were of very high magnitude (such as 17.26 and -16.26 for two neighbors).

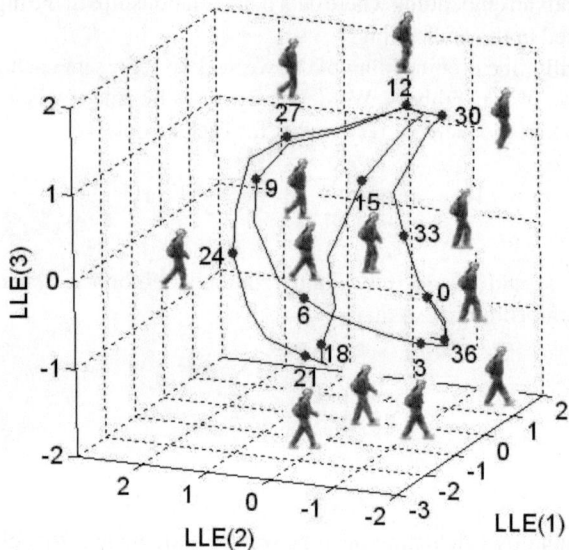

Fig. 6. Low dimensional representation of the walking sequence using Locally Linear Embeddings(LLE)

Figures 5 and 6 illustrate the projection of a walking sequence onto a low dimensional manifold using the LLE technique. Figure 7 shows the embeddings of sequences of a person walking on a treadmill obtained from the CMU MoBo database.

4.2 System Dynamics Identification in Manifold Space

Once the low dimensional manifold has been found, the dynamics governing the motion there can be found using the identification approach outlined in Section 3, by simply using as data the projection w_k on the manifold, rather than the actual high dimensional feature y_k (see Figure 4).

Figure 8 illustrates the use of Caratheodory-Fejer (CF) interpolation to learn the temporal evolution of the points on an embedding. In this example, CF interpolation was applied to one of the embeddings shown in Fig. 7 corresponding to a sequence of 160 frames. The dynamics of the points on this embedding was learned from its first 80 points, assuming an impulse signal as the input. Figure 8 (top) shows the close agreement between the temporal evolution of the coordinates of the points on the embedding and the positions predicted by the CF identified dynamics. An alternative view of these results is given in Fig. 8 (bottom) where the predicted and actual points on the embedding are shown.

4.3 Learning View Correspondences

After obtaining low dimensional representations of a set of video sequences, we want to learn correspondences between views across sequences. One way to learn this correspondence is to align the embeddings so that corresponding views map to the same low

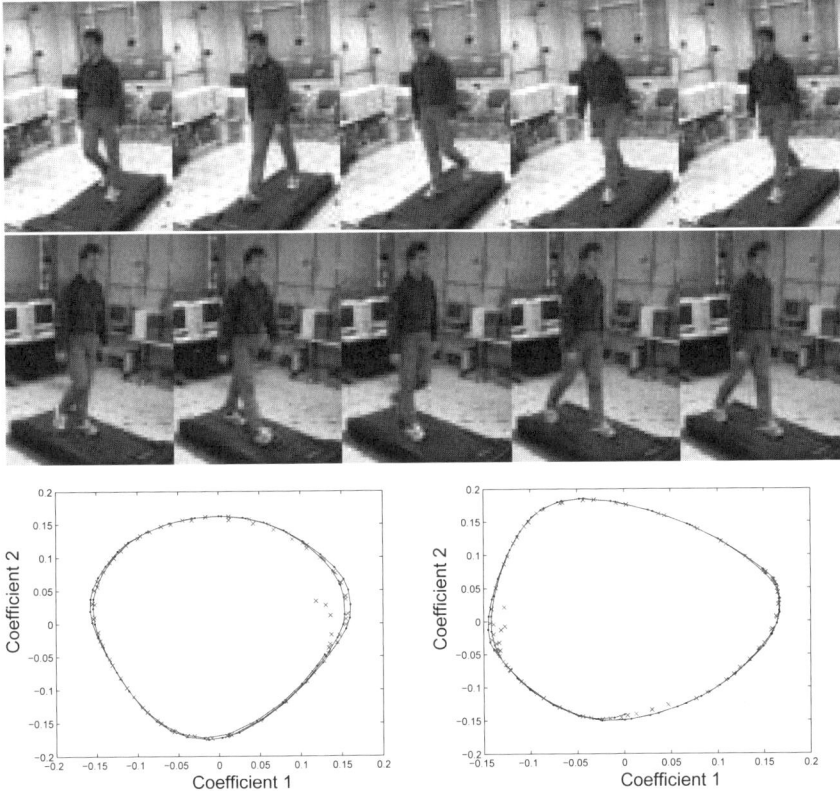

Fig. 7. Top: Sample images. Bottom: Embeddings of two sequences found by LLE. Blue and red points are training and test image embedding coordinates, respectively.

dimensional coordinates. Another option is to model correspondence as an input-output LTI system, where the embedding coordinates of one view are the input to the system and the corresponding image embedding coordinates are the output. These approaches are described in more detail next.

Correspondences By Embedding Alignment.   Finding correspondences between views of two video sequences X^1 and X^2 becomes trivial if their corresponding manifolds are aligned – i.e. if corresponding views $x_i^1 \in X^1$ and $x_j^2 \in X^2$ have *identical* low dimensional embedding representations $y_i^1 = y_j^2$. In general one-to-one correspondences between all training views are not available, since the cameras may not be synchronized or one camera may be occluded at times. However, it is not unreasonable to assume that *some* correspondences might be available. In this case, the method proposed in [34; 35] can be used to align the manifolds.

First we divide the data sets into subsets for which we know correspondences and for which we do not. Let X_c^1 and X_c^2 contain the same number of samples each, where x_i^1 corresponds to x_i^2. Similarly X_u^1 and X_u^2 contain the samples from each sequence for

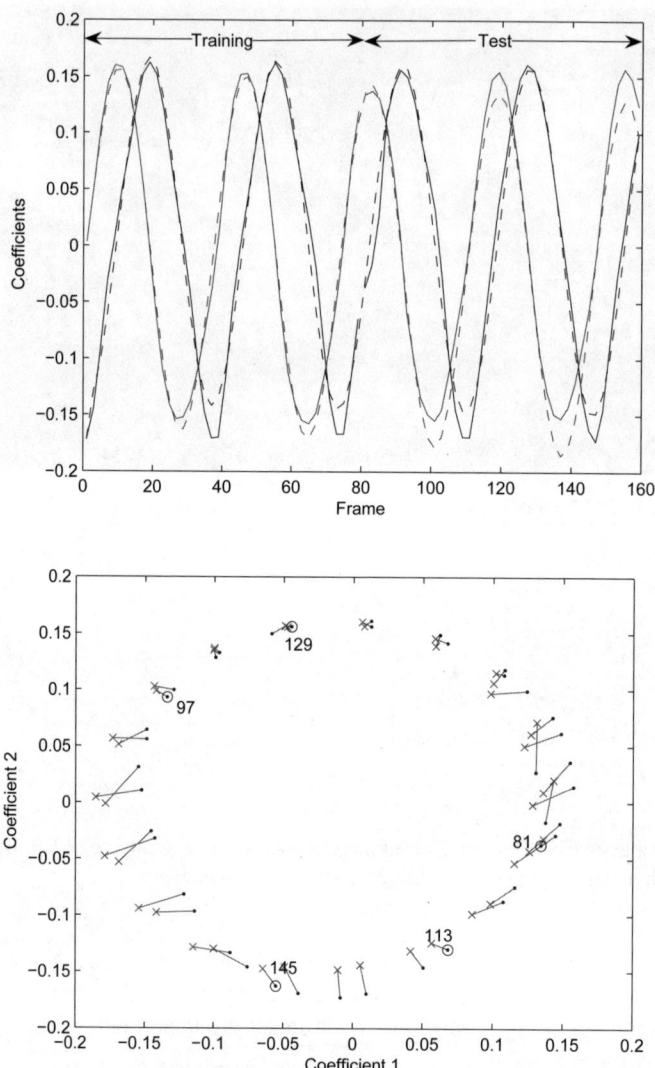

Fig. 8. Learning temporal dynamics. Top: First two coefficients of sequence 2 as time progresses. Solid and dotted lines show actual and interpolated coefficients, respectively. Bottom: The predicted(red) and actual(blue) points on the embedding.

which we do not know correspondences (X_u^1 and X_u^2 can be empty and do not necessarily have the same number of samples).

To align two data sets where we know the correspondence of some or all of the samples, we first compute L^1 and L^2 as shown in Equation 6, where $X^1 = \begin{bmatrix} X_c^1 & X_u^1 \end{bmatrix}$ and $X^2 = \begin{bmatrix} X_c^2 & X_u^2 \end{bmatrix}$. We can then split each L^k into corresponding and non-corresponding parts:

$$L^k = \begin{bmatrix} L_{cc}^k & L_{cu}^k \\ L_{uc}^k & L_{uu}^k \end{bmatrix}.$$

To find the embedding where $Y_c^1 = Y_c^2$ is a hard constraint, we let

$$L = \begin{bmatrix} L_{cc}^1 + L_{cc}^2 & L_{cu}^1 & L_{cu}^2 \\ L_{uc}^1 & L_{uu}^1 & 0 \\ L_{uc}^2 & 0 & L_{uu}^2 \end{bmatrix}$$

and we then find the eigenvalues and eigenvectors for the solution. Once the embedding is computed, we can then map a new sample x_{new}^1 into the embedding using the method described above to get y_{new}^1, which we assume is equal to y_{new}^2 since the embeddings are aligned for the two sequences. We can then generate the second image by mapping from y_{new}^2 to x_{new}^2. The results of this approach are illustrated in Fig. 9 where the embeddings from Fig. 7 are now aligned using LLE.

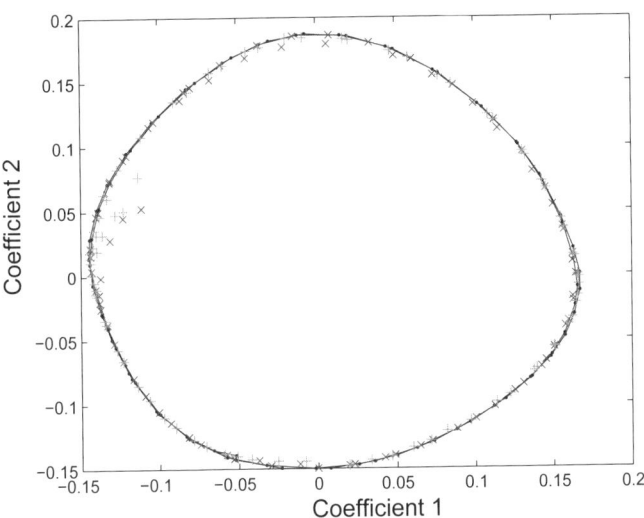

Fig. 9. Embeddings aligned using LLE. Blue dots: training embeddings. Red X: test sequence 2 embeddings. Green +: test sequence 5 embeddings.

Correspondences by System Identification. An alternative approach to finding view correspondences is to capture the temporal correlations between sequences with a LTI operator that generates as output the points on the manifold from one camera when it is excited with a sequence of points from the manifold of the other camera as an input. This operator can be easily identified with the CF interpolation technique described in Section 3, by setting in Equation (1) f and e to the coordinates of sets of points in the first and second manifold, respectively[4]. This approach is illustrated in Figure 10. Figure 11 shows plots of the temporal evolution of the coordinates of the points on two embeddings,

---

[4] Note that the number of points in f and e do not have to be the same.

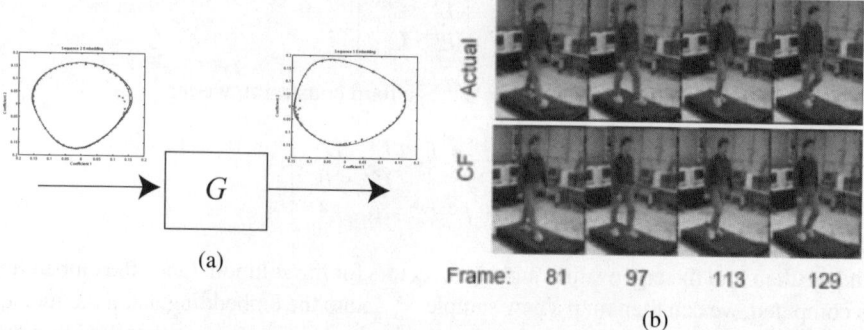

(a)

(b)

Fig. 10. (a) Operator mapping manifolds (b)Actual (top) and Predicted (bottom) correspondences

and the predictions obtained by learning the dynamic relation between them. In this case, f was set to the coordinates of the first 80 points of one embedding and e was set to the coordinates of the corresponding points on the second embedding. The plot on the top of the figure shows the accuracy of the predictions for the next 80 points, obtained using the learned dynamics excited with the coordinates from the second embedding.

4.4 Generating Views

If the correspondences between views and their dynamics are learned using the methods described above, they can be used to generate new views in two situations: (1) when at

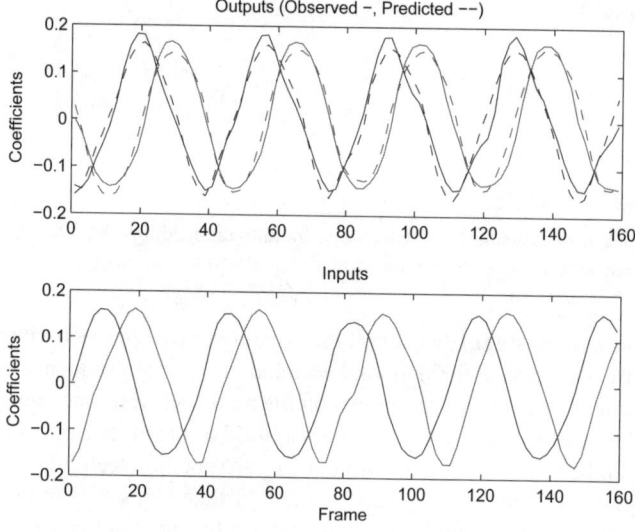

Fig. 11. View correspondences using system dynamics. Top: First two output coefficients as time progresses. Solid and dotted lines show actual and interpolated coefficients, respectively. Bottom: First two coefficients of sequence 2 are the inputs.

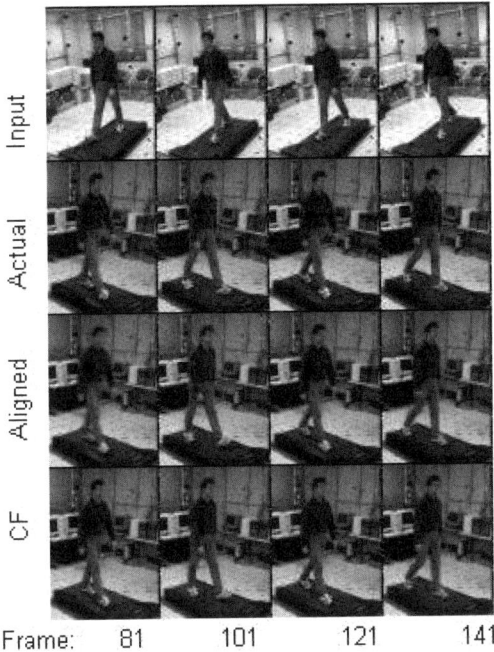

Fig. 12. Generating one sequence from another. Row 1: input. Row 2: actual images. Rows 3 and 4: generated by aligned LLE and CF interpolation, respectively.

time t, we have the image of an object in one view but not in the other, and (2) when we do not have the image of an object in any of the views at time t but we had it in the previous views.

In the first case, we can generate a new image in one of two ways, depending on how the correspondences were learned. If the embeddings were aligned during training by the dimensionality reduction method, then we can simply map the input view x_{in} onto the embedding to get a corresponding y_{in}. Since the embeddings of both views are aligned, $y_{in} = y_{out}$, so we simply map y_{out} into the output space using the neighbors of y_{out} from the output sequence. If the embeddings were aligned using system identification, then y_{in} and y_{out} are not equal, but are related by a dynamic system that we learned. Thus, we can obtain y_{out} from a sequence of inputs from the other manifold using the identified dynamics, and then map it into the high dimensional output space to get a new view. We note that each mapping(to and from) will use different neighboring points in the embedding since the training sequences can be of different sizes and not all images in the sequences are in one-to-one correspondence. Figure 12 illustrates the results of using both methods to generate missing views on the treadmill sequences. We conducted our experiments on the first 160 frames of the *slowWalk* image sequence from the CMU MoBo database [44]. The first 80 images were used to train our embeddings and the last 80 were used for testing the reconstruction of the views. One sequence (top row) is used as input to generate the other (row 2). Both methods are very effective at reconstructing the actual views.

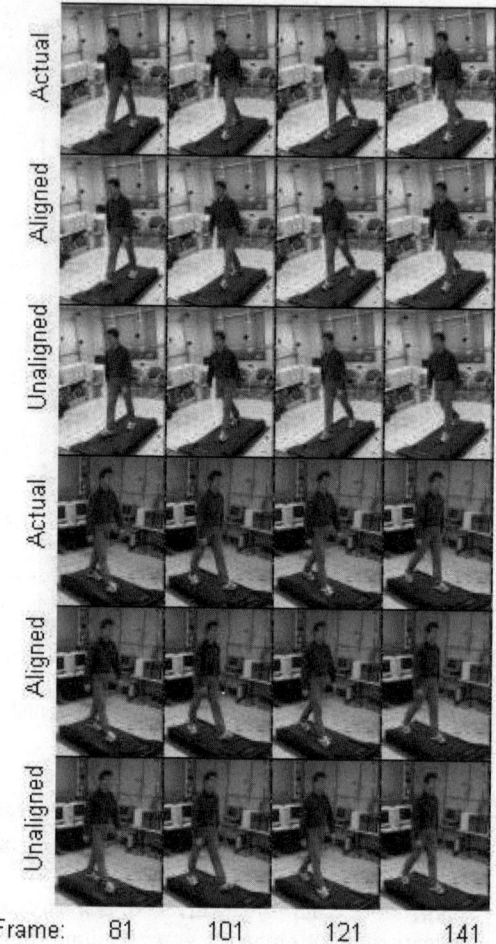

Fig. 13. Generated and actual images generated by predicting position on embedding

In the second case, we can predict new views in one of two ways, again depending on how the correspondences were learned. If correspondences were learned as part of the dimensionality reduction step, there is only one embedding for all images. The temporal dynamics of the low dimensional coordinates along the embedding can then be learned and used to predict where on the low dimensional embedding a view will be in the future, y_{future}. From that point, we can generate the high-dimensional views by mapping into the spaces of each of the input sequences. Similarly, if system identification was used to learn correspondences, the embeddings will be separate for each view, so the dynamics will be learned for each embedding separately and used to generate a new position on each embedding from which a new view can be constructed. Figure 13 illustrates the result of predicting views using both methods. We used the first 80 frames to learn the low dimensional embeddings and then learned the temporal dynamics of the coefficients of the low dimensional embeddings to predict the next 80 views.

5   Experimental Validation

5.1   Preprocessing

To model correspondences between person appearance in multiple views, the objects first need to be extracted and normalized so that they can be compared in a meaningful way. First, we use foreground segmentation methods such as background subtraction and morphological operations to smooth the resulting binary images. After thresholding for size, only the blobs corresponding to persons remain in the image. These are then resized to a standard size for each frame. Figure 14 illustrates one example of preprocessing multiple views of a scene containing two persons. The appearance templates are then transformed into column vectors that are then used for manifold learning and system identification steps.

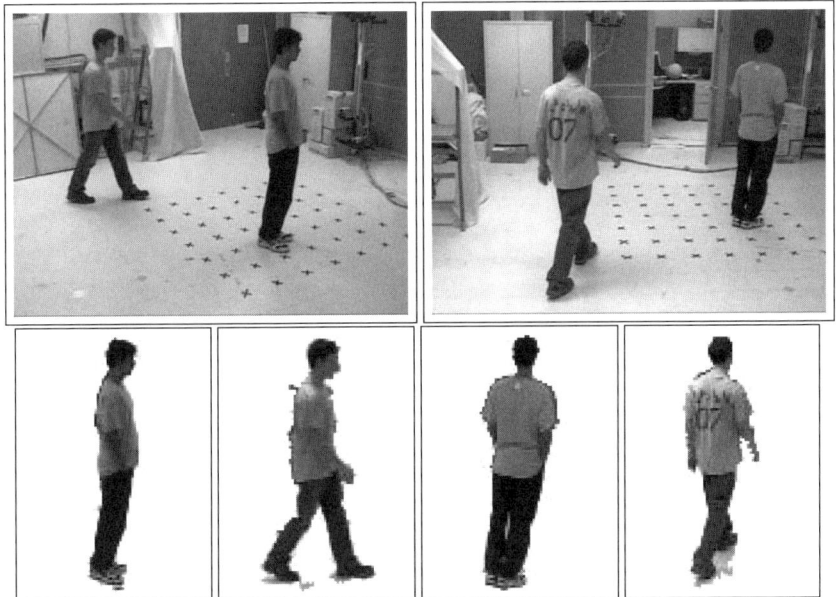

Fig. 14. Example of tracking in two views. Row 1: The input images. Row 2: Normalized person appearance.

For our experiments, we implemented a tracker that extracts persons from multi-camera views and, given an initial manual labeling, tracks the persons and their appearance throughout the sequence, while maintaining their correct identities. For the foreground segmentation, we used the Codebook Background Subtraction algorithm [45]. During the training period, we tracked each person using the blob tracker described by Argyros and Lourakis [46] and extracted the appearance template for each person. During the occlusion periods, the appearance templates could no longer be extracted in one of the videos. However, we used one of our proposed methods, alignment

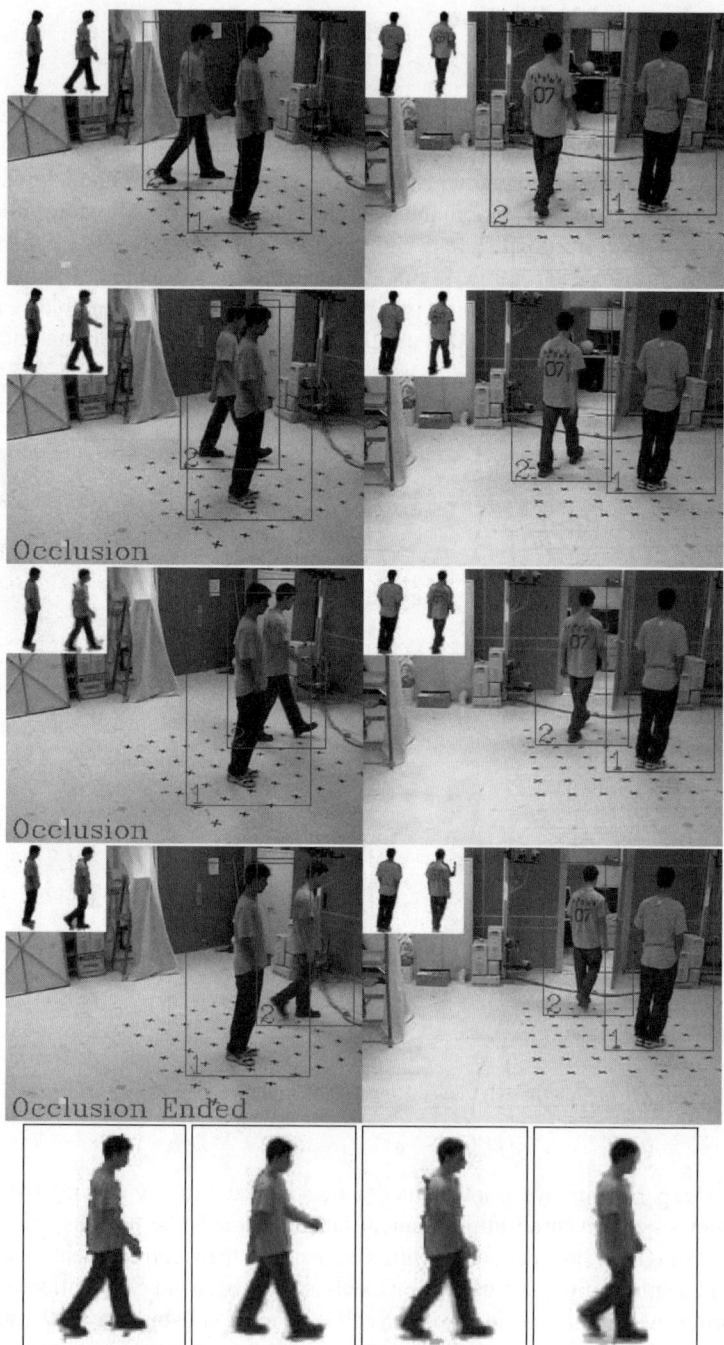

Fig. 15. Learned correspondence is used to generate appearance of occluded person and to maintain identity. Top: tracker views. Bottom: templates of occluded person.

of embeddings through LLE, to create the views of each person despite the occlusion. When the occlusion period ends, we compare the two extracted templates with our generated templates to make sure that the identities are correct, and relabel if necessary. We note that the persons had very similar appearance – both persons were wearing yellow shirts and jeans and both persons were of approximately the same build. Thus, methods that normally depend on such appearance characteristics as color would not be able to maintain correct identities. Figure 15 shows selected frames before, during, and after the occlusion period. In the corner of each view are the templates maintained by the tracker. The templates for person 2, which are generated during the occlusion are provided at the bottom of the figure. Additional results and the corresponding videoclips are available at http://www.umiacs.umd.edu/ morariu and http://robustsystems.ee.psu.edu.

6   Conclusions

Dynamic vision – the confluence of control and computer vision – is uniquely positioned to enhance the quality of life for large segments of the general public. Aware sensors endowed with tracking and scene analysis capabilities can prevent crime, reduce time response to emergency scenes and allow elderly people to continue living independently. Moreover, the investment required to accomplish these goals is relatively modest, since a large number of imaging sensors are already deployed and networked. For instance, the number of outdoor surveillance cameras in public spaces is already large (10,000 in Manhattan alone), and will increase exponentially with the introduction of camera cell phones capable of broadcasting and sharing live video feeds in real time. The challenge now is to develop a theoretical framework that allows for *robustly* processing this vast amount of information, within the constraints imposed by the need for real time operation in dynamic, partially stochastic scenarios. In this chapter we showed that efficient camera coordination leading to robust tracking in the presence of occlusion and clutter can be accomplished by exploiting a combination of identification and manifold discovery tools. The main idea is to exploit the high degree of spatio–temporal correlation of the data to project it, via nonlinear dimensionality tools, to a low order manifold where the underlying dynamics are approximately linear. Once in this manifold, tracking is accomplished by using robust identification tools to extract a compact model of the dynamics that can be used to predict the next position of the target, thus assisting in overcoming occlusion and disambiguating targets with similar appearance. Efficient camera coordination is accomplished by having the sensors share the low order data and associated models in these manifolds, rather than raw video streams. These results were illustrated with several examples. Research is currently under way seeking to reduce even further the amount of data to be shared among sensors by exploiting concepts from Information Based Complexity to eliminate redundancies.

Acknowledgments

We thank the University of Maryland for allowing us to use the Keck Laboratory and the Codebook Background Subtraction code. Also, the blob tracking code was written by the first author at the Navy Center for Applied Research in Artificial Intelligence.

References

[1] Cai, Q., Aggarwal, J.K.: Tracking human motion in structured environments using a distributed camera system. PAMI **22** (2000) 1241–1247

[2] Chang, T.H., Gong, S.: Tracking multiple people with a multi-camera system. In: ICCV. (2001)

[3] Nummiaro, K., Koller-Meier, E., Svoboda, T., Roth, D., Gool, L.V.: Color-based object tracking in multi-camera environments. In: DAGM. Springer LNCS 2781 (2003) 591–599

[4] Comaniciu, D., Berton, F., Ramesh, V.: Adaptive resolution system for distributed surveillance. Real Time Imaging **8** (2002) 427–437

[5] Black, M., Ellis, T.: Multiple camera image tracking. In: PETS. (2001)

[6] A.Mittal, Davis, L.S.: M2tracker: A multi-view approach to segmenting and tracking people in a cluttered scene. IJCV **51** (2003)

[7] Collins, R., Amidi, O., Kanade, T.: An active camera system for acquiring multi-view video. In: ICIP. Volume I. (2002) 517–520

[8] Dockstader, S.L., Tekalp, A.M.: Multiple camera tracking of interacting and occluded human motion. In: Proceedings of the IEEE. Volume 89. (2001) 1441–1455

[9] Lee, L., Romano, R., G.Stein: Monitoring activities from multiple video streams: Establishing a common frame. PAMI **22** (2000) 758–767

[10] Lee, L., Stein, G.: Monitoring activities from multiple video streams: Establishing a common coordinate frame. PAMI **22** (2000) 758–767

[11] Caspi, Y., Irani, M.: A step towards sequence-to-sequence alignment. In: cvpr. (2000)

[12] Khan, S., Shah, M.: Consistent labeling of tracked objects in multiple cameras with overlapping fields of view. PAMI **25** (2003) 1355–1360

[13] Isard, M., Blake, A.: CONDENSATION – conditional density propagation for visual tracking. IJCV **29** (1998) 5–28

[14] Julier, S., Uhlmann, J., Durrant-Whyte, H.F.: A new approach for filtering nonlinear systems. In: Proceedings of the 1995 American Control Conference. (1995) 1628–1632

[15] Kalman, R.E., Bucy, R.S.: New results in linear filtering and prediction theory. Trans. ASME Ser. D: J. Basic Eng. **83** (1961) 95–108

[16] North, B., Blake, A., Isard, M., Rittscher, J.: Learning and classification of complex dynamics. PAMI **22** (2000) 1016–1034

[17] Camps, O.I., Lim, H., Mazzaro, C., Sznaier, M.: A caratheodory-fejer approach to robust multiframe tracking. In: ICCV. (2003) 1048–1055

[18] Parrilo, P.A., Pena, R.S.S., Sznaier, M.: A parametric extension of mixed time/frequency domain based robust identification. IEEE Trans. Autom. Contr. **44** (1999) 364–369

[19] Paganini, F., Feron, E.: LMI methods for robust $\mathcal{H}_2$ analysis: A survey with comparisons. In Ghaoui, L.E., Niculescu, S., eds.: Recent Advances on LMI methods in Control. SIAM press (1999)

[20] Turk, M., Pentland, A.: Face Recognition Using Eigenfaces. In: CVPR. (1991) 586–591

[21] Murase, H., Nayar, S.K.: Visual Learning and Recognition of 3-D Objects from Appearance. IJCV **14** (1995) 5–24

[22] Black, M.J., Jepson, A.D.: Eigentracking: Robust matching and tracking of articulated objects using a view-based representation. IJCV **26** (1998) 63–84

[23] la Torre, F.D., Black, M.J.: Robust principal component analysis for computer vision. In: ICCV. (2001) 362–369

[24] la Torre, F.D., Black, M.J.: Robust parameterized component analysis: theory and applications to 2d facial appearance models. CVIU **91** (2003) 53–71

[25] Brand, M.: Subspace mappings for image sequences. In: Workshop Statistical Methods in Video Processing. (2002)

[26] la Torre, F.D., Black, M.J.: Dynamic coupled component analysis. In: CVPR. Volume 2. (2001) 643–650

[27] Doretto, G., Chiuso, A., Wu, Y.N., Soatto, S.: Dynamic textures. IJCV **51** (2003) 91–109

[28] Brand, M.: Charting a manifold. In: NIPS, MIT Press (2003)

[29] Brand, M.: Continuous nonlinear dimensionality reduction by kernel eigenmaps. In: IJCAI. (2003) 547–554

[30] Brand, M.: From subspaces to submanifolds. In: BMVC. (2004)

[31] Elgammal, A.: Nonlinear generative models for dynamic shape and dynamic appearance. 2nd International Workshop on Generative Model-Based Vision (2004)

[32] Elgammal, A., Lee, C.S.: Inferring 3d body pose from silhouettes using activity manifold learning. In: CVPR. (2004) 681–688

[33] Elgammal, A., Lee, C.S.: Separating style and content on a nonlinear manifold. In: CVPR. (2004) 478–485

[34] Ham, J., Lee, D.D., Saul, L.K.: Semisupervised alignment of manifolds. In: Artificial Intelligence and Statistics. (2005)

[35] Ham, J., Lee, D.D., Saul, L.K.: Learning high dimensional correspondences from low dimensional manifolds. In: Workshop on the Continuum from Labeled to Unlabeled Data in Machine Learning and Data Mining at ICML. (2003) 34–39

[36] Saul, L.K., Roweis, S.T.: Think globally, fit locally: unsupervised learning of low dimensional manifolds. Journal on Machine Learning Research **4** (2003) 119–155

[37] Verbeek, J.J., Roweis, S.T., Vlassis, N.A.: Non-linear cca and pca by alignment of local models. In: NIPS. (2003)

[38] Weinberger, K.Q., Sha, F., Saul, L.K.: Learning a kernel matrix for nonlinear dimensionality reduction. In: ICML, ACM Press (2004)

[39] Weinberger, K.Q., Saul, L.K.: Unsupervised learning of image manifolds by semidefinite programming. In: CVPR. (2004) 988–995

[40] Tenenbaum, J.B., de Silva, V., Langford, J.C.: A global geometric framework for nonlinear dimensionality reduction. Science **290** (2000) 2319–2323

[41] Belkin, M., Niyogi, P.: Laplacian eigenmaps for dimensionality reduction and data representation. Neural Computation **15** (2003) 1373–1396

[42] Donoho, D.L., Grimes, C.E.: Hessian eigenmaps: locally linear embedding techniques for high-dimensional data. In: Proceedings of the National Academy of Arts and Sciences. Volume 100. (2003) 5591–5596

[43] Zhang, Z., Zha, H.: Principal manifolds and nonlinear dimension reduction via local tangent space alignment. In: SIAM Journal of Scientific Computing. Volume 26. (2004) 313–338

[44] Gross, R., Shi, J.: The cmu motion of body (mobo) database. Technical Report CMU-RI-TR-01-18, Robotics Institute, Carnegie Mellon University (2001)

[45] Kim, K., Chalidabhongse, T.H., Harwood, D., Davis, L.S.: Real-time foreground-background segmentation using codebook model. Real-Time Imaging **11** (2005) 172–185

[46] Argyros, A., Lourakis, M.I.A.: Real time tracking of multiple skin-colored objects with a possibly moving camera. In: ECCV. Volume 3. (2004) 368–379

A Lagrangian-Based Algorithm for a Combinatorial Motion Planning Problem

Sai K. Yadlapalli, Waqar A. Malik, Swaroop Darbha, and Siva Rathinam

Department of Mechanical Engg., Texas A&M University,
College Station, TX - 77843-3123, USA
Department of Civil and Enviromental Engineering, University of California,
Berkeley, CA - 94720-1710, USA
{kris5372,waqar_am,dswaroop}@tamu.edu, rsiva@berkeley.edu

Abstract. We consider a combinatorial motion planning problem (CMP) that naturally arises in many applications involving unmanned aerial vehicles (UAVs) with fuel and motion constraints. The motion constraint we consider is the inability of a vehicle to turn at an arbitrary yaw rate. The CMP is a generalization of a single Travelling Salesman Problem and is NP-Hard. In this paper, we exploit the combinatorial structure of the problem and provide heuristics with computational results to address the same.

1 Introduction

Motion planning of a collection of unmanned aerial vehicles (UAVs) has significant applications, see [SC1, DR1, CR1, AG1] and the references therein. The problem of motion planning considered for these applications involves the solution of a combinatorial problem, wherein one must determine the set of targets to be visited by each vehicle and the sequence in which they must be visited before returning to its initial location (depot). Equally important is the consideration of motion constraints of the vehicles in the planning. In this paper, we address a combinatorial motion planning problem involving a homogenous collection of vehicles where the motion of each vehicle satisfies a non-holonomic constraint. The non-holonomic constraint we consider is that the yaw rate of the vehicle at any time is upper bounded by a constant. Hence, if the vehicle is travelling at constant speed, this constraint is equivalent to a lower bound on the turning radius of the vehicle. The combinatorial motion planning problem (CMP) we address is the following:

Given a set of m vehicles and n targets on a plane, the heading angles of each target and the initial heading angles of each vehicle, the CMP is to

- choose at most $p(\leq m)$ vehicles,
- assign a set of targets for each chosen vehicle such that each target is visited exactly once,
- find a feasible path (i.e. a path that satisfies the yaw rate constraints) for each chosen vehicle such that the vehicle starts at its initial position, visits its assigned set of targets at their respective heading angles in a specified sequence and returns to its initial position.

M.J. Hirsch et al. (Eds.): Adv. in Cooper. Ctrl. & Optimization, LNCIS 369, pp. 373–387, 2007.
springerlink.com © Springer-Verlag Berlin Heidelberg 2007

The goal is to minimize the sum of the distances travelled by all the chosen vehicles. In many miltary applications, some targets must be visited prior to other targets due to tactical reasons. Hence, in this paper, we also address CMP with precedence constraints on the targets.

The problem of finding the minimum distance path the vehicle must take between any two positions on a plane subject to the constraints on the yaw rate has been solved by Dubins [Du1]. Hence, the CMP can be posed as a multiple depot Asymmetric Travelling Salesman Problem (ATSP). This problem is a generalization of the single TSP and is NP-Hard. The difficulty of this CMP is due to the following reasons:

1. The vehicle-target assignment is not given.
2. Given the vehicle-target assignment, finding the optimal sequence for each vehicle is again a single depot ATSP which is NP-Hard. Several approximation algorithms and heuristics that work well for the single symmetric TSP does not work well for single depot ATSP [GP1].

The following are the main contributions of this paper:

1. We formulate the CMP as an integer program with $(n+m)^2+m$ variables (one variable for each edge joining any two vertices and one variable for each depot). This formulation exploits the fact that the Dubins' distances satisfy triangle inequality.
2. We solve the vehicle-target assignment problem by solving a Lagrangian dual [Fi1] of the formulated integer program. This step involves finding a minimum cost directed spanning tree with a degree constraint. We solve this problem by penalizing this degree constraint if violated and using the approach given in [RG1].
3. Given a set of targets, the Lagrangian heuristic available in [Ti1] is used to find a sequence of targets each vehicle must visit.
4. The Lagrangian dual of the integer program also gives a tight lower bound for the integer program. This lower bound is used in the Branch and Bound solver to find the optimal solution to the integer program.
5. We provide experimental results that compare the cost of the solution produced by the algorithm given in this paper with the optimal cost of the integer program.
6. We show how to extend the results on the lower bound for CMP to the CMP with precedence constraints on the targets.

2   Literature Review

One can refer to [Be2] for an extensive review of the solution procedures for the multiple Travelling Salesman Problem. As previously mentioned in the introduction, CMP is NP-Hard. Unlike the symmetric counterparts that have constant factor approximation[1] algorithms [RS1], the best approximation algorithms available *even* for a single depot ATSP have approximation ratios scale in the order of $\log(n)$ [Bl1, KL1]. One way to address a CMP is to convert to the CMP into a single ATSP and use the algorithms available for ATSP to solve CMP. But this is currently available only for $m = 2$ [Ra1].

---

[1] A polynomial algorithm that returns an approximate solution whose cost is within a guaranteed factor of the optimal solution.

For a general m, Laporte gives a transformation of CMP to a constrained assignment problem. As mentioned in [Be2, Gu1], it is an incomplete transformation due to the presence of non assignment constraints.

Branch and Bound methods can be used to solve CMP [NW1]. In general the effectiveness of a B&B procedure depends on the tightness of the lower and upper bounds that one has at hand. In this paper, we generate tight lower bounds for CMP using Lagrangian Relaxation. This generalizes the results by Held-Karp [HK1] available for the single TSP to the CMP.

The combinatorial motion planning problem addressed in this paper assumes that the heading of each target is known. This allows one to view CMP purely as a combinatorial problem using Dubins [Du1] result. The CMP without this assumption has also received significant attention in the literature [RS1, SFB1, NF1, TO1]. Though motion constraints are an integral part of all these variants of the CMP, it is hard to envision good algorithms or heuristics for the same that do not exploit the combinatorial structure of the problem. This is the main focus of our paper.

3 CMP Formulation

Let D represent the set of depots (initial locations of vehicles), T represent the set of targets and let $V = D \cup T$. The cardinality of D is m and that of T is n. The set of all the edges connecting any two vertices in V is represented by E. An arc $e = (x, y)$ is considered to be directed from x to y. y is called the head and x is called the tail of the arc. Let c_e be the cost of arc e. Basically, c_e is the length of the Dubins path from vertex x to vertex y. Note that the costs, c_e, satisfy the triangle inequality. We will use $\delta(A)$ to indicate the set of edges with their tails in A, $\Delta(A)$ to indicate the set of edges with their heads in A and $E(X), X \subset V$ to indicate the set of edges with both their heads and tails in X. We will let $x_e, e \in E$ and $y_v, v \in D$ to be the binary variables that respectively represent the choice of the edge and the depot in the solution. The **integer program** for the CMP is formulated as follows:

$$CMP^* = \min \sum_{e \in E} c_e x_e, \tag{1}$$

subject to

$$\sum_{e \in \delta(v) \cap \Delta(T)} x_e = y_v, v \in D \tag{2}$$

$$\sum_{e \in \Delta(v) \cap \delta(T)} x_e = y_v, v \in D \tag{3}$$

$$\sum_{e \in \delta(v)} x_e = 1, v \in T \tag{4}$$

$$\sum_{e \in \Delta(v)} x_e = 1, v \in T \tag{5}$$

$$\sum_{e \in E(S)} x_e \leq |S| - 1, \forall S \subset T \tag{6}$$

$$\sum_{e \in E(T)} x_e + \sum_{v \in D} y_v = n \tag{7}$$

$$\sum_{v \in D} y_v \le q \tag{8}$$

$$x_e \in \{0,1\}, \ y_v \in \{0,1\}. \tag{9}$$

Constraints (2) and (3) represent the out-degree and the in-degree constraints on the depots respectively. In particular, if a depot is not chosen, then no edge incident on the depot(incoming or outgoing) can be chosen from the solution as stated by (2) and (3). Constraints (5) and (4) require the in-degree and out-degree of each target equal to one. The constraint (6) eliminates the presence of any cycles among the target vertices. Constraint (7) indicates that if p depots were chosen in the solution, then the graph $(T, E(T))$ must have exactly p components. Constraint (8) requires that any feasible solution must choose at most q depots.

Proposition 1. *The integer program for the CMP is correct (i.e. the optimal solution of the integer program is an optimal solution to the CMP) if the costs, c_e, satisfy triangle inequality.*

Proof. Every feasible solution to the CMP satisfies the constraints (2) through (9). Now, consider an optimal solution to the integer program. Since the indegree and the outdegree of every selected depot vertex and the target vertex is 1, the optimal solution must represent a union of cycles and isolated depots. Clearly, the constraint (6) does not admit a cyclic solution amongst the target cities and hence, it must be the case that every cycle of an optimal solution to CMP must contain at least one depot vertex. It cannot have more than one depot vertex; otherwise, using triangle inequality, additional depot vertices can be short cut to produce a solution to CMP with a smaller cost than the optimal solution. Since the optimal solution to the binary program is a feasible solution to CMP, the integer program formulated for the CMP is correct.

4   A Lagrangian Relaxation of the CMP

In this section, we show how tight lower bounds can be obtained for the integer program stated in the previous section. In later sections, we show how the results in this section are used to develop a heuristic for the CMP. The method here (Lagrangian Relaxation) follows the approach by Held and Karp who used it for solving the symmetric TSP [HK1]. The basic idea in Lagrangian Relaxation is to first identify the constraints that make the integer program difficult to solve. Then, remove these complicating constraints and penalize them in the objective whenever they are violated. A Lagrangian Relaxation of the integer program for CMP is:

$$L(\Pi, \Psi) := \min \sum_{e \in E} c_e x_e + \sum_{v \in T} \pi_v \left(\sum_{e \in \delta(v)} x_e - 1 \right) + \sum_{v \in T} \psi_v \left(\sum_{e \in \Delta(v)} x_e - 1 \right) \tag{10}$$

subject to

$$\sum_{e\in\delta(v)\cap\Delta(T)} x_e = y_v, v \in D$$

$$\sum_{e\in\Delta(v)\cap\delta(T)} x_e = y_v, v \in D$$

$$\sum_{e\in E(S)} x_e \le |S| - 1, \forall S \subset T$$

$$\sum_{e\in E(T)} x_e + \sum_{v\in D} y_v = n$$

$$\sum_{v\in D} y_v \le q$$

$$x_e \in \{0,1\}, y_v \in \{0,1\}$$

where, π_v (ψ_v) is the penalty variable when the out-degree (in-degree) constraint of a target vertex v is violated and Π (Ψ) indicates the vector of penalty variables π_v (ψ_v). Now we show in the following lemma that $L(\Pi,\Psi)$ can be computed using a polynomial time algorithm. Hence, for any given Π and Ψ, computing $L(\Pi,\Psi)$ would yield a lower bound for CMP^*.

Lemma 1. *For any given Π,Ψ, the Lagrangian Relaxation in (10) is solvable in polynomial time.*

Proof. It is sufficient to show that the following program is polynomially solvable for every integer p lying between 1 and q:

$$J_p(\Pi,\Psi) := \min \sum_{e\in E} c_e x_e + \sum_{v\in T} \pi_v \Big(\sum_{e\in\delta(v)} x_e - 1\Big) + \sum_{v\in T} \psi_v \Big(\sum_{e\in\Delta(v)} x_e - 1\Big), \quad (11)$$

subject to

$$\sum_{e\in\delta(v)\cap\Delta(T)} x_e = y_v, v \in D \quad (12)$$

$$\sum_{e\in\Delta(v)\cap\delta(T)} x_e = y_v, v \in D \quad (13)$$

$$\sum_{v\in D} y_v = p, \quad (14)$$

$$\sum_{e\in E(S)} x_e \le |S| - 1, \forall S \subset T, \quad (15)$$

$$\sum_{e\in E(T)} x_e = n - p, \quad (16)$$

$$x_e \in \{0,1\}, y_v \in \{0,1\}. \quad (17)$$

Observe that the variables in constraints (12,13,14), $\{y_v : v \in D\}, \{x_e : e \in \delta(D)\cup \Delta(D)\}$, and the variables in constraints (15,16), $\{x_e : e \in E(T)\}$, are not coupled. Hence the Lagrangian Relaxation can be decoupled into two problems and can be solved separately as follows:

Problem I

$$J_p^1(\Pi,\Psi) := \min \sum_{e\in E(T)} c_e x_e + \sum_{v\in T} \pi_v \sum_{e\in\delta(v)\cap E(T)} x_e + \sum_{v\in T} \psi_v \sum_{e\in\Delta(v)\cap E(T)} x_e, \quad (18)$$

subject to

$$\sum_{e\in E(S)} x_e \leq |S| - 1, \forall S \subset T, \quad (19)$$

$$\sum_{e\in E(T)} x_e = n - p, \quad (20)$$

$$x_e \in \{0,1\}, \, y_v \in \{0,1\}. \quad (21)$$

Problem II

$$J_p^2(\Pi,\Psi) := \min \sum_{e\in E\setminus E(T)} c_e x_e + \sum_{v\in T} \pi_v\Big(\sum_{e\in\delta(v)\cap\Delta(D)} x_e - 1\Big) + \sum_{v\in T} \psi_v\Big(\sum_{e\in\Delta(v)\cap\delta(D)} x_e - 1\Big), (22)$$

subject to

$$\sum_{e\in\delta(v)\cap\Delta(T)} x_e = y_v, v \in D \quad (23)$$

$$\sum_{e\in\Delta(v)\cap\delta(T)} x_e = y_v, v \in D \quad (24)$$

$$\sum_{v\in D} y_v = p, \quad (25)$$

$$x_e \in \{0,1\}, \, y_v \in \{0,1\}. \quad (26)$$

Problem I involves computing a minimum cost, p-component, directed spanning forest $(DMSF_p^*)$ that can be solved using a polynomial time algorithm given in the appendix. The solution to **Problem II** can be found using the following steps:

1. Let the modified cost of each edge e in $\delta(T)$ $(\Delta(T))$ be $c_e + \pi_{v:e\in\delta(v)}$ $(c_e + \psi_{v:e\in\Delta(v)})$. Determine the cheapest incoming edge and outgoing edge incident on every $v \in D$. Let their total cost be t_v.
2. Sort t_v, $v \in D$. The optimal solution, E_p^*, is the set of $2p$ edges corresponding to the p cheapest costs.

The optimal cost of the Lagrangian Relaxation, $L(\Pi,\Psi)$, can be computed as $L(\Pi,\Psi) = \min_p(J_p^1(\Pi,\Psi) + J_p^2(\Pi,\Psi))$.

Now, since for every Π,Ψ, $CMP^* \geq L(\Pi,\Psi)$, we can conclude that

$$CMP^* \geq \max_{\Pi,\Psi} L(\Pi,\Psi). \quad (27)$$

$\max_{\Pi,\Psi} L(\Pi,\Psi)$ is the Lagrangian Dual of the integer program for CMP. Note that $L(\Pi,\Psi)$ is a concave function of Π and Ψ. Details on how to solve this Lagrangian Dual is given in section 7.

5   Primal Feasible Algorithm

To generate a primal feasible solution, we use the p-directed spanning forest $DMSF_p^*$ generated through the Lagrangian relaxation given in the previous section. The primal algorithm that assigns the depots to each component of the $DMSF_p^*$ and forms the feasible p-directed tours is given below:

Primal feasible Algorithm

1. For each $v \in D$ and i^{th} component of $DMSF_p^*$, $i \in \{1, 2, \ldots, p\}$, we compute the cost, A_{vi} to be the total cost of the the the cheapest edge in $\delta(v) \cap \Delta(S_i)$ and the cheapest edge in $\Delta(v) \cap \delta(S_i)$, where S_i is the set of nodes in the ith component of $DMSF_p^*$.
2. Let v_i be the depot assigned to the component corresponding to set of nodes, S_i. Define $V_i = S_i \cup v_i$. Assign a depot to every component in $DMSF_p^*$ such that the total assignment cost $\min_{v_i} \sum_i A_{v_i i}$ is minimum.
3. We transform the problem of finding a directed, feasible tour with nodes in V_i to a problem of finding a feasible tour with symmetric costs by doubling the nodes in V_i as described in [GP1]. The transformation can be simply put as follows: We replace each node n by a pair of nodes n^+, n^- and the define the costs as follows: Let $n_1, n_2 \in S_i$ then $\tilde{c}_i(n_1^+, n_2^-) = c(n_1, n_2)$ and $\tilde{c}_i(n_2^+, n_1^-) = c(n_2, n_1)$. We also set $\tilde{c}_i(n_1^-, n_1^+) = -M$ and all the other costs in $\tilde{c}_i$ to be $+M$, where M is a sufficiently large positive number such that all the arcs whose costs are $+M$ are excluded from all the feasible tours and all the arcs with $-M$ are included in any feasible tour.
4. Now for each modified cost matrix c_i and the node set S_i, we use the Lagrangian heuristics in [Ti1] to get a primal feasible tour.

6   CMP with Precedence Constraints

This section shows how the approach presented in the previous sections can be used to address CMP with a given set of precedence constraints. If target m has to be visited before target n, then the precedence constraint related with targets m and n is written as $m \prec n$. The set of l precedence constraints is denoted by $\mathscr{P} = \{i_1 \prec j_1, i_2 \prec j_2, \cdots, i_l \prec j_l\}$ where $i_o, j_o \in T$ for $o \in \{1, 2, \cdots, l\}$. We assume that the given set of precedence constraints are consistent (i.e. there are no constraints of the form $\{a \prec b, b \prec c, c \prec a\}$). To formulate the CMP with precedence constraints, we introduce a variable t_j that denotes the order in which the jth target is visited. t_j takes values in $\{1, 2...n\}$. So if $t_i \leq t_j$ then the ith target is visited before the jth target. All other notations and variables are the same as used in the problem formulation for CMP (section 3). The integer programming formulation for CMP with precedence constraints is as follows:

$$CMPPC^* = \min \sum_{e \in E} c_e x_e, \tag{28}$$

subject to

$$\sum_{e \in \delta(v) \cap \Delta(T)} x_e = y_v, v \in D \tag{29}$$

$$\sum_{e \in \Delta(v) \cap \delta(T)} x_e = y_v, v \in D \tag{30}$$

$$\sum_{e \in \delta(v)} x_e = 1, v \in T \tag{31}$$

$$\sum_{e \in \Delta(v)} x_e = 1, v \in T \tag{32}$$

$$\sum_{e \in E(S)} x_e \leq |S| - 1, \forall S \subset T \tag{33}$$

$$\sum_{e \in E(T)} x_e + \sum_{v \in D} y_v = n \tag{34}$$

$$\sum_{v \in D} y_v \leq q \tag{35}$$

$$(n-1) + t_j - t_i \geq n x_e, \forall e \in E(T) \tag{36}$$

$$\{e\} = \delta(i) \text{ and } \{e\} = \Delta(j) \tag{37}$$

$$t_j \geq t_i + 1, \forall (i \prec j) \in \mathscr{P} \tag{38}$$

$$t_i \in \{1, 2, \cdots, n\} \forall i \in T \tag{39}$$

$$x_e \in \{0, 1\}, \ y_v \in \{0, 1\}. \tag{40}$$

The precedence constraints are formulated in 7. Constraints (37) relate the ordering variables t_i and x_e. Now, one can obtain tight lower bounds using Lagrangian relaxation in the same way as we previously discussed in section 4. The idea is to dualize the constraints that make the above integer program difficult. A Lagrangian relaxation of the above problem obtained by relaxing constraints (32,33, 37) is as follows:

$$L(\Pi, \Psi, \alpha) = \min \sum_{e \in E} c_e x_e + \sum_{v \in T} \pi_v \left(\sum_{e \in \delta(v)} x_e - 1 \right) +$$
$$\sum_{v \in T} \psi_v \left(\sum_{e \in \Delta(v)} x_e - 1 \right) + \sum_{e \in E(T)} \alpha_e (\gamma_e - n x_e), \tag{41}$$

subject to

$$\sum_{e \in \delta(v) \cap \Delta(T)} x_e = y_v, v \in D$$

$$\sum_{e \in \Delta(v) \cap \delta(T)} x_e = y_v, v \in D$$

$$\sum_{e \in E(S)} x_e \leq |S| - 1, \forall S \subset T,$$

$$\sum_{e \in E(T)} x_e + \sum_{v \in D} y_v = n,$$

$$\sum_{v \in D} y_v \leq q,$$

$$n - 1 + t_j - t_i \geq \gamma_e, \forall e \in E(T),$$
$$\{e\} = \delta(i) \text{ and } \{e\} = \Delta(j)$$
$$t_i + 1 \leq t_j, \forall (i \prec j) \in \mathscr{P}$$
$$t_i \in \{1, 2, \cdots, n\} \forall i \in T \tag{42}$$
$$x_e \in \{0, 1\}, \ y_v \in \{0, 1\}. \tag{43}$$

In the above relaxation, α_e is the variable that penalizes the constraint $\gamma_e = nx_e$ whenever violated. Note that it is relatively easy to solve the Lagrangian relaxation as the variables t_i, γ_e, x_e and y_v are not coupled through any equation (as discussed in section 4) and hence can be optimized independently. The Lagrangian dual that provides the best lower bound to the optimal cost of the CMP with precedence constraints is given by $\max_{\Pi, \Psi, \alpha} L(\Pi, \Psi, \alpha)$.

7 Numerical Results

In this section, we present the implementation details and the overall algorithm accompanied with the simulation results. To calculate the best lower bound discussed in section 4, we need to compute $\max_{\Pi, \Psi} L(\Pi, \Psi)$. This can be computed using a gradient ascent algorithm. Let $[\Pi]^k$ and $[\Psi]^k$ indicate the values of Π and Ψ at the kth iteration respectively. At each iteration k, we compute a new set of penalty parameters ,$[\Pi]^{k+1}, [\Psi]^{k+1}$, from $[\Pi]^k, [\Psi]^k$ respectively through an update scheme where the direction of update is defined through the subgradient. We define the subgradient as follows:

$$gi_v = \sum_{e \in \Delta(v)} x_e - 1, \forall v \in T \tag{44}$$

$$go_v = \sum_{e \in \delta(v)} x_e - 1, \forall v \in T \tag{45}$$

$$gi_v = 0, \forall v \in D \tag{46}$$

$$go_v = 0, \forall v \in D \tag{47}$$

Let $g = [gi \quad go]$ be the vector of all the subgradients stacked together. The new update $[\pi_v]^{k+1}$ is computed as follows:

$$[\Pi]^{k+1} = [\Pi]^k + \beta^k [go]^k \quad \forall v \tag{48}$$

where the size of the step, β at iteration k is computed as

$$\beta^k = \zeta^k \frac{MDMTSP^* - \phi([\Pi, \Psi]^k)}{||[g]^k||} \tag{49}$$

$[\Psi]^{k+1}$ can be computed in a similar fashion as $[\Pi]^{k+1}$. The above expression (49) is commonly referred to as *Polyak rule II*. Since the optimal solution CMP^* is not known, alternatively we use the cost of the best primal solution found so far. A common practice is to start ζ^k with a fixed value and reduce ζ^k by a constant factor after a specified

Table 1. Duality gap for various instances

n	m	CMP^*	$\phi([\Pi]^k,[\Psi]^k)_{k=25}$	% of Duality Gap in k Iterations $[Cprimal^*]^k_{k=25}$	$[\varepsilon]^k_{k=25}$	$\phi([\Pi]^k,[\Psi]^k)_{k=50}$	$[Cprimal^*]^k_{k=50}$	$[\varepsilon]^k_{k=50}$
14	3	1624.8	1566.6	1624.8	3.7163	1568.5	1624.8	3.5959
19	4	2142.9	2142.2	2142.9	0.030342	2142.9	2142.9	0.00046666
22	7	2076.9	2041.8	2204.5	7.968	2049.7	2204.5	7.5535
24	3	2638.1	2637.9	2638.1	0.0068235	2638.1	2638.1	0.00075812
24	7	2352.7	2294.8	2370.9	3.3161	2312.2	2370.9	2.5418
26	3	2833.2	2833.2	2916.8	2.9493	2833.2	2916.8	2.9493
26	6	2678.9	2598.9	2706.7	4.1476	2611.6	2706.7	3.6387
28	5	2824.5	2728.5	2824.6	3.5187	2740.2	2824.6	3.0794
30	4	2872.1	2759.2	2944.6	6.7193	2778.7	2944.6	5.9711
31	4	3333.9	3268	3459.6	5.8622	3268	3459.6	5.8622
32	3	2898.2	2786.1	2940.6	5.545	2786.1	2940.6	5.545
36	3	3271.1	3149.7	3386.6	7.5216	3167.3	3386.6	6.9235
38	4	3497.9	3479.1	3497.9	0.54181	3480	3497.9	0.51408
40	7	3061	2992.4	3061	2.2918	3001.4	3061	1.9854
45	2	3724.6	3685.5	3748.4	1.7061	3692.6	3748.4	1.5119
48	5	3722.1	3681.9	3723.7	1.1342	3688	3723.7	0.96692
50	5	3242.4	3216	3346.2	4.0475	3218.3	3346.2	3.9754

number of iterations or whenever $\phi([\Pi]^k,[\Psi]^k)$ does not increase within specified number of iterations. The iterative procedure can be briefly put as follows:

Dual and Primal algorithm for the CMP

1. Initial step: $k = 0$, Initialize $\zeta^k = \zeta_0$.
2. For the computed $[\Pi]^k$ and $[\Psi]^k$, solve the Lagrangian relaxation $L([\Pi]^k,[\Psi]^k)$.
3. Use the **Primal feasible Algorithm** to generate a primal feasible solution from the dual solution. Let the cost of the best primal feasible solution found so far be $[C^*]^k$.
4. Stopping criterion: If $[\varepsilon]^k \leq \varepsilon^*$ or $k = N^{max}$, go to 6.
5. Compute $[\Pi]^{k+1},[\Psi]^{k+1}$ and set $k = k+1$ and go to 2.
6. Stop the iterative process.

Note that $[\varepsilon]^k$ is the duality gap at iteration k and is defined as $\frac{[C^*]^k - \phi([\Pi]^k,[\Psi]^{k+1})}{\phi([\Pi]^k,[\Psi]^{k+1})} \cdot \varepsilon^*$ is the desired duality gap. The maximum number of iterations allowed is chosen to be 50. ζ^k was chosen to start with a value of 0.5 and is reduced by a factor of 2, if the dual does not improve in 3 successive iterations. The value of ε for the stopping criterion is chosen to be 10^{-4}. In the simulations, we allow all the depots to participate in the tour, i.e, $q = |D|$.

In Table 1, n refers to the number of targets, m is the number of depots available, $[Cprimal^*]^k$ is the cost of the best primal found at iteration k. In Table I, we report the dual gap at iterations $k = 25$ and $k = 50$. CMP^* refers to the optimal cost for that instance. We compute CMP^* using the GNU Linear Programming Kit (GLPK). The code is written in Matlab and YALMIP [Lö1] is used to formulate the problem and also provides the interface to GLPK. In figure 1 the convergence of the dual gap with the

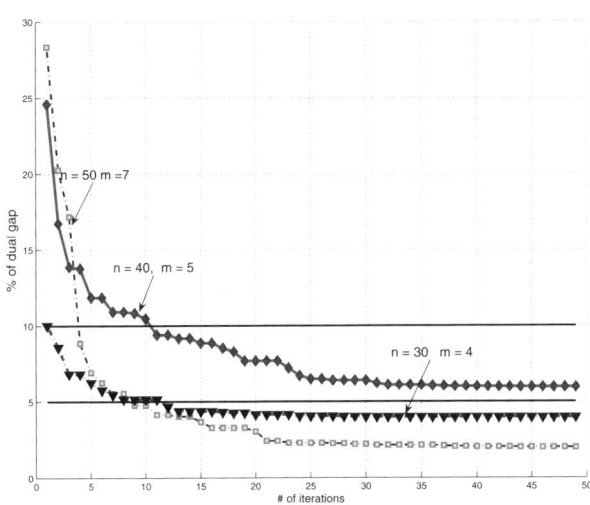

Fig. 1. Convergence of dual gap for random instances

number of iterations is shown for few random instances. The sizes of the instances are as indicated.

In Table 2 the optimal cost to CMPPC for various instances is shown. $LCMPPC^*$ refers to the optimal solution of the linear program obtained by relaxing the integral constraints (39, 40). Clearly, $LCMPPC^* \leq CMPPC^*$. ℓ refers to the gap between $CMPPC^*$ and $LCMPPC^*$, i.e., $\ell = \frac{CMPPC^* - LCMPPC^*}{CMPPC^*} \times 100$.

Table 2. CMP with precedence constraints

n	m	$CMPPC^*$	$LCMPPC^*$	ℓ
13	3	1461.8	1312.2	10.24
16	5	1175.2	938.28	20.16
19	3	1713.7	1577.3	7.9614
22	2	2472.9	2351.0	4.9269
25	2	2623	2541	3.1287
28	4	1849.2	1795.3	4.8542
31	6	1838.5	1616.6	12.0682
34	6	2980.5	2839.1	4.74
37	3	3046.4	2940.8	3.4635
40	6	2998.0	2901.8	3.2094

8   Conclusions

In this chapter, we have considered the problem of motion planning of m Dubins' vehicles through n points in a plane. The location of all the points and the heading at which the vehicles are to arrive and depart is already specified. We provide algorithms to choose at most $q(\leq m)$ Dubin's vehicles and construct sub-optimal tours such that the total cost of the tours of the chosen vehicles is a minimum amongst all the possible choice of vehicles and their tours.

References

[AG1]   Alexopoulos, Griffin, P. M.: "Path Planning for a mobile robot," *IEEE Transactions on Systems, Man and Cybernautics*, Vol. **22**, no. 2, (1992).

[Be1]   Bellman, R. E.: *Dynamic Programming*, Dover Publications, 2003.

[SC1]   Schumacher, C. J., Chandler P. R., Rasmussen S. J.: "Task allocation for wide area search munitions," *Proceedings of the American Control Conference*, (2002), pp. 1917-1922.

[CR1]   Cunningham, C. T., and Roberts, R. S.: "An Adaptive Path Planning Algorithm for Co-operating Unmanned Air Vehicles," *Proceedings of the 2001 IEEE International Conference on Robotics and Automation*,(2001) pp. 3981-3986.

[DR1]   Gross, D., Rasmussen, S., Chandler, P., Feitshans, G.: "Cooperative Operations in UrbaN TERrain (COUNTER)," 2006 SPIE Defense & Security Symposium, Orlando, FL, 17-21 April 2006.

[Va1]   Vazirani, V. V., Approximation Algorithms, Springer 2001.

[GP1]   Gutin, G., Punnen, A. P. (Editors): The Travelling Salesman Problem and its Variations, Kluwer Academic Publishers 2002.

[Ra1]   Rao, M. R.: A note on Multiple Traveling Salesmen Problem," *Operations Research* **28** no.3 (1980) pp. 628-632.

[HK1]   Held, M., Karp, R. M.: The Traveling-Salesman Problem and Minimum Spanning Trees, Operations Research, **18** (1970) pp. 1138-1162.

[HK2]   Held, M., and Karp, R. M.: The travelling-salesman problem and minimum spanning trees: Part II Mathematical Programming, **1** (1971) pp. 6-25.

[Du1]   Dubins, L.E.: "On curves of minimal length with a constraint on average curvature, and with prescribed initial and terminal positions and tangents," *American Journal of Mathematics*, Vol. 79, pp.497-516.

[RS1]   Rathinam, S., Sengupta, R., Darbha, S.: On Resource Allocation Algorithm for Multi-Vehicle Systems with Non-Holonomic Constraints, to appear in IEEE Transactions on Automation Sciences and Engineering, 2006.

[Be2]   Bektas, T.: The Multiple Traveling Salesman Problem: an Overview of Formulations and Solution Procedures. OMEGA: The International Journal of Management Science, **34**(3), (2006) 209-219.

[MR1]   Malik, W., Rathinam, S., Darbha, S.: "A 2-approximation algorithm for a symmetric, generalized, multiple TSP," to appear in *Operations Research Letters*, February 2007.

[Bl1]   Blaser, M.: A new approximation algorithm for the asymmetric TSP with triangle inequality, Proceedings of the fourteenth annual ACM-SIAM symposiumon Discrete algorithms, (2003) pp: 638 - 645.

[KL1]   Kaplan, H., Lewenstein, M., Shafrir, N., Sviridenko, M.: Approximation algorithms for Asymmetric TSP by Decomposing Directed Regular Multigraphs, Proceedings of the 44th Annual IEEE Symposium on Foundations of Computer Science (2003) pp: 56- 65.

[Gu1]   GuoXing Y.: Transformation of multidepot multisalesmen problem to the standard traveling salesman problem. European Journal of Operational Research 81:55760 (1995).

[LN1]   Laporte G, Nobert Y, Taillefer S.: Solving a family of multi-depot vehicle routing and location-routing problems. Transportation Science 22(3):16172, (1988).

[NW1]   Nemhauser G.L., Wolsey L.A.: Integer Programming and Combinatorial Optimization, Wiley-Interscience, 1988.

[NF1]   Ny, J.L., Feron, E.: An approximation algorithm for the curvature constrained traveling salesman problem," Proceedings of the 43rd Annual Allerton Conference on Communications, Control and Computing, 2005.

[TO1]   Tang, Z., Ozguner, U.: Motion planning for multi-target surveillance with mobile sensor agents, IEEE Trans. Robotics, vol. **21** no. 5 (2005) pp. 898-908.

[Fi1]   Fisher, M.: The Lagrangean relaxation method for solving integer programming problems,Management Science **27** 118 (1981).

[RG1]   Ravi, R., Goemans, M. X.: The Constrained Minimum Spanning Tree Problem. Proceedings of the Scandinavian Workshop on Algorithm Theory, Lecture Notes on Computer Science **1097** (1996) p.66-75.

[SFB1] Savla, K., Frazzoli, E., Bullo, F.: "On the point-to-point and traveling salesperson problems for Dubins' vehicle," Proceedings of the American Control Conference, (2005) pp. 786-791.

[Lö1]   Löfberg, J., YALMIP : A Toolbox for Modeling and Optimization in MATLAB, Proceedings of the CACSD Conference 2004.

[Ti1]   Timsjö, S.: An Application of Lagrangian Relaxation to the Traveling Salesman Problem, Technical Report IMa- TOM-1999-02, Department of Mathematics and Physics, Malardalen University, Vasteras, Sweden, 1999.

9   Computing a Constrained, Directed Spanning Forest

Add a root vertex r and join r to each of the vertices in T with a zero cost edge. Now, the problem of finding the minimum cost, p-component directed spanning forest can be posed as a problem of finding the minimum cost, directed spanning tree with a degree constraint on the root vertex as follows:

$$\min \sum_{e \in E(T \bigcup \{r\})} c_e x_e, \tag{50}$$

subject to

$$\sum_{e \in \delta(\{r\})} x_e = p \tag{51}$$

$$\sum_{e \in E(S)} x_e \leq |S| - 1, \forall S \subset T \bigcup \{r\}, \tag{52}$$

$$\sum_{e \in E(T)} x_e = n - p, \tag{53}$$

$$x_e \in \{0, 1\}. \tag{54}$$

$$\tag{55}$$

Removing the zero cost edges from the optimal solution to the above problem would yield the desired minimum cost forest. We first consider the following Lagrangian relaxation of the above problem:

$$L(z) = \min_x \sum_{e \in E(T \bigcup \{r\})} c_e x_e + z\left(\sum_{e \in \delta(\{r\})} x_e - p \right) \tag{56}$$

subject to

$$\sum_{e \in E(S)} x_e \leq |S| - 1, \forall S \subset T \bigcup \{r\},$$

$$\sum_{e \in E(T)} x_e = n - p,$$

$$x_e \in \{0, 1\}.$$

Let (z^*) solve the Lagrangian dual $\max_z L(z)$. If x^* is the unique optimal solution that solves the minimization problem in $L(z^*)$, then using the results in [RG1],[MR1] we can conclude that x^* also satisfies the complicating constraint. We perturb the cost of the edges so that, in practice, we have a unique optimal solution x^*. So, the algorithm we use to find the degree constrained spanning tree is as follows:

Directed spanning forest algorithm

1. Perturb the cost of each edge c_e to $\tilde{c}_e = c_e + u_e$, where $\{u_e : e \in E(T \bigcup \{r\})\}$ represent independent, uniform random variables chosen in the interval[2] $[0, \frac{1}{2(n)}]$.
2. Solve the Lagrangian dual problem (56) corresponding to cost $\tilde{c}$. The solution to the Lagrangian dual problem is the desired optimal solution to problem (50) with probability one.

The following part of the section gives a simple proof as to why the Lagrangian dual problem must have a unique optimal solution with probability one. Specifically, Proposition 2 states why there should be a unique feasible solution and proposition 3 shows why the unique feasible solution is also optimal.

Let x_1 and x_2 be any two feasible solutions that satisfy the constraints in 53 and 54. Let $cost(x, c) = \sum_{e \in E(T \bigcup \{r\})} c_e x_e$.

Proposition 2. *Let* $P(cost(x_1, c + u) = cost(x_2, c + u))$ *indicate the probability that the solutions* x_1 *and* x_2 *have the same cost. Then,* $P(cost(x_1, c + u) = cost(x_2, c + u)) = 0$.

Let S_c^* be the set of all the optimal solutions that solve the minimization problem in (50) corresponding to the cost function c_e.

Proposition 3. *For all* $e \in E(T \bigcup \{r\})$, *let* a_e *be any constant in the interval* $[0, \frac{1}{2(n)}]$. *Then* $S_{(c+a)}^* \subseteq S_c^*$.

Proof. Consider a solution $x^1 \notin S_c^*$ and any $x^* \in S_c^*$. Since all c_e are integers, $cost(x^1, c) - cost(x^*, c) \geq 1$. If all the edges corresponding to x^* are perturbed from c_e to $c_e + a_e$, then $cost(x^*, c + a) \leq cost(x^*, c) + \frac{n}{2(n)} < cost(x^*, c) + 1$. Hence $cost(x^1, c + a) > cost(x^*, c + a)$. Therefore, $S_{(c+a)}^* \subseteq S_c^*$.

---

[2] Assume c_e for all $e \in E(T \bigcup \{r\})$ are integers. If c_e are rational numbers one can always multiply them by appropriate constants to make them integers.

Table II presents the convergence results of this randomized algorithm for computing the minimum cost, directed spanning forest. In Table II, n refers to the number of targets, p^* refers to the desired number of components and i^* is the number of iterations required to compute the optimal directed tree.

Table 3. # of iterations for computing $DMSF^*$

n	p^*	i^*
16	3	9
21	6	15
27	3	21
30	7	20
32	6	12
33	5	20
34	6	13
35	5	2
38	4	8
40	8	13
41	2	15
42	6	14
44	7	16
45	3	13
47	7	8
48	5	8
49	8	8
50	3	8
57	7	14
66	4	2

A Random Keys Based Genetic Algorithm for the Target Visitation Problem

Ashwin Arulselvan[1], Clayton W. Commander[1,2], and Panos M. Pardalos[1]

[1] Center for Applied Optimization
University of Florida
{ashwin,pardalos}@ufl.edu
[2] Air Force Research Laboratory
Eglin Air Force Base
clayton.commander@eglin.af.mil

Abstract. The objective of the TARGET VISITATION PROBLEM is to determine a path for an unmanned aerial vehicle that begins at a point of origin and needs to visit several targets before returning to its starting point. An optimal visitation sequence is one which minimizes the total distance traveled and maximizes the utility of the visitation order. This utility measure is defined for each pair of targets and represents the relative value of visiting a particular target before another. In this chapter, we present the results of a preliminary study investigating the effectiveness of a genetic algorithm for the TARGET VISITATION PROBLEM. The encoding scheme is based on random keys. Numerical results are presented for a set of randomly generated test problems and compared with the optimal solutions as computed by a commercial integer programming package.

1 Introduction

Path planning problems represent an enormous amount of the literature on cooperative control and optimization problems. This is particularly true for those involving military applications [2,3,17]. In this paper, we consider the so-called TARGET VISITATION PROBLEM (TVP), whose objective is to determine a path for an unmanned aerial vehicle (UAV), starting at an origin visiting a set of targets, and then before returning to its starting point. The objective is to determine an optimal path which minimizes the total distance traveled and maximizes the utility of the visitation sequence. The TVP has many military applications including combat search and rescue and disaster relief [12].

To date, the only work on the TVP is the original contribution by Grundel and Jeffcoat [12] in which the problem was first proposed. Here the authors presented the TVP and provided a basic analysis examining the similarities between the TVP and other well-known combinatorial problems. In addition, they described the implementation of a metaheuristic for the TVP based on the greedy randomized adaptive search procedure (GRASP) [19].

In this chapter, we describe the implementation of a genetic algorithm (GA) for the TARGET VISITATION PROBLEM. Genetic algorithms represent a very active

M.J. Hirsch et al. (Eds.): Adv. in Cooper. Ctrl. & Optimization, LNCIS 369, pp. 389–397, 2007.
springerlink.com

area of research in computational optimization and have been applied with great success to a myriad of combinatorial problems [9]. The chapter is organized as follows. In the next section, we present the problem statement and analyze some basic properties of the TVP. In Section 3, we describe in detail the genetic algorithm implementation. Section 4 presents the preliminary results of the GA when tested on a set of randomly generated instances. As a basis of comparison, we examine the performance of the GA to the optimal solutions as computed by a commercial integer programming package. We provide concluding remarks in Section 5 and discuss directions of future research.

2   Problem Description

Before formally defining the problem statement, we introduce the symbols and notations we will employ throughout this paper. We use the symbol "$b := a$" to mean "the expression a defines the (new) symbol b" in the sense of King [15]. Of course, this could be conveniently extended so that a statement like "$(1 - \epsilon)/2 := 7$" means "define the symbol ϵ so that $(1 - \epsilon)/2 = 7$" holds [5]. Also, let $|N|$ denote the cardinality of the set N. Finally, we will use *italics* for emphasis and SMALL CAPS for problem names. Any other locally used terms and symbols will be defined in the sections in which they appear.

An instance of the TARGET VISITATION PROBLEM consists of a set $N = \{1, 2, \ldots, n\}$ of targets located at distinct points in the plane. There is a matrix $\mathbf{D} = \{d_{i,j}\}_{m \times m}$, where $m := n + 1$. The extra node, say node 0 represents the UAV's origin. The values of $d_{i,j}$ represent the distances between each pair of targets. There is also a value $d_{0,j}$ which represents the distance from the UAV's point of origin to target j, for all $j \in N$. We note that the distances need not be symmetric. That is, $d_{i,j} \neq d_{j,i}$ in general. Lastly, the instance consists of a matrix $\mathbf{R} = \{\rho_{i,j}\}_{n \times n}$ where $\rho_{i,j}$ represents the preference or utility of visiting target i before target j. The intuition is that targets for which $\rho_{i,j}$ is relatively large should be visited earlier in the sequence as they are assumed to have a higher "threat level" or "cause of interest".

As Grundel and Jeffcoat mention in [12], the values of $d_{i,j}$ are usually easy to obtain since literal distance measures or other metrics such as travel time are available. However, the values of $\rho_{i,j}$, the value added by visiting target i before target j, are not always so simple to obtain. This is because utility is largely based on personal preference and opinion, both measures which are usually more qualitative than quantitative. To overcome this, there are several methods used by military strategists to arrive at the values of $\rho_{i,j}$. The most common method, and the one we adopt in this paper, is known as "target value reconciliation" [12]. In this method a group of experts offer a set of pair-wise rankings for the targets from which the preference matrix is derived. More specifically, for all targets i and j, each expert is to specify a preference of visiting target i before j [4]. The value of $\rho_{i,j}$ is simply the cumulative number of experts who prefer to visit i before j. For a discussion on other techniques, the reader is referred to [12].

Let π be a permutation of the set of integers $[1, \ldots, n+1] \cap \mathbb{Z}$, such that $j =: \pi(i)$ implies that target j is the ith position of the visitation sequence. Then a feasible solution to the TVP is one in which the UAV leaves its starting point, visits each target exactly once, and returns to the origin. An optimal visitation sequence is one which minimizes the total distance traveled and maximizes the utility of the sequence. With this, we can formulate the TVP as the following combinatorial optimization problem [12].

$$\text{Maximize } Z(\pi) = \left[\sum_{i=1}^{n-1} \sum_{j=i+1}^{n} \rho_{\pi(i),\pi(j)} \right] - \left[d_{0,\pi(1)} + \sum_{k=1}^{n-1} d_{\pi(k),\pi(k+1)} + d_{\pi(n),0} \right].$$

$$(1)$$

In [12], the authors provide a nice discussion of the similarities between the TVP and TRAVELING SALESMAN PROBLEM (TSP) [16] and the LINEAR ORDERING PROBLEM (LOP) [8]. It is easy to see that if there were no added benefits of visiting one target before another, then the components of the utility matrix would all be equal. Hence the contribution of this terms in the objective function would be constant. In this case, the problem would reduce to a TSP since the objective would only be a function of the distance traveled [8]. On the other hand, if the distances were irrelevant, then the TVP reduces directly to a LOP. Grundel and Jeffcoat provide example graphs further illuminating these similarities and the reader is referred to their paper for further discussion [12].

Before proceeding to the description of the heuristic for the TVP, we note a crucial observation first made in [12]. Notice that for a given instance of the TVP it might be the case that the entries in one of the matrices in (1) dominates the other. However, for our consideration both distance and utility should play an equal role in determining an optimal visitation sequence. To circumvent this issue, we adopt a simple heuristic which balances the matrices. Let π_r be a random permutation of the targets to be visited. Define $\gamma \in \mathbb{R}$ such that $\hat{\mathbf{R}} := \gamma \mathbf{R}$. In order to normalize the $\mathbf{D}$ and $\mathbf{R}$ matrices, we adjust the particular value of γ so that

$$\frac{\sum_{i=1}^{n-1} \sum_{j=i+1}^{n} \tilde{\rho}_{\pi_r(i),\pi_r(j)}}{d_{0,\pi_r}(1) + \sum_{i=1}^{n-1} d_{\pi_r(i),\pi_r(i+1)} + d_{\pi_r(n),0}} \approx 1. \qquad (2)$$

We see that the particular value of γ can be adjusted to serve as a weight if it is deemed that the relative importance of the distances and utilities are not equal. For larger values of γ, the solution will tend to favor the utility of the sequence over the total distance traveled [12].

Using the permutation based formulation in (1), we now describe the implementation details for a genetic algorithm for finding near optimal solutions for the TARGET VISITATION PROBLEM.

3   Genetic Algorithm

Genetic algorithms (GAs) receive their name from an explanation of the way they behave. It comes as no surprise that they are based on Darwin's Theory of Natural Selection [7]. GAs store a set of solutions, or a *population*, and the population *evolves* by replacing these solutions with better ones based on certain fitness criteria represented by the objective function value.

In successive iterations, or *generations*, the population evolves by *reproduction*, *crossover*, and *mutation*. Reproduction is the probabilistic selection of the next generations elements determined by their fitness level. Crossover is the combination of two current solutions, called *parents*, which produces one or more solutions, referred to as *offspring*. Finally, mutation is the random modification of the offspring. Mutation is performed as an escape mechanism to avoid getting trapped at a local optimal solution [9]. In successive generations, only those solutions having the best *fitness* are carried to the next generation in a process which mimics the fundamental principle of natural selection, *survival of the fittest* [7]. Figure 1 provides pseudo-code for a standard genetic algorithm. Genetic algorithms were introduced in 1977 by Holland [10], and were greatly invigorated by the work of Goldberg in [9].

We note that though the GA does converge in probability to the optimal solution, it is common to stop the procedure after some "terminating condition" (see line 3) is satisfied. This condition could be one of several things including, a maximum running time, a target objective value, or a limit on the number of generations. For our implementation, we use the latter option and the best solution after `MaxGen` generations is returned.

---

procedure GeneticAlgorithm
1   Generate population P_k
2   Evaluate population P_k
3   **while** terminating condition not met **do**
4   Select individuals from P_k and copy to P_{k+1}
5   Crossover individuals from P_k and put in P_{k+1}
6   Mutate individuals from P_k and put in P_{k+1}
7   Evaluate population P_{k+1}
8        $P_k \leftarrow P_{k+1}$
9        $P_{k+1} \leftarrow \emptyset$
10   **end while**
11   **return** best individual in P_k
end procedure GeneticAlgorithm

Fig. 1. Pseudo-code for generic genetic algorithm

When designing a genetic algorithm for an optimization problem, one must provide a means to encode the population, define the crossover operator, and define the *mutation* operator which allows for random changes in offspring to help prevent the algorithm from converging prematurely. The encoding scheme we propose for our GA is based on random keys and follows exactly as described by Bean [1]. As mentioned in [1], GAs often have a difficult time maintaining feasibility of solutions in successive generations. This problem is overcome by the use of random keys as an encoding mechanism for the population. Random keys work by encoding the solution vector using random numbers. The feasibility issue is then moved into the objective function, and subsequently all offspring produced are guaranteed to be feasible solutions.

For the GA implementation for the TVP, we have the following definitions. As mentioned above, solutions are represented by a random vector. To determine the visitation sequence, a random deviate from a distribution which is uniform onto $(0, 1) \in \mathbb{R}$ is generated for each target. The tour is determined by sorting the random numbers and sequencing the targets in descending order of the sort. For example, suppose there are $n = 3$ targets to visit. Then a chromosome such as

$$(.34, .71, .28)$$

would correspond to the visitation sequence

$$2 \to 1 \to 3.$$

The objective value of the sequence can be evaluated, thus determining the fitness of the chromosome.

In order to evolve the population over successive generations, we use a reproduction method which copies the best individuals in the current generation to the next. We aptly refer to this set the BEST set. This technique ensures that the best solution is monotonically improving in every generation [1]. To breed new solutions, we implement a strategy known as *parameterized uniform crossover* [20]. This method works by selecting two solutions to serve as parents. In our implementation, one parent is chosen at random from the BEST set, and the other is chosen from the entire population (including BEST). Then, for each target to be visited, a biased coin is tossed. If the result is heads, then the allele of the BEST parent is chosen, otherwise the allele is taken from the other parent. The probability that the coin lands on heads is known as CrossProb, and is determined empirically. Figure 2 provides an example of a potential crossover when the number of targets is 5 and CrossProb = 0.65.

Coin Toss	T	H	H	T	H
Parent 1	0.56	0.81	0.22	0.7	0.86
Parent 2	0.29	0.49	0.98	0.12	0.32
Offspring	0.29	0.81	0.22	0.12	0.86

Fig. 2. An example of the crossover operation. In this case, CrossProb = 0.65.

Finally, the mutation operator is defined as follows. Instead of introducing random perturbations to selected offspring, we instead replace a set of individuals having the worst fitness with new solutions generated at random from the same distribution as the original population. This replacement set is referred to as the WORST set. Using this method, we are able to ensure that the GA does not converge prematurely. This is a common method, sometimes referred to as *immigration* and appears throughout the literature [1,11]. An overall pictorial view of the generational evolution of the proposed GA is provided in Figure 3.

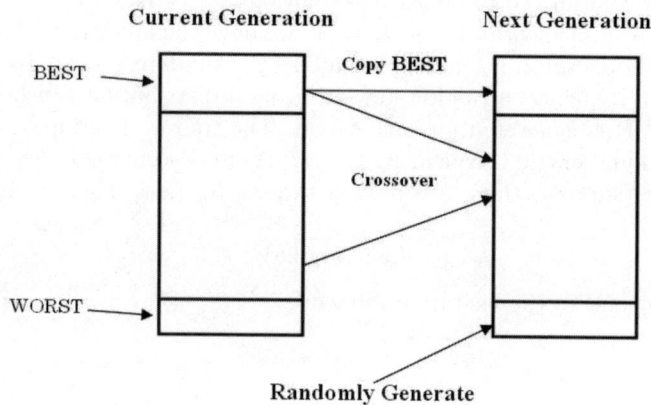

Fig. 3. Graphical representation of generational evolution

4   Computational Results

The proposed heuristic was implemented in the C++ programming language and complied using GNU g++ version 3.4.4, using optimization flags -O2. It was tested on a PC equipped with a 1700MHz Intel® Pentium® M processor and 1.0 gigabytes of RAM operating under the Microsoft® Windows® XP Professional environment.

In order to have a means to compare the results of the GA, we have implemented the integer programming model for the TARGET VISITATION PROBLEM using the CPLEX™ optimization suite from ILOG [6]. CPLEX contains an implementation of the simplex method [13], and uses a branch and bound algorithm [21] together with advanced cutting-plane techniques [14,18]. The CPLEX solver was implemented in the Redhat Linux environment and compiled using g++ version 3.2.3.

The algorithms were tested on a set of randomly generated instances varying in size from 8-16 targets. For each instance, the number of "experts" used to derive the utility matrix was 10. Each of the "expertly defined" pair-wise preferences were generated uniformly. That is, for each target pair (i, j) a random integer in the set $\{0, \dots, 10\}$ was generated. This value represented the number of experts preferring to visit target i before j. Also, the matrices were balanced using the heuristic described in Equation (2) above. For each instance, the maximum distance between the targets varied from 20 to 150 units.

For all of the instances tested, the parameters used for the genetic algorithm (GA) are provided in Table 1.

Table 1. Parameters used for the GA and HGA heuristics

CrossProb $= 0.7$ Population Size $(\texttt{PopSize}) = 2 * \lvert N \rvert$
MaxGen $= 10000$ $\hspace{2em} \lvert \texttt{BEST} \rvert = .1 * \texttt{PopSize}$
$\lvert \texttt{WORST} \rvert = .2 * \texttt{PopSize}$

The comparative results of 250 independent runs of the proposed heuristic on the 25 randomly generated instances are presented in Table 2. The table is organized as follows. The first two columns provide the instance name and number of targets to be visited. The following two columns provide the optimal solutions as computed by CPLEX as well as the required computation time. The heuristic data is listed next. Namely, for each instance we provide the best, worst, and average solutions computed during the 250 runs. The average time to compute the best solution is provided as well as the average deviation from the optimal solution for each instance.

Table 2. The numerical results for a set of 25 randomly generated instances are provided. The optimal solutions are also shown.

Instance		IP Model		Genetic Algorithm				
Name	Targets	Optimal Solution	Execution Time (s)	Max Soln	Min Soln	Avg. Soln	Avg. Time (s)	Avg. Dev (%)
rand8-1	8	60.2766	0.01	60.2766	56.3404	59.8895	0.054	0.642
rand8-2	8	115.944	0.02	115.944	112.653	115.681	0.044	0.27
rand8-3	8	195.333	0.01	195.333	194.96	188.555	0.032	5.006
rand8-4	8	29.0074	0.02	29.0074	25.8592	28.888	0.052	0.412
rand8-5	8	314	0.03	314	314	314	0.008	-
rand10-1	10	157.404	3.01	157.404	140.133	154.084	0.123	2.109
rand10-2	10	208	2.87	208	200	207.36	0.044	0.308
rand10-3	10	520.679	0.01	520.679	437.12	518.698	0.049	0.380
rand10-4	10	532.5	2.45	532.5	489.667	529.891	0.107	0.49
rand10-5	10	365.125	4.87	365.125	303.615	349.457	0.068	4.291
rand12-1	12	124.179	45.37	124.179	106.645	121.022	0.148	2.54
rand12-2	12	318.38	61.17	318.38	266.641	308.668	0.128	3.050
rand12-3	12	420.959	51.89	420.959	341.866	403.31	0.095	4.193
rand12-4	12	594.546	16.44	594.546	487.099	580.956	0.137	2.286
rand12-5	12	472.354	14.68	472.354	409.102	456.735	0.131	3.307
rand14-1	14	137.609	303.55	137.609	110.948	128.208	0.225	6.832
rand14-2	14	405.774	370.01	405.774	334.807	383.21	0.29	5.561
rand14-3	14	631.711	184.82	631.711	508.412	594.765	0.267	5.849
rand14-4	14	176.631	301.71	176.631	146.979	164.377	0.250	6.938
rand14-5	14	679.625	2700.78	679.625	530.617	638.161	0.225	6.101
rand16-1	16	381.934	1353.77	381.934	298.207	351.275	0.351	8.027
rand16-2	16	431.531	2556.77	431.531	333	387.659	0.322	10.167
rand16-3	16	415.338	462.5	415.338	332.868	380.319	0.329	8.431
rand16-4	16	421.658	2810.45	417.109	314.109	386.355	0.397	8.372
rand16-5	16	249.939	3788.49	249.939	187.592	234.192	0.326	6.3

We begin our analysis of the data by noting that for all 6250 runs, the GA computed optimal solutions 95.95% of the time requiring 0.168 seconds on average. This compares favorably with respect to CPLEX which required on average 601.428 seconds to compute the optimal solutions. The CPLEX time was greatly improved for the instances containing 16 targets which we supplied with a feasible solution as a starting point. Without the starting solution, the computation time for these instances was on the order of 75000 seconds. The algorithm also scaled well, averaging less than 0.5 seconds of computing time for each instance. It is reasonable to assume that better performance could be achieved provided the algorithm was able to run for more generations.

5 Conclusion

In this chapter, we described the implementation of a genetic algorithm for the TARGET VISITATION PROBLEM. We began by formally introducing the problem statement. Then we described the details of the proposed heuristic. We presented numerical results comparing the GA solutions with the optimal solutions as computed by the commercial integer programming package CPLEX.

With the current literature on the TVP being slight, there exist many avenues to pursue future research. A natural extension of the work presented here is to augment the GA to include a local search intensification. The resulting, so-called hybrid genetic algorithm is gaining popularity and often adds significant improvement for a minimal amount of computation time. Alternatively, other metaheuristics could be applied as well as advanced cutting plane techniques to try to obtain optimal solutions for larger instances. Lastly, we suggest an investigation of approximation algorithms to produce solutions which have a guaranteed worst-case lower bound.

References

1. J.C. Bean. Genetic algorithms and random keys for sequencing and optimization. *ORSA Journal on Computing*, 6(2):154–160, 1994.
2. S.I. Butenko, R.A. Murphey, and P.M. Pardalos, editors. *Cooperative Control: Models, Applications, and Algorithms*. Springer, 2003.
3. S.I. Butenko, R.A. Murphey, and P.M. Pardalos, editors. *Recent Developments in Cooperative Control and Optimization*. Springer, 2004.
4. B.H. Chiarini, W. Chaovalitwongse, and P.M. Pardalos. A new algorithm for the triangulation of input-output tables in eEconomics. In P. Pardalos, A. Migdalas, and G. Baourakis, editors, *Supply Chain and Finance*. World Scientific, 2004.
5. C.W. Commander, P.M. Pardalos, V. Ryabchenko, and S. Uryasev. The wireless network jamming problem. *Journal of Combinatorial Optimization*, published online, DOI 10.1007/s10878-007-9071-7, 2007.
6. ILOG CPLEX. http://www.ilog.com/products/cplex, Accessed October 2006.
7. C. Darwin. *The Origin of Species*. Murray, sixth edition, 1872.
8. M.R. Garey and D.S. Johnson. *Computers and Intractability: A Guide to the Theory of NP-Completeness*. W.H. Freeman and Company, 1979.

9. D.E. Goldberg. *Genetic Algorithms in Search, Optimization, and Machine Learning*. Addison-Wesley, 1989.
10. D.E. Goldberg. *Genetic Algorithms in Search, Optimization and Machine Learning*. Kluwer Academic Publishers, 1989.
11. J.F. Gonçalves, J.J.M. Mendes, and M.G.C. Resende. A hybrid genetic algorithm for the job scho scheduling problem. *European Journal of Operations Research*, 167:77–95, 2005.
12. D.A. Grundel and D.E. Jeffcoat. Formulation and solution of the target visitation problem. In *Proceedings of the AIAA 1st Intelligent Systems Technical Conference*, 2004.
13. F.S. Hillier and G.J. Lieberman. *Introduction to Operations Research*. McGraw Hill, 2001.
14. R. Horst, P.M. Pardalos, and N.V. Thoai. *Introduction to Global Optimization*, volume 3 of *Nonconvex Optimization and its Applications*. Kluwer Academic Publishers, 1995.
15. J. King. Three problems in search of a measure. *American Mathematical Monthly*, 101:609–628, 1994.
16. E.L. Lawler, J.K. Lenstra, A.H.G. Rinnooy Kan, and D.B. Shmoys, editors. *The Traveling Salesman Problem*. Wiley, 1985.
17. R.A. Murphey and P.M. Pardalos, editors. *Cooperative Control and Optimization*. Springer, 2002.
18. C.A.S. Oliveira, P.M. Pardalos, and T.M. Querido. A combinatorial algorithm for message scheduling on controller area networks. *Int. Journal of Operations Res.*, 1(1/2):160–171, 2005.
19. M.G.C. Resende and C.C. Ribeiro. Greedy randomized adaptive search procedures. In F. Glover and G. Kochenberger, editors, *Handbook of Metaheuristics*, pages 219–249. Kluwer Academic Publishers, 2003.
20. W.M. Spears and K.A. DeJong. On the virtues of parameterized uniform crossover. In *Proceedings of the Fourth International Conference on Genetic Algorithms*, pages 230–236, 1991.
21. L. Wolsey. *Integer Programming*. Wiley, 1998.

Cooperative Rendezvous Between Active Autonomous Vehicles

Yechiel J. Crispin and Marie E. Ricour

Aerospace Engineering Department,
College of Engineering,
Embry-Riddle Aeronautical University,
Daytona Beach, FL 32114, USA
crispinj@erau.edu

Abstract. The rendezvous problem between autonomous vehicles is formulated as an optimal cooperative control problem with terminal constraints. Optimal control problems are often solved by seeking solutions which satisfy the first order necessary conditions for an optimum. Such an approach is based on a Hamiltonian formulation, which leads to a difficult two-point boundary-value problem. We propose a different approach in which the control history is found directly by a genetic algorithm search method. The main advantage of the method is that it does not require the development of a Hamiltonian formulation and consequently, it eliminates the need to deal with an adjoint problem, which leads to a difficult two-point boundary-value problem in nonlinear ordinary differential equations. This method has been applied to the solution of interception and rendezvous problems in an underwater environment, where the direction of the velocity vector is used as the control. We consider the effects of gravity, thrust and viscous drag and treat the rendezvous location as a terminal constraint. We then study cooperative rendezvous problems between spacecraft. We treat the case where the magnitude of the continuous low thrust vector is fixed and the direction of the thrust is used as the control. The spacecraft start from different points on an initial circular orbit and meet at a point on a circular orbit of larger radius, with the same final orbital velocity. The present genetic algorithm was developed to treat complex interception and rendezvous problems involving multiple vehicles.

1   Introduction

Rendezvous problems can be divided into two main classes. The first class includes active-passive rendezvous problems, whereas the second includes cooperative rendezvous problems. In an active-passive rendezvous problem between two vehicles, the passive or target vehicle does not apply any control maneuvers and moves passively along its trajectory. The active or chaser vehicle is controlled or guided such as to meet the passive vehicle at a later time, matching both the location and the velocity of the target vehicle. On the other hand, in a cooperative rendezvous problem, the two vehicles are active and maneuver such as to meet at a later time, at the same location with the same velocity. The two vehicles start the motion from different initial locations and might have different initial velocities. The rendezvous problem consists of finding the control sequences or the guidance laws that are required in order to bring the two vehicles

M.J. Hirsch et al. (Eds.): Adv. in Cooper. Ctrl. & Optimization, LNCIS 369, pp. 399–422, 2007.
springerlink.com

to a final state of rendezvous. In addition to matching final locations and velocities, a rendezvous problem might include additional criteria to be achieved. For example, in a rendezvous problem between two active spacecraft, one might also require the total amount of propellant expended by the vehicles or the time duration to be minimized.

One possible approach to the solution of the rendezvous problem is to formulate it as an optimal control problem in which one is seeking the controls that minimize the differences between the final locations and final velocities of the vehicles in some mathematical sense, for example in the least squares sense. An optimal control problem consists of finding the control histories (control as a function of time) and the state variables of the dynamical system that minimize a performance index. The differential equations of motion of the vehicles are then treated as dynamical constraints.

The methods of approach for solving an optimal control problem include the classical indirect methods and the more recent direct methods. The indirect methods are based on the calculus of variations and its extension to the maximum principle of Pontryagin, which is based on a Hamiltonian formulation. These methods use necessary first order conditions for an optimum, they introduce adjoint variables and require the solution of a two-point boundary value problem (TPBVP) for the state and adjoint variables. Usually, the state variables are subjected to initial conditions and the adjoint variables to terminal or final conditions. TPBVPs are much more difficult to solve than initial value problems (IVP). For this reason, direct methods of solution have been developed which avoid completely the Hamiltonian formulation. For example, a possible approach is to reformulate the optimal control problem as a nonlinear programming (NLP) problem by direct transcription of the dynamical equations at prescribed discrete points or collocation points.

Direct Collocation Nonlinear Programming (DCNLP) is a numerical method that has been used to solve optimal control problems. This method uses a transcription of the continuous equations of motion into a finite number of nonlinear equality constraints, which are satisfied at fixed collocation points. This method was originally developed by Dickmanns and Well (Dickmanns, 1975) and used by Hargraves and Paris (Hargraves, 1987) to solve several atmospheric trajectory optimization problems. Another class of direct methods is based on biologically inspired methods of optimization. These include evolutionary methods such as genetic algorithms (GAs), see for example (Goldberg, 1989) and (Michalewicz, 1994), particle swarm optimization methods (Venter, 2002) and ant colony optimization algorithms (Dorigo, 2004). algorithm mimics the social behavior of swarms of birds or insects, whereas the ant colony optimization algorithms include also chemical communication between the members of the swarm.

Cooperative rendezvous problems between spacecraft have been treated using classical indirect methods. Coverstone-Carroll and Prussing (Coverstone-Carroll, 1992) obtained analytical solutions for a minimum fuel rendezvous between two active power-limited spacecraft. They first studied the rendezvous problem assuming a Hill-Clohessy-Wiltshire linearized gravity field. Then, they obtained a solution for the case of an inverse-square law gravity field, using the DCNLP method mentioned above. A similar problem was studied in which two power-limited spacecraft perform a rendezvous in the linearized Hill-Clohessy-Wiltshire gravity field, in the vicinity of neighboring circular orbits (Coverstone-Carroll, 1993). The same authors (Coverstone-Carroll, 1994) treated a rendezvous problem between two spacecraft in the inverse-square law gravity field where the motion is confined to coplanar orbits. In the case of equal initial power-to-mass ratios and circular initial orbits, cooperative rendezvous resulted in significant savings of propellant when compared to the simpler non-cooperative chaser-target rendezvous.

Pourtakdoust and Jalali (Pourtakdoust, 1995) studied orbital transfer problems for thrust-limited spacecraft. Optimal three dimensional transfer trajectories were obtained using a DCNLP method. In another study an optimal low-thrust chaser-target rendezvous with constraints based on a variational approach, has been conducted (Marinescu, 1976). Recently, Park, Scheeres and Guibout (Park, 2005) introduced a method based on generating functions for the optimal feedback control and rendezvous trajectories for continuous low thrust spacecraft. A Hamiltonian formulation for the state and adjoint variables with split boundary conditions was derived. Generating functions were used to find the optimal solution, treating the TPBVP as a canonical transformation. The main advantage of this method is that it does not require the guess of the initial or terminal values of the adjoint variables to solve the problem. Jezewski (Jezewski, 1992) studied an optimal rendezvous problem subject to arbitrary perturbations and constraints, using primer vector theory as the basis for the optimal control formulation. The solution of the constrained nonlinear parameter problem was found using NLP. However, in this study, the thrust was impulsive and the rendezvous was of the chaser-target type.

Rauwolf and Coverstone-Carroll (Rauwolf, 1996) studied low-thrust spacecraft orbital transfers using GAs. They treated both the case where the thrust is constant as well as the case of variable thrust. The near optimal solutions obtained proved to be accurate enough to be used for preliminary mission planning, or as initial guesses for direct optimization techniques. Other papers treat chaser-target rendezvous problems using GAs. Rendezvous trajectories of the chaser-target type in the presence of disturbing forces were studied by Carpenter and Jackson (Carpenter, 2003). Although the Clohessy-Wiltshire equations provide a linearized approximation for preliminary mission planning, they can lead to significant error in actual use, due to the presence of disturbing forces. Here, a GA was used to minimize the range error after an impulsive maneuver. The Clohessy-Wiltshire equations were used to generate an initial population of solutions for the GA. A similar problem of the chaser-target type has been treated by Kim and Spencer (Kim, 2002), where a minimum fuel solution of the impulsive rendezvous between two spacecraft was obtained. Olsen and Fowler (Olsen, 2004) obtained results using a GA to generate near optimal rendezvous trajectories. Their method provided solutions that closely matched the reference optimal trajectories. Like Rauwolf and Coverstone-Carroll (Rauwolf, 1996), these authors emphasized the importance of obtaining near optimal solutions generated by a GA, since they can be used as an initial guess for a more accurate calculus of variations method.

GAs are a powerful alternative method for solving optimal control problems. They have been used to solve control problems, orbital transfer and rendezvous problems. GAs use a stochastic search method and are robust when compared to gradient methods. They are based on a directed random search which can explore a large region of the design space without conducting an exhaustive search. This increases the probability of finding a global optimum solution to the problem. They can handle continuous or discontinuous variables since they use binary coding. They require only values of the objective function but no values of the derivatives. However, GAs do not guarantee convergence to the global optimum. If the algorithm converges too fast, the probability of exploring some regions of the design space will decrease. Methods have been developed for preventing the algorithm from converging to a local optimum (Fogel, 1995) and (Schraudolph, 1992). These include dynamic parameter encoding, increased probability of mutation, redefinition of the fitness function and other methods that can help maintain the diversity of the population during the genetic search.

2   Cooperative Rendezvous as an Optimal Control Problem

We study trajectory optimization for vehicles moving in an incompressible viscous fluid in a two-dimensional domain. We treat the case where the medium in which the vehicles are moving is incompressible and viscous. The vehicle weight $W = mg\boldsymbol{j}$ acts downward. Here $\boldsymbol{j}$ is a unit vector in the positive y direction. The vehicle has a propulsion system that delivers a thrust $\boldsymbol{T}$ of constant magnitude and is controlled by varying the thrust direction γ. Since the fluid is viscous, a drag force $\boldsymbol{D}$ acts on the vehicle, in the opposite direction of the velocity. The control variable of the problem is the thrust direction $\gamma(t)$. The angle $\gamma(t)$ is measured positive clockwise from the horizontal direction as shown in Figure 1.

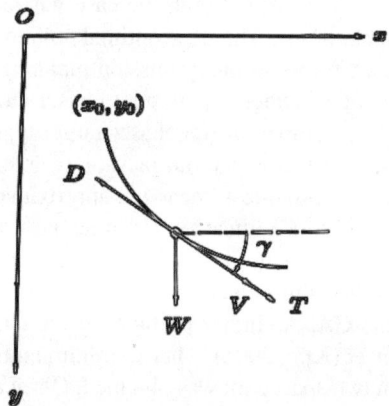

Fig. 1. System of forces acting on an underwater vehicle

We approach the rendezvous problem as an optimal control problem, in which it is required to determine the control functions, or control histories $\gamma(t)$ of the two vehicles, such that they will meet at a prescribed location (x_f, y_f) at the terminal time $t = t_f$. Since GAs deal with discrete variables, we discretize the values of $\gamma(t)$. We assume that the mass m of each vehicle is constant and that the thrust $\boldsymbol{T}$ is always tangent to the trajectory. The motion of the vehicle is governed by Newton's second law of motion and the kinematic relations between velocity and distance:

$$d\left(m\boldsymbol{V}\right)/dt = m\boldsymbol{g} + \boldsymbol{T} + \boldsymbol{D} \tag{2.1}$$

$$dx/dt = V\cos\gamma \tag{2.2}$$

$$dy/dt = V\sin\gamma \tag{2.3}$$

where $\boldsymbol{V}$ is the velocity vector and $\boldsymbol{g} = g\boldsymbol{j}$ is the acceleration of gravity. Since we assumed m is constant,

$$d\boldsymbol{V}/dt = \boldsymbol{g} + \boldsymbol{T}/m + \boldsymbol{D}/m \tag{2.4}$$

Writing this equation for the components of the forces along the tangent to the vehicle's path, we get:

$$dV/dt = g\sin\gamma + T/m - D/m \tag{2.5}$$

Here V, T and D are the magnitudes of the velocity, thrust and drag vectors, respectively. The drag D can be expressed in terms of the drag coefficient:

$$D = \frac{1}{2}\rho V^2 SC_D \tag{2.6}$$

where ρ is the fluid density, S a typical cross-section area of the vehicle and C_D its drag coefficient, which depends on the Reynolds number $Re = \rho Vd/\mu$, where d is a characteristic diameter of the vehicle and μ is the fluid viscosity.

Substituting the drag from Equation (2.6) and writing $T = amg$, where a is the thrust to weight ratio T/mg, Equation (2.5) becomes:

$$dV/dt = g\sin\gamma + ag - \rho V^2 SC_D/2m \tag{2.7}$$

Introducing a characteristic length L_c, time t_c and speed v_c as

$$L_c = 2m/\rho SC_D, \quad t_c = \sqrt{L_c/g}, \quad v_c = \sqrt{gL_c} \tag{2.8}$$

the following nondimensional variables, denoted by an overline (bar) can be defined:

$$x = L_c\bar{x}, \quad y = L_c\bar{y}$$

$$t = (L_c/g)^{1/2}\bar{t}, \quad V = (gL_c)^{1/2}\bar{V} \tag{2.9}$$

Substituting in Equation (2.7), we have:

$$d\bar{V}/d\bar{t} = a + \sin\gamma(t) - \bar{V}^2 \tag{2.10}$$

Similarly, the other equations of motion can be written in nondimensional form as

$$d\bar{x}/d\bar{t} = \bar{V}\cos\gamma(t) \tag{2.11}$$

$$d\bar{y}/d\bar{t} = \bar{V}\sin\gamma(t) \tag{2.12}$$

For each vehicle the initial conditions are:

$$V(0) = V_0, \quad x(0) = x_0, \quad y(0) = y_0 \tag{2.13}$$

For a single vehicle, an optimal control problem can be defined as follows. Starting with the given initial conditions, determine the control function $\gamma(t)$ such as to maximize the horizontal distance traveled in a given time t_f, provided the vehicle arrives at a prescribed depth y_f, i.e.

$$\text{maximize } \overline{x}(\overline{t}_f)$$

subject to the terminal constraint

$$\overline{y}(\overline{t}_f) = \overline{y}_f = y_f/L_c \tag{2.14}$$

where the nondimensional final time is given by $\overline{t_f} = t_f/\sqrt{L_c/g}$.

We now define a rendezvous problem between two vehicles. We denote the variables of the first vehicle by a subscript 1 and those of the second vehicle by a subscript 2. We will now drop the bar notation indicating nondimensional variables. The two vehicles might have different thrust to weight ratios, which we denote by a_1 and a_2, respectively. The equations of motion for the system of two vehicles are:

$$dV_i/dt = a_i + \sin \gamma_i(t) - V_i^2 \quad i = 1, 2 \tag{2.15}$$

$$dx_i/dt = V_i \cos \gamma_i(t) \quad i = 1, 2 \tag{2.16}$$

$$dy_i/dt = V_i \sin \gamma_i(t) \quad i = 1, 2 \tag{2.17}$$

The vehicles can start the motion from different locations and at different speeds. The initial conditions are given by:

$$V_i(0) = V_{i0}, \quad x_i(0) = x_{i0}, \quad y_i(0) = y_{i0} \quad i = 1, 2 \tag{2.18}$$

The cooperative rendezvous problem consists of finding the control functions $\gamma_1(t)$ and $\gamma_2(t)$ such that the two vehicles arrive at a given terminal location (x_f, y_f) and at the same speed in the given time t_f. The terminal constraints are then given by:

$$x_1(t_f) = x_f, \quad x_2(t_f) = x_f, \quad y_1(t_f) = y_f$$

$$y_2(t_f) = y_f, \quad V_1(t_f) = V_2(t_f) \tag{2.19}$$

We can also define an interception problem, of the target-chaser type, in which one vehicle is passive and the chaser vehicle maneuvers such as to match the location of the target vehicle, but not its velocity. Consistent with the above terminal constraints, we define the following objective function for the optimal control problem:

$$\text{minimize } f(\boldsymbol{x}_j(t_f)) = \sum_{j=1}^{N_v} \left\| \boldsymbol{x}_j(t_f) - \boldsymbol{x}_f \right\|^2 \tag{2.20}$$

where N_v is the number of vehicles and $\boldsymbol{x}_f = (x_f, y_f)$ is the prescribed interception or rendezvous point.

We use standard numerical methods for integrating the differential equations. The time interval t_f is divided into N time steps of duration $\Delta t = t_f/N$. The discrete time is $t_i = i\Delta t$. We used a second-order Runge-Kutta method with fixed time step.

We also tried a fourth-order Runge-Kutta method and a variable time step and found that the results were not sensitive to the method of integration. The control function $\gamma(t)$ is discretized to $\gamma(i) = \gamma(t_i)$ according to the number of time steps N used for the numerical integration. Depending on the accuracy of the desired solution, we can choose the number of bits n^i for encoding the value of the control $\gamma(i)$ at each time step i. The size n^i used for encoding $\gamma(i)$ and the number of time steps N will have an influence on the computational time. Therefore n^i and N must be chosen carefully, in order to obtain an accurate enough solution in a reasonable time. The total length of the chromosome is given by:

$$L_{ch} = n^i N N_v \qquad (2.21)$$

For this problem, we were able to increase the rate of convergence of the algorithm by introducing heuristic arguments. For instance, having noticed that $\gamma(t)$ is a monotonically decreasing function of time, we were able to speed up the algorithm by choosing a function with such a property, a priori. Therefore, instead of waiting for the algorithm to converge towards a monotonous $\gamma(t)$, we can sort the values of γ of each individual solution in decreasing order, before calculating its fitness. We also use smoothing of the control function by fitting a third or fourth-order polynomial to the discrete values of γ. The values of the polynomial at the N discrete time points are then used as the current values of γ and are used in the integration of the differential equations.

Appropriate ranges for γ would be either $\gamma \in [0, \pi/2]$ or $\gamma \in [-\pi/2, \pi/2]$ according to the specific interception or rendezvous case at hand. We choose $N = 30$ as a reasonable number of time steps. We now need to choose the parameters associated with the GA. First, we select the lengths of the "genes" for encoding the discrete values of γ. A choice of $n^i = 8$ bits for $\forall i \in [0, N-1]$ was made. The interval between two consecutive possible values of γ is given by:

$$\Delta\gamma = (\gamma_{max} - \gamma_{min})/(2^n - 1) \approx 0.0062\,\text{rad} = 0.35\,\text{deg}$$

For two vehicles $N_v = 2$ and $N = 30$ time steps, the length of a chromosome is given by $L_{ch} = n^i N N_v = 480$ bits.

The problem of choosing a population size and other parameters for the GA has been treated in the early work by De Jong (De Jong, 1975) who tested the GA on a suite of test functions. Numerical experiments show that a reasonable size for the population of solutions is typically in the range $n_{\text{pop}} \in [50, 100]$, see also (Mitchell, 1996). For this problem, there is no need for a particularly large population, so we select $n_{\text{pop}} = 50$. The probability of mutation is set to a value of $p_{\text{mut}} = 0.05$.

3 Simplified Test Cases

Before treating the cooperative rendezvous problem, we checked the method by solving two simplified problems for which we know beforehand what kind of solution to expect. In these cases, the two vehicles start from the same initial conditions and end the motion at the same terminal conditions. Basically, we are reducing the problem to that of a single vehicle, since the trajectories of the two vehicles should be identical, the vehicles moving side by side from the initial point to the terminal point. In both cases, the solution for the motion of a single vehicle was recovered as expected. Since these are degenerate cases, they are mentioned here because they serve as test cases for checking the method, but because of lack of space, we do not show all the details. Detailed results are shown for the more interesting cases where the trajectories of the two cooperating vehicles are significantly different.

In the first test case, the simplified problem consists of finding the control functions $\gamma_1(t)$ and $\gamma_2(t)$ such as to maximize the horizontal distances $x_1(t_f)$ and $x_2(t_f)$ traveled in a given time t_f, provided both vehicles arrive at a prescribed depth y_f, while also matching their horizontal location x_f. Upon testing the algorithm, we expect to obtain the same control for the two vehicles $\gamma_1(t) = \gamma_2(t)$ and the same trajectories. The two vehicles have the same thrust to weight ratios $a_1 = a_2 = a = 0.05$. The initial and final conditions are given in Table 3.1. The problem can be written as

$$\text{maximize } x_1(t_f) + x_2(t_f)$$

subject to the initial and terminal conditions given in Table 1. The following objective, or fitness function is minimized. We use a weight $w = 0.5$:

$$f[x_1(t_f),\, x_2(t_f),\, y_1(t_f),\, y_2(t_f)] =$$
$$= -w[x_1(t_f) + x_2(t_f)] + (1 - w)[(y_1(t_f) - y_f)^2 + (y_2(t_f) - y_f)^2] \qquad (3.1)$$

The parameters for this test case are summarized in Table 1. The algorithm converged in 200 generations and equal control functions $\gamma_1(t) = \gamma_2(t)$ and trajectories were obtained.

Table 1. Parameters, initial and final conditions for the first test case

N_v	n_{pop}	n^i	N	p_{mut}	N_{gen}	$\gamma_{\min}$	$\gamma_{\max}$
2	50	8	30	0.05	200	0	$\pi/2$
$a_1,\, a_2$	t_0	t_f	(x_{01}, y_{01})	(x_{02}, y_{02})	(V_{01}, V_{02})	x_f	$y_1(t_f), y_2(t_f)$
0.05	0	5	$(0,0)$	$(0,0)$	$(0,0)$	max	2

In the second test case, the degenerate cooperative rendezvous problem consists of finding the control functions $\gamma_1(t)$ and $\gamma_2(t)$ such that the two vehicles, starting from the same initial conditions, arrive at the same terminal point (x_f, y_f) in a given time t_f. The two vehicles have the same thrust to weight ratios $a_1 = a_2 = a = 0.05$. Upon testing the algorithm, we expect to obtain the same control functions $\gamma_1(t) = \gamma_2(t)$ for the two vehicles as well as the same trajectories. The final velocities are also required to be equal $V_1(t_f) = V_2(t_f)$. The fitness function is given by:

$$f[x_1(t_f),\, x_2(t_f),\, y_1(t_f),\, y_2(t_f),\, V_1(t_f),\, V_2(t_f)] =$$
$$= (x_1(t_f) - x_f)^2 + (x_2(t_f) - x_f)^2 + (y_1(t_f) - y_f)^2 + (y_2(t_f) - y_f)^2 + (V_1(t_f) - V_2(t_f))^2 \qquad (3.2)$$

The parameters, initial and final conditions for this test case are summarized in Table 2. The algorithm converged in 200 generations and equal control functions $\gamma_1(t) = \gamma_2(t)$ and trajectories were obtained.

Table 2. Parameters and conditions for the second test case

N_v	n_{pop}	n^i	N	p_{mut}	N_{gen}	γ_{min}	γ_{max}
2	50	8	30	0.05	200	$-\pi/2$	$\pi/2$
a_1, a_2	t_0	t_f	(x_{01}, y_{01})	(x_{02}, y_{02})	(V_{01}, V_{02})	$x_1(t_f), x_2(t_f)$	$y_1(t_f), y_2(t_f)$
0.05	0	5	$(0,0)$	$(0,0)$	$(0,0)$	2.8	2

4   Interception and Rendezvous Between Two Underwater Vehicles

In this section we treat the more interesting cases where the trajectories of the two vehicles are significantly different. In section 4.1 we present results for an interception problem between a maneuvering active chaser vehicle and a passive target vehicle whose trajectory is known a priori. The rendezvous problem between two active cooperative vehicles is treated in section 4.2.

4.1   Chaser-Target Interception

In this section, we study a chaser-target interception problem between two vehicles. In this case the first vehicle is active and the second is passive and moves along a straight line at a constant depth $y_2 = y_f$, constant speed V_2 and constant trajectory angle $\gamma_2 = 0$. The two vehicles start from different points and the interception occurs at the known depth of the target vehicle $y_2 = y_f$. The horizontal distance x_f to the interception point is free. We first check the case where the target moves at low speed and can easily be captured by the active chaser vehicle. The thrust to weight ratios are $a_1 = a = 0.05$ for the chaser vehicle and $a_2 = 3a = 0.15$ for the target.

Since this is an interception problem, we do not require matching between the final velocities. In order to match the final locations, the following objective or fitness function is defined:

$$f[x_1(t_f), x_2(t_f), y_1(t_f)] = (x_1(t_f) - x_2(t_f))^2 + (y_1(t_f) - y_f)^2 \qquad (4.1)$$

The GA parameters, initial and final conditions are summarized in Table 3.

Table 3. Parameters and conditions for the interception problem at low target speed

N_v	n_{pop}	n^i	N	p_{mut}	N_{gen}	γ_{1min}	γ_{1max}
2	50	8	30	0.05	50	0	$\pi/2$
a	t_0	t_f	(x_{01}, y_{01})	(x_{02}, y_{02})	(V_{01}, V_{02})	x_f	y_f
0.05	0	5	$(0,0)$	$(0,2)$	$(0, \sqrt{3a})$	free	2

The results of the low target speed interception problem are presented in Figures 2-4. Figure 2 shows the control function $\gamma_2 = 0$ of the target vehicle (straight line with dots) required for maintaining a straight line trajectory at constant speed and fixed depth, see also Figure 3. The curve with the points denoted by circles is the control function $\gamma_1(t)$ of the chaser vehicle required to achieve the interception of the target in the given

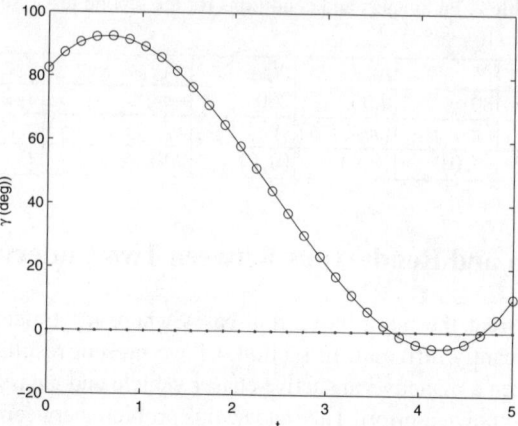

Fig. 2. Control functions $\gamma(t)$ for underwater chaser-target interception with prescribed target depth and low target speed. The curve with circles is for the chaser vehicle. The target vehicle moves at a fixed depth, so $\gamma_2 = 0$.

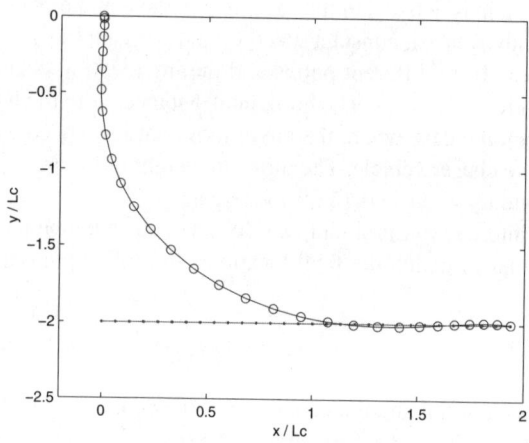

Fig. 3. Trajectories for underwater chaser-target interception with prescribed target depth and low target speed

time. Note that at the interception point $\gamma_1(t_f = 5) > \gamma_2$ because the velocities are not required to be equal.

The trajectories are displayed in Figure 3. It can be seen that the chaser vehicle is first diving vertically in order to build up speed and then starts maneuvering, eventually intercepting the target at $t = t_f$. This speed build up phase can also be seen in Figure 4, which shows the normalized kinetic energy of the chaser vehicle as a function of time. The chaser vehicle then decelerates in order to complete the interception. The kinetic energy of the target vehicle is fixed and is represented by a single dot near the final point at $t = t_f$.

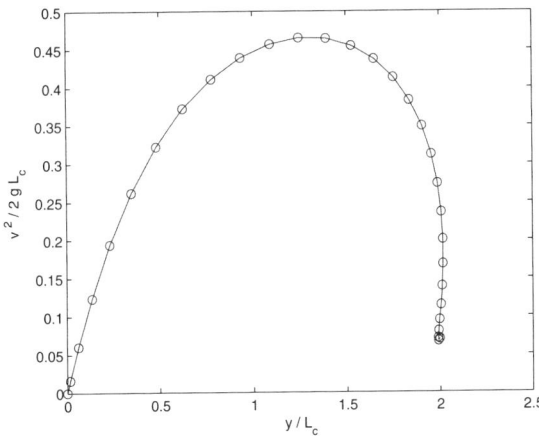

Fig. 4. Kinetic energy as a function of depth for underwater chaser-target interception with prescribed target depth and low target speed

Table 4. Parameters and conditions for the underwater interception problem at moderate speed

N_v	n_{pop}	n^i	N	p_{mut}	N_{gen}	γ_{1min}	γ_{1max}
2	50	8	30	0.05	50	0	$\pi/2$
a	t_0	t_f	(x_{01}, y_{01})	(x_{02}, y_{02})	(V_{01}, V_{02})	x_f	y_f
0.05	0	5	$(0,0)$	$(0,2)$	$(0, \sqrt{5a})$	free	2

Next, we study the same chaser-target interception problem between two vehicles as before, but now the target moves at a moderate speed and can still be captured by the chaser active vehicle. For this case, the thrust to weight ratios are chosen as $a_1 = a = 0.05$ for the chaser and $a_2 = 5a = 0.25$ for the target. The fitness function is given by Equation (4.1). The parameters and conditions for this interception case are summarized in Table 4.

The results of the moderate target speed interception problem are presented in Figures 5-7. Figure 5 shows the control function $\gamma_2 = 0$ of the target vehicle as a straight line with dots. The curve with the points denoted by circles is the control function $\gamma_1(t)$ of the chaser vehicle. It can be seen that at the interception point $\gamma_1(t_f = 5) > \gamma_2$ since the velocity vectors of the chaser and target are not required to be equal.

The trajectories for the case of moderate target speed are displayed in Figure 6. The chaser vehicle starts maneuvering right away in order to compensate for the higher speed of the target, eventually intercepting the target "at the last moment" $t = t_f$. No interception can be obtained at higher target speeds in the given limited time t_f. The normalized kinetic energy of the chaser vehicle is shown in Figure 7, which displays a similar behaviour as in the previous case of low target speed. Here also, the kinetic energy of the target vehicle is fixed and is represented by a single dot near the final point at $t = t_f$.

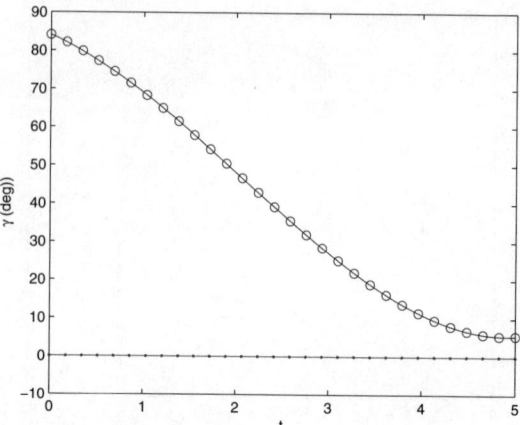

Fig. 5. Control functions $\gamma_1(t)$ and $\gamma_2(t)$ for underwater chaser-target interception with prescribed target depth at moderate target speed

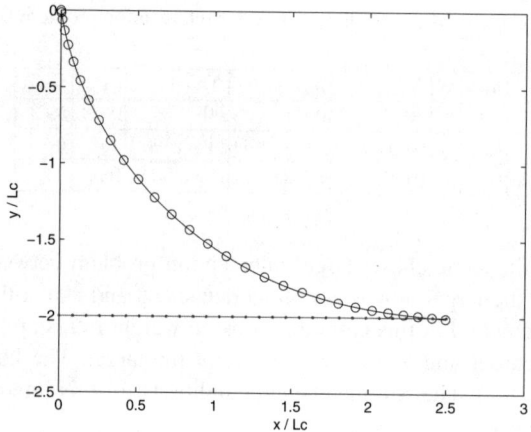

Fig. 6. Trajectories for underwater chaser-target interception with prescribed target depth at moderate target speed

4.2 Rendezvous Between Two Active Underwater Vehicles

We next treat a rendezvous problem with two vehicles. The two vehicles start from different points and rendezvous at a given point (x_f, y_f) in a given time t_f. The vehicles have the same thrust to weight ratios $a_1 = a = 0.05$ and $a_2 = a = 0.05$. In this case, the two vehicles are also required to have the same terminal speed. Therefore, the objective or fitness function is given by:

$$f[x_1(t_f), x_2(t_f), y_1(t_f), y_2(t_f), V_1(t_f), V_2(t_f)] = (x_1(t_f) - x_f)^2 +$$
$$+(y_1(t_f) - y_f)^2 + (x_2(t_f) - x_f)^2 + (y_2(t_f) - y_f)^2 + (V_1(t_f) - V_2(t_f))^2 \quad (4.2)$$

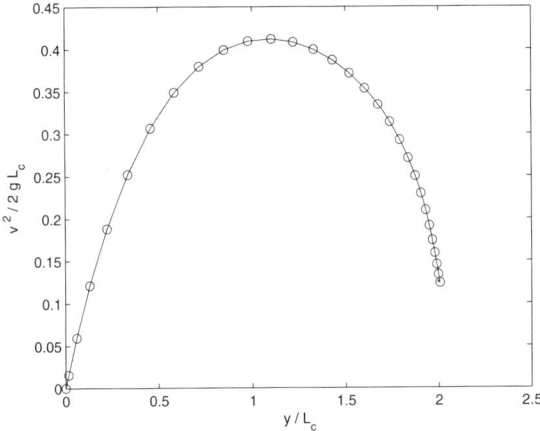

Fig. 7. Kinetic energy as a function of depth for underwater chaser-target interception with pre-scribed target depth at moderate target speed

Table 5. Parameters and conditions for the underwater rendezvous problem

N_v	n_{pop}	n^i	N	p_{mut}	N_{gen}	γ_{min}	γ_{max}
2	50	8	30	0.05	200	0	$\pi/2$
a	t_0	t_f	(x_{01}, y_{01})	(x_{02}, y_{02})	(V_{01}, V_{02})	x_f	y_f
0.05	0	5	$(0,0)$	$(0.5, 0)$	$(0,0)$	2.8	2

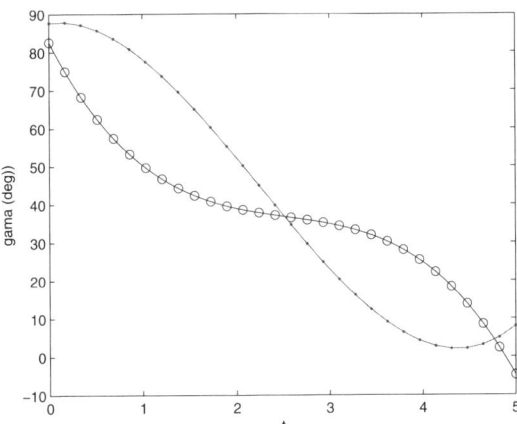

Fig. 8. Control functions $\gamma_1(t)$ and $\gamma_2(t)$ for the rendezvous between two underwater vehicles with prescribed terminal point

The parameters and conditions for this test case are summarized in Table 5. The control functions $\gamma_1(t)$ and $\gamma_2(t)$ are given in Figure 8 and the trajectories in Figure 9. The first vehicle starts at the origin (0,0) and the second vehicle starts at (0.5,0), see

Table 5 and Figure 9. The trajectory of the first vehicle is close to a trajectory required to achieve maximum horizontal distance in the given time t_f. This trajectory can be divided into three segments: first, a dive to acquire speed, then a dive along a straight line at an almost constant angle γ_1, followed by a third segment during which the angle $\gamma_1(t)$ varies more rapidly, see the curve with the circles in Figures 8 and 9. The trajectory of the second vehicle has a much steeper dive segment at the beginning of the maneuver, since the second vehicle is closer to the rendezvous point, see the curves with dots in figures 8 and 9.

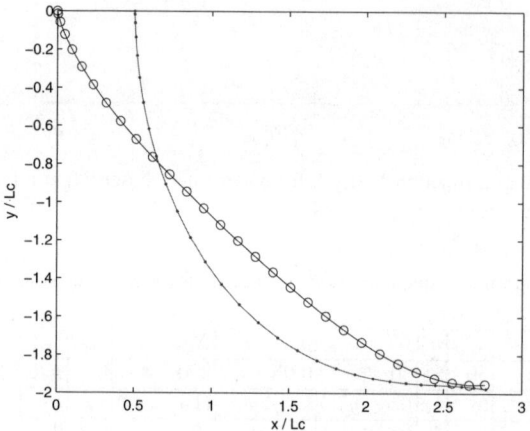

Fig. 9. Trajectories for the rendezvous between two underwater vehicles with prescribed terminal point

5   Cooperative Rendezvous Between Low-Thrust Spacecraft

In this section the rendezvous problem between spacecraft is formulated as an optimal control problem, in which it is required to find the controls of all active spacecraft such as to achieve the common goal of the rendezvous in a prescribed time. In section 5.1 a mathematical formulation of the rendezvous problem is developed. The results are discussed in section 5.2.

5.1   Formulation of the Spacecraft Rendezvous Problem

Two spacecraft, starting from the same circular orbit around an attracting body have to meet on a larger prescribed circular orbit in a given time. The location of a vehicle is defined by the polar coordinates (r, ν) in an inertial frame of reference, as shown in Figure 10.

The state variables of a vehicle are its distance $r(t)$ from the center of attraction, its true anomaly $\nu(t)$, measured counterclockwise with respect to the x-axis, its radial component of velocity $u(t)$ and its velocity component $v(t)$ perpendicular to $r(t)$. The thrust T acts at an angle θ measured from the direction of the velocity component v, positive clockwise from the direction of v, see Figure 10.

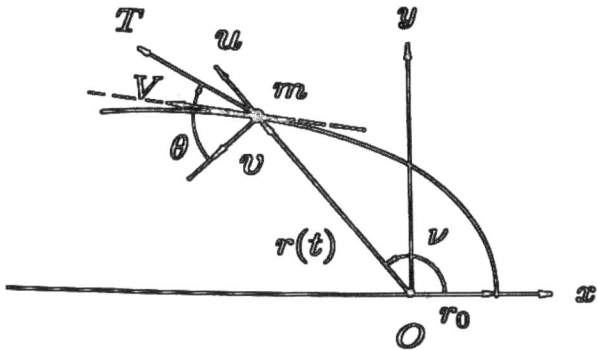

Fig. 10. Notation for the spacecraft problem

We treat spacecraft that use continuous low-thrust rockets, for example spacecraft equipped with electric propulsion devices. Problems involving impulsive maneuvers have been studied extensively and are often easier to solve than their continuous thrust counterpart problems. After an impulsive thrust maneuver, a spacecraft moves passively on a Keplerian orbit, which might be easier to determine. In our case, all spacecraft are continuously active over the time duration of the maneuver. We assume that there is no perturbation of the gravity field such as oblateness of the attracting body or the influence of a third body. We also assume that there is no mutual gravitational force between the spacecraft as it is negligible with respect to the gravitational pull of the main attracting body. Also, without loss of generality, we assume that the two spacecraft are identical, meaning that the spacecraft initial masses are the same, as well as their propulsion devices and thrust magnitudes, which are assumed constant. Furthermore, the constant propellant mass flow rates $\dot{m}$ are the same for all spacecraft.

The spacecraft dynamics are governed by the following equations of motion, see for example Bryson (Bryson, 1999):

$$dr/dt = u \tag{5.1}$$

$$mdu/dt = mv^2/r - m\mu/r^2 + T\sin\theta \tag{5.2}$$

$$mdv/dt = -muv/r + T\cos\theta \tag{5.3}$$

$$d\nu/dt = v/r \tag{5.4}$$

where $\mu = GM_s$ is the gravitational constant of the sun, G is the universal gravitational constant and M_s is the mass of the sun. The equations are then written in non-dimensional form, using the following characteristic (or reference) parameters:

$$t^* = \sqrt{r_0^3/\mu}, \quad V_0 = \sqrt{\mu/r_0}, \quad T_0 = \mu m_0/r_0^2 \tag{5.5}$$

where t^*, V_0 and T_0 are the characteristic time, speed and thrust, respectively. The radius r_0 is the radius of the initial orbit and m_0 is the initial mass of each spacecraft. The reference speed V_0 is the initial orbital speed of each spacecraft, when they are moving passively without thrust on the initial circular orbit. Therefore, there are no radial components u of the spacecraft velocities before starting the cooperative maneuver. We develop a non-dimensional form of the equations using the following non-dimensional variables, denoted by an overline (bar):

$$r = r_0\bar{r}, \quad m = m_0\bar{m}, \quad t = \bar{t}t^* = \bar{t}\sqrt{r_0^3/\mu}$$

$$u = \bar{u}V_0 = \bar{u}\sqrt{\mu/r_0}, \quad v = \bar{v}V_0 = \bar{v}\sqrt{\mu/r_0}, \quad T = T_0\tau = (\mu m_0/r_0^2)\tau \quad (5.6)$$

Here τ is the non-dimensional thrust. Introducing the new variables, the first two equations of motion become:

$$d\bar{r}/d\bar{t} = \bar{u} \quad (5.7)$$

$$\bar{m}d\bar{u}/d\bar{t} = \bar{m}\bar{v}^2/\bar{r} - \bar{m}/\bar{r}^2 + \tau\sin\theta \quad (5.8)$$

The mass of the vehicles varies as a function of time:

$$m(t) = m_0 + \dot{m}t$$

where $\dot{m} < 0$, i.e., the mass of the vehicle is decreasing as propellant is expended by the rocket propulsion device. Rewriting this equation in non-dimensional form, we get

$$\bar{m}\left(\bar{t}\right) = 1 - \left(|\dot{m}|/m_0\right)\sqrt{r_0^3/\mu}\,\bar{t} \quad (5.9)$$

Introducing the nondimensional parameter:

$$B = |\dot{m}|\,t^*/m_0 = \left(|\dot{m}|/m_0\right)\sqrt{r_0^3/\mu} \quad (5.10)$$

Equation (5.9) becomes:

$$\bar{m}\left(\bar{t}\right) = 1 - B\bar{t} \quad (5.11)$$

Substituting Equation (5.11) into Equation (5.8) and rearranging, we get

$$d\bar{u}/d\bar{t} = \bar{v}^2/\bar{r} - 1/\bar{r}^2 + \tau\sin\theta(\bar{t})/(1 - B\bar{t}) \quad (5.12)$$

Similarly, Equations (5.3) and (5.4) are written as:

$$d\bar{v}/d\bar{t} = -\bar{u}\bar{v}/\bar{r} + \tau\cos\theta(\bar{t})/(1 - B\bar{t}) \quad (5.13)$$

$$d\nu/d\bar{t} = \bar{v}/\bar{r} \quad (5.14)$$

The initial conditions for each spacecraft j are:

$$\bar{r}_j(0) = 1, \quad \bar{u}_j(0) = 0, \quad \bar{v}_j(0) = 1, \quad \nu_j(0) = \nu_{j0} \quad (5.15)$$

To achieve a rendezvous within a prescribed time $\bar{t}_f$, the motion is subjected to the following final conditions:

$$\bar{r}_j\left(\bar{t}_f\right) = \bar{r}_f, \quad \bar{u}_j\left(\bar{t}_f\right) = \bar{u}_f = 0, \quad \bar{v}_j\left(\bar{t}_f\right) = \bar{v}_f = 1/\sqrt{\bar{r}_f} \quad (5.16)$$

where $\overline{u_f}$ and $\overline{v_f}$ correspond to the radial and tangential components of the velocities, respectively, that the spacecraft must have in order to maintain circular motion along the final orbit of radius $\overline{r_f}$.

There are different ways of formulating a rendezvous problem. We can minimize the transfer time t_f knowing the final radius r_f and true anomaly ν_f. Equivalently, we can maximize the radius r_f in a given time t_f. Here r_f and t_f are prescribed, and we minimize the difference $r_j(t_f) - r_f$, where $j \in [1, N_v]$ and minimize the difference $\nu_{j_1}(t_f) - \nu_{j_2}(t_f)$ for all

$$(j_1, j_2) \in \{j_1 \in [1, N_v], j_2 \in [1, N_v]; j_1 < j_2\}$$

That is, we minimize the error between the actual final radii of the spacecraft and the prescribed final radius r_f, as well as the difference between the final true anomalies of the vehicles. This ensures that all vehicles have to be on the prescribed final orbit and that they are all close to each other. For rendezvous problems, the final tangential and radial components of the spacecraft velocities are also prescribed, so that upon shutting off the rocket engines, the vehicles will keep moving on the circular orbit defined by the final radius. We minimize the differences $u_j(t_f) - u_f$ and $v_j(t_f) - v_f$ for all $j \in [1, N_v]$. The objective function can then be written as:

$$f = w_1 \sum_{j=1}^{N_v} [r_f - r_j(t_f)]^2 + w_2 \sum_{j=1}^{N_v} [u_f - u_j(t_f)]^2 + w_3 \sum_{j=1}^{N_v} [v_f - v_j(t_f)]^2 +$$

$$+ w_4 \sum_{j_1=2}^{N_v} \sum_{j_2=1}^{j_1-1} [\nu_{j_1}(t_f) - \nu_{j_2}(t_f)]^2 \tag{5.17}$$

The w_k, for $k \in [1, 4]$, are weights corresponding to the variables r, u, v and ν respectively. The rendezvous problem can now be formulated as follows:

Find the control functions $\theta_j(t)$ for $t \in [0, t_f]$ and $j \in [1, N_v]$, such as to minimize the objective function f defined in Equation (5.17), subject to the state equations for each vehicle j:

$$d\overline{r}/d\overline{t} = \overline{u} \tag{5.18}$$

$$d\overline{u}/d\overline{t} = \overline{v}^2/\overline{r} - 1/\overline{r}^2 + \tau \sin\theta(\overline{t})/(1 - B\overline{t}) \tag{5.19}$$

$$d\overline{v}/d\overline{t} = -\overline{uv}/\overline{r} + \tau \cos\theta(\overline{t})/(1 - B\overline{t}) \tag{5.20}$$

$$d\nu/d\overline{t} = \overline{v}/\overline{r} \tag{5.21}$$

subject to the initial conditions for each vehicle j:

$$\overline{r_j}(0) = 1, \quad \overline{u_j}(0) = 0, \quad \overline{v_j}(0) = 1, \quad \nu_j(0) = \nu_{j0} \tag{5.22}$$

and subject to the final conditions:

$$\overline{r_j}(\overline{t_f}) = \overline{r_f}, \quad \overline{u_j}(\overline{t_f}) = 0, \quad \overline{v_j}(\overline{t_f}) = 1/\sqrt{\overline{r_f}} \tag{5.23}$$

We integrate the equations of motion using a Runge-Kutta 4th order method, for the time interval between $t_0 = 0$ to a given t_f for the given initial conditions. In order to use a GA to search for rendezvous trajectories, each control function $\theta(t)$ is represented by N discrete values θ_i, in order to form the binary chromosomes needed for the algorithm. As the solution evolves, the binary values of the chromosomes are converted to real discrete values of $\theta(t)$ and the objective function is evaluated. Before integrating the differential equations for each generation of solutions, the real discrete values of θ_i, which might contain some noise from the genetic operators, are smoothed by a polynomial approximation. During the integration, the values of θ at any given time t can be calculated from the polynomial coefficients. We represent a continuous control function $\theta(t)$ by N discrete values θ_i at N discrete times t_i uniformly distributed in the interval $[0, t_f]$. We also need to select the number of bits n^i used to encode each value θ_i. As an example, we present reference parameters corresponding to a transfer from Earth to Mars:

$$r_0 = 1\,\mathrm{au} = 1.4959787 \times 10^8\,\mathrm{km}$$

where r_0, the radius of the orbit of the earth about the sun, is defined as 1 au, one astronomical unit or one canonical distance unit. The gravitational constant μ for the sun is given by:

$$\mu = \mu_s = GM_s = 1.3271244 \times 10^{11}\,km^3/s^2$$

where G is the universal gravitational constant and M_s is the mass of the sun. For the initial mass of the vehicles, we use an example given by (Bryson, 1999) where:

$$m_0 = 4536\,kg$$

For the given initial orbit, the circular orbit of the earth around the sun, we determine the characteristic time t^* and orbital speed V_0 as:

$$t^* = \sqrt{r_0^3/\mu} = 5.022643 \times 10^6\,s = 58.13\,days$$

$$V_0 = \sqrt{\mu/r_0} = 29.7847\,km/s$$

The time t^* is also defined as one canonical time unit. The characteristic thrust is given by:

$$T_0 = \mu m_0/r_0^2 = 0.0269\,kgkm/s^2 = 26.9\,N$$

A typical value for the actual constant low thrust of an 4536 kg spacecraft is on the order of a few newtons, see (Bryson, 1999):

$$T = 3.778\,N$$

It follows that the nondimensional thrust is:

$$\tau = T/T_0 = 0.1405$$

We now obtain an estimate the mass flow rate $\dot{m}$. The thrust is given by:

$$T = \dot{m}u_e$$

where u_e is the exhaust speed of the propellant at the exit of the rocket's nozzle, which can be estimated from the specific impulse:

$$I_{sp} = u_e/g$$

For a spacecraft with a specific impulse $I_{sp} \approx 5700\,s$ (Bryson, 1999), we have:

$$u_e = I_{sp}g = 55917\,m/s$$

The mass flow rate can be estimated from the thrust and the exhaust speed

$$\dot{m} = T/u_e = 6.7564 \times 10^{-5}\,kg/s$$

We then obtain the parameter B as:

$$B = |\dot{m}|\,t^*/m_0 = 0.0748$$

We can now estimate values for the final conditions. For a transfer maneuver from Earth to Mars, the average distance from the sun of the Martian orbit r_f is:

$$r_f = 1.5237\,au = 2.2794 \times 10^8\,km$$

$$\overline{r_f} = 1.5237$$

$$u_f = 0$$

$$\overline{v_f} = 1/\sqrt{\overline{r_f}} = 0.8101$$

We need to use a reasonable rendezvous time $\overline{t_f}$. In (Bryson, 1999), the total time of transfer of one vehicle from the Earth to Mars orbit is set to 3.3155 canonical units. In our case, we choose a larger value of $\overline{t_f}$, since the rendezvous problem is more complex. We use a longer time:

$$\overline{t_f} = 5.5$$

We have now all the information required to initialize the GA. We drop the non-dimensional overline (bar) notation in the tables appearing in the following sections.

5.2   Rendezvous Between Two Spacecraft

We study a rendezvous maneuver between two spacecraft, starting from two locations on the orbit of the Earth about the sun. The initial true anomalies for the two spacecraft are ν_{10} and ν_{20}. They rendezvous at a point on Mars orbit with no prescribed value for their final true anomaly. The range for $\theta(t) \in [\theta_{min}, \theta_{max}]$ is the largest possible range: $\theta \in [-\pi, \pi]$, since we do not know beforehand the shape of the control functions $\theta(t)$ of the two spacecraft. We use $N = 40$ discrete values of θ_i to represent a solution $\theta(t)$ for the GA. The set of parameters for this first case is given in Table 6.

The average CPU time was 18 minutes (average of 30 simulations on a Pentium 4, 2 GHz computer). Since the initial population is generated randomly, the rate of convergence may vary dramatically when running the same case many times, because the initial populations are generated randomly and the GA operations are also stochastic

Table 6. Parameters and conditions for the rendezvous between two spacecraft

N_v	n_{pop}	n^i	N	p_{mut}	f_{stop}	θ_{min}	θ_{max}	w_1	w_2	w_3	w_4
2	80	6	40	0.05	.0002	$-\pi$	π	1	0.3	1.5	1.5
τ	t_0	t_f	r_{01}, r_{02}	ν_{10}	ν_{20}	u_{01}, u_{02}	v_{01}, v_{02}	r_f	u_f	v_f	
0.1405	0	5.5	1	0	$\pi/2$	0	1	1.5237	0	0.8101	

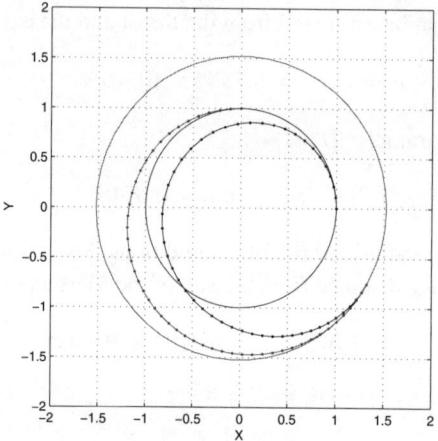

Fig. 11. Trajectories for a rendezvous between two spacecraft. Spacecraft 1 starts at (1,0) and spacecraft 2 starts at (0,1). The sun is located at the origin. The inner circle is the orbit of Earth and the outer circle is the orbit of Mars. Distances are in astronomical units (au). Similar results were obtained with the Chebyshev polynomials approximation.

processes. The rendezvous trajectories of the two spacecraft are given in Figure 11. The initial and final circular orbits and the starting locations of vehicle 1 and 2 are also shown in the figure. The control functions $\theta_1(t)$ and $\theta_2(t)$ of the two spacecraft are shown in Figure 12.

The results of the solution found by the GA are summarized in Table 7. The errors in r_f for both spacecraft are less than 0.5%. The difference between the two spacecraft true anomalies $\nu_1(t_f) - \nu_2(t_f)$ is 0.0076 radians, which is less than 0.5 degree. The maximum error in the tangential velocities is less than 2.13%.

Table 7. Final conditions for the rendezvous between two spacecraft

N_{gen}	CPU (s)	r_f	$r_1(t_f)$	$r_2(t_f)$	$\nu_1(t_f)$	$\nu_2(t_f)$
60	1080	1.5237	1.5184	1.5311	5.9824	5.9748
u_f	$u_1(t_f)$	$u_2(t_f)$	v_f	$v_1(t_f)$	$v_2(t_f)$	
0	-0.02150	-0.01155	0.8101	0.8163	0.8274	

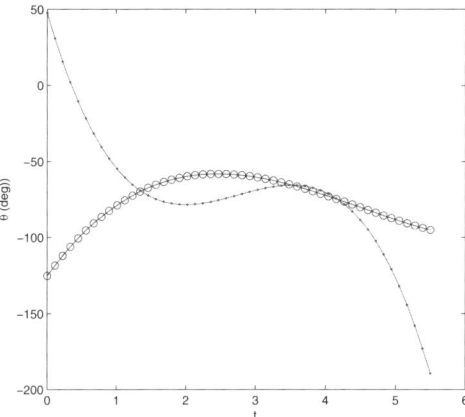

Fig. 12. Thrust direction angles $\theta_1(t)$ and $\theta_2(t)$ for the rendezvous between two spacecraft using the discrete approach. The curve with the circles is for spacecraft 1 and the curve with the dots belongs to spacecraft 2. Similar results were obtained with the Chebyshev polynomials approximation.

5.3   Approximation of the Variables by Chebyshev Polynomials

The results presented in section 5.2 were obtained using discrete values of the control function $\theta(t)$ for the GA implementation. A relatively large number ($N = 40$) of discrete values θ_i were required. The main disadvantage is that the corresponding chromosomes are large. To reduce the size of the chromosomes as well as to use an exact representation of the variations of θ, we also used the coefficients of an approximating polynomial as the design variables. The Chebyshev polynomials $T_c(t)$ can be used to approximate the control functions $\theta(t)$:

$$\theta(t) = \sum_{c=0}^{N_c} A_c T_c(t)$$

Chebyshev polynomials are orthogonal polynomials used in function approximation. They have been described in the numerical analysis literature, see for example a discussion of function approximation in (Press, 1992). For a given N_C number of Chebyshev polynomials, a continuous function $p(x)$ of a real variable x can be approximated by:

$$p(x) = \sum_{c=0}^{N_C} A_c T_c(x) \text{ for } x \in [-1, 1] \tag{5.24}$$

In order to represent the control function $\theta(t)$ in the time interval $t \in [0, t_f]$ for values of $\theta(t)$ in the range $\theta_{min} \leq \theta \leq \theta_{max}$, the following transformation can be used:

$$x = 2(t - t_0)/(t_f - t_0) - 1 = 2t/t_f - 1 \tag{5.25}$$

The use of Chebyshev polynomials will allow a substantial decrease of the length of the chromosome depending on the choice of N_c and n^i (recall that n^i is the number of bits used to encode each design variable). For given N_c and n^i, the length of a chromosome is $L_{ch} = n^i \, N_c \, N_v$. For the same number of bits n^i, a chromosome will be shorter because the number of polynomials N_c required will be much smaller than the number of discrete values N required to represent a solution accurately in the discrete formulation of the GA.

We now determine a range for the values of the coefficients A_c. Since we do not have a priori knowledge about the behavior of $\theta(t)$, we must rely on experience obtained from our numerical results using the discrete approach. We determined the polynomial in Equation (5.24) together with the transformation given by Equation (5.25) that best fits the control history $\theta(t)$ found in the previous section using the discrete approach. We use the first five Chebyshev polynomials to represent the variations of $\theta(t)$, so the variables are A_0, A_1, A_2, A_3 and A_4 and the functions $\theta(t)$ are approximated by polynomials of degree 4. Then, using the maximum and minimum values of the computed variables A_c, we obtained an estimate of the minimum and maximum values of the Chebyshev coefficients as:

$$(A_c)_{min} = -1.4863 \text{ and } (A_c)_{max} = 0.4140$$

We then used a rounded largest absolute value of $[(A_c)_{min}, (A_c)_{max}]$ as the value of the upper and lower limit for the variables of the GA:

$$(A_c)_{min} = -1.5 \text{ and } (A_c)_{max} = 1.5$$

A chromosome is made of N_v binary strings, each containing a sequence of N_c binary strings of length n^i, encoding values of $A_c \in [A_{min}, A_{max}]$. The parameters used to initialize the GA are given in Table 8. The rendezvous final conditions are summarized in Table 9.

Table 8. Parameters and conditions for the spacecraft rendezvous using Chebyshev polynomials

N_v	n_{pop}	n^i	N_c	p_{mut}	f_{stop}	A_{min}	A_{max}	w_1	w_2	w_3	w_4
2	80	12	5	0.05	0.0002	−1.5	1.5	1	0.3	1.5	1.5
τ	t_0	t_f	r_0	ν_{10}	ν_{20}	u_0	v_0	r_f	u_f	v_f	
0.1405	0	5.5	1	0	$\pi/2$	0	1	1.5237	0	0.8101	

After $N_{gen} = 192$ generations, the fitness function decreased below $f_{stop} = 0.0002$. The error in the final radius is $|r_1(t_f) - r_f|/r_f = 0.0035$ for the first spacecraft and $|r_2(t_f) - r_f|/r_f = 0.0049$ for the second. The difference between the final true anomalies is $|\nu_1(t_f) - \nu_2(t_f)| = 0.0073 \, \text{rad} = 0.42 \, \text{deg}$.

Table 9. Final conditions for the spacecraft rendezvous using Chebyshev polynomials

N_{gen}	CPU (s)	r_f	$r_1(t_f)$	$r_2(t_f)$	$\nu_1(t_f)$	$\nu_2(t_f)$
192	810	1.5237	1.5184	1.5311	6.0367	6.0294
u_f	$u_1(t_f)$	$u_2(t_f)$	v_f	$v_1(t_f)$	$v_2(t_f)$	
0	-0.02319	-0.007	0.8101	0.8209	0.8108	

The solution found by the GA using the Chebyshev polynomial approximation is very close to the one found using the discrete approach.

6   Conclusion

The rendezvous problem between two active autonomous vehicles moving in an underwater environment has been treated using an optimal control formulation with terminal constraints. The two vehicles have fixed thrust propulsion systems and use the direction of the velocity vector for steering and control. We use a genetic algorithm to determine directly the control histories of both vehicles by evolving populations of possible solutions. In order to test the method on a simplified case, we treat an interception problem, where one vehicle moves along a straight line with constant velocity and the second vehicle acts as a chaser, maneuvering such as to capture the target in a given time. It was found that the chaser can capture the target within the prescribed time as long as the target speed is below a critical speed. We then treated the rendezvous problem between two active vehicles where both the final positions and velocities are matched.

The rendezvous problem between active autonomous low-thrust spacecraft has been treated using a similar formulation. The spacecraft have propulsion systems which provide a continuous thrust of constant magnitude and use the direction of the thrust vector for steering and control. As the initial distance between the two vehicles is increased, it becomes more difficult to solve the problem and the genetic algorithm requires more generations to converge to a near optimal solution. The approximate solutions obtained by the GA can be used as initial guesses in more accurate numerical methods such as finite difference methods, collocation methods or variational methods.

References

Bryson, A. E., 1999: Dynamic Optimization, Addison Wesley Longman.

Carpenter, B. and Jackson, B., 2003: "Stochastic optimization of spacecraft rendezvous trajectories", Journal of Guidance and Control, Vol. 113, pp. 219-232.

Coverstone-Carroll, V. and Prussing, J. E., 1992: "Optimal cooperative power limited rendezvous with propellant constraints", AIAA-92-4508-CP, Washington, American Institute of Aeronautics and Astronautics,p. 246-255.

Coverstone-Carroll, V. and Prussing, J. E., 1993: "Optimal cooperative power limited rendezvous between neighboring circular orbits", Journal of Guidance, Control, and Dynamics, Vol. 16, No. 6, pp. 1045-1054

Coverstone-Carroll, V. and Prussing, J. E., 1994: "Optimal cooperative power limited rendezvous between coplanar circular orbits", Journal of Guidance, Control, and Dynamics, Vol. 17, No. 5, pp. 1096-1102

Coverstone-Carroll, V., 1997: "Near-Optimal Low-Thrust Trajectories via Micro-Genetic Algorithms", Journal of Guidance, Control, and Dynamics, Vol. 20, No. 1, pp. 196-198

Crispin, Y., 2005: "Cooperative control of a robot swarm with network communication delay", The First International Workshop on Multi-Agent Robotic Systems (MARS 2005), Setubal, Portugal

Crispin, Y., 2006: "An evolutionary approach to nonlinear discrete-time optimal control with terminal constraints", in Informatics in Control, Automation and Robotics I, pp.89-97, Springer, Dordrecht, Netherlands.

Crispin, Y., 2007: "Evolutionary computation for discrete and continuous time optimal control problems", in Informatics in Control, Automation and Robotics II, Springer, Dordrecht, Netherlands.

De Jong, K.A., 1975: An analysis of the behavior of a class of genetic adaptive systems, Ph.D. thesis, University of Michigan, Ann Arbor, MI.

Dickmanns, E. D. and Well, H., 1975: "Approximate solution of optimal control problems using third-order hermite polynomial functions," Proceedings of the 6th Technical Conference on Optimization Techniques, Springer-Verlag, New York.

Dorigo, M. and Stutzle, T., 2004, Ant Colony Optimization, The MIT Press, Cambridge, MA.

Fogel, D.B., 1995: Evolutionary Computation, IEEE Press, New York.

Goldberg, D.E., 1989: Genetic Algorithms in Search, Optimization and Machine Learning, Addison-Wesley Publishing Company

Hargraves, C. R. and Paris, S. W., 1987: "Direct Trajectory Optimization Using Nonlinear Programming and Collocation", AIAA Journal of Guidance, Control and Dynamics, Vol. 10, pp. 338-342

Hartmann, J.W., 1999: "Low thrust trajectory optimization using stochastic optimization methods", Master of Science in Aeronautical and Astronautical Engineering Thesis

Jezewski, D.J., 1992: "Optimal rendezvous trajectories subject to arbitrary perturbations and constraints", AIAA-92-4507-CP, Washington, American Institute of Aeronautics and Astronautics, p. 235-245

Kim, Y.H. and Spencer, D.B., 2002: "Optimal spacecraft rendezvous using genetic algorithms", Journal of Spacecraft and Rockets, Vol. 39, No. 6, pp. 859-865

Marinescu, Al., 1976: "Optimal low-thrust orbital rendezvous", Journal of spacecraft, Vol. 13, No. 7, pp. 385-398.

Michalewicz, Z., 1994: Genetic Algorithms + Data Structures = Evolution Programs, Second, Extended Edition, Springer-Verlag, Berlin, Heidelberg, New York.

Mitchell, M., 1996: An Introduction to Genetic Algorithms, MIT Press, Cambridge, MA.

Olsen, C. and Fowler W., 2005: "Characterization of the relative motions of rendezvous between vehicles in proximate, highly elliptic orbits", Advances in the Astronautical Sciences. Vol. 119, Part I: Spaceflight Mechanics 2004, pp. 879-895

Park, C., Scheeres, D.J. and Guibout, V., 2006: "Solving optimal continuous thrust rendezvous problem with generating functions", Journal of Guidance, Control, and Dynamics, Vol.29, No.2, pp. 321-331

Pourtakdoust, S.H. and Jalali, M. A., 1995: "Thrust limited optimal three dimensional spacecraft trajectories", AIAA-95-3325-CP, Washington, American Institute of Aeronautics and Astronautics, p. 1395-1404

Press, W.H., Teukolsky, S.A., Vetterling, W.T. and B.P. Flannery, 1992: Numerical Recipes in Fortran, The Art of Scientific Computing, Second Edition, Cambridge University Press, New York, Chapter 5, section 5.8.

Rauwolf, G.A. and Coverstone-Carroll, V., 1996: "Near optimal low-thrust orbit transfers generated by a genetic algorithm", Journal of Spacecraft and Rockets, Vol. 33, No. 6, pp. 859-862

Rauwolf, G.A. and Friedlander, A., "Near-optimal solar sail trajectories generated by a genetic algorithm", Advances in the Astronautical Sciences. Vol. 103, part 1.

Schraudolph, N.N. and R.K. Belew, 1992: "Dynamic parameter encoding for genetic algorithms", Machine Learning, Vol. 9, pp. 9-21.

Venter, G. and Sobieszczanski-Sobieski, J., 2002: "Particle swarm optimization", 43rd AIAA/ASME/ASCE/AHS/ASC Structures, Structural Dynamics and Materials Conference.

Author Index

Printing: Mercedes-Druck, Berlin
Binding: Stein+Lehmann, Berlin

Lecture Notes in Control and Information Sciences

Edited by M. Thoma, M. Morari

Further volumes of this series can be found on our homepage:
springer.com